Springer Complexity

Springer Complexity is a publication program, cutting across all traditional disciplines of sciences as well as engineering, economics, medicine, psychology and computer sciences, which is aimed at researchers, students and practitioners working in the field of complex systems. Complex Systems are systems that comprise many interacting parts with the ability to generate a new quality of macroscopic collective behavior through self-organization, e.g., the spontaneous formation of temporal, spatial or functional structures. This recognition, that the collective behavior of the whole system cannot be simply inferred from the understanding of the behavior of the individual components, has led to various new concepts and sophisticated tools of complexity. The main concepts and tools – with sometimes overlapping contents and methodologies – are the theories of self-organization, complex systems, synergetics, dynamical systems, turbulence, catastrophes, instabilities, nonlinearity, stochastic processes, chaos, neural networks, cellular automata, adaptive systems, and genetic algorithms.

The topics treated within Springer Complexity are as diverse as lasers or fluids in physics, machine cutting phenomena of workpieces or electric circuits with feedback in engineering, growth of crystals or pattern formation in chemistry, morphogenesis in biology, brain function in neurology, behavior of stock exchange rates in economics, or the formation of public opinion in sociology. All these seemingly quite different kinds of structure formation have a number of important features and underlying structures in common. These deep structural similarities can be exploited to transfer analytical methods and understanding from one field to another. The Springer Complexity program therefore seeks to foster cross-fertilization between the disciplines and a dialogue between theoreticians and experimentalists for a deeper understanding of the general structure and behavior of complex systems.

The program consists of individual books, books series such as "Springer Series in Synergetics", "Institute of Nonlinear Science", "Physics of Neural Networks", and "Understanding Complex Systems", as well as various journals.

T0155678

Springer Series in Synergetics

SSSyn – An Interdisciplinary Series on Complex Systems

The success of the Springer Series in Synergetics has been made possible by the contributions of outstanding authors who presented their quite often pioneering results to the science community well beyond the borders of a special discipline. Indeed, interdisciplinarity is one of the main features of this series. But interdisciplinarity is not enough: The main goal is the search for common features of self-organizing systems in a great variety of seemingly quite different systems, or, still more precisely speaking, the search for general principles underlying the spontaneous formation of spatial, temporal or functional structures. The topics treated may be as diverse as lasers and fluids in physics, pattern formation in chemistry, morphogenesis in biology, brain functions in neurology or self-organization in a city. As is witnessed by several volumes, great attention is being paid to the pivotal interplay between deterministic and stochastic processes, as well as to the dialogue between theoreticians and experimentalists. All this has contributed to a remarkable cross-fertilization between disciplines and to a deeper understanding of complex systems. The timeliness and potential of such an approach are also mirrored – among other indicators – by numerous interdisciplinary workshops and conferences all over the world.

Didier Sornette

Critical Phenomena in Natural Sciences

Chaos, Fractals,
Selforganization and Disorder:
Concepts and Tools

Second Edition
With 102 Figures

 Springer

Professor Didier Sornette

Chair of Entrepreneurial Risks
Department of Management,
Technology and Economics (D-MTEC)
ETH Zentrum
8092 Zürich, Switzerland

and

Laboratoire de Physique de la Matière Condensée
CNRS UMR6622
Université de Nice-Sophia Antipolis
Faculté des Sciences, B.P. 71
06108 Nice Cedex 2, France

Library of Congress Control Number: 2006920906

2nd Printing of the Hardcover Edition with ISBN 3-540-40754-5

ISSN 0172-7389

ISBN-10 3-540-30882-2   2nd Edition   Springer Berlin Heidelberg New York
ISBN-13 978-3-540-30882-9   2nd Edition   Springer Berlin Heidelberg New York

Springer is a part of Springer Science+Business Media

springeronline.com

© Springer-Verlag Berlin Heidelberg 2006
Printed in Germany

Typesetting: by the author
Production: LE-TEX Jelonek, Schmidt & Vöckler GbR, Leipzig
Cover design: Erich Kirchner, Heidelberg
Printed on acid-free paper   54/3100/YL   5 4 3 2 1

To Anne, Jaufray and Paul

Preface

Since its first edition, the ideas discussed in this book have expanded significantly as a result of very active research in the general domain of complex systems. I have also seen with pleasure different communities in the geo-, medical and social sciences becoming more aware of the usefulness of the concepts and techniques presented here.

In this second edition, I have first corrected, made more precise and expanded a large number of points. I have also added a significant amount of novel material which I describe briefly below.

Chapter 1 has been expanded by inclusion of stricter and more in-depth discussions of the differences between objective and subjective (Bayesian) probabilities with, in particular, the addition of the Dutch book argument. A presentation of the Gnedenko–Pickands–Balkema–de Haan theorem for the distribution of peaks-over-threshold has been added, which derives the generalized Pareto distribution of the asymptotic distribution for independent random variables from the extreme value distributions. I have also added a formal treatment of the expectation of the sum over the maximum of random variables, for fat-tailed and non-fat-tailed probability distribution functions (pdf's).

In Chap. 2, I have added a section on the extraction of model equations from experimental data.

In Chap. 4, the explicit representation of stable Lévy distributions is given in terms of Fox functions and the useful properties of the generalized Mittag–Leffler exponentials and of Fox functions are described. Chapter 4 also contains additional information on the expectation of the sum over the maximum of random variables for fat-tailed pdf's.

Chapter 5 contains a new section on the multifractal random walk (MRW), a recently introduced stochastic process that generalized the fractional Brownian motion by having an exact multifractal structure in the continuous limit.

Chapter 6 contains a new section on conditional power law distributions with application to "fractal plate tectonics" and a significant development on Wilk statistics of embedded hypothesis testing to compare the relative merits of power law versus stretched exponential distributions. A novel embedding of the family of power law pdf's within the family of stretched exponentials is presented and the general formulas for the covariance of the estimators are given.

New figures have been added to Chap. 7 to clarify and enhance the discussion on the relevance of the concept of temperature to out-of-equilibrium systems. Chapter 7 also contains a new section on the Beck–Cohen superstatistics which provides a dynamical origin of non-extensive Tsallis-type statistics.

Chapter 8 contains a new presentation of fractional diffusion equations, their relationship with Lévy laws and the associated anomalous diffusion.

Chapter 10 contains applications of the critical precursors and critical dynamics to explain for instance the way our internal hearing organ, the cochlea, works.

Chapter 11 has been significantly expanded to include a section of functional reconstruction of approximants based on renormalization group ideas, which have been shown to be an improvement over the Padé approximants. Chapter 11 also contains a new section on the Weierstrass and Weierstrass-type functions and concludes with recalling Anderson's message "more is different."

New figures have been added to Chap. 13 to clarify and enhance the discussion of quasi-dynamical rupture models.

Chapter 14, already a favorite in the first edition, has been significantly enhanced by including several other mechanisms for the generation of power law distributions. The discussion of the Kesten process in terms of multiplicative noise has been expanded. A new section presents the class of growth models with preferential attachment, which has a wide range of applications. A new section discusses the superposition of log-normal pdf's. Another section presents the coherent-noise models and their limits for the application to earthquakes.

Chapter 15 expands on the mechanism of self-organized criticality in terms of the feedback of the order parameter onto the control parameter. A new section also presents the linear fractional stable motions for extremal dynamics.

Chapter 16 contains a new section reviewing the fundamental Kolmogorov's theorem on fragmentation models which played a fundamental role in attracting the attention on the importance of log-normal distributions for general multiplicative processes.

I would have liked to enrich this second edition much more and remain frustrated by the limits of its achievements. Nevertheless, I hope that the readers, and especially the "students" in the extraordinary rich fields of complex dynamical systems, will find this new edition valuable.

In addition to the many collaborators and colleagues mentioned in the preface of the first edition and who contributed to my understanding, this second edition owes a lot to V.F. Pisarenko who provided numerous comments and suggestions on the first edition, as well as detailed explanations on subtle points in the field of mathematical statistics. The errors remain mine. T.P. O'Brien has been also very stimulating in his penetrative questions and remarks. In addition to my colleagues saluted in the first edition, I acknowledge inspiring exchanges with Y. Ageon, S. Gluzman, A. Helmstetter, K. Ide, Y.Y. Kagan, T. Lux, Y. Malevergne, M.E.J. Newman, A. Saichev, H. Takayasu, H.J. Viljoen, V.I. Yukalov, and W.-X. Zhou.

UCLA and Nice,
October 2003 Didier Sornette

Preface to the First Edition:
Variability and Fluctuations

Life is fundamentally risky, reflecting the pervasive out-of-equilibrium nature of the surrounding world. Risk is synonymous with uncertainty about the future, leading not only to potential losses and perils, but also to gains. This uncertainty results from the numerous dynamical factors entering our life, giving it spice and color as well as its dangerous flavor. Life consists of a succession of choices that have to be made with often limited knowledge and in a complex and changing environment. These choices result in a sequence of often unpredictable outcomes, whose accumulation defines the specific trajectory characterizing each individual, somewhat similar to the apparent random trajectory of a leaf carried by a turbulent wind. The notion of risk is probably one of the most general concepts pervading all the facets of our life [285, 794].

Risk is a companion to most of our daily activities, professional or private. Crossing a street or driving a car involves risk that is quantified by the statistics of traffic accidents and police reports and which impacts on our insurance premium. Staying at home is also risky: falling, burning, electrocution, plane crash, earthquakes, hurricanes, etc. Risk is present in the choice of a career, in the selection of a college and university program as well as in the effect of social interactions on the development of children. Any choice is intrinsically risky, since the existence of a choice implies several alternatives that are all thought to be possible outcomes, albeit with possibly different likelihood. In industry, companies have to face a multitude of risks: R&D, choice of a niche, capital, production, sales, competition, etc., encompassing all types of risks that, ideally, have to be optimized at each instant. The apparent random nature of price variations in both organized and emerging stock markets leads to risky investment choices, with impact on the global economy and our welfare (retirement funds).

The Earth provides its share of risks, partly overcome with the development of technology, but hurricanes, earthquakes, tsunamis, volcanic eruptions and meteorites bring episodic destruction each year, constituting as many Damocles' swords over our heads. Neither is biological risk negligible, with endemic epidemics and the emergence of novel diseases. Human society, with its technical development and population growth, introduces new risks: unemployment, strike, dysfunction of cities, rupture of sensitive technological structures (hydroelectric dams, chemical plants, oil tankers, nuclear plants, etc.). Scientific and technical development and the growing interaction between the different organizational levels of human society introduce an increasing complexity, leading often to an enhanced vulnerability. The weight of human activity has developed to a point where there are growing concerns about new planetary risks such as global warming, ozone-layer depletion, global pollution, demographic overcrowding, and the long-term agricultural and economic sustainability of our finite planet. Paling's little book [715] provides an interesting and stimulating synopsis in which a logarithmic scale is used to quantify all the risks that we have to face, from the largest, which are not always those we think about, to the smallest. This logarithmic scale (similar to the earthquake magnitude scale) reflects the extremely large variability of risk sizes. The concept of risk thus covers the notion of variability and uncertainty.

Our main goal in this book is to present some of the most useful modern theoretical concepts and techniques to understand and model the large variability found in the world. We present the main concepts and tools and illustrate them using examples borrowed from the geosciences. In today's rapidly evolving world, it is important that the student be armed with concepts and methods that can be used outside his/her initial specialization for a better adaptation to the changing professional world. It is probably in the everyday practice of a profession (for instance as an engineer or a risk-controler in a bank) that the appreciation of variabilities and of the existence of methods to address it will be the most useful.

These ideas are of utmost importance in the advancement of the traditional scientific disciplines and it is in their context that this book is presented. The notions of variability, fluctuations, disorder, and non-reproducibility, on a deep conceptual level, progressively penetrate the traditional disciplines, which were initially developed using the concepts of averages, or more generally, of representative elements (as in thermodynamics, mechanics, acoustics and optics, etc.). Modern physics deals, for instance, with heterogeneous composite systems and new materials, chaotic and self-organizing behaviors in out-of-equilibrium systems, and complex patterns in the growth and organization of many structures (from that of the universe at the scale of hundreds of megaparsecs to the minute branchings of a snowflake). It is clear that these phenomena are all deeply permeated by the concepts of variability, fluctuations, self-organization and complexity. In the context of natural evolution, let us mention the remarkable illustrations (evolution and baseball) presented by S.J. Gould [358], in which the full distribution (and not only the average) of all possible outcomes/scenarios provides the correct unbiased description of reality. This is in contrast with the usual reductionist approach in terms of a few indicators such as average and variance.

The physical sciences focus their attention on a description and understanding of the surrounding inanimate world at all possible scales. They address the notion of risk as resulting from the intrinsic fluctuations accompanying any possible phenomenon, with chaotic and/or quantum origins. Mathematics has developed a special branch to deal with fluctuations and risk, the theory of probability, which constitutes an essential tool in the book. We begin with a review of the most important notions to quantify fluctuations and variability, namely *probability distribution* and *correlation.* "Innocuous" Gaussian distributions are contrasted with "wild" heavy-tail power law distributions. The importance of characterizing a phenomenon by its full distribution and not only by its mean (which can give a very distorted view of reality) is a recurrent theme. In many different forms throughout the book, the central theme is that of *collective* or *cooperative* effects, i.e. the whole is more than the sum of the parts. This concept will be visited with various models, starting from the sum of random variables, the percolation model, and self-organized criticality, among others.

The first six chapters cover important notions of statistics and probability and show that collective behavior is already apparent in an ensemble of uncorrelated elements. It is necessary to understand those properties that emerge from the law of large numbers to fully appreciate the additional properties stemming from the interplay between the large number of elements and their interactions/correlations. The second part (Chaps. 7–15) discusses the behavior of many correlated elements, including bifurcations, critical transitions and self-organization in out-of-equilibrium systems which constitute the modern concepts developed over the last two decades to deal with complex natural systems, characterized by collective self-organizing behaviors with long-range correlations and sometimes frozen heterogeneous structures. The last two chapters, 16 and 17, provide an introduction to the physics of frozen heterogeneous systems in which remarkable and non-intuitive behaviors can be found.

The concepts and tools presented in this book are relevant to a variety of problems in the natural and social sciences which include the large-scale structure of the universe, the organization of the solar system, turbulence in the atmosphere, the ocean and the mantle, meteorology, plate tectonics, earthquake physics and seismo-tectonics, geomorphology and erosion, population dynamics, epidemics, bio-diversity and evolution, biological systems, economics and so on. Our emphasis is on the concepts and methods that offer a unifying scheme and the exposition is organized accordingly. Concrete examples within these fields are proposed as often as possible. The worked applications are often very simplified models but are meant to emphasize some basic mechanisms on which more elaborate constructions can be developed. They are also useful in illustrating the path taken by progress in scientific endeavors, namely "understanding", as synonymous with "simplifying". We shall thus attempt to present the results and their derivations in the simplest and most intuitive way, rather than emphasize mathematical rigor.

This book derives from a course taught several times at UCLA at the graduate level in the department of Earth and Space Sciences between 1996 and 1999. Essentially aimed at graduate students in geology and geophysics offering them an introduction to the world of self-organizing collective behaviors, the course attracted graduate students and post-doctoral researchers from space physics, meteorology, physics, and mathematics. I am indebted to all of them for their feedback. I also acknowledge the fruitful and inspiring discussions and collaborations with many colleagues over many years, including J.V. Andersen, J.-C. Anifrani, A. Arneodo, W. Benz, M. Blank, J.-P. Bouchaud, D.D. Bowman, F. Carmona, P.A. Cowie, I. Dornic, P. Evesque, S. Feng, U. Frisch, J.R. Grasso, Y. Huang, P. Jögi, Y.Y. Kagan, M. Lagier, J. Laherrère, L. Lamaignère, M.W. Lee, C. Le Floc'h, K.-T. Leung, C. Maveyraud, J.-F. Muzy, W.I. Newman, G. Ouillon, V.F. Pisarenko, G. Saada, C. Sammis, S. Roux, D. Stauffer, C. Vanneste, H.-J. Xu, D. Zajdenweber, Y.-C. Zhang, and especially A. Johansen, L. Knopoff, H. Saleur, and A. Sornette. I am indebted to M.W. Lee for careful reading of the manuscript and to F. Abry and A. Poliakov for constructive comments on the manuscript.

UCLA and Nice,   *Didier Sornette*
April 2000

Contents

1. **Useful Notions of Probability Theory** 1
   1.1 What Is Probability? 1
       1.1.1 First Intuitive Notions 1
       1.1.2 Objective Versus Subjective Probability 2
   1.2 Bayesian View Point 6
       1.2.1 Introduction 6
       1.2.2 Bayes' Theorem 7
       1.2.3 Bayesian Explanation for Change of Belief 9
       1.2.4 Bayesian Probability and the Dutch Book 10
   1.3 Probability Density Function........................... 12
   1.4 Measures of Central Tendency 13
   1.5 Measure of Variations from Central Tendency 14
   1.6 Moments and Characteristic Function 15
   1.7 Cumulants... 16
   1.8 Maximum of Random Variables and Extreme Value Theory. 18
       1.8.1 Maximum Value Among N Random Variables 19
       1.8.2 Stable Extreme Value Distributions 23
       1.8.3 First Heuristic Derivation
             of the Stable Gumbel Distribution 25
       1.8.4 Second Heuristic Derivation
             of the Stable Gumbel Distribution 26
       1.8.5 Practical Use and Expression of the Coefficients
             of the Gumbel Distribution 28
       1.8.6 The Gnedenko–Pickands–Balkema–de Haan Theorem
             and the pdf of Peaks-Over-Threshold 29

2. **Sums of Random Variables, Random Walks
   and the Central Limit Theorem**........................... 33
   2.1 The Random Walk Problem 33
       2.1.1 Average Drift 34
       2.1.2 Diffusion Law 35
       2.1.3 Brownian Motion as Solution of a Stochastic ODE . 35
       2.1.4 Fractal Structure 37

	2.1.5	Self-Affinity	39
2.2		Master and Diffusion (Fokker–Planck) Equations	41
	2.2.1	Simple Formulation	41
	2.2.2	General Fokker–Planck Equation	43
	2.2.3	Ito Versus Stratonovich	44
	2.2.4	Extracting Model Equations from Experimental Data	47
2.3		The Central Limit Theorem	48
	2.3.1	Convolution	48
	2.3.2	Statement	50
	2.3.3	Conditions	50
	2.3.4	Collective Phenomenon	51
	2.3.5	Renormalization Group Derivation	52
	2.3.6	Recursion Relation and Perturbative Analysis	55

3. Large Deviations 59
3.1		Cumulant Expansion	59
3.2		Large Deviation Theorem	60
	3.2.1	Quantification of the Deviation from the Central Limit Theorem	61
	3.2.2	Heuristic Derivation of the Large Deviation Theorem (3.9)	61
	3.2.3	Example: the Binomial Law	63
	3.2.4	Non-identically Distributed Random Variables	64
3.3		Large Deviations with Constraints and the Boltzmann Formalism	66
	3.3.1	Frequencies Conditioned by Large Deviations	66
	3.3.2	Partition Function Formalism	68
	3.3.3	Large Deviations in the Dice Game	70
	3.3.4	Model Construction from Large Deviations	73
	3.3.5	Large Deviations in the Gutenberg–Richter Law and the Gamma Law	76
3.4		Extreme Deviations	78
	3.4.1	The "Democratic" Result	78
	3.4.2	Application to the Multiplication of Random Variables: a Mechanism for Stretched Exponentials	80
	3.4.3	Application to Turbulence and to Fragmentation	83
3.5		Large Deviations in the Sum of Variables with Power Law Distributions	87
	3.5.1	General Case with Exponent $\mu > 2$	87
	3.5.2	Borderline Case with Exponent $\mu = 2$	90

4. **Power Law Distributions** 93
   4.1   Stable Laws: Gaussian and Lévy Laws 93
         4.1.1   Definition 93
         4.1.2   The Gaussian Probability Density Function 93
         4.1.3   The Log-Normal Law 94
         4.1.4   The Lévy Laws 96
         4.1.5   Truncated Lévy Laws 101
   4.2   Power Laws... 104
         4.2.1   How Does One Tame "Wild" Distributions? 105
         4.2.2   Multifractal Approach 110
   4.3   Anomalous Diffusion of Contaminants
         in the Earth's Crust and the Atmosphere................ 112
         4.3.1   General Intuitive Derivation 113
         4.3.2   More Detailed Model of Tracer Diffusion in the Crust 113
         4.3.3   Anomalous Diffusion in a Fluid 115
   4.4   Intuitive Calculation Tools
         for Power Law Distributions 116
   4.5   Fox Function, Mittag–Leffler Function
         and Lévy Distributions................................ 118

5. **Fractals and Multifractals** 123
   5.1   Fractals 123
         5.1.1   Introduction 123
         5.1.2   A First Canonical Example: the Triadic Cantor Set . 124
         5.1.3   How Long Is the Coast of Britain?............... 125
         5.1.4   The Hausdorff Dimension 127
         5.1.5   Examples of Natural Fractals 127
   5.2   Multifractals... 141
         5.2.1   Definition 141
         5.2.2   Correction Method for Finite Size Effects
                 and Irregular Geometries 143
         5.2.3   Origin of Multifractality and Some Exact Results... 145
         5.2.4   Generalization of Multifractality:
                 Infinitely Divisible Cascades 146
   5.3   Scale Invariance 148
         5.3.1   Definition 148
         5.3.2   Relation with Dimensional Analysis.............. 150
   5.4   The Multifractal Random Walk 153
         5.4.1   A First Step: the Fractional Brownian Motion 153
         5.4.2   Definition and Properties
                 of the Multifractal Random Walk................ 154
   5.5   Complex Fractal Dimensions
         and Discrete Scale Invariance 156
         5.5.1   Definition of Discrete Scale Invariance 156
         5.5.2   Log-Periodicity and Complex Exponents 157

5.5.3   Importance and Usefulness
of Discrete Scale Invariance 159
5.5.4   Scenarii Leading to Discrete Scale Invariance 160

6.   Rank-Ordering Statistics and Heavy Tails 163
6.1   Probability Distributions 163
6.2   Definition of Rank Ordering Statistics.................... 164
6.3   Normal and Log-Normal Distributions 166
6.4   The Exponential Distribution 167
6.5   Power Law Distributions 170
6.5.1   Maximum Likelihood Estimation 170
6.5.2   Quantiles of Large Events 173
6.5.3   Power Laws with a Global Constraint:
"Fractal Plate Tectonics" 174
6.6   The Gamma Law 179
6.7   The Stretched Exponential Distribution 180
6.8   Maximum Likelihood and Other Estimators
of Stretched Exponential Distributions 181
6.8.1   Introduction 182
6.8.2   Two-Parameter Stretched Exponential Distribution   185
6.8.3   Three-Parameter Weibull Distribution 194
6.8.4   Generalized Weibull Distributions 196

**7.   Statistical Mechanics: Probabilistic Point of View
and the Concept of "Temperature"** 199
7.1   Statistical Derivation of the Concept of Temperature....... 200
7.2   Statistical Thermodynamics 202
7.3   Statistical Mechanics as Probability Theory
with Constraints 203
7.3.1   General Formulation 203
7.3.2   First Law of Thermodynamics 206
7.3.3   Thermodynamic Potentials 207
7.4   Does the Concept of Temperature Apply
to Non-thermal Systems?................................ 208
7.4.1   Formulation of the Problem 208
7.4.2   A General Modeling Strategy 210
7.4.3   Discriminating Tests 211
7.4.4   Stationary Distribution with External Noise 213
7.4.5   Effective Temperature Generated
by Chaotic Dynamics 214
7.4.6   Principle of Least Action
for Out-Of-Equilibrium Systems.................. 218
7.4.7   Superstatistics 219

8. Long-Range Correlations 223
    8.1   Criterion for the Relevance of Correlations............... 223
    8.2   Statistical Interpretation 226
    8.3   An Application: Super-Diffusion in a Layered Fluid
          with Random Velocities 228
    8.4   Advanced Results on Correlations 229
          8.4.1   Correlation and Dependence 229
          8.4.2   Statistical Time Reversal Symmetry 231
          8.4.3   Fractional Derivation and Long-Time Correlations . 236

9. Phase Transitions: Critical Phenomena
    and First-Order Transitions.............................. 241
    9.1   Definition 241
    9.2   Spin Models at Their Critical Points 242
          9.2.1   Definition of the Spin Model 242
          9.2.2   Critical Behavior 245
          9.2.3   Long-Range Correlations of Spin Models
                  at their Critical Points 246
    9.3   First-Order Versus Critical Transitions 248
          9.3.1   Definition and Basic Properties 248
          9.3.2   Dynamical Landau–Ginzburg Formulation 250
          9.3.3   The Scaling Hypothesis: Dynamical Length Scales
                  for Ordering 253

10. Transitions, Bifurcations and Precursors 255
    10.1 "Supercritical" Bifurcation 256
    10.2 Critical Precursory Fluctuations........................ 258
    10.3 "Subcritical" Bifurcation 262
    10.4 Scaling and Precursors Near Spinodals 264
    10.5 Selection of an Attractor in the Absence
          of a Potential 265

11. The Renormalization Group 267
    11.1 General Framework 267
    11.2 An Explicit Example: Spins on a Hierarchical Network 269
          11.2.1 Renormalization Group Calculation............... 269
          11.2.2 Fixed Points, Stable Phases and Critical Points 273
          11.2.3 Singularities and Critical Exponents 275
          11.2.4 Complex Exponents
                  and Log-Periodic Corrections to Scaling 276
          11.2.5 "Weierstrass-Type Functions"
                  from Discrete Renormalization Group Equations ... 279
    11.3 Criticality and the Renormalization Group
          on Euclidean Systems 283

11.4   A Novel Application to the Construction
       of Functional Approximants 287
       11.4.1   General Concepts 287
       11.4.2   Self-Similar Approximants 288
11.5   Towards a Hierarchical View of the World 291

12. The Percolation Model 293
12.1   Percolation as a Model of Cracking 293
12.2   Effective Medium Theory and Percolation 296
12.3   Renormalization Group Approach to Percolation
       and Generalizations 298
       12.3.1   Cell-to-Site Transformation 299
       12.3.2   A Word of Caution
                on Real Space Renormalization Group Techniques .. 301
       12.3.3   The Percolation Model
                on the Hierarchical Diamond Lattice 303
12.4   Directed Percolation 304
       12.4.1   Definitions 304
       12.4.2   Universality Class 306
       12.4.3   Field Theory: Stochastic Partial Differential Equation
                with Multiplicative Noise 308
       12.4.4   Self-Organized Formulation of Directed Percolation
                and Scaling Laws 309

13. Rupture Models 313
13.1   The Branching Model 314
       13.1.1   Mean Field Version or Branching
                on the Bethe Lattice 314
       13.1.2   A Branching–Aggregation Model
                Automatically Functioning at Its Critical Point 316
       13.1.3   Generalization of Critical Branching Models 317
13.2   Fiber Bundle Models and the Effects
       of Stress Redistribution 318
       13.2.1   One-Dimensional System
                of Fibers Associated in Series 318
       13.2.2   Democratic Fiber Bundle Model (Daniels, 1945).... 320
13.3   Hierarchical Model 323
       13.3.1   The Simplest Hierarchical Model of Rupture 323
       13.3.2   Quasi-Static Hierarchical Fiber Rupture Model..... 326
       13.3.3   Hierarchical Fiber Rupture Model
                with Time-Dependence............................ 328
13.4   Quasi-Static Models in Euclidean Spaces 330
13.5   A Dynamical Model of Rupture Without Elasto-Dynamics:
       the "Thermal Fuse Model" 335

13.6   Time-to-Failure and Rupture Criticality 339
        13.6.1   Critical Time-to-Failure Analysis 339
        13.6.2   Time-to-Failure Behavior
                 in the Dieterich Friction Law 343

14. **Mechanisms for Power Laws** 345
     14.1   Temporal Copernican Principle
            and $\mu = 1$ Universal Distribution of Residual Lifetimes 346
     14.2   Change of Variable 348
            14.2.1   Power Law Change of Variable Close to the Origin . 348
            14.2.2   Combination of Exponentials 354
     14.3   Maximization of the Generalized Tsallis Entropy 356
     14.4   Superposition of Distributions........................... 359
            14.4.1   Power Law Distribution of Widths................. 359
            14.4.2   Sum of Stretched Exponentials (Chap. 3).......... 362
            14.4.3   Double Pareto Distribution by Superposition
                     of Log-Normal pdf's 362
     14.5   Random Walks: Distribution of Return Times to the Origin. 363
            14.5.1   Derivation 364
            14.5.2   Applications 365
     14.6   Sweeping of a Control Parameter Towards an Instability.... 367
     14.7   Growth with Preferential Attachment 370
     14.8   Multiplicative Noise with Constraints 373
            14.8.1   Definition of the Process 373
            14.8.2   The Kesten Multiplicative Stochastic Process 374
            14.8.3   Random Walk Analogy 375
            14.8.4   Exact Derivation, Generalization and Applications.. 378
     14.9   The "Coherent-Noise" Mechanism 381
     14.10 Avalanches in Hysteretic Loops and First-Order Transitions
            with Randomness.. 386
     14.11 "Highly Optimized Tolerant" (HOT) Systems 389
            14.11.1 Mechanism for the Power Law Distribution of Fire
                     Sizes 390
            14.11.2 "Constrained Optimization with Limited Deviations"
                     (COLD)... 393
            14.11.3 HOT versus Percolation 393

15. **Self-Organized Criticality** 395
     15.1   What Is Self-Organized Criticality? 395
            15.1.1   Introduction 395
            15.1.2   Definition 397
     15.2   Sandpile Models... 398
            15.2.1   Generalities..................................... 398
            15.2.2   The Abelian Sandpile 398
     15.3   Threshold Dynamics 402

15.3.1   Generalization 402
15.3.2   Illustration of Self-Organized Criticality
Within the Earth's Crust 404
15.4   Scenarios for Self-Organized Criticality 406
15.4.1   Generalities.................................. 406
15.4.2   Nonlinear Feedback of the "Order Parameter"
onto the "Control Parameter" 407
15.4.3   Generic Scale Invariance 409
15.4.4   Mapping onto a Critical Point 414
15.4.5   Mapping to Contact Processes 422
15.4.6   Critical Desynchronization...................... 424
15.4.7   Extremal Dynamics............................. 427
15.4.8   Dynamical System Theory of Self-Organized Criticality 435
15.5   Tests of Self-Organized Criticality in Complex Systems:
the Example of the Earth's Crust..................... 438

16. **Introduction to the Physics of Random Systems** 441
16.1   Generalities.. 441
16.2   The Random Energy Model........................... 445
16.3   Non-Self-Averaging Properties 449
16.3.1   Definitions 449
16.3.2   Fragmentation Models 451

17. **Randomness and Long-Range Laplacian Interactions**...... 457
17.1   Lévy Distributions from Random Distributions of Sources
with Long-Range Interactions 457
17.1.1   Holtsmark's Gravitational Force Distribution 457
17.1.2   Generalization to Other Fields
(Electric, Elastic, Hydrodynamics)................ 461
17.2   Long-Range Field Fluctuations Due to Irregular Arrays
of Sources at Boundaries 463
17.2.1   Problem and Main Results 463
17.2.2   Calculation Methods........................... 464
17.2.3   Applications 471

References 477

Index 525

1. Useful Notions of Probability Theory

The true logic of this world is in the calculus of probabilities.
James Clerk Maxwell

The theory of probability is an important branch of mathematics and our goal is not to review it extensively but only to briefly present useful key concepts and results.

1.1 What Is Probability?

1.1.1 First Intuitive Notions

The intuitive notion of a probability is clear in coin-tossing or roulette games: the probability is defined by the relative frequency of a given outcome when repeating the game, ideally, an infinite number of times. Probabilities reflect our partial ignorance, as in the outcome of the coin-tossing game. Meteorology (what the weather will be like tomorrow or in one week) and hazard estimations of natural catastrophes such as earthquakes and volcanic eruptions also rely on probabilities. For instance, the working group on California earthquake probabilities estimated an 80%–90% probability of a magnitude $m \geq 7$ earthquake within southern California between 1994 and 2024 [450].

This probability estimation is in fact not as obvious as it seems to be. The usual frequency interpretation of probability would lead us to consider ten planet Earths (a hundred or more would be better) carrying ten southern Californias. Then, the probability statement would mean that only about one or two out of the ten would not have an $m \geq 7$ earthquake between 1994 and 2024. This is obviously nonsense because there is a single southern California and there will be only two outcomes: either one or more $m \geq 7$ earthquakes will occur or none. In fact, what is underlying the statement of a probability, in these circumstances where the "game" cannot be repeated, is that there may be many paths leading to the same result and many other paths leading to different outcomes. The probability of an outcome is then a statement about the fraction of paths leading to this outcome.

A specific path involves, for instance, the sequence of smaller precursory earthquakes redistributing stress in southern California as well as all other

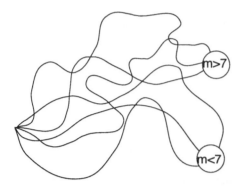

Fig. 1.1. Schematic representation of possible paths or scenarios for the evolution of the system leading to the outcome of an earthquake of magnitude larger or smaller than 7. Note the possible intricate interwoven nature of the set of paths

facts that may have to bear on the outcome of an $m \geq 7$ earthquake. It is reasonable to assume that many such paths can lead to the presence of an $m \geq 7$ earthquake between 1994 and 2024. Many other paths will lead to the absence of such an earthquake. The statement of probability is, thus, our estimation of the fraction of those paths that lead to an $m \geq 7$ earthquake between 1994 and 2024 among all possible ones. It does not embody a hard fact (such as a frequency) but rather our own limited understanding of the possible scenarios and their degree of validity as well as the uncertainty in our understanding of present conditions. The same comments hold for a statement about the probability of the weather in the future. In all these problems, the dynamical evolution of the stress field in the earth or the meteorological variables in the atmosphere is governed by highly non-linear equations exhibiting the property of sensitivity with respect to initial conditions. This sensitivity is responsible for chaotic behaviors and justifies our description of the time evolution in statistical terms.

Another example is the Gutenberg–Richter law, which gives the relative fraction of earthquake magnitudes, in other words, the probability that an earthquake is of a given magnitude. In this case, we measure many earthquakes (at least for the small and intermediate ones) and the Gutenberg–Richter is a frequency statement.

The probability is a number between 0 and 1, where the two extremes correspond to a certainty (i.e. impossible and true, respectively). A probabilistic description is the only tool when we do not know everything in a process, and even if we knew everything, it is still a convenient representation if the process is very complicated. See the chapter on probability in Feynman's Lecture Notes in Physics [299] for an intuitive introduction to the use of probability.

1.1.2 Objective Versus Subjective Probability

The following discussion is based on suggestions by V.F. Pisarenko.

The notion of probability has two aspects: mathematical and applied. As to the mathematical aspect, there is no disagreement among mathematicians.

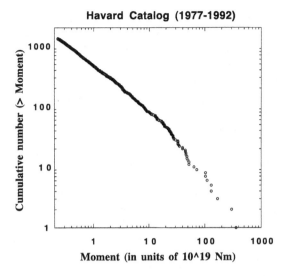

Havard Catalog (1977-1992)

Fig. 1.2. Log–log plot of the number of earthquakes with seismic moment (approximately proportional to the energy release) of the largest shallow earthquakes (depth < 70 km) in the Harvard worldwide catalog. See [903] for more information on the Gutenberg–Richter distribution and its domain of validity

According to the Kolmogorov's axiomatics [519], the probability is a non-negative measure normalized to unity on a σ-algebra of elementary events. But, with respect to the applied aspects of probability, there is no consensus among experts. There are two main approaches (with many variations):

- the so-called "objective approach," named sometimes "frequency approach," see e.g. [186, 240, 292, 304, 631, 682], and
- the "subjective approach," see e.g. [462, 524, 653, 813].

In the objective approach, using a probability requires an idealized model of the experiment that can be, at least mentally, repeated an arbitrary number of times. The objective approach thus uses the notions of "population" and of "ensemble of realizations," and therefores excludes unique events automatically. For instance, it is difficult in the objective approach to interpret what is the meaning of probability in the statement: "The probability that our Universe was created as a result of the Big Bang approximately 14×10^9 years ago equals 0.75." The Universe was created by one or another unique way. It is difficult to invent an "ensemble of realizations" here.

There are events that can be imbedded in an ensemble (e.g., the ensemble of planets with living conditions close to ours), but such imbedding would be empty if we are interested just in the unique member of the ensemble (the Earth). In the introduction in Sect. 3 of his first volume [292], W. Feller describes the objective approach to the applied notion of probability as follows: "We shall be concerned further not with methods of inductive conclusions, but with such entities that can be called physical, or statistical probability. Roughly speaking, we can characterize this notion by saying that our probabilities refer to outcomes of a mental experiment, and not to opinions. In our system, there is no place for hypotheses using the probability of sunrise tomorrow. Before speaking about such probability, we should stipulate that some (idealized) model of experiment would exist. Such model could look as follows: 'We select in a random way one of an infinite ensemble of planet systems.' A little imagination is enough in order to construct such a model, but it turns out to be both uninteresting and senseless."

In the objective approach, the probability of some event is thought as a stable frequency of its appearance in a long series of repeated experiments, conducted in identical conditions and independently. It should be noted, however, that some statistical conclusions can be inferred concerning unique events or objects. Such conclusions are connected with so-called confidence regions for unknown (and non-random) parameters. There is no contradiction in such inference with the objective frequency approach since a confidence region (in particular, a confidence interval for a one-dimensional parameter) is a domain with random boundaries derived from a random sample. Thus, the assertion "The parameter x has a confidence interval C with probability $1 - p$" has a clear statistical meaning and the probability $1 - p$ corresponds to some repeated experiment.

The objective frequency point of view on applied aspects of probability was adopted by such eminent mathematicians as Kolmogorov, Feller, Doob, Cramer, Mises, Fisher, Neyman.

In the subjective approach, a probability can be assigned to any event or assertion. It expresses a measure of likelihood of occurrence of this event. Such measure has often a subjective character [462] (see as well "personal" probabilities introduced by Savage [813] and "intuitive" probabilities [524]). This approach does not necessarily require ensembles, and probabilities can be assigned to such events as e.g.: the city of Rome was founded by Romulus; the Riemann hypothesis concerning non-trivial zeros of the ζ-function is true; "the Iron Mask" was the brother of Louis XIV [186]; there exists an organic life on the planet of Mars [186], etc.

Adherents of the frequency approach do not generally deny the possible usefulness of subjective opinions with regard to the likelihood of unique events of the type mentioned above, or with regard to the usefulness of the Bayes theorem, discussed in the next section, that handles these opinions. Sometimes, these opinions are necessary for practical purposes, because there may exist no corresponding device, or no objective measurement methods. An example is found in sport gymnastic, in figure skating, in wine competitions and so on. It is natural to use statistical methods (the Bayes theorem in particular) in treating personal marks in such cases. Let us mention for instance the well-known practice of throwing out the highest and lowest marks before averaging the remaining marks given by referees in figure skating. This approach is justified by the statistical theory of robust estimation.

However, unchecked use of subjective probabilities can lead sometimes to misleading or false results because they reflect not only the objective reality of a phenomenon under study but as well the level of knowledge of a particular group of experts. There is thus a "noise" or bias introduced by experts, and sometimes it is difficult to estimate its level. Let us consider two possible subjective estimations of probabilities. Perhaps, an "expert" opinion concerning our solar system before Copernicus and Galileo would have assigned a probability, say 0.9999, to the assertion that the Sun turns around the Earth and not conversely. Similarly, an "expert" opinion concerning mechanical motion before Einstein would have assigned a probability, say 0.9999, to the possibility of relative speeds between two bodies larger than the speed of light. We know now that both assertions turned out to be false.

It is difficult to delineate definitely the domains where subjective opinions and probabilities are useful from those where they are dangerous. For instance, the construction of such a boundary is a serious problem in medicine. What can be asserted as a rule is that the use of a statistical terminology in the subjective approach is more fuzzy and vague with often indefinite meaning. Inferences based on subjective probabilities should not be regarded as objective results confirmed by the theory of probability.

A.N. Kolmogorov defined the applied aspects of probability as follows (see for more details [521–523]): "Probabilities are objective characteristics of those phenomena which, due to some their inherent properties, possess frequency stability; the probability of an event does not depend on any subjective opinion."

The subjective approach is based on Bayes' theorem discussed in the next sections. In the context of this general discussion, let us contrast the content of Bayes' theorem with the objective approach. Bayes' theorem, or the Bayes' formula of a posteriori probability, has as "input" an a priori distribution of some parameter, which defines the hypothesis. As "output," this theorem provides an a posteriori distribution based on an observed sample. An a priori distribution can be sometimes suggested on the basis of past experiments. The prior as it is called can in particular be constructed on some statistical frequency estimates. However, in most cases, this a priori distribution is unknown. In such cases, one has to set a priori distributions in an arbitrary way in order to be able to apply Bayes' theorem. One then refers to "equally probable a priori chances," or one uses "for the lack of the better" a uniform distribution in some region whose choice is often arbitrary. Such references are mostly ill-justified and unconvincing. The main fact justifying the application of an arbitrary a priori distribution (continuous and positive in the whole domain of definition) consists in a theorem of equivalence between Bayesian estimates and maximum likelihood estimates (see Chap. 3, Sect. 53, Theorem 1 in [103]). For the sake of fairness, it should be noted that the least favorable a priori distribution in the Bayesian approach is quite reasonable and can be justified from the point of view of the minimax criterion, which minimizes the largest possible risk. This approach is related to the probability of a false decision in hypothesis testing and provides some measure of deviation of parameter estimates from the true value.

An inference or estimate based on subjective a priori distributions often turns out to be quite sensible and useful. However, such an inference needs very careful checking in practice. Let us finish with an exemplary historical case. The famous Maxwell–Boltzmann statistic of classical physics (see Sect. 3.3 and Chap. 7) appears as the result of "natural" or "random" distributions of particles over phase space cells. However, when physicists tried to apply this statistic to quantum mechanical systems, it was found that no known particle or systems of particles did obey the Maxwell–Boltzmann statistic. Thus, one had to replace the "natural," random Boltzmann distribution by the Bose–Einstein distribution (applied to photons and some other boson particles), and by the Fermi–Dirac distribution (applied to electrons and other fermion particles). In this case, an intuitively clear uniform distribution of particles at the microscopic level failed when applied in quantum mechanics.

Thus, there is no ground to oppose the Bayesian to the frequency approach, as well as to consider it as a more fruitful one. Both have value when used carefully with full knowledge and appreciation of their domain of application and limitations.

1.2 Bayesian View Point

1.2.1 Introduction

We take inspiration from [193] for this brief presentation. Traditionally, the various contributions to uncertain knowledge are classified in terms of "statistical" and "systematic" uncertainties, which reflect the sources. Statistical uncertainties vanish, in general, if the number of observations becomes very large (except for certain systems said to possess non-averaging properties as occurs in random media [622]; see Chaps. 16–17). On the other hand, it is not possible to treat "systematic" uncertainties coherently in the frequence framework, as there are no universally accepted prescriptions for how to combine the "statistical" and "systematic" uncertainties (linear or nonlinear or partial addition, etc.). The only way to deal with these and related problems in a consistent way is to abandon the frequence interpretation of probability introduced at the beginning of this century, and to recover the intuitive concept of probability as *degree of belief*. Stated differently, one needs to associate the idea of probability with the lack of knowledge, rather than with the outcome of repeated experiments. This has been recognized also by the International Organization for Standardization (ISO) which assumes the subjective definition of probability in its *Guide to the Expression of Uncertainty in Measurement* [448].

The three different definitions of a probability are:

1. *combinatorial* – the ratio of the number of favorable cases to the number of all cases;

2. *frequence* – the ratio of the times the event occurs in a test series to the total number of trials in the series;
3. *Bayesian* – a measure of the degree of belief that an event will occur.

Indeed, the concept of "probable" arises in reasoning when the concept of "certain" is not applicable. When it is impossible to state firmly if an event or proposition, relative to past, present or future, is *true* or *false*, we just say that this is possible or probable. Different events may have different levels of probability, depending on whether we think that they are more likely to be true or false. This is the definition found in books on Bayesian analysis [83, 212, 462, 747, 1023] and used in the ISO *Guide to Expression of Uncertainty in Measurement* [448].

Bayesian statistics is based on the subjective definition of probability as "degree of belief" and on Bayes' theorem, the basic tool for assigning probabilities to hypotheses combining a priori judgements and experimental information. This was the original point of view of Bayes, Bernoulli, Gauss and Laplace, and contrasts with later conventional definitions of probabilities, which implicitly presuppose the concept of frequences, as we already mentioned. The Bayesian approach is useful for data analysis and for assigning uncertainties to the results of measurements. Let us summarize its main properties.

- The Bayesian definition is natural and general, and can be applied to any thinkable event, independent of the feasibility of making an inventory of all (equally) possible and favorable cases, or of repeating the experiment under conditions of equal probability.
- It avoids the problem of having to distinguish "scientific" probability from "non-scientific" probability used in everyday reasoning.
- As far as measurements are concerned, it allows us to talk about the probability of the *true value* of a physical quantity. In the frequence frame, it is only possible to talk about the probability of the *outcome* of an experiment, as the true value is considered to be a constant.
- It is possible to make a general theory of uncertainty which can take into account any source of statistical and systematic error, independently of their distribution.

1.2.2 Bayes' Theorem

We briefly sum up the most important rules for conditional probabilities from which Bayes' theorem can be derived and we give a few illustrations. The expression of the conditional probability $P(E|H)$ that event E occurs under the condition that hypothesis H occurs is

$$P(E|H) = \frac{P(E \cap H)}{P(H)} \qquad (P(H) \neq 0).$$ (1.1)

Thus

$$P(E \cap H) = P(E|H)P(H),\tag{1.2}$$

and by symmetry

$$P(E \cap H) = P(H|E)P(E).\tag{1.3}$$

Two events are called *independent* if

$$P(E \cap H) = P(E)P(H).\tag{1.4}$$

This is equivalent to saying that $P(E|H) = P(E)$ and $P(H|E) = P(H)$, i.e. the knowledge that one event has occurred does not change the probability of the other. If $P(E|H) \neq P(E)$, then the events E and H are *correlated*. In particular:

- if $P(E|H) > P(E)$ then E and H are *positively* correlated;
- if $P(E|H) < P(E)$ then E and H are *negatively* correlated.

Let us think of all the possible, mutually exclusive, hypotheses H_i which could condition the event E. The problem is to determine the probability of H_i under the hypothesis that E has occurred. This is a basic problem for any kind of measurement: having observed an *effect*, to assess the probability of each of the *causes* which could have produced it. This intellectual process is called *inference* and is a fundamental building block of the construction of scientific knowledge.

In order to calculate $P(H_i|E)$, let us rewrite the joint probability $P(H_i \cap E)$, making use of (1.2) and (1.3), in two different ways:

$$P(H_i|E)P(E) = P(E|H_i)P(H_i),\tag{1.5}$$

obtaining

$$P(H_i|E) = \frac{P(E|H_i)P(H_i)}{P(E)}\tag{1.6}$$

or

$$\frac{P(H_i|E)}{P(H_i)} = \frac{P(E|H_i)}{P(E)}.\tag{1.7}$$

Since the hypotheses H_i are mutually exclusive (i.e. $H_i \cap H_j = \emptyset, \forall i, j$) and exhaustive (i.e. $\bigcup_i H_i = \Omega$), E can be written as $E \cup H_i$, the union of E with each of the hypotheses H_i. It follows that

$$P(E) = P\left(E \cap \bigcup_i H_i\right) = P\left(\bigcup_i (E \cap H_i)\right),$$
$$= \sum_i P(E \cap H_i),$$
$$= \sum_i P(E|H_i)P(H_i),\tag{1.8}$$

where we have made use of (1.2) again in the last step. It is then possible to rewrite (1.6) as

$$P(H_i|E) = \frac{P(E|H_i)P(H_i)}{\sum_j P(E|H_j)P(H_j)} \, . \tag{1.9}$$

This is the standard form of *Bayes' theorem*. The denominator of (1.9) is nothing but a normalization factor, such that $\sum_i P(H_i|E) = 1$.

- $P(H_i)$ is the *initial*, or *a priori*, probability (or simply *"prior"*) of H_i, i.e. the probability of this hypothesis with the information available before the knowledge that E has occurred;
- $P(H_i|E)$ is the *final*, or *"a posteriori"*, probability of H_i after the new information;
- $P(E|H_i)$ is called the *likelihood*.

1.2.3 Bayesian Explanation for Change of Belief

To better understand the terms "initial", "final", and "likelihood", let us formulate the problem in a way closer to the mentality of scientists, referring to *causes* and *effects*: the causes can be all the physical sources which may produce a certain *observable* (the effect). The likelihoods are – as the word says – the likelihoods that the effect follows from each of the causes.

Let us assume that a natural catastrophe (a hurricane such as Andrew [748]) occurs at a rate of once in 100 years, but there is a concern over the possible effect of human impact, such as "global warming", in modifying this natural rate. Let us assume that the public and scientific community believes a priori that the existence H_e and nonexistence H_{ne} of the anthropogenic effect is equally probable, that is $P(H_e) = P(H_{ne}) = 0.5$. In the absence of "global warming", we assume that $P(\text{rate} \leq 1 \text{ per century}|H_{ne}) = 0.99$ while $P(\text{rate} > 1 \text{ per century}|H_{ne}) = 0.01$. Suppose that we are able to estimate the impact "global warming" may have on the rate of hurricanes; global warming is assumed to increase the likelihood of large hurricanes to the value $P(\text{rate} > 1 \text{ per century}|H_e)$ that we take equal to 0.5. Thus, $P(\text{rate} \leq 1 \text{ per century}|H_e) = 0.5$ also (these numbers are given only for the sake of the example and do not represent genuine scientific calculations). Suppose then that two catastrophic hurricanes of the magnitude of Andrew occurred in our century. Then, (1.9) gives the probability for the hypothesis that global warming is present:

$$P(H_e|\text{rate} > 1) =$$
$$\frac{P(\text{rate} > 1 \text{ per century}|H_e)P(H_e)}{P(\text{rate} > 1 \text{ per century}|H_e)P(H_e) + P(\text{rate} > 1 \text{ per century}|H_{ne})P(H_{ne})}$$
$$= \frac{0.5 \times 0.5}{0.5 \times 0.5 + 0.01 \times 0.5} \approx 98\% \, . \tag{1.10}$$

With these numbers, the occurrence of just one additional catastrophic hurricane would make the existence of "global warming" very plausible. The public will not think that the second catastrophic hurricane was simply bad luck. See [541] for other applications in engineering and [434] for a quantification of how a skeptical belief (with prior belief of 10^{-3}) can be turned into an a posteriori probability of 93%, given the evidence provided by the data.

1.2.4 Bayesian Probability and the Dutch Book

A very interesting view expounded in [150] is that Bayesian probabilities are "consistent" subjective quantification of the degree of uncertainty on a system. The following exposition borrows from Caves et al. [150]. The key idea is to develop an operational definition of the degrees of belief or uncertainty by Bayesian probabilities using decision theory [814], i.e., the theory of how to decide in the face of uncertainty. The Bayesian approach captures naturally the notion that probabilities can change when new information is obtained. Recall that the fundamental Bayesian probability assignment is to a single system or a single realization of an experiment. As we already stressed, Bayesian probabilities are defined without any reference to the limiting frequency of outcomes in repeated experiments. Bayesian probability theory does allow one to make (probabilistic) predictions of frequencies. Rather, frequencies in past experiments provide valuable information for updating the probabilities assigned to future trials. Despite this connection, probabilities and frequencies are strictly separate concepts.

The simplest operational definition of Bayesian probabilities is in terms of *consistent betting behavior*, which is decision theory in a nutshell. Consider a bookie who offers a bet on the occurrence of outcome E in some situation. The bettor pays in an amount px – the *stake* – up front. The bookie pays out an amount x – the *payoff* – if E occurs and nothing otherwise. Conventionally this is said to be a bet at *odds* of $(1 - p)/p$ to 1. For the bettor to assign a probability p to outcome E means that he is willing to accept a bet at these odds with an arbitrary payoff x determined by the bookie. The payoff can be positive or negative, meaning that the bettor is willing to accept either side of the bet. We call a probability assignment to the outcomes of a betting situation *inconsistent* if it forces the bettor to accept bets in which he incurs a sure loss; i.e., he loses for every possible outcome. A probability assignment will be called *consistent* if it is not inconsistent in this sense.

Remarkably, consistency alone implies that the bettor must obey the standard probability rules in his probability assignment: (i) $p \geq 0$, (ii) $p(A \vee B) = p(A) + p(B)$ if A and B are mutually exclusive, (iii) $p(A \wedge B) = p(A|B)p(B)$, and (iv) $p(A) = 1$ if A is certain. $A \vee B$ means A or B; $A \wedge B$ means A and B. Any probability assignment that violates one of these rules can be shown to be inconsistent in the above sense. This is the so-called *Dutch-book argument* [206, 259]. We stress that it does not invoke expectation values or averages in repeated bets; the bettor who violates the probability rules suffers a sure loss in a single instance of the betting situation.

For instance, to show that $p(A \vee B) = p(A) + p(B)$ if A and B are mutually exclusive, assume that the bettor assigns probabilities p_A, p_B, and p_C to the three outcomes A, B, and $C = A \vee B$. This means he will accept the following three bets: a bet on A with payoff x_A, which means the stake is $p_A x_A$; a bet on B with payoff x_B and thus with stake $p_B x_B$; and a bet on C with payoff x_C and thus with stake $p_C x_C$. The net amount the bettor receives is

$$R = \begin{cases} x_A(1 - p_A) - x_B p_B + x_C(1 - p_C) & \text{if } A \wedge \neg B \\ -x_A p_A + x_B(1 - p_B) + x_C(1 - p_C) & \text{if } \neg A \wedge B \text{ ;} \\ -x_A p_A - x_B p_B - x_C p_C & \text{if } \neg A \wedge \neg B \end{cases} \quad (1.11)$$

$\neg B$ means non-B. The outcome $A \wedge B$ does not occur since A and B are mutually exclusive. The bookie can choose values x_A, x_B, and x_C that lead to $R < 0$ in all three cases unless

$$0 = \det \begin{pmatrix} 1 - p_A & -p_B & 1 - p_C \\ -p_A & 1 - p_B & 1 - p_C \\ -p_A & -p_B & -p_C \end{pmatrix} = p_A + p_B - p_C . \quad (1.12)$$

The probability assignment is thus inconsistent unless $p(A \vee B) = p_C = p_A + p_B$.

As scientists working in the natural or social sciences, it is important to accept and embrace the notion that subjective probabilities receive their *only* operational significance from decision theory, the simplest example of which is the Dutch-book argument in which probabilities are *defined* to be betting odds. In the Dutch-book approach, the structure of probability theory follows solely from the requirement of consistent betting behavior. There is no other input to the theory. For example, normalization of the probabilities for exclusive and exhaustive alternatives is not an independent assumption, so obvious that it needs no justification. Instead normalization follows from probability rules (ii) and (iv) above and thus receives its sole justification from the requirement of consistent betting behavior.

The only case in which consistency alone leads to a particular numerical probability is the case of certainty, or *maximal information*. If the bettor is certain that the outcome E will occur, the probability assignment $p < 1$ means he is willing to take the side of the bookie in a bet on E, receiving an amount px up front and paying out x if E occurs, leading to a certain loss of $x(1 - p) > 0$. Consistency thus requires that the bettor assign probability $p = 1$. More generally, consistency requires a particular probability assignment only in the case of maximal information, which classically always means $p = 1$ or 0.

Caves et al. [150] have used the Dutch book approach to show that, despite being prescribed by a fundamental law, probabilities for individual quantum systems can be understood within the Bayesian approach. This approach quantifying the lack of knowledge may also be instrumental in the broad questions of model validation for decision making, as occurs for instance in climate modeling (global warming or not?), reliability of a rocket (launch or not?), or of a nuclear stockpile in the context of nuclear stewardship, to cite a few example.

1.3 Probability Density Function

Consider a process X whose outcome is a real number. The probability density function (pdf) $P(x)$ (also called probability distribution) of X is such that the probability that X is found in a small interval Δx around x is $P(x) \Delta x$. The probability that X is between a and b is given by the integral of $P(x)$ between a and b:

$$\mathcal{P}(a < X < b) = \int_a^b P(x)\, \mathrm{d}x \ . \tag{1.13}$$

The pdf $P(x)$ depends on the units used to quantity the variable x and has the dimension of the inverse of x, such that $P(x) \Delta x$, being a probability i.e. a number between 0 and 1, is dimensionless. In a change of variable, say $x \to y = f(x)$, the probability is invariant. Thus, the invariant quantity is the probability $P(x) \Delta x$ and not the pdf $P(x)$. We thus have

$$P(x) \Delta x = P(y) \Delta y \ , \tag{1.14}$$

leading to $P(y) = P(x)|\mathrm{d}f/\mathrm{d}x|^{-1}$, taking the limit of infinitesimal intervals. We will repeatedly use this expression (1.14).

By definition, $P(x) \geq 0$. It is normalized

$$\int_{x_{\min}}^{x_{\max}} P(x)\, \mathrm{d}x = 1 \ , \tag{1.15}$$

where $x_{\min}$ and $x_{\max}$ are the smallest and largest possible values for x, respectively. We will, from now on, take the upper and lower bounds as respectively $-\infty$ and $+\infty$ by putting $P(x) = 0$ for $-\infty < x \leq x_{\min}$ and $x_{\max} \leq x < +\infty$. We all know how to plot the pdf $P(x)$ with the horizontal axis scaled as a graded series for the measure under consideration (the magnitude of the earthquakes, etc.) and the vertical axis scaled for the number of outcomes or measures in each interval of horizontal value (the earthquakes of magnitude between 1 and 2, between 2 and 3, etc.). This implies a "binning" into small intervals. If the data is sparse, the number of events in each bin becomes small and fluctuates making a poor representation of the data. In this case, it is useful to construct the cumulative distribution $P_\leq(x)$ defined by

$$\mathcal{P}_\leq(x) = \mathcal{P}(X \leq x) = \int_{-\infty}^x P(y)\, \mathrm{d}y \ , \tag{1.16}$$

which is much less sensitive to fluctuations. $\mathcal{P}_\leq(x)$ gives the fraction of events with values less than or equal to x. $\mathcal{P}_\leq(x)$ increases monotonically with x from 0 to 1. Similarly, we can define $\mathcal{P}_>(x) = 1 - \mathcal{P}_\leq(x)$.

For random variables which take only discrete values $x_1, x_2, \ldots, x_n$, the pdf is made of a discrete sum of Dirac functions $(1/n)[\delta(x - x_1 + \delta(x - x_2) + \ldots + \delta(x - x_n)]$. The corresponding cumulative distribution function (cdf) $\mathcal{P}_\leq(x)$ is a staircase. There are also more complex distributions made of continuous cdf but which are singular with respect to the Lebesgue measure $\mathrm{d}x$. An example is the Cantor distribution constructed from the Cantor set (see Chap. 5). Such singular cdf is continuous but has its derivative which is zero almost everywhere: the pdf does not exist (see e.g. [293], Chapt. 5, Sect. 3a).

1.4 Measures of Central Tendency

What is the typical value of the outcome of a chance process? The answer is given by the so-called "central tendency". There are three major measures of central tendency. The mean or average noted $\langle x \rangle$ is obtained by adding all the values and dividing by the number of cases, or in continuous notation

$$\langle x \rangle = \int_{-\infty}^{\infty} x P(x)\, \mathrm{d}x \ . \tag{1.17}$$

The median $x_{1/2}$ is the halfway point in a graded array of values, in other words half of all values are below $x_{1/2}$ and half are above:

$$\mathcal{P}_\leq(x_{1/2}) = \frac{1}{2} \ . \tag{1.18}$$

For instance, the IQ median is 100. Finally, the most probable value or mode is the value x_{mp} that maximizes $P(x)$:

$$\frac{\mathrm{d}P(x)}{\mathrm{d}x}\Big|_{x=x_{\mathrm{mp}}} = 0 \ . \tag{1.19}$$

If several values satisfy this equation, the most probable is the one with the largest $P(x)$.

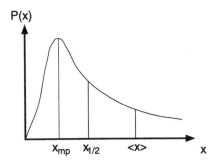

P(x)

x_{mp} $x_{1/2}$ $\langle x \rangle$ x

Fig. 1.3. Distribution skewed to the right with a thick tail, where $x_{\mathrm{mp}}, x_{1/2}$ and $\langle x \rangle$ are represented

For a unimodal symmetric pdf such as the Gauss distribution, all these three quantities are equal. However, for a skewed pdf, they differ and all the more so if the pdf exhibits a thick tail to one side. As an extreme illustration, let us write the pdf of (normalized) earthquake energies as

$$P(E) = \mu E^{-(1+\mu)} , \qquad \text{for } 1 \leq E < \infty , \qquad (1.20)$$

where μ is usually taken close to 2/3. Then, $E_{\text{mp}} = 1$, $E_{1/2} = 2^{1/\mu} \approx 2.8$ and $\langle E \rangle = \infty$, because $\mu < 1$. The diverging average betrays the fact that the average is controlled by the largest event ever measured and not by the crowd of small events (see Chap. 4). For pdf's which are skewed, say to the right as in this example, we have in general $x_{\text{mp}} \leq x_{1/2} \leq \langle x \rangle$. The difference between $x_{\text{mp}}, x_{1/2}$ and $\langle x \rangle$ has an important significance: depending on the type of measurement, the "typical" behavior will be different. For a few trials, the most probable values will be sampled first and their typical value will be not far from x_{mp}. An average made on a few such measures will thus be close to x_{mp}. On the other hand, if many measurements are made, their average will drift progressively to the true $\langle x \rangle$ as their number increases. A clear discussion of this cross-over is given in [764] for the log-normal distribution which is also discussed in Chap. 4. This evolution of the apparent average as a function of the sampling size has obvious important consequences and should be kept in mind.

1.5 Measure of Variations from Central Tendency

When repeating a measurement or an observation several times, one expects them to be within an interval anchored at the central tendency (when well-defined) of a certain width. This width is a measure of the variations. There are several ways to measure the width of the variations. A first measure is the average of the absolute value of the spread (D_{abs}) defined by

$$D_{\text{abs}} = \int_{-\infty}^{\infty} |x - x_{1/2}| P(x) \, dx . \qquad (1.21)$$

D_{abs} does not always exist, such as for pdf's with thick tails decaying as (or slower than) $1/x^2$ for large x.

The *variance* σ^2 is the square of the standard deviation σ which provides a second measure and is defined by the average of the square of the distance to the mean:

$$\sigma^2 = \langle x^2 \rangle - \langle x \rangle^2 = \int_{-\infty}^{+\infty} (x - \langle x \rangle)^2 P(x) \, dx . \qquad (1.22)$$

The variance does not always exist, such as for pdf's with thick tails decaying as (or slower than) $1/x^3$ for large x.

1.6 Moments and Characteristic Function

The moments are defined by the average of the powers of x:

$$m_n = \langle x^n \rangle = \int_{-\infty}^{+\infty} x^n P(x)\, dx . \tag{1.23}$$

The mean $\langle x \rangle$ is the first moment m_1 and the variance is related to the second moment: $\sigma^2 = m_2 - m_1^2$. For this definition (1.23) to be meaningful, the integral on the right-hand side must be convergent, i.e. $P(x)$ must decay sufficiently fast for large $|x|$.

A necessary condition for the existence of a moment m_n of order n is that the pdf $P(x)$ decays faster than $1/|x|^{n+1}$ for $x \to \pm\infty$. This is trivially obeyed for pdf's which vanish outside a finite region.

Consider a pdf whose asymptotic shape for large $|x|$ is a power law with exponent μ,

$$P(x) \sim \frac{C_\pm}{|x|^{1+\mu}} \text{ for } x \to \pm\infty . \tag{1.24}$$

This pdf does not have moments of order $n \geq \mu$. For instance, this pdf does not have a variance if $\mu \leq 2$ and does not have a mean if $\mu \leq 1$, as already seen. Only positive μ's need to be considered to ensure the normalization of $P(x)$.

Statisticians often work with moments (or cumulants, see below) because they replace the difficult problem of determining a full functional behavior (the pdf) with the estimation of a few numbers. In principle, the knowledge of all the moments is (almost) equivalent to that of the pdf. This is not strictly correct as there are examples where two different distributions have the same moments. For instance, the pdf

$$\frac{1}{\sqrt{2\pi}} x^{-1} \exp\left[-\frac{1}{2} (\ln x)^2\right] [1 + a \sin(2\pi \ln x)] ,$$

with $-1 < a < 1$, has exactly the same moments as the log-normal distribution

$$\frac{1}{\sqrt{2\pi}} x^{-1} e^{-(1/2)(\ln x)^2}$$

(see [293], Vol. II, p. 227). This constitutes a warning that it is always preferable to work with the complete pdf. The strict equivalence between the knowledge of all the moments and the pdf is obtained under additional analyticity conditions of the characteristic function in the neighborhood of the origin [500]. Specifically, if the characteristic function (as defined in (1.25) below) admits an expansion in integer powers of ik up to order q included, then all moments up to order q and vice-versa [575].

The calculation of an increasing number of moments thus offers a succession of improving approximations for the determination of the pdf. In practice however, the determination of the moments of high order is generally unstable, which makes their use delicate for the description of empirical data. For instance, the intermittent nature of hydrodynamic turbulence is often characterized by calculating the moments of order smaller than 10, while higher orders are found to be unreliable even for the largest available time series.

The Fourier transform of $P(x)$ (in the continuous case) defines the characteristic function $\hat{P}(k)$:

$$\hat{P}(k) = \int_{-\infty}^{\infty} \exp\left(\mathrm{i}kx\right) P(x)\,\mathrm{d}x \ . \tag{1.25}$$

Inversely, we have

$$P(x) = \frac{1}{2\pi} \int_{-\infty}^{\infty} \exp\left(-\mathrm{i}kx\right) \hat{P}(k)\,\mathrm{d}k \ . \tag{1.26}$$

The normalization condition on $P(x)$ is equivalent to $\hat{P}(0) = 1$. The moment of the pdf can be obtained from the successive derivatives of the characteristic function at $k = 0$:

$$m_n = (-\mathrm{i})^n \left.\frac{\mathrm{d}^n}{\mathrm{d}k^n}\hat{P}(k)\right|_{k=0} \ . \tag{1.27}$$

Provided some analyticity conditions are obeyed and all moments exist, the characteristic function reads

$$\hat{P}(k) = \sum_n^{\infty} \frac{m_n}{n!}(\mathrm{i}k)^n \ . \tag{1.28}$$

1.7 Cumulants

The probability of a set of independent events is the product of their probabilities. It is thus convenient to work with $\ln P(x)$ (which is additive) and define the analog of (1.27) and (1.28). This leads to the introduction of the *cumulants* c_n of a pdf, defined as the derivatives of the *logarithm* of its characteristic function:

$$c_n = (-\mathrm{i})^n \left.\frac{\mathrm{d}^n}{\mathrm{d}k^n} \ln \hat{P}(k)\right|_{k=0} \ . \tag{1.29}$$

In other words, we can write the characteristic function as

$$\hat{P}(k) = \exp\left[\sum_n^{\infty} \frac{c_n}{n!}(\mathrm{i}k)^n\right] \ . \tag{1.30}$$

A cumulant c_n is a combination of moments m_l of orders $l \leq n$, as can be easily seen by expansion of the exponential in the right-hand side of (1.30). We get for the six first cumulants [937]

$$c_1 = m_1 \,, \tag{1.31a}$$

$$c_2 = m_2 - m_1^2 \,, \tag{1.31b}$$

$$c_3 = m_3 - 3m_2 m_1 + 2m_1^3 \,, \tag{1.31c}$$

$$c_4 = m_4 - 4m_3 m_1 - 3m_2^2 + 12m_2 m_1^2 - 6m_1^4 \,, \tag{1.31d}$$

$$c_5 = m_5 - 5m_4 m_1 - 10m_3 m_2 + 20m_3 m_1^2 + 30m_2^2 m_1$$
$$- 60m_2 m_1^3 + 24m_1^5 \,, \tag{1.31e}$$

$$c_6 = m_6 - 6m_5 m_1 - 15m_4 m_2 + 30m_4 m_1^2 - 10m_3^2$$
$$+ 120m_3 m_2 m_1 - 120m_3 m_1^3 + 30m_2^3$$
$$- 270m_2^2 m_1^2 + 360m_2 m_1^4 - 120m_1^6 \,. \tag{1.31f}$$

Cumulants enjoy remarkably useful properties. For instance, the cumulants of the pdf of the sum of two independent random variables are just the sum of the respective cumulants of the pdf's of each variable. To derive this result, take $x = x_1 + x_2$, where x_1 and x_2 are independent and distributed with the pdf's P_1 and P_2 respectively. Then, the pdf of x is given by

$$P(x) = \int_{-\infty}^{\infty} dx_1 P_1(x_1) \int_{-\infty}^{\infty} dx_2 P_2(x_2) \delta(x - x_1 - x_2)$$
$$= \int_{-\infty}^{\infty} dx_1 P_1(x_1) P_2(x - x_1) \,, \tag{1.32}$$

showing that $P(x)$ is the convolution of P_1 and P_2. Now, the Fourier transform of the product of convolution is the product of the Fourier transform:

$$\hat{P}(k) = \hat{P}_1(k)\hat{P}_2(k). \tag{1.33}$$

Using (1.30) in this expression and identifying term by term the coefficient of k^n, we get the announced result

$$c_n = c_n^{(1)} + c_n^{(2)} \,. \tag{1.34}$$

Another important property is that cumulants of order larger than two offer natural measures of the deviation from normality, with an increasing sensitivity to the largest fluctuations as the order n increases. Indeed, the normal (or Gauss) law has all its cumulants of order larger than 2 identically zero.

Normalized cumulants are defined by

$$\lambda_n \equiv \frac{c_n}{\sigma^n} \,. \tag{1.35}$$

$\lambda_3 \equiv c_3/\sigma^3$ is called the skewness and

$$\kappa \equiv \lambda_4 = \frac{\langle (x - \langle x \rangle)^4 \rangle}{\sigma^4} - 3 \,, \tag{1.36}$$

is called the *excess kurtosis*. The (normal) kurtosis, defined as m_4/m_2^2, is simply $\kappa + 3$ for symmetric pdf's. For symmetric distributions, κ quantifies the first correction to the Gaussian approximation. A Gaussian pdf has zero excess kurtosis and normal kurtosis equal to 3.

1.8 Maximum of Random Variables and Extreme Value Theory

Central values and typical fluctuations are not enough to characterize natural systems which exhibit rare but extreme events often dominating the long term balance:

- the largest earthquake in California accounts for a significant fraction, maybe a third, of the total long-term energy released by the crust;
- a centenial or millenial flood has often by itself more impact on erosion and landscape shaping than the cumulative effect of all other erosion mechanisms;
- the largest volcanic eruptions bring in major meteorological perturbations that may lead to important modifications of the biosphere;
- the largest hurricane (such as Andrews for the twentieth century in terms of impact on human properties [748]) as well as a large earthquake in Los Angeles or Tokyo may have a major impact on the economy of the country.

It is thus important to determine the statistical properties of such rare but extreme events, with applications to the calculation of the largest risks that insurance companies and governments must face, to the construction of wave and tsunami barriers, to the definition of engineering building codes and so on.

The theory of extreme events is called extreme value theory (EVT). The two classic references are Gumbel [381] and Galambos [326]. A more recent book [274] provides a synthetis of recent results on EVT, with application to finance and insurance. We also refer to other general references of interest that cover other developments and applications [148, 269, 327, 380, 504]. The EVT evaluates the probable size of the largest event among a population, or the probability of exceedance, i.e. that the size of the event be larger than some value. The asumption of independence can be relaxed [274]. EVT has also been worked out for processes both in discrete and continuous time, with or without independence and/or stationarity assumptions (see [274] and references therein).

Heterogeneous systems often exhibit large fluctuations and their physics is often controlled by some rare and large fluctuations. There is a close link between the different classes of extreme value distributions discussed below and some of the techniques developed to calculate the properties of random interacting systems as discussed in Chap. 16. For instance, Parisi's "replica symmetry breaking" scheme [622] needed to obtain the correct low temperature properties of various random systems, such as spin glasses, can be understood from the point of view of extreme value theory. Indeed, at low temperature, a disordered system preferentially occupies its extremely low energy states, which are random variables because of the disordered nature of the systems. We refer to [114] for a first step in a categorization of the universality classes that appear in the physics of random systems and which are analogous to classifications presented below.

In this section, we first show how to estimate the size of the largest value. We then recall some important results on the stable laws of extreme event distributions and how they can be derived. The exposition is non-rigorous in order to provide an intuitive understanding.

1.8.1 Maximum Value Among N Random Variables

Definition and Properties. The N independent realizations of the same random process can be for instance the observation of N earthquakes in a given area, assumed independent as a first approximation. This assumption of independence is known to be wrong in reality since, on the large time scale, the earthquake must accomodate between themselves, and some possible creep, the global plate tectonic motion. This provides a constraint on the first moment of the Gutenberg–Richter distribution. Also aftershocks and the Omori law are the signature of short-time coupling.

By the law of large numbers which dictates that the sample average approximates the mathematical expectation and from the very definition of a probability, an event whose probability is p occurs typically Np times. One should thus expect to encounter in these N realizations only one event whose probability is of the order of $1/N$, while it would be surprising to see an event with a probability much less than $1/N$. This simple reasoning allows us to get the estimation of the typical value of the largest value λ observed among N realizations:

$$\mathcal{P}_{\geq}(\lambda) = 1/N \ . \tag{1.37}$$

It is possible to be more precise and specify the full pdf of the maximum value among N observations. Consider N random independent identically distributed (iid) variables X_i with pdf $P(x)$. We define the random variable $X_{\max} = \max\{X_i; i = 1, \ldots, N\}$. Its cumulative distribution $\Pi_{<}(\lambda)$, defining the probability that $X_{\max} < \lambda$, is obtained from the fact that the maximum of the X_i is smaller than λ if all variables X_i are smaller than λ. From the condition of independence, the probability for this to occur is simply the product of the probabilities:

$$\Pi_{<}(\lambda) = [\mathcal{P}_{<}(\lambda)]^N \ . \tag{1.38}$$

This expression holds with no further restrictions on $P(x)$. Replacing $\mathcal{P}_{<}(\lambda)$ by $1 - \mathcal{P}_{\geq}(\lambda)$, we can write

$$\Pi_{<}(\lambda) = [1 - \mathcal{P}_{\geq}(\lambda)]^N = \exp\{N \ln[1 - \mathcal{P}_{\geq}(\lambda)]\} \ , \tag{1.39}$$

which for large N yields

$$\Pi_{<}(\lambda) \simeq \exp\{-N\mathcal{P}_{\geq}(\lambda)\} \ . \tag{1.40}$$

We have expanded the logarithm for small values of $\mathcal{P}_>(\lambda)$, which are the only cases keeping the exponential appreciable for large N. The median of X_{max} is the value $\lambda_{1/2}$ such that $\Pi_<(\lambda_{1/2}) = 1/2$, and is thus the solution of the following equation in terms of the initial pdf $\mathcal{P}_\geq$:

$$\mathcal{P}_\geq(\lambda_{1/2}) = 1 - \left(\frac{1}{2}\right)^{1/N} \simeq \frac{\ln 2}{N} . \tag{1.41}$$

More generally, the value λ_p of the maximum which will not be exceeded with probability p satisfies

$$\mathcal{P}_\geq(\lambda_p) \simeq \frac{\ln(1/p)}{N} . \tag{1.42}$$

With these results, the intuitive formula (1.37) can be put on a firm basis. This estimation of the maximum provided by (1.37) is the value that is not exceeded with probability $p = 1/e \approx 0.37$.

The median value of the largest among N variables depends on the pdf $P(x)$ only through its asymptotic behavior at large x. For a pdf with an asymptotic exponential decay $\mathcal{P}_\geq(x) \sim \exp(-x/a)$, the maximum is typically

$$\lambda \simeq a \ln N \tag{1.43}$$

which grows very slowly with N. This result can be refined using (1.42). For instance, for an exponential pdf, we get

$$\lambda(p) \approx a \ln \frac{N}{\ln(1/p)} . \tag{1.44}$$

Let us take a symmetric pdf $P(x) = (1/2)\exp(-|x|)$ with standard deviation $\sqrt{2}$. The median value of the maximum of 1000 variables is only 4.7 standard deviations and for 10 000 it is 6.3 standard deviations. For "gentle" pdf's (i.e. decaying faster than any power law), the maximum is a slowly increasing function of N. The Gaussian pdf is the archetype of a gentle function for which the maximum grows like $\sqrt{\ln N}$.

Table 1.1. Comparison of characteristic scales of the exponential e^{-x} and power law $1.5/x^{1+1.5}$ for $x \geq 1$ distributions. Note that, notwithstanding their comparable central values $x_{mp}, x_{1/2}$, and $\langle x \rangle$, their extreme fluctuations are very different: in the exponential case, the maximum value $\lambda(p)$ grows slowly (logarithmically) with the number N of variables and with the confidence level p. In contrast, the maximum value $\lambda(p)$ grows like $[N/\ln(1/p)]^{1/\mu}$ for a power law

$P(x)$	x_{mp}	$x_{1/2}$	$\langle x \rangle$	$\lambda(37\%, N = 10^3)$	$\lambda(37\%, 10^4)$	$\lambda(99\%, 10^3)$	$\lambda(99\%, 10^4)$
$\exp(-x)$	0	0.69	1	6.9	9.2	11.5	13.8
$1.5/x^{1+1.5}$	1	1.6	3	100	460	2150	10 000

For pdf's $P(x)$ with asymptotic power law dependence (1.24), the situation is very different. In this case,

$$P_\geq(x) \simeq \frac{C_+}{\mu x^\mu} , \qquad \text{for } x > 0 , \qquad (1.45)$$

and the typical value of the maximum is

$$\lambda \sim (C_+ N)^{1/\mu} . \qquad (1.46)$$

As a numerical example, take a pure power law defined between 1 and $+\infty$ (i.e. with $C_+ = \mu$) with $\mu = 1.5$. Then the largest value among 1000 or 10 000, that will not be exceeded with probability $p = 1/e \approx 37\%$ is 100 or 460, respectively. For a larger confidence level of $p = 99\%$, the largest value among 1000 or 10 000 is 2150 or 10 000, respectively. These values must be compared with $x_{mp} = 1$, $x_{1/2} = 2^{1/\mu} = 1.6$, and $\langle x \rangle = \mu/(\mu - 1) = 3$. These extremely large fluctuations are above all the measures of central tendency. Intuitively, we now understand better why the variance has no meaning for $\mu \leq 2$ (recall that we found it mathematically infinite) as fluctuations as large as thousands of times larger than the mean can occur!

Another quantity of important use is the exceedance. Fix a threshold u and estimate the size of the events beyond this level u. This is given from the conditional probability $P_<(X - u \leq x | X > u)$, i.e. the conditional probability that, given that there is an event larger than u, the exceedance $X - u$ is no bigger than some level x. For sufficient data, an estimate of this probability is obtained from the events $X_1 \geq X_2 \geq \ldots \geq X_n$ ordered from the largest event such that $X_n \geq u \geq X_{n-1}$. For insufficient data, we need to find a suitable model or approximation for the pdf to calculate this conditional probability. Formally, the mean exceedance

$$\langle X \rangle|_{>u} = \frac{\int_u^{+\infty} x \, dP_<(x)}{\int_u^{+\infty} dP_<(x)} \qquad (1.47)$$

measures the expected value (conditionned to be above u) above u. It is interesting to note that $\langle X \rangle|_{>u} - u$ is constant for an exponential distribution:

$$\langle X \rangle|_{>u} - u = \frac{1}{\theta} \quad \text{for} \quad P_<(x) \approx 1 - \exp\left(e^{-x/\theta}\right) \quad \text{with } 0 \leq x . \quad (1.48)$$

This property is related to the absence of memory of the exponential Poisson law. For a Gaussian distribution, $\langle X \rangle|_{>u}$ is very close to u showing that the Gaussian pdf is characterized by small fluctuations. For power laws with exponent $\mu > 1$, $\langle X \rangle|_{>u} = (\mu/(\mu - 1))u$ and grows with u, a signature of the absence of characteristic scales.

Relative Weight of the Maximum X_{max} in a Sum $X_1 + \ldots + X_N$.
A very natural question is to ask what is the relative importance of the largest term X_{max} in the sum $X_1 + \ldots + X_N$. Is the largest value contributing a finite fraction of the sum (a situation in which X_{max} could be called a "king")? Or is its contribution of the same order $1/N$ as most of the others (a more

"democratic" case)? The following presentation borrows from exchanges with V.F. Pisarenko.

To answer this question, we again use the notation $P(x)$ for the pdf of the random variable X and $\mathcal{P}_<(x)$ for its cumulative distribution. The pdf of $X_{\max}$ is

$$\frac{d[\mathcal{P}_<(x)]^N}{dx} = NP(x)[\mathcal{P}_<(x)]^{N-1} . \tag{1.49}$$

The conditional pdf of X_1 under fixed $X_{\max}$, denoted as $P(x|X_{\max})$ is

$$P(x|X_{\max}) = \frac{1}{N}\delta(x - X_{\max}) + \left(1 - \frac{1}{N}\right)\frac{P(x)\ H(X_{\max} - x)}{\mathcal{P}_<(X_{\max})} , \tag{1.50}$$

where H is the Heaviside step-wise function. The first term in the r.h.s. of (1.50) corresponds to the event $X_1 = X_{\max}$ which occurs with probability $1/N$. The complementary event $X_1 < X_{\max}$ occurs with probability $1 - 1/N$ and the conditional pdf in this case is as shown in (1.50).

Now, the ratio $X_1/X_{\max}$ has the following cumulative distribution function denoted $\mathcal{G}(z)$:

$$\begin{aligned}
\mathcal{G}(z) &= \Pr\{X_1/X_{\max} < z\} \\
&= \int_0^{+\infty} \Pr\{X_1/X_{\max} < z|X_{\max}\}\, d[\mathcal{P}_<(X_{\max})]^N \\
&= \int_0^{+\infty} \Pr\{X_1 < zX_{\max}|X_{\max}\}\, d[\mathcal{P}_<(X_{\max})]^N . \tag{1.51}
\end{aligned}$$

The corresponding pdf $G(z)$ of $X_1/X_{\max}$ is obtained by differentiating (1.51):

$$G(z) = \int_0^{+\infty} X_{\max}P(zX_{\max}|X_{\max})\, d[\mathcal{P}_<(X_{\max})]^N . \tag{1.52}$$

We express $P(zX_{\max}|X_{\max})$ using (1.50):

$$G(z) = \frac{1}{N}\delta(z-1) + (N-1)\int_0^{+\infty} dy\, yP(zy)P(y)[\mathcal{P}_<(y)]^{N-2} , \tag{1.53}$$

for $0 \le z \le 1$. Finally, the expectation of the ratio of the sum $X_1 + \ldots + X_N$ divided by $X_{\max}$ reads

$$\mathrm{E}\left[\frac{X_1 + \ldots + X_N}{\max\{X_1, \ldots, X_N\}}\right] = N\int_0^1 dz\, zG(z)$$

$$= 1 + N(N-1)\int_0^{+\infty} dy\, yP(y)[\mathcal{P}_<(y)]^{N-2}\int_0^1 dz\, zP(yz) . \tag{1.54}$$

Using the fact that $\int_0^1 dz\, zP(yz) = (1/y)[\mathcal{P}_<(y) - T(y)]$ where

$$T(y) \equiv (1/y)\int_0^y du\, \mathcal{P}_<(u) , \tag{1.55}$$

we finally obtain

$$E\left[\frac{X_1 + \ldots + X_N}{\max\{X_1, \ldots, X_N\}}\right] = N - N \int_0^{+\infty} T(y)\, \mathrm{d}[\mathcal{P}_<(y)]^{N-1} . \tag{1.56}$$

This expression (1.56) is valid for any pdf of positive random variables. In Sect. 4.2.1, we use this expression to discuss the case of pdf with power law tails.

1.8.2 Stable Extreme Value Distributions

Under widely applicable conditions, the cumulative distribution of the largest observation of an i.i.d. sample $X_1, X_2, \ldots, X_N$ can be approximated by a member of the following class of extreme value cumulative distributions:

$$H_{\xi,m,a}(x) = \exp\left[-\left(1 + \xi\frac{x - m}{a}\right)_+^{-1/\xi}\right] , \tag{1.57}$$

with the notation $y_+ = \max(y, 0)$. This three parameter family of distributions $H_{\xi,m,a}(x)$ has

1. a location parameter $-\infty < m < +\infty$,
2. a scale parameter $a > 0$,
3. a shape parameter $-\infty < \xi < +\infty$.

- The Gumbel distribution is obtained from (1.57) by taking the limit $\xi \to 0$:

$$H_{0,m,a}(x) = \exp\left[-\exp\left(-\frac{x - m}{a}\right)\right] , \tag{1.58}$$

defined for $-\infty < x < +\infty$.
- The Fréchet distribution

$$H_{\xi>0,m,a}(x) = \exp\left(-\frac{1}{[1 + \xi(x - m)/a]_+^{1/\xi}}\right) , \tag{1.59}$$

defined for $m - a/\xi < x < +\infty$, is obtained for $\xi > 0$.
- The Weibull distribution

$$H_{\xi<0,m,a}(x) = \exp\left[-\left(\frac{m + (a/|\xi|) - x}{a}\right)_+^{1/|\xi|}\right] , \tag{1.60}$$

defined for $-\infty < x < m + a/|\xi|$, is obtained for $\xi < 0$. The Weibull distribution has a finite right endpoint $x_F = m + (a/|\xi|)$.

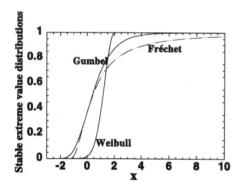

Fig. 1.4. Comparison of the three stable distributions. The parameters are $m = 0$, $a = 1$ and $\xi = 1/2$ for the Fréchet, $\xi = -1/2$ for the Weibull distribution. The Gumbel distribution is defined over $-\infty < x < +\infty$. For these parameters, the Fréchet distribution is defined over $-2 < x < +\infty$ and the Weibull distribution is defined over $-\infty < x < 2$. Notice the long power law tail of the Fréchet distribution and its slow convergence to 1

The remarkable result of EVT is that for any pdf of N random variables, the distribution $\Pi_<(\lambda)$ of their maximum given by (1.38) tends asymptotically for large N to one of the three EV distribution $H_{\xi,m,a}(x)$ (1.58, 1.59, 1.60). The Gnedenko EV theorem states that, if, after an adequate centering and normalization using two N-dependent parameters a_N, b_N, the distribution of λ converges to a *non-degenerate* distribution as N goes to infinity, this limit distribution is then necessarily the Generalized Extreme Value distribution defined with a N-independent number ξ by

$$\Pi_<(\lambda) \to_{N\to+\infty} H_{\xi,m,a}\left(\frac{\lambda - b_N}{a_N}\right) . \tag{1.61}$$

The two N-dependent numbers a_N and b_N can be calculated explicitly as a function of N and of the parameters of the pdf of the random variables. In the same way as the Gaussian and Lévy stable laws for the sum of N random variables, which will be discussed later on, the three EV distributions $H_{\xi,m,a}$ have each their domain of attraction:

- Gumbel: any pdf with a tail falling faster than a power law will have its EV distribution converging to the Gumbel distribution;
- Fréchet: any pdf with a tail falling as a power law $X^{-1-\mu}$ for $X \to +\infty$ will have its EV distribution converging to the Fréchet distribution with $\xi = 1/\mu$;
- Weibull: any pdf with a finite right endpoint x_F and with a dependence close to this right endpoint proportional to $(x_F - x)^{1/|\xi|}$ times a slowly varying function of x will have its EV distribution converging to the Weibull distribution. The Weibull distribution is often used in engineering and materials science under different forms, as discussed in Chap. 6.

We now give two different heuristic derivations of this result for the case of the Gumbel distribution, which has the largest domain of attraction since it applies to all distributions with infinite domain of definition and which decay asymptotically faster than power laws.

1.8.3 First Heuristic Derivation
of the Stable Gumbel Distribution

Consider N realizations $X_1, X_1, \ldots, X_N$ of a random variable and define

$$\hat{X}_N^{(1)} \equiv \max(X_1, X_2, \ldots, X_N) . \tag{1.62}$$

Denote $P_<(x)$ the cdf of the X_i's, i.e. the probability to get a value less than x. Then, the probability $G_<(x)$ that the maximum $\hat{X}_N^{(1)}$ is less than x is obviously

$$G_<(x) = [P_<(x)]^N , \tag{1.63}$$

as the different realizations are assumed uncorrelated. Consider another set of N realizations $X_{N+1}, X_{N+2}, \ldots, X_{2N}$ and define

$$\hat{X}_N^{(2)} \equiv \max(X_{N+1}, X_{N+2}, \ldots, X_{2N}) . \tag{1.64}$$

Then obviously,

$$\max(X_1, X_2, \ldots, X_{2N}) = \max(\hat{X}_N^{(1)}, \hat{X}_N^{(2)}) . \tag{1.65}$$

This suggests that the distribution of the maximum value should obey some kind of stability condition, which is usually stated as

$$[G_<(x)]^n = G_<(x + b_n) . \tag{1.66}$$

The same holds true for $G_>(x) = 1 - G_<(x)$. We thus drop the index $<$ or $>$ for the time being. Taking (1.66) to the power m yields

$$[G(x)]^{nm} = [G(x + b_n)]^m = G(x + b_n + b_m) , \tag{1.67}$$

where the second equality is derived by applying (1.66) to $[G(x + b_n)]^m$. But applying (1.66) to the first term also gives

$$G(x + b_{nm}) = G(x + b_n + b_m) , \tag{1.68}$$

and thus

$$b_{nm} = b_n + b_m . \tag{1.69}$$

The solution of (1.69) is

$$b_n = \theta \ln n , \tag{1.70}$$

where θ is an arbitrary constant. Now, taking the logarithm of (1.66) twice, we write

$$\ln n + \ln[-\ln G(x)] = \ln[-\ln G(x + \theta \ln n)] , \tag{1.71}$$

where the minus sign have been inserted because the cdf G is smaller than one and thus its logarithm is negative. Calling $h(x) \equiv \ln[-\ln G(x)]$, we see that $h(x)$ is solution of

$$\ln n + h(x) = h(x + \theta \ln n) , \tag{1.72}$$

whose solution is the affine function $h(x) = h(0) + x/\theta$. This provides finally the expression of the extreme value we were looking for:

$$G(x) = \exp\left(-p_0 e^{x/\theta}\right) , \qquad (1.73)$$

where we have noted $p_0 = e^{h(0)}$. For $x > 0$ large, $G_<(x) \to 1$ and for $x < 0$ with $|x|$ large, $G_<(x) \to 0$. We thus have

$$G_<(x) = 1 - \exp\left(-p_0 e^{x/\theta}\right) . \qquad (1.74)$$

p_0 has a simple interpretation when small. For $x < 0$, $e^{x/\theta} < 1$, we can thus expand

$$\exp\left(-p_0 e^{x/\theta}\right) = 1 - p_0 \exp\left(x/\theta\right) , \qquad (1.75)$$

showing that p_0 is simply the probability to get a negative value.

1.8.4 Second Heuristic Derivation of the Stable Gumbel Distribution

We start again from (1.63) and use the following parameterization

$$P_<(x) = 1 - e^{-f(x)} , \qquad (1.76)$$

where $f(x)$ goes from zero to $+\infty$ when x goes from $-\infty$ to $+\infty$. This leads to

$$G_<(x) = \left(1 - e^{-f(x)}\right)^N = \exp\left[N \ln(1 - e^{-f(x)})\right]$$
$$\approx \exp\left(-N e^{-f(x)}\right) , \qquad (1.77)$$

where the last expression becomes a better and better approximation as N increases and x is larger so that $e^{-f(x)}$ becomes smaller and smaller.

The function $\exp\left(-N e^{-f(x)}\right)$ is very small for small or negative x and becomes very close to one for large x. This transition from 0 to 1 occurs more and more abruptly when N increases. The location of this transition is at the inflection point defined by $d^2 G_<(x)/dx^2 = 0$. For simplicity, we consider unimodal pdf's such that $f(x)$ is monotonically increasing. This ensures the existence of a unique inflection point x^* of $G_<(x)$ given by

$$N e^{-f(x^*)} = 1 - \frac{f''_{x=x^*}}{(f'_{x=x^*})^2} , \qquad (1.78)$$

noting $f' = df/dx$ and $f'' = d^2 f/dx^2$. The right-hand side of (1.78) is very close to 1 since, for distributions $1 - P_<(x)$ which decay to zero for large x faster than any power law, $f''_{x=x^*}/(f'_{x=x^*})^2$ goes asymptotically to zero for large x. In contrast, for power law tails, this term approaches a constant for large x. The typical interval Δx over which the transition from 0 to 1 occurs is such that

$$\Delta x \left. \frac{dG_<(x)}{dx} \right|_{x=x^*} \sim 1 \; , \tag{1.79}$$

leading to

$$\Delta x \sim \frac{1}{(dG_<(x)/dx)|_{x=x^*}} \; . \tag{1.80}$$

We verify that $\Delta x \ll x^*$ as soon as $f(x) \gg \ln x$, i.e. the tail of the distribution decays faster than any power law. It is thus possible to expand $f(x)$ around this inflection point x^* to derive the Gumbel distribution:

$$f(x) = f(x^*) + (x - x^*)f'_{x=x^*} + \frac{1}{2}(x - x^*)^2 f''_{x=x^*} + \dots \; . \tag{1.81}$$

This expansion is not valid for power laws which lead to the different (Fréchet) stable distribution for extremes. The fundamental origin of this difference is that the width Δx is proportional x^*, which invalidates the expansion.

Inserting (1.81) in (1.77) yields

$$G_<(x) = \exp \left\{ -Ne^{-f(x^*)} \exp \left[-(x - x^*)f'_{x=x^*} - \frac{1}{2}(x - x^*)^2 f''_{x=x^*} \right] \right\}$$

$$= \exp \left(-Ne^{-f(x^*)} \exp \left\{ -(x - x^*)f'_{x=x^*} \left[1 + \frac{1}{2}(x - x^*)\frac{f''_{x=x^*}}{f'_{x=x^*}} \right] \right\} \right) . \tag{1.82}$$

The Gumbel distribution corresponds to neglecting the term

$$\frac{1}{2}(x - x^*)\frac{f''_{x=x^*}}{f'_{x=x^*}}$$

in the right-hand side of (1.82). This is justified for large N (leading to large x^*) since $x - x^*$ is typically of order Δx and thus

$$(x - x^*)\frac{f''_{x=x^*}}{f'_{x=x^*}} \approx \frac{f''_{x=x^*}}{(f'_{x=x^*})^2} \frac{e^{f(x^*)}}{NG_<(x^*)}$$

$$= -\left. \frac{d\left[1/(df/dx)\right]}{dx} \right|_{x=x^*} \frac{e^{f(x^*)}}{NG_<(x^*)} \; . \tag{1.83}$$

The first order-of-magnitude equality is derived by using (1.80). The last term in (1.83) decays to zero for large x^*, leading to the convergence of the distribution of the maximum to the Gumbel distribution. It is possible to be more precise and control in a similar way all higher order terms in the expansion (1.81) of $f(x)$. For this, we refer first to the initial work of R. Fisher and L. Tippet [?] which contains an heuristic derivation, lacking however strict mathematic rigor. The first rigorous proof is due to the

Russian statistician Boris Gnedenko. He published his paper in French during the second World War [349]. A simplified proof of this result can be
found in [1011]. A half-page sketch of the proof can be found in [274],
pp. 121–122.

The speed of convergence, i.e. how well $G_<(x)$ is approached by the suitable normalized Gumbel distribution for a given N, is obtained by taking the
ratio of $G_<(x)$ to the corresponding normalized Gumbel distribution $G_<^G(x)$:

$$\frac{G_<(x)}{G_<^G(x)} = [G_<^G(x)]^\alpha , \tag{1.84}$$

where

$$\alpha = \exp\left[-\frac{1}{2}(x - x^*)^2 f''_{x=x^*}\right] - 1 . \tag{1.85}$$

Consider as an illustration the power-exponential pdf for which $f(x) = x^c$
with $c > 0$. We find $x^* \sim (\ln N)^{1/c}$, $\Delta x \sim (x^*)^{1-c}$, leading to $\Delta x/x^* \sim (x^*)^{-c}$
and $\alpha \approx -c(c - 1)/2 \ln N$. This shows a very slow logarithmic convergence
of the EV distribution to the Gumbel law. In general, the rate of convergence depends very much on the tail of the distribution of the random variables. For instance, Gaussian random variables have their EV distribution
converging to the Gumbel law only as $\mathcal{O}(1/\ln N)$ as above. For random variables distributed according to the exponential pdf, the convergence rate is
much faster, as $\mathcal{O}(1/N^2)$. In addition, the situation is complicated by the
fact that the convergence rate depends on the precise choice of normalising
constants. This is much more complex and distribution dependent than the
situation for the convergence to the Gaussian for the sum of random numbers
as given by the central limit theorem discussed in Chap. 2. For the Fréchet
and Weibull laws, the convergence rates can be anything. See [214, 771]
for more informations on the rate of convergence: the general mathematical result is that the weak convergence is uniform whenever the limit is
continuous.

1.8.5 Practical Use and Expression of the Coefficients of the Gumbel Distribution

For practical implementation, we do not recommend the determination of
the distribution of extreme values by using the EV distribution. This procedure which is often advocated does not in fact use the full data set. It is
preferable to determine the underlying pdf and use it in (1.38) or to use the
so-called "peak over threshold", where one models the marked point process
of exceedances over high thresholds. This is also intimately related to the
rank-ordering method [326, 381, 903]. These methods work well for relatively
small samples (a few tens of measurements) as shown in [903].

It is interesting to express explicitly the constants m and a in (1.58) as
a function of the distribution $P_<(x)$ of the individual variables. It can be

shown that a suitable choice is

$$a \rightarrow \frac{P_>[x_n(\gamma)]}{p[x_n(\gamma)]} \, , \tag{1.86}$$

$$m \rightarrow x_n(\gamma) - \frac{a}{\ln \gamma} \, , \tag{1.87}$$

where p is the pdf of the individual variables and $x_n(\gamma)$ is defined by

$$\{P_<[x_n(\gamma)]\}^n = e^{-\gamma} \, , \tag{1.88}$$

where γ is a positive real number. Optimizing γ will lead to a faster convergence to the Gumbel distribution.

1.8.6 The Gnedenko–Pickands–Balkema–de Haan Theorem and the pdf of Peaks-Over-Threshold

For practical applications, it is inconvenient to work with extremes. Indeed, one has only a single largest event for each given series of random events. Thus, in order to construct empirical statistics of extremes, the initial series of N random variables must be partitioned into a possibly large number n of groups (which become necessarily of rather small sizes N/n), of which the largest element can be defined, yielding n samples of extreme values. It is however obvious that this approach is suboptimal [518] compared with the analysis in terms of the pdf of peaks-over-threshold, that we now describe.

The main relevant mathematical tool is the Gnedenko–Pickands–Balkema–de Haan (GPBH) theorem which describes how the distribution of the large events conditioned to be larger than some threshold can be characterized by the Generalized Pareto Distribution (GPD). The following borrows from [738]. The GPD denoted as $G(x/\xi, a)$ is derived from the distribution $H_{\xi,a}(x)$ of the largest value given by (1.57), thus showing the link with Extreme Value Theory:

$$G(x/\xi, a) = 1 + \ln(H_{\xi,a}(x)) = 1 - \left(1 + \xi \frac{x}{a}\right)_+^{-1/\xi} \, , \tag{1.89}$$

where the two parameters (ξ, a) are such that $-\infty < \xi < +\infty$ and $a > 0$. We have dropped the position parameter m for notational simplicity. For $\xi \geq 0$, $x \geq 0$ and for $\xi < 0$, $0 \leq x \leq -a/\xi$. In order to state the GPBH theorem, we define the right end-point x_P of the cdf $P_<(x)$ as

$$x_P = \sup\{x : P_<(x) < 1\} \, , \tag{1.90}$$

which can often be considered to be infinite in many practical applications. We also define the excess distribution $P_u(x)$:

$$P_u(x) = \text{Proba}\{X - u < x | X > u\} \, , \qquad x \geq 0 \, . \tag{1.91}$$

Gnedenko–Pickands–Balkema–de Haan theorem. Suppose $P_<(x)$ is a cdf with excess distribution $P_u(x)$, with $u > 0$. Then, for $-\infty < x < +\infty$, $P_<(x)$ belongs to the the the Maximum Domain of Attraction of $H_{\xi,a}(x)$ if and only if there exists a positive function $a(u)$ such that

$$\lim_{u \to x_P} \sup_{0 \le x \le x_P - u} |\bar{P}_u(x) - \bar{G}(x/\xi, a(u))| = 0 . \tag{1.92}$$

We use the standard notation $\bar{P}_u(x) = 1 - P_u(x)$. Therefore, by definition, $\bar{P}_<(x) = 1 - P_<(x) = P_>(x)$. Other names for $\bar{P}$ and $\bar{G}$ are the "complementary cumulative" distribution or "survivor" function.

Intuitively, the statement (1.92) means that the tail $\bar{P}_<(x)$ of the distribution is asymptotically given by the GPD defined by (1.89) (to be precise, by one minus the expression in (1.89)) as x approaches the very end of the tail of the distribution. The strength of the GPBH theorem is that it is not a statement only on the largest value of a data set, as is the case for the Extreme Value Theory leading to the limit distributions (1.57). As already mentioned, Knopoff and Kagan [518] showed that using Extreme Value Theory to constraint the shape of the tail of the distribution is sub-optimal as one effectively discards a significant part of the data which leads to unreliable results. In contrast, the analysis of the tail provided by the GPBH theorem makes full use of all the data present in the tail.

Let us denote by n_u the number of observations exceeding a threshold u and by $y_1, \dots, y_{n_u}$ the observations decreased by u: $y_i = x_{j(i)} - u$ where $x_{j(i)} > u$. The GPBH theorem yields an approximation to the tail $\bar{P}_<(x)$ by a GPD as a tail estimator:

$$\bar{P}_<(x) = \frac{n_u}{N} \bar{G}(x/\hat{\xi}, \hat{a}) . \tag{1.93}$$

The estimates of the two parameters $\hat{\xi}, \hat{a}$ can be obtained through the Maximum Likelihood estimation (ML). The log-likelihood L is given by

$$L = -n_u \ln a - \left(1 + \frac{1}{\xi}\right) \sum_{i=1}^{n_u} \ln \left(1 + \frac{\xi y_i}{a}\right) . \tag{1.94}$$

Maximization of the log-likelihood $\ln L$ can be done numerically. The limit standard deviations of the ML-estimates as $n_u \to +\infty$ can be easily obtained [274]:

$$\sigma_\xi = \frac{1 + \xi}{\sqrt{n_u}} ; \quad \sigma_a = \sqrt{2(1 + \xi)/n_u} . \tag{1.95}$$

In practice, one usually replaces the unknown parameters in (1.95) by their estimates. One can also calculate the q-quantile estimator

$$x_q = u + \frac{\hat{a}}{\hat{\xi}} \left(\left(\frac{n_u(1-q)}{N}\right)^{\hat{\xi}} - 1\right) , \tag{1.96}$$

which is the value not overpassed by the random variable with probability q.

The shape parameter ξ is of great interest in the analysis of the tails. When x becomes large and $\xi > 0$, the tail of the cdf in (1.89) approaches the power law

$$\bar{G}(x/\xi, a) \simeq \left(\frac{a}{\xi x}\right)^{1/\xi} . \tag{1.97}$$

$1/\xi$ is therefore the asymptotic exponent of the survivor distribution function. It corresponds asymptotically to the exponent μ for the Pareto law. Thus, the GPD is asymptotically scale invariant. For $1/\xi = 0$, the power law tail is replaced by an exponential decay $\sim \exp(-x/a)$. We refer to [584, 738, 739] for recent developments on these issues with applications to the distributions of earthquake sizes and financial returns.

It should be noted that the scale parameter $a = a(u)$ depends on the threshold u, while the shape parameter ξ is in theory independent of u and solely determined by the pdf $P(x)$ of the data points. Thus, one can hope to find a reasonable GPD fit to the tail if it is possible to take a sufficiently high threshold u and to keep a sufficiently large number of excesses over it. Of course, this is not always possible. The dependence $a(u)$ can be illustrated in the case of Weibull distributions discussed further in Chap. 6. For this family of distributions, the conditional cdf $\bar{P}_u(x)$ reads

$$\bar{P}_u(x) = \exp\left(u^c - (u+x)^c\right) . \tag{1.98}$$

Then, the GPBH theorem states that

$$\lim_{u \to \infty} \sup_x \left| \exp\left(u^c - (u+x)^c\right) - \exp\left(-x/a(u)\right) \right| = 0 . \tag{1.99}$$

Thus, for any u it is possible to find a value $a(u)$ minimizing the difference in (1.99). This minimal difference tends to 0 as u tends to infinity. To find explicitly $a(u)$ and the corresponding minimal difference is not easy but it is possible to estimate these values using asymptotic expansions for transcendental equations resulting from (1.99). What is perhaps counter-intuitive is that (1.99) holds also for $c < 1$, for which the Weibull cdf has a fatter tail than its corresponding asymptotic GPD.

2. Sums of Random Variables, Random Walks and the Central Limit Theorem

Why do we care about sums of random variables? The answer is that everywhere around us the processes that we see often depend on the accumulation of many contributions or are the result of many effects. The pressure in a room, measured on a surface, is the sum of the order of 10^{23} momentum exchanges between the air molecules and the surface. The large time tectonic deformation is the (tensorial) sum of the deformation associated with the myriad of earthquakes. Errors and/or uncertainties in measurements are often the aggregation of many sources and are in many cases distributed according to a Gaussian law (see below). In fact, it is hard to find an observation that is *not* controlled by many variables. Studying the sum of random variables allows us to grasp the fundamental notion of collective behavior without the need for further complications.

In general however, the different contributions may be correlated in a complicated way, sometimes even showing wildly varying scales. It is useful to first investigate the simplest case where the variables which constitute the sum are uncorrelated and well-behaved. This will be a useful benchmark against which all other processes can be compared.

2.1 The Random Walk Problem

Consider the 1D-problem of a random walker on a line [53]: starting at time zero from the origin, the random walker is at $x(t)$ at time t. It then makes a random step of length $l(t)$ between time t and $t + \tau$ to reach the position $x(t + \tau)$ at time $t + \tau$. We assume that $l(t)$ is distributed according to a pdf $\Pi(l)$. The random walk is thus described by the equation

$$x(t + \tau) = x(t) + l(t) . \tag{2.1}$$

It is assumed for the time being that the variables $l(t)$ are i.i.d. (independent identically distributed), i.e. $\langle l(t) l(t') \rangle = \delta_{tt'} \langle l(t) \rangle^2$, with $\delta_{tt'} = 1$ if $t = t'$ and zero otherwise. The equation (2.1) is a simple example of a discrete stochastic equation. By using a procedure which amounts to applying the central limit theorem (CLT) explained below, one can show that there is a unique continuous limit of (2.1), which is called the Wiener process [1034]. By taking the continuous limit, any infinitesimal increment dx can be decomposed into an infinite number of infinitesimal steps, and by the value of the CLT, this shows that the distribution of the increments dx is Gaussian, whatever the distribution of the infinitesimal increments one starts from to construct the process. The miracle appears by taking the limit $l \to 0$ together with $\tau \to 0$ while keeping constant the ratio l^2/τ. The Wiener process is the simplest example of what is known in physics as a Langevin equation [780].

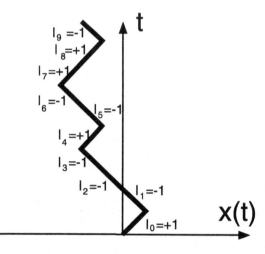

Fig. 2.1. Construction of the random walk in the space–time representation in the simple case where all step size of the same length but with random signs

The solution of (2.1) is obviously

$$x(t) = l(t - \tau) + l(t - 2\tau) + \ldots + l(\tau) + l(0) , \tag{2.2}$$

where t is taken as a multiple of the elementary unit time τ. The expression (2.2) defines the variable $x(t)$ as the sum of $N \equiv t/\tau$ random variables. Let us describe a few properties of this sum and use it as a way to introduce a number of concepts.

2.1.1 Average Drift

$$\langle x(t) \rangle = \sum_{i=1}^{N} \langle l_i \rangle = N \langle l \rangle , \tag{2.3}$$

for i.i.d. variables, where we have denoted $l_i \equiv l((i-1)\tau)$. $\langle x(t) \rangle$ represents the average position taken over a large assembly of random walkers. The expression (2.3) defines an average drift with velocity

$$v = \frac{\langle l \rangle}{\tau} . \tag{2.4}$$

If the average step length $\langle l \rangle = 0$, then the random walker remains on average at the point where he started from.

2.1.2 Diffusion Law

The variance $\langle x(t)^2 \rangle - \langle x(t) \rangle^2$ of the position characterizes the typical size of the excursion of the random walk around its average position vt:

$$\langle x(t)^2 \rangle - \langle x(t) \rangle^2 = \sum_{i=1}^{t} \sum_{j=1}^{t} \left[\langle l_i l_j \rangle - \langle l_i \rangle \langle l_j \rangle \right] = \sum_{i=1}^{t} \sum_{j=1}^{t} C_{ij} , \qquad (2.5)$$

where we denote

$$C_{ij} = \langle l_i l_j \rangle - \langle l_i \rangle \langle l_j \rangle , \qquad (2.6)$$

the correlation function of l_i and l_j estimated at the same time. Since the l_i's are assumed uncorrelated, $C_{ij} = [\langle l^2 \rangle - \langle l \rangle^2] \delta_{ij}$ and we get

$$\langle x(t)^2 \rangle - \langle x(t) \rangle^2 = N[\langle l^2 \rangle - \langle l \rangle^2] \equiv N\sigma^2 , \qquad (2.7)$$

where

$$\sigma^2 = \langle l^2 \rangle - \langle l \rangle^2 \qquad (2.8)$$

is the variance of the pdf of the l-variables. Defining the diffusion coefficient

$$D \equiv \frac{\sigma^2}{2\tau} , \qquad (2.9)$$

we get the well-known diffusion or Fick's law relating the variance $\langle (x(t))^2 \rangle - \langle x(t) \rangle^2$ of the sum of t/τ random i.i.d. variables to the individual variance σ^2:

$$\langle (x(t))^2 \rangle - \langle x(t) \rangle^2 = 2Dt . \qquad (2.10)$$

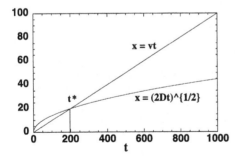

Fig. 2.2. This figure shows how the standard deviation $\sqrt{\langle x(t)^2 \rangle - \langle x(t) \rangle^2} = \sqrt{2Dt}$ of the random walk position compares to a constant drift $\langle x(t) \rangle = vt$. The parameters are $D = 1$ and $v = 0.1$. Note that for times $t < t^* = 2D/v^2 = 200$, the diffusion is faster, meaning that a constant drift will not be felt, while at large times, the constant velocity drift dominates

2.1.3 Brownian Motion as Solution of a Stochastic ODE

The usual description of continuous stochastic processes is based on the Brownian motion $W(t)$ written in mathematical terms as

$$W(t) = \int_0^t d_u W(u) , \qquad (2.11)$$

with

$$\langle d_t W(t) \rangle = 0 , \tag{2.12}$$

and

$$\mathrm{Var}\,[d_t W(t)] = dt , \tag{2.13}$$

where Var denotes the variance. $W(t)$ is called a Wiener process. Alternatively, $W(t)$ can be defined as the solution of the following stochastic *ordinary* differential equation (SODE):

$$\frac{dW(t)}{dt} = \eta(t) . \tag{2.14}$$

$\eta(t)$ is a Gaussian noise, characterized by the following covariance

$$\mathrm{Cov}\,[\eta(t), \eta(t')] = \delta(t - t') , \tag{2.15}$$

where δ designates the Dirac distribution and Cov denotes the covariance. The expression (2.14) describes a particle at position W which is incessantly subjected to random velocity impulses leading to random variations η of its position. The solution of (2.14) is formally

$$W(t) = \int_0^t dv\, \eta(v) . \tag{2.16}$$

This shows that $dt\, \eta(t)$ is simply a notation for $d_t W(t)$, and, as usual, this has mathematical meaning only under the integral representation. From (2.16), we can calculate easily the covariance

$$\mathrm{Cov}\,[W(t), W(t')] = \int_0^t dv \int_0^{t'} dv'\, \mathrm{Cov}\,[\eta(v),\ \eta(v')]$$

$$= \int_0^t dv \int_0^{t'} dv'\, \delta(v - v') = \mathrm{Min}(t, t') . \tag{2.17}$$

This result (2.17) expresses the fact that the correlation between $W(t)$ and $W(t')$ is due to the set of noise contributions $\{\eta\}$ that are common to both of them. We define $d_t W(t)$ as the limit of $[W(t + \delta t) - W(t)]$ when the small but finite increment of time δt becomes the infinitesimal dt. Using (2.17), we get $\mathrm{Var}\,[W(t + \delta t) - W(t)] = \delta t$ which recovers (2.13) in the infinitesimal time increment limit.

The definition of the Brownian motion as the solution of a SODE is very useful to generalize to other processes. Maybe the simplest extension is the well-known Ornstein–Uhlenbeck (O–U) process $U(t)$ which can be defined as the solution of the following SODE:

$$\frac{dU(t)}{dt} = -\kappa U(t) + \eta(t) , \tag{2.18}$$

where κ is a positive constant. In addition to the random variations η of the velocity, the Brownian particle is now subjected to a restoring force tending to bring it back to the origin (mean reversal term). The solution reads

$$U(t) = \int_0^t dv\, \eta(v) e^{-\kappa(t-v)} \ . \tag{2.19}$$

Its covariance is

$$\mathrm{Cov}\left[U(t),\ U(t')\right] = \frac{1}{2\kappa}\left(e^{-\kappa|t-t'|} - e^{-\kappa(t+t')}\right) , \tag{2.20}$$

which is essentially $(1/2\kappa)e^{-\kappa|t-t'|}$ at large times. We have used $2\,\mathrm{Min}(t,t') = t+t'-|t-t'|$. By adding more terms in (2.18), more complex stochastic processes can be easily generated.

2.1.4 Fractal Structure

It is straightforward to generalize the random walk to a space of dimension d. One can then show that (2.10) is transformed into

$$\langle(x(t))^2\rangle - \langle x(t)\rangle^2 = 2dDt \ , \tag{2.21}$$

by the simple rule of additivity of the variance of each projection of the random walk position over each dimension. The same square-root law for the standard deviation holds in any dimension, only the prefactor is modified. Note that this expression (2.21) expresses the fact that the line trajectory of a particular random walk forms a very convoluted object of fractal dimension exactly equal to 2 in any dimension.

As we will elaborate in Chap. 5, a fractal is an object which enjoys the property of self-similarity, namely arbitrary sub-parts are statistically similar to the whole provided a suitable magnification is performed along all directions.

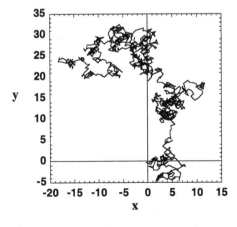

Fig. 2.3. A random walk of 10^3 steps in a plane generated with step lengths l_x, l_y along x and y uniformly taken in the interval $[-1,+1]$. This random walk started at the origin and arrived at $(-15.4, 22.5)$ at $t = 1000$. The radius of gyration equal to 25.5 quantifies the typical excursion of the random walk. One can observe the multiple self-crossings which are characteristic of a $d_{\mathrm{f}} = 2$ fractal set embedded in a plane

By definition of the fractal dimension [291, 592], the "mass" M of an object measured within a sphere of radius R with a resolution ϵ is

$$M \propto \epsilon^d \left(\frac{R}{\epsilon} \right)^{d_f} . \tag{2.22}$$

This formula describes two ways for measuring the fractal (capacity) dimension d_f:

1. for a fixed resolution (or stick length) ϵ, the mass of an object of radius of gyration R increases as R^{d_f};
2. for a fixed object of macroscopic size R, the "observable mass" decreases as ϵ^{d-d_f} when the resolution becomes better ($\epsilon \to 0$). This simply reflects the fact that the fractal object becomes more and more tenuous in comparison to the space of dimension d in which it is embedding, when its dimension d_f is less than d.

This formula (2.22) is equivalent to the definition [592]

$$d_f = \frac{\ln N(r)}{\ln(1/r)} , \tag{2.23}$$

where $N(r)$ is the number of "elements" seen at the scale r, by the identification $r \to \epsilon/R$ and $N \to M/\epsilon^d$.

To apply these notions to a random walk and derive its fractal dimension, one must define what one means by its "mass". A natural measure is its length $\langle l \rangle t/\tau$, for which one thinks of the step length as massive hard sticks. By (2.21), this mass is the square of the typical radius of gyration of the random walk, measured by its standard deviation $\sqrt{\langle (x(t))^2 \rangle - \langle x(t) \rangle^2}$. Therefore, $d_f = 2$. For a random walk in one dimension, this means that the random walk which is intrinsically a two-dimensional fractal object has been "folded" many times to fit within a one-dimensional space. In other words, the random walker comes back an infinite number of times on its previous steps. It does so marginally within a plane and only a finite number of times in three and higher dimensions.

Consider two objects of dimensions d_1 and d_2 imbedded in a space of dimension d. It is a well-known result that the intersection of the two objects has dimension $d_1 + d_2 - d$ with probability one. For instance, two planes in space intersect generically along a line ($2 + 2 - 3 = 1$). A plane and a line in space intersect generically at a point ($2 + 1 - 3 = 0$). For a random walk with $d_f = 2$, we thus see that we need to go to a space of dimension $d = 4$ for the number of intersections to constitute a set of zero dimension, i.e. for the set of crossing to become almost vanishing. In other words, in a space of four or more dimensions, a random walk has very little chance to cross itself and this explains why four dimensions plays a special role in theories of interacting fields, such as spin models that we will study later on. At and above four dimensions, these theories are well-described by so-called mean-field approaches while below four dimensions, the large number of crossings of a random walk make the role of fluctuations important and lead to complex behaviors. This results from the existence of deep connections between these theories and problems of interacting random walks [297].

2.1.5 Self-Affinity

The projections on the x and y axis of the random walk presented in Fig. 2.3 are shown in Fig. 2.4. These curves exhibit the property of "self-affinity" with a self-affine exponent $\zeta = 1/2$. Self-affinity is a generalization of the notion of fractality, in which a subpart of the system is statistically similar to the whole under a dual magnification by a factor λ for the time axis and a different factor λ^ζ along the other direction. Curves with $\zeta = 0$ are essentially flat and do not roughen as the magnification increases. Curves with $\zeta = 1$ correspond to tilted line with constant non-zero average slope, i.e. with a non-vanishing drift.

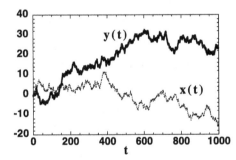

Fig. 2.4. The random walk shown in Fig. 2.3 is decomposed into its time evolution $x(t)$ and $y(t)$ projected onto the x and y axis. Since $x(t)$ and $y(t)$ are independent, this graph gives an intuition of the degree of variability of random walks

Since the standard deviation of x and y grows as $t^{1/2}$, as seen from (2.10), this confirms the value $\zeta = 1/2$ of the self-affine exponent of the curves $x(t)$ and $y(t)$. Self-affinity describes approximately the one-dimensional transection of a mountainous landscape. It turns out that recent measurements on mountains in the United States have found a value close to $\zeta = 1/2$ [678, 918].

Let us now discuss the transition from fractality to self-affinity. If we measure the fractal dimension of the random walk trajectory $x(t)$ or $y(t)$ in the space–time diagram, we get $d_f = 1$. This is because the length of the wiggly line in the space–time diagram is simply proportional to t/τ lengths of the yard-stick used to measure it. This must be contrasted to the value $d_f = 2$ of the random walk in space.

Let us now consider a generalization of the random walk trajectory. In the previous random walk, we have

$$D \sim cl \,, \tag{2.24}$$

where l is the step length (also called mean free path) and c is the ballistic velocity between two successive changes of direction. Since

$$l \sim c\tau ,$$ (2.25)

where τ is the time step, i.e. the time interval between successive steps, we get

$$D \sim c^2\tau \sim \frac{l^2}{\tau} .$$ (2.26)

The standard deviation of the random walk is then

$$\sigma(t) \sim \sqrt{Dt} \sim l \left(\frac{t}{\tau}\right)^{1/2} .$$ (2.27)

This recovers the characteristic distance travelled during τ is l, which is the obvious definition of the step length.

We now generalize this problem to a curve $y(t)$ such that we add a large prefactor $C \gg 1$;

$$\sigma(t) \sim Cl \left(\frac{t}{\tau}\right)^{1/2} .$$ (2.28)

This is no longer a random walk but still a self-affine curve. The novel property is that at small scales the curve becomes a genuine fractal in the following sense.

To simplify the notation, we call $\sigma_0 \equiv Cl$, so that expression (2.28) reads

$$\sigma(t) \sim \sigma_0 \left(\frac{t}{\tau}\right)^{1/2} .$$ (2.29)

Then, the unfolded length of the new random walk in the space–time diagram can be estimated in order of magnitude by the arc length joining the starting point to the ending point:

$$L \sim \sqrt{t^2 + [\sigma(t)]^2} \sim t\sqrt{1 + \left(\frac{\sigma}{t}\right)^2} .$$ (2.30)

For $t < \sigma_0^2/\tau$, $(\sigma/t)^2$ is much larger than one and

$$L \sim t\sqrt{\left(\frac{\sigma}{t}\right)^2} \sim t\sqrt{\frac{\sigma_0^2}{\tau t}} \sim t^{1/2} .$$ (2.31)

If we interpret t as the resolution ϵ in (2.22), then we get

$$d_f = \frac{3}{2} .$$ (2.32)

At larger scales however, $(\sigma/t)^2$ becomes much smaller than one, then $L \sim t$ and

$$d_f = 1 .$$ (2.33)

There are two lessons to get from this calculation.

1. The random walk problem can be seen in several ways with different scaling laws and fractal properties.
2. A self-affine line (or more generally surface) of self-affine exponent ζ can be interpreted as a fractal structure of fractal dimension $d - \zeta$ in an embedding space of d dimensions, below a certain characteristic length scale while, at large length scales, the dimension recovers the value $d - 1$.

Generalizing, if $L(t) \sim t^H$ for t small but H is unknown a priori and one has access to the expansion $L(t) \sim t + \alpha/t + \mathcal{O}(t^2)$ at large times, the technique of construction of approximants by self-similar renormalization [344–348, 1037–1040] allows one to determine H as a function of α (see Chap. 11 for more on the self-similar renormalization method).

2.2 Master and Diffusion (Fokker–Planck) Equations

2.2.1 Simple Formulation

There is a general correspondence between stochastic (Langevin) microscopic equations of type (2.1) with so-called Master and Fokker–Planck equations. In contrast to the stochastic (Langevin) microscopic equation (2.1) which describes the evolution of the position of a given random walker, the Master and Fokker–Planck equations give the time evolution of the pdf $P(x, t)$ of the position of the random walker as a function of time, in other words, they describe the behavior of a large population of walkers and the fraction of this population at position x at time t.

The Master equation corresponding to (2.1) is

$$P(x, t + \tau) = \int_{-\infty}^{+\infty} \Pi(l) P(x - l, t) \mathrm{d}l \ . \tag{2.34}$$

It simply states that, in order to be at x at time $t + \tau$, a walker was at some position $x - l$ at time t and has then just made the step of the correct length l and direction to reach x at time $t + \tau$. The integral sums over all possible scenarii. This expression uses the identity $P(A \text{ and } B) = P(A)P(B)$, i.e. the probability for two uncorrelated events to occur is the product of the corresponding probabilities.

Let us now consider the continuous limit where $\tau \to 0$ and the step lengths l, in a certain sense to be described below, also go to zero. The continous limit of the Master equation gives us the Fokker–Planck equation which, in the context of the random walk problem, is nothing but the diffusion equation. Let us expand $P(x - l, t)$ up to second order

$$P(x - l, t) = P(x, t) - l \left.\frac{\partial P}{\partial x}\right|_{(x,t)} + \frac{1}{2} l^2 \left.\frac{\partial^2 P}{\partial x^2}\right|_{(x,t)} + \mathcal{O}(l^3) \ , \tag{2.35}$$

where $\mathcal{O}(l^3)$ represents the higher order terms in powers of l. We also write $P(x, t + \tau) = P(x, t) + \tau\, \partial P(x, t)/\partial t + \mathcal{O}(\tau^2)$. Replacing in (2.34), we get

$$\frac{\partial P(x,t)}{\partial t} = -\frac{\langle l \rangle}{\tau} \left.\frac{\partial P}{\partial x}\right|_{(x,t)} + \frac{1}{2} \frac{\langle l^2 \rangle}{\tau} \left.\frac{\partial^2 P}{\partial x^2}\right|_{(x,t)} + \mathcal{O}\left(\frac{l^3}{\tau}\right) + \mathcal{O}(\tau) \,. \quad (2.36)$$

The continuous limit is taken such that $\langle l \rangle / \tau$ becomes a constant that we call the velocity v and $(\langle l^2 \rangle - \langle l \rangle^2)/2\tau$ also becomes a constant that we call the diffusion coefficient D. Notice that these two limits can be taken with no contradiction as they refer to two different aspects of the random walk behavior, its average drift and the fluctuations around it. The two parameters v and D are thus independent. We can thus write the coefficient $\langle l^2 \rangle / 2\tau$ of the second derivative of P in the r.h.s. of (2.36) under the form: $D + \langle l \rangle^2 / 2\tau$. Since $\langle l \rangle \to v\tau$, this implies that $\langle l \rangle^2 / 2\tau \to v^2\tau/2 \to 0$ and $\langle l^2 \rangle / 2\tau \to D$, in the continuous limit. The expression (2.36) becomes

$$\frac{\partial P(x,t)}{\partial t} = -\frac{\partial j(x,t)}{\partial x} = -v\frac{\partial P(x,t)}{\partial x} + D\frac{\partial^2 P(x,t)}{\partial x^2} \,, \quad (2.37)$$

where $v = \langle l \rangle / \tau$ and $D = (1/2\tau)[\langle l^2 \rangle - \langle l \rangle^2]$ are proportional to the first two cumulants of $\Pi(l)$. Notice that the continuous limit has a physical meaning if the limits $\tau \to 0$ and the step sizes $\to 0$ are taken such that $[\langle l^2 \rangle - \langle l \rangle^2]/\tau$ goes to a constant. Physically, this amounts to imposing that the diffusion law (2.10) remains valid when the microscopic scale shrinks to zero. This constraint implies that the characteristic scale of the fluctuation of the step length is $\propto \sqrt{\tau}$, ensuring that the higher order terms are of order or smaller than $\sqrt{\tau} \to 0$ in the continuous limit. v is non-zero if $\langle l \rangle \neq 0$. In this case, it is also easy to check that the higher order terms are negligible for distributions $\Pi(l)$ with a finite second moment. For a more systematic derivation, including extensions in terms of the derivation of the Kramers–Moyal expansion, we refer to Hänggi and Thomas [398].

Let us give an illustration and consider, for the sake of simplicity, the Gaussian pdf translated from the origin

$$\Pi(l) = \frac{1}{\sqrt{2\pi}\sigma}\mathrm{e}^{-(l-l_0)^2/2\sigma^2} \,. \quad (2.38)$$

The continuous limit corresponds to taking $l_0 \to 0$ and $\sigma \to 0$, with $l_0/\tau = v$ and $\sigma^2/\tau = 2D$ fixed (when τ also goes to zero). Consider the third moment appearing in the expansion (2.35):

$$\langle l^3 \rangle = \int_{-\infty}^{\infty} dl\, \Pi(l)l^3 = \int_{-\infty}^{\infty} dX\, (X + l_0)^3 \frac{1}{\sqrt{2\pi}\sigma}\mathrm{e}^{-X^2/2\sigma^2}$$
$$= 3\sigma^2 l_0 + l_0^3 \,. \quad (2.39)$$

Thus,

$$\frac{\langle l^3 \rangle}{\tau} \to 3v(\sigma^2 + l_0^2) \to 0 \quad (2.40)$$

in the continuous limit, showing its consistency. The same check can easily be made for the higher order terms.

$j(x,t)$ is the flux of random walkers defined by

$$j(x,t) = vP(x,t) - D\frac{\partial P(x,t)}{\partial x} \; . \tag{2.41}$$

The expression (2.37) states the conservation of probability. It can be shown that this description (2.37) is generic in the limit of narrow Π distributions: the details of Π are not important for the large t behavior; only its first two cumulants control the results [780]. The parameters v and D introduce a characteristic "length" $x^* = D/|v|$ as already discussed in Fig. 2.2.

The pdf $P(x,t)$ for a set of random walkers obeys the same diffusion equation as the temperature T or the concentration C of a chemical species diffusing according to the heat equation and Fick's law respectively. The fundamental reason is that the variables T and C are the macroscopic manifestations of a large number of microscopic degrees of freedom undergoing random walk motions. For the temperature, the microscopic degrees of freedom are the phonons colliding with the surrounding heterogeneities which make them follow a kind of random walk. For the concentration, the microscopic degrees of freedom are the molecules undergoing a brownian motion.

The solution of the diffusion equation (2.37), for a population of random walkers all at the origin at time 0, is called the Green function of the diffusion equation, which takes the form of the Gaussian distribution

$$P_G(x,t) = \frac{1}{\sqrt{2\pi}}\frac{1}{\sqrt{2Dt}}e^{-(x-vt)^2/4Dt} \; . \tag{2.42}$$

If $P(x,t=0)$ is given (and in general different from a delta function), the solution of the diffusion equation reads

$$P(x,t) = \int \mathrm{d}x' P_G(x-x',t)P(x',t=0) \; . \tag{2.43}$$

The concentration of random walkers at a given point x at time t is the sum of the diffusion resulting from all sources.

2.2.2 General Fokker–Planck Equation

In the presence of space varying forces, the most general expression of the Fokker–Planck equation reads [780]

$$\frac{\partial P(y,t)}{\partial t} = -\frac{\partial}{\partial y}\left(A(y)P\right) + \frac{1}{2}\frac{\partial^2}{\partial y^2}\left(B(y)P\right) \; . \tag{2.44}$$

The stationary solution of this equation is known:

$$P^s(y) = \frac{\text{const}}{B(y)}\exp\left[2\int_0^y \frac{A(y')}{B(y')}\mathrm{d}y'\right] \; . \tag{2.45}$$

P(x,t)

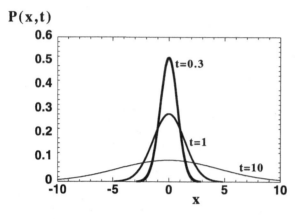

Fig. 2.5. Distribution $P_G(x,t)$ given by (2.42) for three times $t = 0.3, 1$ and 10, with $D = 1, v = 0$

The functions A and B can be estimated from measurements as follows. Let $P(y,t|y_0,t_0)$ for $t \geq t_0$ be the solution which at t_0 reduces to $\delta(y - y_0)$. Take a later time $t = t_0 + \Delta t$. The centered moments of $y - y_0 = \Delta y$ are then

$$\frac{\langle \Delta y \rangle}{\Delta t} = A(y_0) \,, \tag{2.46a}$$

$$\frac{\langle (\Delta y - \langle \Delta y \rangle)^2 \rangle}{\Delta t} = B(y_0) \,, \tag{2.46b}$$

$$\frac{\langle (\Delta y - \langle \Delta y \rangle)^\nu \rangle}{\Delta t} = 0 \,, \quad \text{for all } \nu > 2 \,. \tag{2.46c}$$

These properties suggest a strategy to construct the appropriate Fokker–Planck equation in various contexts. Suppose that we observe some stochastic process. Pick a time Δt so small that y does not vary much. Compute $\langle \Delta y \rangle_y$ and $\langle (\Delta y)^2 \rangle_y$ to first order in Δt. This provides $A(y)$ and $B(y)$ in the Fokker–Planck equation.

2.2.3 Ito Versus Stratonovich

This presentation borrows from exchanges with I. Procaccia. Consider the stochastic differential equation that generalizes (2.1, 2.14), by introducing a systematic position-dependent drift $A(y)$ and a position-dependent step length $C(y)$:

$$\dot{y} = A(y) + C(y)\eta(t) \,. \tag{2.47}$$

This equation as it stands has no meaning, and we do not know how to associate to it a Fokker–Planck equation. The reason is that the δ-function character of the correlation of $\eta(t)$ causes jumps in $y(t)$ such that the value of y at time t is not defined. Accordingly $C(y)$ is also not defined. Basically, the problem comes from the fact that the equation does not specify what value of $C(y)$ should be used. Think of the discrete version of (2.47). We can evaluate $C(y)$ before, after or as a mean of the values before and after the jump. Which one is correct? This has important consequences for numerical simulations of random walk equations with space-dependent force and diffusion coefficients. There are two ways to give sense to (2.47), the Ito and the Stratonovich interpretations.

The Ito Interpretation. In this interpretation, the value of y is taken *before* the jump. Explicitly

$$y(t + \Delta t) - y(t) = A(y(t)) \, \Delta t + C(y(t)) \int_t^{t+\Delta t} \eta(t') \, dt' \ . \tag{2.48}$$

From this, we can compute the first and second moment of Δy, and read the Fokker–Planck equation:

$$\frac{\partial P(y, t)}{\partial t} = -\frac{\partial}{\partial y} (A(y)P) + \frac{1}{2} \frac{\partial^2}{\partial y^2} (C^2(y)P) \ . \tag{2.49}$$

The Stratonovich Interpretation. In this interpretation, we take the mean of $y(t)$ before and after the jump:

$$y(t + \Delta t) - y(t) = A(y(t)) \, \Delta t$$
$$+ C \left(\frac{y(t) + y(t + \Delta t)}{2} \right) \int_t^{t+\Delta t} \eta(t') \, dt' \ . \tag{2.50}$$

The expectation of $y(t + \Delta t) - y(t)$ is

$$\langle y(t + \Delta t) - y(t) \rangle = \left[A(y(t)) + \frac{1}{2} C(y(t)) \left. \frac{\partial C(y)}{\partial y} \right|_{y=y(t)} \right] \Delta t \ . \tag{2.51}$$

Expression (2.51) comes from the expansion

$$C \left(\frac{y(t) + y(t + \Delta t)}{2} \right) = C(y(t)) + \frac{y(t + \Delta t) - y(t)}{2} \left. \frac{\partial C(y)}{\partial y} \right|_{y=y(t)} , \tag{2.52}$$

which introduces a term proportional to $\int_t^{t+\Delta t} \eta(t'') \, dt''$ which multiplies the other term $\int_t^{t+\Delta t} \eta(t') \, dt'$ in (2.50). Then using $\langle \int_t^{t+\Delta t} \eta(t') \, dt' \int_t^{t+\Delta t} \eta(t'') \times dt'' \rangle = \Delta t$, we obtain (2.51). The first moment of $\Delta y = y(t+\Delta t) - y(t)$ is thus given by (2.51) and the second centered moment is $[C(y(t))]^2$. Inserting these two values in (2.46a) and (2.46b) gives, after some rearranging, a Fokker–Planck equation different from that obtained in the Ito case:

$$\frac{\partial P(y, t)}{\partial t} = -\frac{\partial}{\partial y} (A(y)P) + \frac{1}{2} \frac{\partial}{\partial y} \left(C(y) \frac{\partial}{\partial y} (C(y)P) \right) \ . \tag{2.53}$$

One notes that, under a nonlinear transformation $\bar{y} = \phi(y)$, the coefficients of (2.53) transform like

$$\bar{A}(\bar{y}) = A(y) \frac{d\phi}{dy} , \quad \bar{C}(\bar{y}) = C(y) \frac{d\phi}{dy} \ . \tag{2.54}$$

If one applies the same transformation to (2.47), one obtains the same transformation laws. This means that the relation between (2.47) and (2.53) is invariant to nonlinear transformations.

This is not so for the Ito interpretation. The same transformation applied to (2.49) yields

$$\bar{A}(\bar{y}) = A(y)\frac{\mathrm{d}\phi}{\mathrm{d}y} + \frac{1}{2}C^2(y)\frac{\mathrm{d}^2\phi}{\mathrm{d}y^2} \ , \quad \bar{C}(\bar{y}) = C(y)\frac{\mathrm{d}\phi}{\mathrm{d}y} \ . \tag{2.55}$$

Relation Between Ito and Stratonovich. Given an equation like (2.47) interpreted à la Ito, we can write another stochastic equation that under the Stratonovich interpretation will give the same stochastic process. The following is known: if we write the two equations in the two interpretations as

$$\mathrm{d}_\mathrm{I}Y_i = b_i^\mathrm{I}\,\mathrm{d}t + \sigma_i\,\mathrm{d}z \ , \tag{2.56}$$

$$\mathrm{d}_\mathrm{S}Y_i = b_i^\mathrm{S}\,\mathrm{d}t + \sigma_i\,\mathrm{d}z \ , \tag{2.57}$$

then the condition for these two equations to describe the same stochastic process is

$$b_i^\mathrm{I} = b_i^\mathrm{S} + \frac{1}{2}\sigma_j\frac{\partial\sigma_i}{\partial Y_j} \ . \tag{2.58}$$

This has important practical applications. Consider for example the motion of a Brownian particle subjected to a position-dependent diffusion coefficient $D(Y) = \sigma(Y)^2/2$ and a constant position-dependent force $F(Y)$. The dependence $D(Y)$ may stem for instance from a position-dependent friction coefficient due to the presence of a wall. The random displacement $\ell(Y)$ of a random walker along $0Y$ during a microscopic time interval τ is then the sum of three terms [270]:

$$\ell(Y) = \pm\sqrt{2D(Y)\tau} + \frac{1}{2}\frac{\mathrm{d}D(Y)}{\mathrm{d}Y}\,\tau + \frac{D(Y)}{k_\mathrm{B}T}\,F(Y)\tau \ . \tag{2.59}$$

The first term in the r.h.s. of (2.59) recovers the usual definition $\langle l^2\rangle = 2D\tau$ of the variance of a random walk. The second term is the Ito correction (2.58) which must be incorporated when performing a numerical simulation in which the value of the step length is locally defined. The last term is the drift term due to the systematic force. The second term may be intuitively rederived by noting that the differential increment of the amplitude ℓ_0 of the first term

$$\mathrm{d}\ell_0 \equiv \mathrm{d}\left[\sqrt{2D(Y)\tau}\right] = \tau\frac{\mathrm{d}D}{\mathrm{d}Y}\frac{\mathrm{d}Y}{\ell_0} \tag{2.60}$$

is equal to the second term in the r.h.s. of (2.59) for $\mathrm{d}Y = \ell_0/2$, which is precisely the Stratonovich interpretation to consider the step length as the average of the starting and ending point.

Another way to state the same thing is to consider the differential of a function $F(W(t))$ of the random walk $W(t)$:

$$dF(W(t)) = \frac{dF(W(t))}{dW(t)}dW(t) + \frac{1}{2}\frac{d^2F(W(t))}{dW(t)^2}\text{Var}\left[dW(t)\right] . \qquad (2.61)$$

Since, by definition, $\text{Var}\left[dW(t)\right] = dt$, we see that, up to first-order in dt, one must keep the two terms in the r.h.s. of (2.61) to be consistent with usual derivatives. This is the essence of Ito's term.

2.2.4 Extracting Model Equations from Experimental Data

Let us present briefly a general data-driven method for formulating suitable model equations for nonlinear complex systems [314], based on the Langevin formalism. We thus consider the class of dynamic systems which can be described by the following differential equation

$$\frac{d}{dt}\mathbf{X}(t) = \mathbf{g}(\mathbf{X}(t), t) + \mathbf{h}(\mathbf{X}(t), t)\Gamma(t) , \qquad (2.62)$$

where $\mathbf{g}$ (resp. $\mathbf{h}\Gamma$) represents the deterministic (resp. stochastic) part. $\mathbf{X}(t)$ denotes the time dependent d-dimensional stochastic vector which characterises the system completely. $\Gamma(t)$ stands for terms of δ-correlated Gaussian white noises and the $d \times d$-matrix $\mathbf{h}$ fixes the dynamic influence of the noise on the system. For the functionals $\mathbf{g}$ and $\mathbf{h}$, no further assumptions have to be made; $\mathbf{g}$ can be nonlinear, and therefore also deterministic chaos can be formulated by (2.62). The question addressed here is to find the deterministic and stochastic laws solely by data analysis. If this only condition, the describability of the system's dynamics by an evolution equation like (2.62), is given, no further assumptions or pre-knowledge have to be included in the following analysis. Deterministic and noisy parts of the dynamics can be separated and quantified, and model equations for the dynamics can be set up by the data-driven method.

The considered class of dynamic systems (2.62) is characterised by the absence of memory effects, i.e., for the time development of the system we need to know only the state of one vector $\mathbf{X}(t)$ at a given time t and not its evolution in the past. In other words, the system is Markovian. The conditional probability density $p(x_{n+1}, t + \tau | x_n, t; x_{n-1}, t - \tau; \ldots)$ describes the probability of states x_{n+1} of the system's variable $\mathbf{X}$ at time $t + \tau$ under the condition that the system is in state x_n at time t, has been in state x_{n-1} at time $t - \tau$ and so on. The Markovian property of a system can be tested by comparing the multiple conditional probability densities with the one-step conditional probability densities $p(x_{n+1}, t + \tau | x_n, t)$. If both expressions agree, the time development of the probability density depends only on the present state and not on its evolution in the past. The assumed qualities of the driving noise terms Γ as being Gaussian white noise functions can be validated, as well, by looking at the conditional probability density distributions.

The central ideas of the construction of the model equations consist in the following. First, stationary dynamics shall be assumed, i.e., the deterministic and stochastic parts $\mathbf{g}$ and $\mathbf{h}$ are not explicitly time dependent (this restriction can be removed by using a moving window technique). Every time t_i the system's trajectory meets an arbitrary but fixed point $\mathbf{x}$ in state space, the localisation of the trajectory at time $t_i + \tau$ is determined by the deterministic function $\mathbf{g}(\mathbf{x})$, which is constant for fixed $\mathbf{x}$, and by the stochastic function $\mathbf{h}(\mathbf{x})\Gamma(t_i)$ with constant $\mathbf{h}$ for fixed $\mathbf{x}$ and Gaussian distributed white noise $\Gamma(t)$. With these pre-requisites, the following relationships have been proved in a strict mathematical way using Ito's definitions for stochastic integrals [780]:

$$\mathbf{g}(\mathbf{x}) = \lim_{\tau \to 0} \frac{1}{\tau} \langle \mathbf{X}(t+\tau) - \mathbf{x} \rangle|_{\mathbf{X}(t)=\mathbf{x}} \, , \tag{2.63}$$

$$\mathbf{h}(\mathbf{x})\mathbf{h}^{\mathrm{T}}(\mathbf{x}) = \lim_{\tau \to 0} \frac{1}{\tau} \langle (\mathbf{X}(t+\tau) - \mathbf{x}) \, (\mathbf{X}(t+\tau) - \mathbf{x})^{\mathrm{T}} \rangle|_{\mathbf{X}(t)=\mathbf{x}} \, . \tag{2.64}$$

Under the condition that the system's trajectory meets the point $\mathbf{x}$ at time t, i.e., $\mathbf{X}(t) = \mathbf{x}$, the deterministic part $\mathbf{g}(\mathbf{x})$ of the dynamics can be evaluated for small τ by the difference of the system's state at time $t + \tau$ and the state at time t, averaged over an ensemble, or in the regarded stationary case, averaged over all $t = t_i$ of the whole time series with $\mathbf{X}(t_i) = \mathbf{x}$. The limit $\tau \to 0$ can be reached by extrapolation. In a similar way, the stochastic influences can be determined from the average of the quadratic terms as shown in (2.64). For every point $\mathbf{x}$ in state space, that is visited statistically often by the trajectory, deterministic and stochastic parts of the dynamics can be estimated numerically. As final step, analytic functions can be fitted to $\mathbf{g}$ and $\mathbf{h}$ in order to formulate model equations for the investigated system (see [314] for applications to tremor data from patients suffering from Parkinson's disease).

2.3 The Central Limit Theorem

The central limit theorem is a fundamental result with broad applications covering many fields, from statistical physics to signal processing.

2.3.1 Convolution

Consider $X = X_1 + X_2$ where X_1 and X_2 are two random variables with pdf $P_1(x_1)$ and $P_2(x_2)$. The probability that X be equal to x (to within dx) is given by the sum of all probabilities of the events that can give $X = x$, and thus corresponds to all combinations of $X_1 = x_1$ and $X_2 = x_2$ with $x_1 + x_2 = x$. Since X_1 and X_2 are assumed to be independent, the joint probability of $X_1 = x_1$ and $X_2 = x - x_1$ is the product of the respective probabilities:

$$P(x) = \int_{-\infty}^{\infty} dx_1 \, P_1(x_1) \int_{-\infty}^{\infty} dx_2 \, P_2(x_2) \delta(x - (x_1 + x_2))$$

$$= \int_{-\infty}^{\infty} dx_1 \, P_1(x_1) P_2(x - x_1) \ . \tag{2.65}$$

The last term of the r.h.s. defines the convolution operation between $P_1(x)$ and $P_2(x)$, that we note $P = P_1 \star P_2$. For a sum of N independent random variables

$$X = X_1 + X_2 + \ldots + X_N \ , \tag{2.66}$$

with pdf $P_i(x_i)$ for variable X_i, we get the pdf of the sum as

$$P(x) =$$

$$\int dx_1 \, P_1(x_1) \int dx_2 \, P_2(x_2) \ldots \int dx_N \, P_N(x_N) \delta(x - (x_1 + x_2 + \ldots + x_N)) =$$

$$\int dx_1 \, P_1(x_1) \int dx_2 \, P_2(x_2) \ldots \int dx_{N-1}$$

$$\times P_{N-1}(x_{N-1}) P_N(x - x_1' - \ldots - x_{N-1}') \ . \tag{2.67}$$

The assumption of independence between the variables allows one to derive the complete knowledge of the pdf of the sum from the sole knowledge of the pdf of the individual variables.

It is convenient to use the characteristic function $\hat{P}(k)$. Indeed, in the Fourier representation, the convolution operation becomes a product: if $P = P_1 \star P_2$, then

$$\hat{P}(k) = \hat{P}_1(k) \hat{P}_2(k) \ . \tag{2.68}$$

This is easily seen directly from (2.65) by the integration of the delta function. Similarly, from (2.67), we get

$$\hat{P}(k) = \hat{P}_1(k) \hat{P}_2(k) \ldots \hat{P}_j(k) \ldots \hat{P}_N(k) = \left(\hat{P}_1(k) \right)^N , \tag{2.69}$$

where the last equality occurs for i.i.d. variables with $\hat{P}_j = \hat{P}_1$. In principle, all is known of the pdf of the sum by taking the inverse Fourier transform of $\left(\hat{P}_1(k) \right)^N$. In practice, this may not be very illuminating!

Let us provide an intuitive approach to this analysis. It relies upon the cumulant expansion (1.30) for the pdf $P_1(x)$ of the i.i.d. variables contributing to the sum. Putting (1.30) in (2.69), we get

$$\hat{P}(k) = \exp \left\{ \sum_{n=1}^{\infty} \frac{N c_n}{n!} (ik)^n \right\} . \tag{2.70}$$

When $N \to +\infty$, the sum X goes to infinity with a drift $\sim N$ and a standard deviation $\sim N^{1/2}$. Since the drift can be zero or can be put to zero by a change of frame moving with the drift velocity, we see that the relevant scale is that of the fluctuations, namely the standard deviation $\sim N^{1/2}$. The corresponding range of k is simply its inverse $\sim N^{-1/2}$, since x and k are conjugate in the Fourier transform and thus have opposite dimensions.

As a consequence, the n-th term in the cumulant expansion (2.70) scales as

$$\frac{Nc_n}{n!}k^n \sim N^{1-n/2} . \tag{2.71}$$

We see that for $n = 2$, the exponent is zero and the cumulant of the pdf of the sum remains invariant while all higher cumulants approach zero as $N \to +\infty$. This shows that, asymptotically, only the second (and the first) cumulant will remain, characterizing a Gaussian pdf. Alternatively, the same result is obtained by studying the normalized cumulants that decay as $\lambda_n \propto N^{1-n/2}$ for $n > 2$. This is the essence of the central limit theorem.

2.3.2 Statement

The precise formulation of the central limit theorem is the following.

The sum, normalized by $1/\sqrt{N}$ of N random independent and identically distributed variables of zero mean and finite variance σ^2, is a random variable with a pdf converging to the Gaussian distribution with variance σ^2. The convergence is said to hold in the measure sense, i.e. the probability that the above normalized sum falls in a given interval converges to that calculated from the Gaussian distribution.

The normalization by $1/\sqrt{N}$ ensures that the variable has a stationary scale, since the characteristic scale of the fluctuations of the sum is of order $\sqrt{N}$.

It is important to realize that the central limit theorem only applies in the "center" of the distribution. Large deviations can occur in the tail of the pdf of the sum, whose weight shrinks as N increases. In Chap. 3, we will discuss the large deviation regime that refines the central limit theorem, for which precise statement can be made. This large deviation regime is relevant to the understanding of the distribution of intermittent velocity bursts in hydrodynamic turbulence and in fragmentation processes, for instance.

2.3.3 Conditions

The main conditions under which the central limit theorem holds are the following.

- The X_i's must be independent. This condition can be relaxed and the CLT still holds for weakly correlated variables. We will come back to this case in Chap. 8, when we will investigate what types of correlations can modify the distribution of the sum away from the Gaussian law.

- The convergence of the sum to a Gaussian pdf also holds if the variables have different pdf's with finite variance of the same order of magnitude: this is in order for one variable not to dominate, as the variance of the sum is the sum of the variances.
- Strictly speaking, the central limit theorem is applicable in the limit of infinite N. In practice, the Gaussian shape is a good approximation of the center of the pdf for the sum if N is sufficiently large. How large is large depends on the range of interest for the sum and on the initial pdf's of the contributing variables.
- The central limit theorem does not say anything about the behavior of the tails for finite N. Only the center is well-approximated by the Gaussian law. The center is a region of width at least of the order of $\sigma \sim \sqrt{N}$ around the average of X. This width depends on detailed properties of the pdf's of the constituting variables. For "gentle" pdf's which have all their cumulants finite, it is of size $\sim N^{2/3}\sigma$ if $c_3 \neq 0$ and $\sim N^{3/4}\sigma$ if $c_3 = 0$ and $c_4 \neq 0$ (see Chap. 3). For power law pdf's with $\mu > 2$, the size of the region over which the Gaussian law holds can become as small as $\sim \sigma\sqrt{N \ln N}$.
- The condition that the variance of the pdf's of the constituting variables be finite can be somewhat relaxed to include the case of power laws with tail exponent $\mu = 2$, as defined in (1.20). In this case, the normalizing factor is no longer $1/\sqrt{N}$ but can contain logarithmic corrections. Technically, this generalization can be written as follows. A pdf belongs to the basin of attraction of the Gaussian law if and only if the following condition holds [350]:

$$\lim_{x \to \infty} x^2 \frac{P_<(-x) + P_>(x)}{\int_{|x'|<x} x'^2 P(x') \, dx'} = 0 \; . \tag{2.72}$$

This condition is always satisfied for a pdf with finite variance and it also include marginal cases such as power laws with $\mu \geq 2$. Pdf's decaying slower than $|x|^{-3}$ for large $|x|$ do not converge to the Gaussian law but to another class of stable distributions, called Lévy laws.
- We refer to [755] for a detailed description of the rate of convergence to the normal pdf of the sum of i.i.d. random variables, with special emphasis on the way the asymmetry of the initial pdf decays when the number of variables in the sum increases.

2.3.4 Collective Phenomenon

"In fact, all epistemologic value of the theory of probability is based on this: that large-scale random phenomena in their collective action create strict, non random regularity." B.V. Gnedenko and A.N. Kolmogorov [350].

The central limit theorem is the expression of a collective phenomenon. It is the most basic among those that we will discuss in this book: the contributions of many random variables lead to a global behavior of the sum which becomes both simple and universal. The individual details of the pdf's of the contributing variables are progressively washed out to give rise to this universal Gaussian shape. The central limit theorem is very important as it can be thought of as the cornerstone for understanding collective phenomena. Indeed, the collective phenomena that we will study further in our exposition can be thought of as complications of this simple beautiful result. The complications brought out by the study of natural systems involve correlations, nonlinearity, external driving, etc., which modify the result in many possible ways, but the surviving fact is the emergence of a macroscopic coherent behavior with, often but not always, well-defined universal behavior. The importance of the Gaussian law also relies on the fact that we can often use it as a starting point in the study of more complicated situations.

2.3.5 Renormalization Group Derivation

We will now derive the central limit theorem, using the technique of the *renormalization group* (RG) theory. This theory has been invented to tackle critical phenomena, a class of behaviors characterized by structures on many different scales [1021] and power law dependences of measurable quantities on the control parameters. In fact, the random walk (and equivalently the sum of random variables) is a critical phenomenon, arguably the simplest of all: up to time t, the standard deviation $\langle [X(t)]^2 \rangle^{1/2} \sim t^{1/2}$ of a symmetric random walk scales as a power law of t, where the "control parameter" $1/t$ can be interpreted as the distance to criticality. The only relevant scale of the problem is $\sim t^{1/2}$ and all scales smaller than $t^{1/2}$ are present with a relative weight given by the scale free power spectrum $\sim k^{-2}$. In the limit of large times, the random walk becomes a critical system (see [213] for a fruitful exploitation of this idea in the context of polymers).

The RG analysis, introduced in field theory and in critical phase transitions, is a very general mathematical tool, which allows one to decompose the problem of finding the "macroscopic" behavior of a large number of interacting parts into a succession of simpler problems with a decreasing number of interacting parts, whose effective properties vary with the scale of observation. Technically, this is done as we will see in Chap. 11 by defining a mapping between observational scale and distance (time) from the critical point. The term "observational scale" usually refers to the physical size of an observation. In the random walk context, the observational scale refers to the number of terms in the sum. The RG approach works best when the system possesses the properties of scale invariance and self-similarity of the observables at the critical point. The purpose of the RG is to translate in mathematical language the concept that the sum is the aggregation of an ensemble of arbitrarily defined sub-sums, each sub-sum defined by the sum of sub-sub-sums, and so on.

We can carry out this program for the sum (2.66) and perform the two main transformation of the RG, decimation and rescaling.

Decimation. The first step of the RG is to "decimate" the degrees of freedom to transform the problem into a simpler one, i.e. with fewer degrees of freedom. Starting from (2.66), one way to do it is to group the terms in the sum by pairs and rewrite it as

$$X = X_1' + X_2' + \ldots + X_{N/2}' \, , \tag{2.73}$$

where $X_1' = X_1 + X_2$, $X_2' = X_3 + X_4$, $\ldots$, $X_{N/2}' = X_{N-1} + X_N$, with $N = 2^m$. This specific choice is not a restriction since we are interested in the limit of large N and the way with which we reach this limit (in the present case by taking $m \to \infty$) is of no consequence.

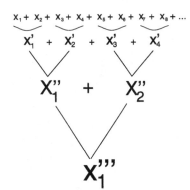

Fig. 2.6. Schematic representation of the decimation procedure, consisting of grouping the variables in pairs

The expression (2.73) is also a sum of random variables but with $N/2$ of them instead of N. Obviously, the X_i' are i.i.d. variables if the X_i are i.i.d variables. If we can deduce their pdf P' from that of the initial pdf P of the X_i, the problem is now to determine the pdf of the sum X of $N/2$ i.i.d. variables with known pdf P'. If we apply successively m times the pairing process, we will go from N variables in the sum to $N/2$, then $N/4, \ldots, 4$, 2 and finally only 1 final variable. This final unique variable cannot be anything else but X itself. We have thus "climbed" up the "staircase" of scales to ascend from the microscopic degrees of freedom to the macroscopic variable X. Now, this decimation of the degrees of freedom comes with some practical difficulties in its implementation, namely the calculation of the mapping allowing us to go from P to P' and then to $P'', \ldots, P^{(n)}, P^{(n+1)}, \ldots$, to end with $P^{(m)}$ which is indeed the pdf of the sum X. It is clear that this mapping is the same at each generation since we use the same pairing algorithm. We already know its expression from (2.65):

$$P^{(n+1)}(x) = \int_{-\infty}^{\infty} dx_1 \, P^{(n)}(x_1) P^{(n)}(x - x_1) \, , \tag{2.74}$$

valid for i.i.d. variables. The characteristic functions thus obey

$$\hat{P}^{(n+1)}(k) = \left[\hat{P}^{(n)}(k) \right]^2 . \tag{2.75}$$

Using the cumulant expansion (1.30), we obtain a relationship between the l-th cumulant of $P^{(n+1)}(x)$ and that of $P^{(n)}(x)$:

$$c_l^{(n+1)} = 2c_l^{(n)} . \tag{2.76}$$

One cannot think of a simpler result!

Since a pdf is completely determined from the knowledge of all its cumulants (barring some special cases mentioned above), we can thus write

$$P(x, N, c_1, c_2, c_3, \ldots, c_l, \ldots) = P\left(x, \frac{N}{2}, 2c_1, 2c_2, 2c_3, \ldots, 2c_l, \ldots\right) . \tag{2.77}$$

$P(x, N, c_1, c_2, c_3, \ldots, c_l, \ldots)$ denotes the pdf of the sum X, characterized by the cumulants $c_1, c_2, c_3, \ldots, c_l, \ldots$ of the initial pdf of the N variables in the sum. Since the sum does not depend on the way we group its individual constituants, its pdf is the same if we look at it at two different scales, provided we correctly scale the properties associated with these two different scales.

Rescaling. Inherent to a pdf is the notion of a scale, as exemplified by the structure around x_{mp}. When we form the pdf for the sum of two i.i.d. random variables, its pdf may display differences to that of the pdf we started from. We compensate for this by the scale factor s for the sum. This leads to the second step of the RG, which is to rescale the X_i' so that the problem involves the same scale as the initial one. The idea of the RG is to decimate the degrees of freedom, while rescaling so as to keep the same scale. We guess that this involves changing X_i' into X_i'/s, where s is to be determined. It is useful to make $s = 2^\alpha$ and the problem is to determine the right choice for the exponent α. We can intuitively infer what the correct value of α will be from the diffusive behavior of the sum of random variables $X(t) \sim t^{1/2}$. For a sum of $t = 2$ terms, this leads to $s = 2^{1/2}$ and thus $\alpha = 1/2$. We will provide a formal derivation of this guess.

With a rescaling of the X_i variables, the cumulants are also rescaled and c_l has to be multiplied by the factor $2^{-\alpha l}$. This stems from the fact that c_l has the dimension k^{-l}, i.e. the dimension of X^l. Under this change of variable, the conservation of probabilities $p(y)\,dy = p(x)\,dx$ introduces a factor $2^{-\alpha}$ in front of the pdf on the r.h.s. of (2.77). We thus obtain

$$P(x, N, c_1, c_2, c_3, \ldots, c_l, \ldots) =$$
$$\frac{1}{2^\alpha} P\left(\frac{x}{2^\alpha}, \frac{N}{2}, 2^{1-\alpha}c_1, 2^{1-2\alpha}c_2, 2^{1-3\alpha}c_3, \ldots, 2^{1-l\alpha}c_l, \ldots\right) . \tag{2.78}$$

We see from (2.78) that the particular choice $\alpha = 1/2$ makes the factor $2^{1-2\alpha}$ of the second cumulant equal to 1, i.e. this choice keeps the pdf in a frame with a constant width as $N \to \infty$. In this limit, all higher cumulants decrease to zero, while the first cumulant increases, reflecting its larger and larger effect. If the first cumulant (equal to the average) is not zero, this means that the sum is dominated by it and grows like N. The standard deviation scales as $N^{1/2}$. In the frame of scale where the standard deviation is constant, this implies that the mean of the sum scales itself as $N^{1/2}$, hence the factor $2^{1-\alpha} = 2^{1/2}$ in front of the first cumulant.

Iteration of the Decimation and Rescaling. Iterating the RG procedure (involving both the decimation and rescaling) m times, (2.78) gives

$$P\left(x, N, c_1, c_2, c_3, \ldots, c_l, \ldots\right) =$$
$$\frac{1}{2^{m\alpha}} P\left(\frac{x}{2^{m\alpha}}, \frac{N}{2^m} = 1, 2^{m(1-\alpha)} c_1, 2^{m(1-2\alpha)} c_2,\right.$$
$$\left. 2^{m(1-3\alpha)} c_3, \ldots, 2^{m(1-l\alpha)} c_l, \ldots\right) . \tag{2.79}$$

If $\alpha = 1/2$, then for $m \to +\infty$, (2.79) becomes

$$P(x, N, c_1, c_2, c_3, \ldots, c_l, \ldots) \to$$
$$\frac{1}{\sqrt{N}} P\left(\frac{x}{\sqrt{N}}, 1, c_1 \sqrt{N}, c_2, c_3 = 0, \ldots, c_l = 0, \ldots\right) , \tag{2.80}$$

for $N \to \infty$, which is a function only of $N = 2^m$, c_1 and c_2. We have thus obtained the asymptotic result that the pdf of the sum for $N \to \infty$ has only its two first cumulant non zero, hence it is a Gaussian law, thus recovering the central limit theorem. Applying the Fourier transform on both sides of (2.80) gives

$$\hat{P}(k, N) = \exp\left\{N\left(\mathrm{i}kc_1 - k^2 \frac{c_2}{2}\right)\right\}, \tag{2.81}$$

which is the Fourier transform of the Gaussian law of mean Nc_1 and variance Nc_2.

This exercise has the merit of introducing the powerful RG concepts and technique. It also shows explicitly the multiple scales involved in the sum of a large number of random variables and thus its intrinsic critical nature. In addition, it demonstrates the convergence to the Gaussian law, now understood as the attractive *fixed point* of the RG process.

The RG has many applications, first of all in allowing the calculation of the critical exponents (among other quantities) of magnetic and fluid systems at their Curie point. It has also been applied to a large variety of problems, such as, among others, material rupture, fragmentation, earthquakes, transport in heterogeneous media (conductivity, permeability, elastic properties...), in the characterization of the transitions to chaotic behavior and in turbulence. We will review some of these problems in Chap. 11.

2.3.6 Recursion Relation and Perturbative Analysis

We continue our strategy to expose modern methods and concepts using the simple context of the sum of random uncorrelated variables. Here, we develop a perturbation approach to the central limit theorem.

Instead of the "bottom-up" decimation of the degrees of freedom performed in the renormalization group method, we now take a "top-to-bottom" approach which amounts to doubling the number of variables at each step of a recursive process $N_m = 2^m \rightarrow N_{m+1} = 2^{m+1}$. The corresponding recursion relationship of the pdf's of the average $x \equiv (x_1 + x_2 + \ldots + x_N)/N$ is simply obtained from (1.32):

$$P_{m+1}(x) = 2 \int_{-\infty}^{\infty} \mathrm{d}x_1 \, P_m(x_1) P_m(2x - x_1) \,, \tag{2.82}$$

where $P_m(x)$ denotes the pdf of the mean of $N_m = 2^m$ variables. The expression (2.82) is an operator transforming $P_m(x)$ into $P_{m+1}(x)$. The properties that we have analyzed previously translate themselves as follows in the present description.

Fixed Point. The Gaussian law $P_m^{\mathrm{G}}(x) = \left(1/\sqrt{2\pi\sigma_m^2}\right) \exp\left(-x^2/2\sigma_m^2\right)$ is a fixed point of this transformation (2.82) in the space of functions, i.e. it is transformed into $P_{m+1}^{\mathrm{G}}(x) = \left(1/\sqrt{2\pi\sigma_{m+1}^2}\right) \exp\left(-x^2/2\sigma_{m+1}^2\right)$, where $\sigma_m^2 = 2^{-m}\sigma_0^2$ using the addition rule on the variance leading to $\sigma_{m+1}^2 = (1/2)\sigma_m^2$. Notice that, in order to call the Gaussian law a fixed point, we have to use a reduced variable. This generalizes to the functional space the concept of fixed points of the renormalization group flow of *control parameters* that we will discuss in Chap. 11 for critical points.

Perturbation. A pdf which is close to the Gaussian law $P_m^{\mathrm{G}}(x)$ can be written as

$$P_m(x) = P_m^{\mathrm{G}}(x) + \frac{\epsilon}{\sigma_m}\phi_m\left(\frac{x}{\sigma_m}\right) \,, \tag{2.83}$$

where the second term of the r.h.s. corresponds to a small perturbation turned on by the small parameter ϵ and where the natural scale free variable x/σ_m has been used. The choice of this scaling ensures that the perturbation has a width comparable to that of the Gaussian law, i.e. that the relative size of the perturbation is small over the whole x range.

It is useful to take the perturbation function $\phi_m(x)$ to be such that $\int_{-\infty}^{+\infty} \mathrm{d}x \, x^n \phi_m(x) = 0$ for $n = 0, 1$ and 2. This condition for $n = 0$ ensures that $P_m(x)$ is still normalized to 1. The condition for $n = 1$ and $n = 2$ express the absence of trend and vanishing second moment. The perturbation is thus acting on the higher cumulants which, as we have seen, distinguish a pdf from a Gaussian.

Putting the expression (2.83) in the recursion relation (2.82) and keeping only the terms proportional to ϵ (and assuming that higher order terms are smaller), we arrive at

$$\sigma_{m+1} = \frac{1}{2}\sigma_m \,, \tag{2.84}$$

and

$$\phi_{m+1} = \mathcal{L}\left(\phi_m\right), \tag{2.85}$$

where $\mathcal{L}$ is a linear operator which can be explicitly written:

$$\phi_{m+1}(x) = 2\sqrt{2} \int_{-\infty}^{+\infty} \frac{dy}{\sqrt{2\pi}} \phi_m(x\sqrt{2} - y)e^{-y^2/2} . \tag{2.86}$$

Taking the Fourier transform of (2.86) gives the form $\hat{\mathcal{L}}$ of the operator $\mathcal{L}$ in Fourier space:

$$\hat{\phi}_{m+1}(k) = 2\hat{\phi}_m\left(\frac{k}{\sqrt{2}}\right) e^{-k^2/4} . \tag{2.87}$$

One verifies that the eigenfunctions of $\hat{\mathcal{L}}$, defined by $\hat{\phi}_{m+1}^{(n)}(k) = E_n\hat{\phi}_m^{(n)}(k)$, are the Hermite polynomials which in Fourier space are $\hat{\phi}_m^n(k) = k^n e^{-k^2/2}$ with eigenvalues $E_n = 2^{1-n/2}$. This result shows that all E_n for $n > 2$ are less than 1. Therefore, upon iteration, the weight of the corresponding eigenfunctions in the expansion of a perturbation on the eigenfunctions decreases approaching zero, like a decay. This ensures the convergence to the Gaussian. This calculation precisely quantifies the rate of convergence to the Gaussian law by specifially how the initial perturbation decays to zero upon one convolution (corresponding to the application of the operator $\mathcal{L}$ once): the application of $\mathcal{L}$ transforms $\sum_{n=2}^{+\infty} a_n\phi_m^{(n)}$ into $\sum_{n=2}^{+\infty} a_n E_n\phi_m^{(n)}$, where a_n are arbitrary coefficients.

In a remarkable effort (see [487] and references therein), Jona-Lasinio shows how to start from this approach valid for independent variables to show that, very generally, the renormalization group has an interesting probabilistic interpretation which clarifies the deep statistical significance of critical universality.

3. Large Deviations

The central limit theorem states that the Gaussian law is a good description of the center of the pdf of a sum of a large number N of random variables with finite variance and that the weight (in probability) of the tail goes to zero for large N. We now make more precise what is meant by the "center" of the pdf. This section is slightly more technical than the previous ones, even though we emphasize a non-rigorous intuitive presentation. The purpose of this chapter is to show that there is a lot of "action" going on in the tails, beyond the central Gaussian region. This must be kept in mind for practical applications and data analysis.

3.1 Cumulant Expansion

If X is the sum of N random i.i.d. variables with average $\langle x \rangle$ and variance σ^2, we define the rescaled variable

$$z = \frac{X - N\langle x \rangle}{\sqrt{N}\sigma} , \tag{3.1}$$

which, according to the central limit theorem, tends to a Gaussian variable of zero mean and unit variance. The convergence also applies to the tail; more precisely, for any z,

$$\lim_{N \to \infty} \mathcal{P}_>(z) \equiv g(z) \tag{3.2}$$

where $g(z)$ is the probability weight of the tail:

$$g(z) = \int_z^\infty \frac{\mathrm{d}x}{\sqrt{2\pi}} \exp(-x^2/2) = \frac{1}{2}\mathrm{erfc}(z/\sqrt{2}) , \tag{3.3}$$

and $\mathrm{erfc}(z)$ is the complementary error function. It is important to realize that the convergence is not uniform as the minimum value of N for which $\mathcal{P}_>(z) \simeq g(z)$ depends on z. In other words, for N fixed, this approximation is valid only for $|z| \ll z_0(N)$.

It is possible to estimate $z_0(N)$ for gentle initial pdf's, i.e. when the pdf's decay faster than any power law at large $|x|$. In this case, all cumulants exist and one can develop a systematic expansion in powers of $N^{-1/2}$ of the difference $\mathcal{P}_>(z) - g(z)$ [350]:

$$
\begin{aligned}
&\mathcal{P}_>(z) - g(z) \\
&= \frac{\exp(-z^2/2)}{\sqrt{2\pi}} \left(\frac{Q_1(z)}{N^{1/2}} + \frac{Q_2(z)}{N} + \ldots + \frac{Q_k(z)}{N^{k/2}} + \ldots \right) ,
\end{aligned} \tag{3.4}
$$

where $Q_k(x)$ are polynomials that are parameterized by the normalized cumulants λ_n of the initial pdf defined in (1.35). For the sake of illustration, we give the first two polynomials

$$
Q_1(x) = \frac{1}{6}\lambda_3(1 - x^2) , \tag{3.5}
$$

and

$$
Q_2(x) = \frac{1}{72}\lambda_3^2 x^5 + \frac{1}{8}\left(\frac{1}{3}\lambda_4 - \frac{10}{9}\lambda_3^2 \right) x^3 + \left(\frac{5}{24}\lambda_3^2 - \frac{1}{8}\lambda_4 \right) x . \tag{3.6}
$$

Obviously, if a pdf has all its cumulants of order larger than 2 identically zero, then all the $Q_k(x)$ are also zero as the pdf $P(x)$ is a Gaussian.

For an arbitrary asymmetric pdf, c_3 is non-vanishing in general and the leading correction is $Q_1(x)$. The deviation from the Gaussian law is negligible for z of order 1 if this correction is small, i.e. if $N \gg \lambda_3^2$. Since the deviation increases with z, the Gaussian approximation remains valid if the relative error remains small compared to 1. For large z, this relative error is obtained by dividing (3.4) by $g(z) \simeq (1/z\sqrt{2\pi})\exp(-z^2/2)$. We thus obtain the width of the pdf over which the Gaussian law holds [293]:

$$
\lambda_3 z^3 \ll N^{1/2} \quad \text{leading to} \quad |X - N\langle x \rangle| \ll \sigma \lambda_3^{-1/3} N^{2/3} , \tag{3.7}
$$

using definition (3.1). The standard deviation of the sum scales as $\sigma N^{1/2}$ and the "central" region in which the Gaussian law holds extends further up to a distance $\sim N^{2/3}$.

A symmetric pdf has $c_3 \equiv 0$ and the excess kurtosis $\kappa = \lambda_4$ provides the first leading correction to the central limit theorem. The Gaussian law is valid if $N \gg \lambda_4$ and

$$
\lambda_4 z^4 \ll N \quad \text{leading to} \quad |X - N\langle x \rangle| \ll \sigma \lambda_4^{-1/4} N^{3/4} , \tag{3.8}
$$

using definition (3.1). The standard deviation of the sum scales as $\sigma N^{1/2}$ and the "central" region in which the Gaussian law holds extends to a distance $\sim N^{3/4}$.

3.2 Large Deviation Theorem

Our presentation is inspired from [316, 549]. The large deviation theorem goes beyond the previous results and determines the probability for the sum X of N random variables X_i, in the limit $N \to \infty$, to take a value Nx, where

x can be different from $\langle x \rangle$. It can be shown that this probability can be expressed as

$$P[X \simeq Nx] \sim e^{Ns(x)} \, dx, \qquad \text{for } x \text{ finite and } N \to \infty , \qquad (3.9)$$

where $s(x)$ is the so-called Cramér function (also called "rate function"). It is obtained from the characteristic function of the pdf of the initial variables contributing to the sum and is given by the formulas (3.17,3.20) derived below.

3.2.1 Quantification of the Deviation from the Central Limit Theorem

The expression (3.9) contains the central limit theorem as a special case. Indeed, a large class of functions are quadratic close to their maximum. Thus, if $x_{\max}$ is the value of x that maximizes $s(x)$, we can then write

$$s(x) = s(x_{\max}) + \frac{1}{2} s''(x_{\max})(x - x_{\max})^2$$
$$+ \frac{1}{3!} s'''(x_{\max})(x - x_{\max})^3 + \frac{1}{4!} s''''(x_{\max})(x - x_{\max})^4 + \dots , \qquad (3.10)$$

where s'', s''', s'''' are the second, third and fourth order derivative. For x close to $x_{\max}$, $s(x)$ is quadratic and therefore $P[X]$ is essentially Gaussian with small contributions from the higher order terms at large N.

The expression (3.9) allows us to recover the regime of validity of the Gaussian law. The deviation from the Gaussian law is given by

$$\frac{P[X] - P_G[X]}{P_G[X]} = \exp\left(N \left[\frac{1}{3!} s'''(x_{\max}) \left(\frac{1}{N} X - x_{\max} \right)^3 \right. \right.$$
$$\left. \left. + \frac{1}{4!} s''''(x_{\max}) \left(\frac{1}{N} X - x_{\max} \right)^4 + \dots \right] \right) - 1 . \qquad (3.11)$$

For a non-symmetric law, $s'''(x_{\max}) \neq 0$, we find $(P[X] - P_G[X])/P_G[X] \ll 1$ as long as $|X - Nx_{\max}| \ll N^{2/3}$. For a symmetric law, the term $s''''(x_{\max})$ dominates and one finds $(P[X] - P_G[X])/P_G[X] \ll 1$ as long as $|X - Nx_{\max}| \ll N^{3/4}$. We thus recover in a simple way the power law dependences $\sim N^{2/3}$ of (3.7) and $\sim N^{3/4}$ of (3.8).

3.2.2 Heuristic Derivation of the Large Deviation Theorem (3.9)

Let us assume that the pdf of the initial variables decays faster than an exponential for large $|X_i|$. This ensures the validity of the characteristic function, which we now define with an imaginary k, such that $ik = \beta$ is real

$$Z(\beta) \equiv \langle e^{-\beta X_i} \rangle \equiv \int_{-\infty}^{+\infty} dX_i \, e^{-\beta X_i} P(X_i) . \qquad (3.12)$$

In order to calculate the Cramér function $s(x)$, we construct

$$Z^n(\beta) = \left\langle e^{-\beta \sum_{j=1}^{N} X_j} \right\rangle , \tag{3.13}$$

and take the limit of large N. If formula (3.9) is true, $\sum_{j=1}^{N} X_j = Nx$ with a probability $\sim \mathrm{d}x\, e^{Ns(x)}$. We can thus rewrite

$$Z^N(\beta) \sim \int \mathrm{d}x\, e^{Ns(x)} e^{-\beta Nx} . \tag{3.14}$$

The Jacobian stemming from the change of variable $X_i \rightarrow x$ provides a proportionality factor. The integral over x can be evaluated by the Laplace method [69]. This method is sometimes referred to as "steepest descent", an inadequate terminology when $f(x)$ is not analytic. This method here amounts to have the integral determined by the value of its integrand in a small neighborhood of x that maximizes $F(x) = -\beta x + s(x)$, i.e. the value x which is solution of the equation

$$s'(x^*) = \beta , \tag{3.15}$$

where s' denotes the first derivative of s. This approximation is valid if the width of the maximum is small compared to its value. In this case, we obtain

$$Z^N(\beta) \sim e^{N(s(x^*) - \beta x^*)} . \tag{3.16}$$

Therefore, up to logarithmic corrections, $s(x)$ is determined from the following equations:

$$s(x) = \ln Z(\beta) + \beta x , \tag{3.17}$$

$$\frac{\mathrm{d}s}{\mathrm{d}x} = \beta . \tag{3.18}$$

These two expressions indicate that $s(x)$ is the Legendre transform of $\ln Z(\beta)$. Therefore, in order to determine the Cramér function $s(x)$ from the calculation of the characteristic function $Z(\beta)$, we must find the value β which corresponds to a given x, knowing $Z(\beta)$. First, we construct $s'(x)$ by differentiating (3.17) with respect to x:

$$s'(x) = \frac{\mathrm{d}\ln Z(\beta)}{\mathrm{d}\beta}\frac{\mathrm{d}\beta}{\mathrm{d}x} + x\frac{\mathrm{d}\beta}{\mathrm{d}x} + \beta . \tag{3.19}$$

Then, we use (3.18) and assume that $\mathrm{d}\beta/\mathrm{d}x$ is non-zero. This last condition is correct when the saddle-node approximation is valid. Indeed, from (3.18), one sees that $s''(x) = \mathrm{d}\beta/\mathrm{d}x$. If $\mathrm{d}\beta/\mathrm{d}x$ vanishes, $1/\sqrt{-s''(x^*)}$ diverges and the condition for the validity of the saddle-node approximation is not obeyed.

We obtain the equation determining $\beta(x)$ knowing $\ln Z(\beta)$ by putting (3.15) into (3.19):

$$\frac{\mathrm{d}\ln Z(\beta)}{\mathrm{d}\beta} = -x . \tag{3.20}$$

This leads to $s(x) = \ln Z(\beta(x)) + \beta(x)x$, where $\beta(x)$ is the solution of (3.20).

3.2.3 Example: the Binomial Law

The binomial law is

$$\mathcal{P}(j) = \binom{N}{j} p^j (1-p)^{N-j} , \tag{3.21}$$

and corresponds to the special case where the initial variables can take only two values X_1 and X_2 with probability p and $1-p$ respectively. The probability that, among a sample of N such random variables, j have the value X_1 and $N-j$ takes the value X_2 is given by (3.21).

If $X_1 =$ tails and $X_2 =$ heads, with $p = 1/2$, expression (3.21) gives the probability to observe j tails among N throws of a coin. In the random walk problem in which a right and left step occur with probability p and $1-p$ respectively, (3.21) gives the probability that the walker has made j right steps and $N-j$ left steps. The position of the random walker, given by the sum $\sum_{i=1}^{N} X_i$, is then $j - (N-j) = 2j - N$.

The large deviation theorem can be obtained using the Stirling formula applied to $\binom{N}{j}$. We give the expression for the case $X_i = 1$ (success) with probability p and $X_i = 0$ (failure), with probability $1-p$, for which the expression is the simplest:

$$\mathcal{P}(j = xN) = e^{Ns(x)} , \tag{3.22}$$

with

$$s(x) = x \ln p + (1-x) \ln(1-p) - x \ln x - (1-x) \ln(1-x) , \tag{3.23}$$

for $0 < x < 1$,

$$s(x) = -\infty \qquad \text{otherwise} . \tag{3.24}$$

Notice that $s(x)$ reduces to

$$s_p(x) = -\frac{(x-p)^2}{2p(1-p)} \tag{3.25}$$

close to its maximum $x = p$, a result which recovers the Gaussian law. However, one can observe large deviations from the Gaussian law for $|x|$ different from p, as seen in Figs. 3.1 and 3.2. Finite variations of x from p correspond to the regime of large deviations as the sum has deviations of order N (exactly $(x-p)N$) away from its mean, instead of the most probable deviations of order $\sqrt{N}$ according to the central limit theorem.

Consider the case shown in Fig. 3.2 where $p = 0.95$ ($X_i = 1$), corresponding to a 5% probability to loose ($X_i = 0$). Take $x = 0.9$, corresponding to a finite deviation from the most probable value 0.95 for the average. Expression (3.23) gives $s(x) = -0.0206$ while its parabolic approximation yields $s_p(x = 0.9) = -0.0264$. The corresponding probabilities are $e^{-0.0206N} \approx 13\%$ for $N = 100$ using the Cramér expression and $e^{-0.0264N} \approx 7\%$ for $N = 100$

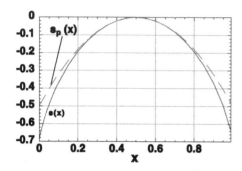

Fig. 3.1. The Cramér function $s(x)$ given by (3.23) and its parabolic approximation $s_p(x)$ as a function of x for $p = 0.5$

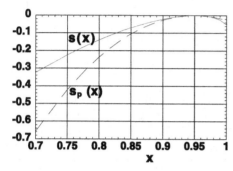

Fig. 3.2. The Cramér function $s(x)$ given by (3.23) and its parabolic approximation $s_p(x)$ as a function of x for $p = 0.95$

in the Gaussian approximation. The Gaussian approximation thus underestimates by almost a factor of two the probability of such a large deviation scenario. The difference becomes more overwhelming when N increases.

To sum up, it is not possible in general to capture the degree of uncertainty in the estimation of the sum of N random variables by the single variance parameter. It is the Cramér function which fully encodes this uncertainty. Applications to finance and portolio theory in the presence of large risks can be found in [882].

3.2.4 Non-identically Distributed Random Variables

When the variables constituting the sum are non-identically distributed, the large deviation theorem still holds with expression (3.9) but the Cramér function $s(x)$ must now be expressed in terms of all the pdf's $P_j(X_j)$ for $j = 1$ to N. Let us study the tail for large positive X, where we use the capital letter X to describe the sum, generalizing the notation $X = Nx$ of the previous section. We construct the characteristic function $\hat{P}_X(N\beta) \equiv \int_0^\infty dX\, e^{-N\beta X} P_X(X)$. If we assume (3.9) to hold and use the Laplace method, we obtain

$$\hat{P}_X(N\beta) \propto \exp[-N\,\mathrm{Inf}_X(\beta X - s(X))] \,, \tag{3.26}$$

where $\mathrm{Inf}_X(\beta X - s(X))$ is the evaluation of this function at the value of the argument X that minimizes it. From the property of convolution of pdf's,

$\hat{P}_X(N\beta)$ can be written in the form $\hat{P}_X(N\beta) = \prod_{j=1}^{N} \hat{P}_j(N\beta)$. Comparing with (3.9), we obtain $s(X)$ as a Legendre transform [316, 549]

$$s(X) = \text{Inf}_\beta \left(\frac{1}{N} \sum_{j=1}^{N} \ln \hat{P}_j(N\beta) + \beta X \right) . \tag{3.27}$$

This result (3.27) with the expression (3.9) provides the general formula for the pdf of arbitrary large values X, i.e. beyond the Gaussian approximation, which is recovered for small variations. This is seen by expanding the Laplace transform $\hat{P}_j(N\beta)$ in powers of β and truncating at the second order in β. The minimization becomes quadratic in β and yields a quadratic dependence in X and thus a Gaussian law is obtained for the pdf. Expression (3.27) generalizes the expressions (3.17, 3.20) while recovering them when all random variables are i.i.d.

It is also worthwhile to mention a more general limit theorem than the central limit theorem, which is valid for the sum of independent but not necessarily identically distributed stochastic variables. This limit theorem is due to Khintchine [350] and states that the sum of N independent random variables converge in probability to a pdf belonging to the class of infinitely divisible pdf's [293]. An infinitely divisible pdf is defined by a characteristic function $\hat{P}(k)$ obeying the following functional equation

$$\hat{P}(k) = [\hat{P}_N(k)]^N , \qquad \text{for all } N \qquad \text{with } \hat{P}_N(0) = 1 , \tag{3.28}$$

and $\hat{P}_N(k)$ continuous. As seen from (2.69), this simply means that a variable taken from an infinitely divisible pdf can be viewed as the sum of an arbitrary number N of contributions, with the pdf of the contributions depending possibly on their number. The importance of infinitely divisible pdf's has been stressed in the development of statistical models of hydrodynamic turbulence [316, 837, 838]. The class of infinitely divisible stochastic process is a much broader class than that of the stable distributions (Gaussian and Lévy). In addition to the Gaussian and Lévy laws, Poisson and Gamma distributions are examples of infinitely divisible pdfs. The class of infinitely divisible stochastic processes selects a subset of all possible stochastic processes. An example of a distribution outside this class is the stretched exponential distribution. As we show below, the sum of N variables, each distributed according to the same stretched exponential distribution, is dominated for extreme deviations by the largest of the N variables and is of the same order of magnitude. This prevents the application of Khintchine's theorem as well as of the extreme theorem discussed below. Indeed, Khintchine's theorem relies essentially on the constraint that the N independent random variables summing up to the random variable X must be infinitesimal (in the limit $N \to \infty$), in other words there is not a single stochastic variable among the N that dominates the sum X.

3.3 Large Deviations with Constraints and the Boltzmann Formalism

3.3.1 Frequencies Conditioned by Large Deviations

For simplicity and convenience of notation, we now assume that each X_i can take only a finite set of discrete values that we call $v_1, v_2, \ldots, v_{n-1}, v_n$. For the coin toss problem, $n = 2$ and in a game with a single die, $n = 6$. Taking $n \to \infty$ recovers the limit of a continuous variable. This assumption does not lead to a loss of generality.

The N realizations of a random variable are drawn from the pool of the n possible values. So in N trials, there are n^N conceivable outcomes. We use the word "result" for a single trial, while "outcome" refers to the experiments as a whole. Thus, one outcome consists of an enumeration of N results, including their order. For instance, five tosses of a die ($n = 6$, $N = 5$) might have the outcome "65133". Each outcome yields a set of sample numbers $\{N_i\}$ and relative frequencies $\{f_i = N_i/N, i = 1, \ldots, n\}$.

The classical law of large numbers states that the frequency with which one measures a given value v_l among n possible values of the same random variable V converges towards its probability. This idea is in fact at the basis of the notion of probability in the frequentistic approach.

In many situations, the outcome of a random experiment is not known completely. One does not know the order in which the individual results occurred, and often one does not even know all n relative frequencies $\{f_i\}$ but only a smaller number m ($m < n$) of linearly independent constraints

$$\sum_{i=1}^{n} G_a^i \, f_i = g_a , \quad a = 1, \ldots, m . \tag{3.29}$$

As an example of a constraint, suppose that the measured mean of these N realizations deviates from the theoretical mean. In this limit of large N, we consider a deviation of the mean which survives even in this limit. We are thus in the large deviation regime of the previous sections.

What can we say about the frequencies of each value v_l taken by the N outcomes which created this deviation? As we will show, the answer is that the frequencies of each value v_l do converge to a well-defined number in the limit of large N, but this number is different from its asymptotic probability!

The fundamental reason for this is that there is a close relationship between the existence of the deviation of the mean from its theoretical value and the existence of frequencies that are different from their theoretical probabilities. The theory of large deviations allows one to compute precisely these anomalous behaviors in the limit of large N. This provides an independent signature of the existence of large deviations.

To quantify these statements, recall that $X \equiv \sum_{j=1}^{N} X_j$ may also be written

$$X = N \sum_{l=1}^{n} f_l v_l , \tag{3.30}$$

where $f_l = N_l/N$ is the observed frequency for the N_l times the value v_l was found in the N realizations. In this expression, the N realizations have been partitioned into groups of identical values, the first group contains N_1 variables each equal to v_1, the second group contains N_2 variables each equal to v_2, and so on, such that $N_1 + N_2 + \ldots + N_n = N$. The law of large numbers states that $f_l = N_l/N \to P(v_l) \equiv p_l$, when $N \to \infty$, where $P(v_l) \equiv p_l$ is the probability of v_l. For a large observed value of $X/N = x$, what can we say about f_l? More precisely, what are the values taken by f_l, conditioned by the observation of $X/N = x$?

The following exposition benefits from exchanges with V. Pisarenko. A natural way to address this question is to use some functional $R(P; P_x)$ measuring the "distance" between the pdf P and the desired modified pdf P_x. Then, one can minimize this distance $R(P; P_x)$ under the given conditions. The problem is that there is a certain degree of arbitrariness in the choice of the distance $R(P; P_x)$, that lead to different solutions.

If one takes the Kullback Distance 1 representing average log-likelihood $\ln[P_x(v)/P(v)]$ of P_x against P [535]

$$R_1(P; P_x) = \int dv \, P_x(v) \ln[P_x(v)/P(v)] , \tag{3.31}$$

one gets directly (using the Lagrange multiplier method) the Gamma distribution

$$P_x(v) = P(v)e^{a-bv} , \tag{3.32}$$

where the constants a and b are determined by the constraints.

However, there are other "distances" that are a priori as justifiable as the Kullback Distance 1 and which lead to different results. The Kullback Distance 2 is the average log-likelihood of P against P_x

$$R_2(P; P_x) = \int dv \, P(v) \ln[P(v)/P_x(v)] , \tag{3.33}$$

which leads to the modified solution

$$P_x(v) = \frac{P(v)}{a + bv} , \tag{3.34}$$

where the constants a and b are again determined by the constraints.

Another example is the Kullback Distance 3, quantifying the "divergence" between P and P_x

$$R_3(P; P_x) = R_1(P; P_x) + R_2(P; P_x) , \tag{3.35}$$

which leads to the following solution

$$P_x(v) = \frac{P(v)}{G(a+bv)} , \qquad (3.36)$$

where $G(z)$ is the inverse function of $g(z) = z + \ln z$.

Thus, as a consequence of the existence of some degree of arbitrariness in the choice of distance $R(P; P_x)$, this approach does not select the law (3.32) as the unique solution of the two conditions 1 and 2 of the previous section.

The approach that we now describe provides a natural way to avoid the arbitrariness in the choice of distance between P and P_x, based on fixing the random sample mean of observed events. This constraint could be seen as too restrictive, since it selects among all realizations of possible sequences, only those where the sample mean is exactly equal to a specified value. Our point is that this constraint can be fixed to a specific value by an independent global measurement. As an illustration, think of the Gutenberg–Richter distribution of earthquakes in which the constraint on the average corresponds to the cumulative strain obtained by geodetic or satellite techniques, thus providing an estimation of the cumulative released moment (neglecting difficulties associated with the tensorial nature of the problem). Thus, we propose to condition the modified pdf on those specific random realizations that are consistent with the global measurement. This rather specific constraint will not apply in all circumstances.

3.3.2 Partition Function Formalism

The probability to observe $f_1, f_2, \ldots, f_n$ from N realizations is simply

$$P(f_1, f_2, \ldots, f_n) = \frac{N!}{(Nf_1)!(Nf_2)! \ldots (Nf_n)!} \prod_{l=1}^{n} [P(v_l)]^{Nf_l} , \qquad (3.37)$$

where $Nf_l = N_l$ and $(Nf_l)!$ is the factorial $Nf_l(Nf_l - 1)(Nf_l - 2) \ldots 4 \times 3 \times 2 \times 1$ of Nf_l. Using the Stirling formula

$$x! \approx \sqrt{2\pi x} \, x^x e^{-x} , \qquad (3.38)$$

we find

$$P(f_1, f_2, \ldots, f_n) \simeq \frac{\sqrt{2\pi N}}{\prod_{j=1}^{n} \sqrt{2\pi f_j N}} \, e^{NH(f_1, f_2, \ldots, f_n)} , \qquad (3.39)$$

where the "entropy" is defined by

$$H(f_1, f_2, \ldots, f_n) = -\sum_{l=1}^{n} f_l \ln \frac{f_l}{p_l} , \quad \text{with} \ \ p_l \equiv P(v_l) . \qquad (3.40)$$

Montroll [645] has similarly discussed the appearance of the notion of entropy in sociotechnical systems.

Consider two different data sets leading to the two sets $\{f_i\}$ and $\{f_i'\}$. The ratio of their respective probabilities is given by

$$\frac{\text{prob}(f|p, N)}{\text{prob}(f'|p, N)}$$

$$= \sqrt{\prod_i \frac{f_i'}{f_i}} \, \exp[N(H(f_1, f_2, \ldots, f_n) - H(f_1', f_2', \ldots, f_n'))] \,. \qquad (3.41)$$

As the prefactor $\sqrt{\prod_i \frac{f_i'}{f_i}}$ is independent of N, for large N and for closely similar distributions $f' \approx f$, the variation of $\text{prob}(f|p, N)/\text{prob}(f'|p, N)$ is completely dominated by the exponential:

$$\frac{\text{prob}(f|p, N)}{\text{prob}(f'|p, N)} \approx \exp[N(H(f_1, f_2, \ldots, f_n) - H(f_1', f_2', \ldots, f_n'))] \,. \quad (3.42)$$

Hence the probability with which any given frequency distribution f is realized is essentially determined by the entropy $H(f_1, f_2, \ldots, f_n)$. The larger this quantity, the more likely is the frequency distribution.

In the absence of constraints other than the normalization condition $\sum_{l=1}^{n} f_l = 1$ and for N large, the frequencies f_l converge towards the values that maximize $H(f_1, f_2, \ldots, f_n)$. We thus recover the law of large numbers

$$f_l \to_{N \to \infty} p_l \,. \qquad (3.43)$$

However, if we observe $X/N \equiv \sum_{l=1}^{n} f_l v_l = x$, the frequencies are those that maximize the function

$$H(f_1, f_2, \ldots, f_n) - \lambda_1 \left(\sum_{l=1}^{n} f_l - 1 \right) - \lambda_2 \left(\sum_{l=1}^{n} f_l v_l - x \right) \,, \qquad (3.44)$$

where λ_1 and λ_2 are two Lagrange parameters determined by the constraint of normalization and the observation of the large deviation x. Recall that the technique of Lagrange multipliers is very useful in solving an optimization problem in the presence of addition constraints. Briefly, the idea is to incorporate the constraints in the function to minimize, with multiplicative factors. Then, the solution depends on these factors, which are then eliminated by imposing the constraints on the solution. See [86] for further information.

The solution is

$$f_l = P(v_l) \frac{e^{-\beta v_l}}{Z(\beta)}, \qquad (3.45)$$

where $\beta \equiv -\lambda_2$ is determined as a function of x by the equation (3.20) and $Z(\beta)$ is defined by (3.12) and reads

$$Z(\beta) = \sum_{l=1}^{n} p_l e^{-\beta v_l} \,. \qquad (3.46)$$

The expression (3.45) gives the frequencies of the values of the random variables, conditioned by the existence of a large deviation of the mean. Notice that for $x = 0$ (no deviation), we recover $f_l = p_l$ asymptotically since then $\beta(x = 0) = 0$.

It is not fortuitous that the expressions (3.45, 3.46) bear a strong similarity with the statistical mechanics formulation [762] of systems composed of many elements, where $Z(\beta)$ is the partition function, β is the inverse temperature and $-\ln Z(\beta)$ is proportional to the free energy. The fact that the constraint is seen as a "high temperature" ($\beta \to 0$) perturbation is clear: the constraint is analogous to an "energy" added to the "free energy" $-\ln Z(\beta)$, which in the absence of constraint is solely controlled by the "entropy" H, i.e. by statistics. The relative importance of entropy and energy is weighted by the temperature, with the entropy dominating at high temperatures.

The probability that N trials will yield a frequency distribution with n values that satisfy the m constraints (3.29) and whose entropy H differs from $H^{\max}$ by more than ΔH is given for large N ($N \gg s/\Delta H$) by [762]

$$\text{prob}(H < (H^{\max} - \Delta H)|m \text{ consts.})$$
$$\approx \frac{1}{\Gamma(s+1)}(N\,\Delta H)^s \exp(-N\,\Delta H) \,, \qquad (3.47)$$

where $s = (n - m - 3)/2$. One recognizes the Poisson formula $p(k, \lambda) = (\lambda^k/k!)\,e^{-\lambda}$, which gives for instance the probability of finding exactly k events within a fixed interval of specified length [293]. As the number N of trials increases, this probability rapidly tends to zero for any finite ΔH. Thus, it becomes virtually certain that the unknown frequency distribution has an entropy H very close to $H^{\max}$. Hence, not only does the maximal point represent the frequency distribution that is the most likely to be realized (cf. (3.42)), but in addition, as N increases, all other – theoretically allowed – frequency distributions become more and more concentrated near this maximal point. Any frequency distribution other than the maximal point becomes highly atypical of those allowed by the constraints.

Another problem with important applications is to estimate the validity of models from finite imperfect data. If we are given a particular family of parametric models (Gaussians, power laws or stretched exponentials for example), the task of modeling the true distribution is reduced to parameter estimation, which is a relatively well-understood, though difficult, problem. Much less is known about the task of model family selection – for example, how do we choose between a family of power laws and a family of stretched exponentials as a model for the true distribution based on the available data? Several recent works have addressed this question by building from the Bayesian approach and Jeffreys theory [462], a statistical theory similar to field theory, from which systematic approximations can be obtained [4, 50, 87].

3.3.3 Large Deviations in the Dice Game

Let us roll a die N times. The probabilities f_l, $l = 1$ to 6, correspond to the frequencies with which each of the six faces of the die occurs. According to

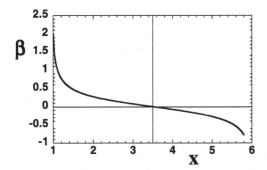

Fig. 3.3. The "inverse temperature" β is shown as a function of x in the allowed range $1 \leq x \leq 6$. Note that $|\beta| \rightarrow \infty$ (zero "temperature" = perfect order) for $x \rightarrow 1$ and $x \rightarrow 6$ which can only be attained when all draws are either 1 or 6. The value $\beta = 0$ is recovered for $x = (1/6) \times 1 + (1/6) \times 2 + (1/6) \times 3 + (1/6) \times 4 + (1/6) \times 5 + (1/6) \times 6 = 3.5$, which is the unconditional average

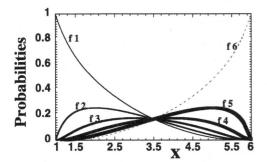

Fig. 3.4. The probabilities $f_1, f_2, f_3, f_4, f_5, f_6$ given by (3.45) are shown as a function of x in the allowed range $1 \leq x \leq 6$. All f_i's are equal to $1/6 = 0.166\ldots$ only when x is equal to the mean 3.5, where all curves cross. For $x > 3.5$, the probabilities f_i with $i > x$ are increased above $1/6$ while those with $i < x$ are decreased below $1/6$. For $x < 3.5$, the probabilities f_i with $i < x$ are increased above $1/6$ while those with $i > x$ are decreased below $1/6$

the law of large numbers, the six f_l tend to $1/6 \simeq 0.166$ for large N and the mean $f_1 + 2f_2 + 3f_3 + 4f_4 + 5f_5 + 6f_6$ tends to 3.5.

Let us now assume that we have observed a mean $x = 4$. The formula (3.45) predicts that the frequencies that contributed to this deviation are not the same anymore. We get $f_1 \rightarrow 0.103$, $f_2 \rightarrow 0.123$, $f_3 \rightarrow 0.146$, $f_4 \rightarrow 0.174$, $f_5 \rightarrow 0.207$, $f_6 \rightarrow 0.247$ in the limit of large N. Observe that the large values become more frequent, as can be expected since the outcomes are biased by the observation of a large mean.

As another example, consider the problem of a loaded die [762]. Let us assume that observations of die rolls have shown that 6 occurs twice as often as 1. Nothing peculiar was observed for the other faces. Given this information and nothing else, i.e., not making use of any additional information that we might get from inspection of the die or from past experience with dice in general, all we know is a single constraint of the form (3.29) $g_1 = 0$ with

$$G_1^i = \begin{cases} 2 : i = 1 \\ 0 : i = 2, \ldots, 5 \\ -1 : i = 6 \end{cases} \tag{3.48}$$

What estimates should we make of the relative frequencies $\{f_i\}$ with which the different faces appeared? Taking the a priori probability distribution – assigned to the various faces before one has asserted the die's imperfection – to be uniform, $\{p_i = 1/6\}$, the best estimate for the frequency distribution reads

$$f_i^{\max} = \begin{cases} Z^{-1} \exp(-2\lambda^1) : i = 1 , \\ Z^{-1} : i = 2, \ldots, 5 , \\ Z^{-1} \exp(\lambda^1) : i = 6 , \end{cases} \tag{3.49}$$

with only a single Lagrange parameter λ^1 and

$$Z = \exp(-2\lambda^1) + 4 + \exp(\lambda^1) . \tag{3.50}$$

The Lagrange parameter is readily determined from

$$\frac{\partial}{\partial \lambda^1} \ln Z = -g_1 = 0 , \tag{3.51}$$

with solution

$$\lambda^1 = (\ln 2)/3 . \tag{3.52}$$

This in turn gives the numerical estimates

$$f_i^{\max} = \begin{cases} 0.107 : i = 1 , \\ 0.170 : i = 2, \ldots, 5 , \\ 0.214 : i = 6 , \end{cases} \tag{3.53}$$

with an associated entropy

$$H^{\max} = \ln(1/6) + \ln Z = -0.019 . \tag{3.54}$$

This negative value of the entropy $H^{\max} = -0.019$ makes $H^{\max}$ smaller than zero, where zero is obtained for the unbiased priors $f_i = 1/6 = 0.166\ldots$ and can be seen as a kind of organization resulting from the bias (the entropy is smaller). In absence of the bias (3.48), the frequencies $f_i^{\max}$ given by (3.53) would be observed with the probability (3.42)

$$\frac{\text{prob}(f_i^{\max} | p = 1/6, N)}{\text{prob}(f = 1/6 | p = 1/6, N)} \approx \exp\left(-\frac{N}{52}\right) . \tag{3.55}$$

In absence of bias, these frequencies (3.53) would become very unlikely for $N \gg 50$. This means that one cannot hope to detect the bias where the 6 appears twice a many times as the 1 when the number of die casts is less than 50 or so. To establish the existence of the bias, one needs to reject the null hypothesis according to which the anomalous observations could occur just by chance. Choosing the significant statistical level of 95%, (3.55) implies that a sample of more than 150 die casts is needed.

3.3.4 Model Construction from Large Deviations

We now present a summary inspired from [762] of Jaynes' analysis of Wolf's die data, which illustrates vividly how hypotheses testing can be used iteratively by enlarging the set of constraints to improve the model [461].

Rudolph Wolf (1816–1893), a Swiss astronomer, had performed a number of random experiments, presumably to check the validity of statistical theory. In one of these experiments, a die was tossed 20 000 times in a way that precluded any systematic favoring of any face over any other. The observed relative frequencies $\{f_i\}$ and their deviations $\{\Delta_i = f_i - p_i\}$ from the a priori probabilities $\{p_i = 1/6\}$ are given in Table 3.1. Associated with the observed distribution is

$$\Delta H = -0.006769 . \tag{3.56}$$

Our "null hypothesis" H0 is that the die is ideal and hence that there are no constraints needed to characterize any imperfection ($m = 0$); the deviation of the experimental distribution from the uniform distribution is supposed to be merely a statistical fluctuation. However, the probability that statistical fluctuations alone yield a ΔH-difference as large as 0.006769 is practically zero. Using (3.47) with $N = 20\,000$ and $s = 3/2$ ($n = 6$ and $m = 0$), we find

$$\text{prob}(\Delta H | H0 \text{ and no constraints}) \sim 10^{-56} . \tag{3.57}$$

Therefore, the null hypothesis is rejected and the frequencies must be biased in some way.

Similar results found in similar experiments [758] (see in particular the random number experiments [820, 821, 925] and the ensueing controversy

Table 3.1. Wolf's die data: frequency distribution f and its deviation Δ from the uniform distribution

i	f_i	Δ_i
1	0.16230	−0.00437
2	0.17245	+0.00578
3	0.14485	−0.02182
4	0.14205	−0.02464
5	0.18175	+0.01508
6	0.19960	+0.02993

in *Physics Today* [76, 243, 244, 585, 926]) are sometimes put forward by certain investigators to suggest that some parapsychological effect has thus been discovered [76, 78, 161, 243, 244, 585, 926]. Belief in the "supernatural" and/or in parapsychology is still widely spread in our modern society [374, 830], even among educated students [993], probably due to the complexity of the psychology of the human mind [187, 931, 1001].

In addition to rigorous statistical hypotheses testing, the principle of Occam's razor [959, 960] is particularly useful to distinguish between competing hypotheses. This principle is attributed to the 14th century logician and Franciscan monk William of Occam which states that "Entities should not be multiplied unnecessarily." The most useful statement of the principle for scientists is "when you have two competing theories which make exactly the same predictions, the one that is simpler is the better." Occam's razor is used to cut away metaphysical concepts. The canonical example is Einstein's theory of special relativity compared with Lorentz's theory that ruler's contract and clocks slow down when in motion through the Ether. Einstein's equations for transforming space–time are the same as Lorentz's equations for transforming rulers and clocks, but Einstein recognised that the Ether could not be detected according to the equations of Lorentz and Maxwell. By Occam's razor, it had to be eliminated.

Thus, between the parapsychological hypothesis and the possibility that the die has some defect leading to the observed systematic bias, we first investigate the later as a potentially "simpler" explanation. Not knowing the mechanical details of the die, we can still formulate and test hypotheses as to the nature of its imperfections. Jaynes argued that the two most likely imperfections are:

- a shift of the center of gravity due to the mass of ivory excavated from the spots, which being proportional to the number of spots on any side, should make the "observable"

$$G_1^i = i - 3.5 \tag{3.58}$$

  have a nonzero average $g_1 \neq 0$;
- errors in trying to machine a perfect cube, which will tend to make one dimension (the last side cut) slightly different from the other two. It is clear from the data that Wolf's die gave a lower frequency for the faces (3,4); and therefore that the (3–4) dimension was greater than the (1–6) or (2–5) ones. The effect of this is that the "observable"

$$G_2^i = \begin{cases} 1 : i = 1, 2, 5, 6 \\ -2 : i = 3, 4 \end{cases} \tag{3.59}$$

  has a nonzero average $g_2 \neq 0$.

Our hypothesis H2 is that these are the only two imperfections present. More specifically, we conjecture that the observed relative frequencies are characterized by just two constraints ($m = 2$) imposed by the measured averages

$$g_1 = 0.0983 \quad \text{and} \quad g_2 = 0.1393 \; . \tag{3.60}$$

As a consequence, we can use the technique developed above to fit the observed relative frequencies with the maximal distribution

$$f_i^{\max(\text{H2})} = \frac{1}{Z} \exp\left(-\sum_{a=1}^{2} \lambda^a \, G_a^i\right) \; . \tag{3.61}$$

In order to test our hypothesis, we determine

$$Z = \sum_{i=1}^{6} \exp\left(-\sum_{a=1}^{2} \lambda^a \, G_a^i\right) \; , \tag{3.62}$$

fix the Lagrange parameters by requiring

$$\frac{\partial}{\partial \lambda^a} \ln Z = -g_a \; , \tag{3.63}$$

and then calculate

$$H_p^{\max(\text{H2})} = \ln(1/6) + \ln Z + \sum_{a=1}^{2} \lambda^a \, g_a \; . \tag{3.64}$$

With this algorithm, Jaynes found

$$H_p^{\max(\text{H2})} = -0.006534 \; , \tag{3.65}$$

and thus

$$\Delta H^{\text{H2}} = H_p^{\max(\text{H2})} - H_p(f) = 0.000235 \; . \tag{3.66}$$

The probability for such an H_p-difference to occur as a result of statistical fluctuations is (with now $s = 1/2$)

$$\text{prob}(H_p < (H_p^{\max} - \Delta H^{\text{H2}})|2 \, \text{constraints}) \approx 2.5\% \; , \tag{3.67}$$

much larger than the previous 10^{-56} but still below the usual acceptance bound of 5%. The more sophisticated model H2 is therefore a major improvement over the null hypothesis H0 and captures the principal features of Wolf's die; yet there are indications that an additional very tiny imperfection may have been present.

Jaynes' analysis of Wolf's die data furnishes a useful paradigm for the experimental method in general. Modern geophysical and astrophysical research yield data in the form of frequency distributions over discrete "bins" for each of the various measurements. The search for interesting signals in the data (new mechanisms, new interactions, new correlations, etc.) essentially proceeds in the same manner in which Jaynes revealed the imperfections of Wolf's die: by formulating physically motivated hypotheses and testing them against the data. Such a test is always statistical in nature. Conclusions (say, about the presence of life in Martian meteorites [612], or about the presence of a certain imperfection of Wolf's die) can never be drawn with absolute certainty but only at some – quantifiable – confidence level.

3.3.5 Large Deviations in the Gutenberg–Richter Law and the Gamma Law

It is well-known that the Gutenberg–Richter power law distribution of earthquake seismic moment releases has to be modified for large seismic moments due to energy conservation and geometrical reasons. Several models have been proposed, either in terms of a second power law with a larger b-value beyond a cross-over magnitude, or based on a "hard" magnitude cut-off or a "soft" magnitude cut-off using an exponential taper. Since the large scale tectonic deformation is dominated by the very largest earthquakes and since their impact on loss of life and properties is huge, it is of great importance to constrain the shape of their distribution as much as possible.

The above probabilistic theoretical approach provides a simple framework to handle the constraints on global tectonic deformations. It is easy to show that the Gamma distribution is the best model, under the two hypothesis that the Gutenberg–Richter power law distribution holds in absence of any condition and that one or several constraints are imposed, either based on conservation laws or on the nature of the observations themselves. The selection of the Gamma distribution does not depend much on the specific nature of the constraint. This approach has been illustrated with two constraints [912], namely the existence of a finite moment release rate and the observation of the size of a maximum earthquake in a finite catalog. The predicted "soft" maximum magnitudes compare favorably with those obtained by Kagan [491] for the Flinn–Engdahl regionalization of subduction zones, collision zones and mid-ocean ridges.

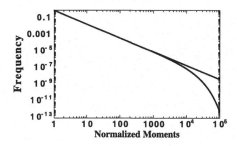

Fig. 3.5. Comparison between the pure Gutenberg–Richter law $1/M^{1+2/3}$ and the Gamma distribution $\exp\left(-\beta(x)M\right)/M^{1+2/3}$, where M are seismic moments normalized in units of $M_t = 5 \times 10^{17}$ N m corresponding to a magnitude 5.8

Figure 3.5 shows the comparison between the pure Gutenberg–Richter law $1/M^{1+2/3}$ and the Gamma distribution $\exp\left(-\beta(x)M\right)/M^{1+2/3}$, where M are seismic moments normalized in units of $M_t = 5 \times 10^{17}$ N m corresponding to a magnitude 5.8 (in other words, we consider large earthquakes of magnitudes larger than 5.8). The global constraint can be expressed as

$$x = \frac{\dot{M}\,\Delta t}{M_t N} \,, \tag{3.68}$$

where $\dot{M}$ is the geological rate of deformation (in $N\,m$/year) of the region under consideration and $N/\Delta t$ is the yearly number of earthquakes with moments above the threshold M_t (taken from the Harvard catalog over a time interval of $\Delta t = 18.5$ years). Taking a typical value $x = 40$ yields $\beta(40) = 8.9 \times 10^{-5}$ which corresponds to a characteristic magnitude of 8.5 controlling the cross-over from the power law behavior to the exponential tail. A magnitude 8.5 corresponds to a "great" earthquake with an energy release of about 3×10^{17} joules (to be compared with the energy 10^{12} joules released by Hiroshima's atomic bomb). Such a great earthquake typically involves a fault slip of many meters over several hundred kilometers.

Figure 3.5 is typical of the cross-over from pure power law to exponential decay, for most of the regions in the world that include subduction zones, collision zones and mid-ocean ridges [491, 912]. We should stress however that the most interesting (and difficult) problem is to go beyond this description of the balance of deformations in terms of moment scalars towards the estimation of the probability distribution for displacements in brittle material due to earthquakes, thus capturing the vectorial nature of the motion of the Earth crust. For this, we need to consider multidimensional statistics and the variables of interest are vector-valued or tensor-valued (displacement, strain, stress) (see [489, 490] for pioneering work in this direction).

Interestingly, the distribution $P(X_1|S)$ of any single term X_1 in the sum $S = X_1 + \ldots + X_N$ conditioned on a fixed value of S is showing a behavior quite different from the exponential taper shown in Fig. 3.5. According to Bayes' rule, we have

$$P(X_1|S) = \frac{P_N(S|X_1)P(X_1)}{P_N(S)} \ , \tag{3.69}$$

where $P_N(S|X_1)$ is the pdf of the sum S conditioned on the value of X_1 and $P_N(S)$ is the unconditional pdf of S. Now, $P_N(S|X_1) = P_{N-1}(S - X_1)$ by definition, for $X_1 < S$. Thus, $P(X_1|S) \sim P_{N-1}(S - X_1)P(X_1)$ since the denominator $P_N(S)$ can be included in the normalizing factor. For pdf's with heavy tails for which Nagaev's theorem hold (see expression (3.100) below), $P_{N-1}(S - X_1) \sim (N - 1)P(S - X_1)$, for $S - X_1 > N^{1/2+\epsilon}$ with arbitrary positive ϵ (which can also be read as $X_1 < S - N^{1/2+\epsilon}$). This leads to

$$P(X_1|S) \sim P(S - X_1)\,P(X_1) \ , \tag{3.70}$$

since we can include the factor $N - 1$ into the normalizing factor. The first factor $P(S - X_1)$ is the result of the conditioning and can thus be considered as a "taper" acting on the initial (unconditional) pdf $P(X_1)$. It is interesting to find that for large $X_1 < S - N^{1/2+\epsilon}$, the taper leads to an increase of the probability for large X_1. If we could use (3.70) in the whole interval $0 < X_1 < S$, then the conditional density (3.70) would take the form of a symmetric U-curve with a minimum at $X_1 = S/2$. Because of the restriction on the argument $X_1 < S - N^{1/2+\epsilon}$, the resulting product (3.70) looks as a U-curve with unequal arms: the left arm is much higher than the right arm, since $P(1|S) \geq P(S - N^{1/2+\epsilon}|S)$ under the condition that $S \sim N$. Thus, the conditional pdf $P(X_1|S)$ given by (3.70) in the range $1 < X_1 < S - N^{1/2+\epsilon}$ (with $S \simeq$ constant $\times N$) behaves practically like the initial unconditional pdf $P(X_1)$ with a small rising (!) taper $P(S - X_1)$. The difference between expression (3.70) and the decreasing taper presented in Fig. 3.5 stems from the fact that the later is obtained by looking simultaneously at all the terms $X_1, X_2, \ldots, X_N$ in the sum, constrained to sum to N. In other words, the pdf shown in Fig. 3.5 is the pdf of the i.i.d. random variable X_i in the sum, while (3.70) gives the pdf of one such random variables over many independent realizations $\{X_1, X_2, \ldots, X_N\}$ of the N variables.

3.4 Extreme Deviations

3.4.1 The "Democratic" Result

The Cramér theorem (3.9) describes large deviations of X/N away from the mean in the limit $N \to \infty$. There is another interesting regime, called *extreme deviation* regime [318]. It corresponds to the different order of limits: N finite and $X/N \to \infty$. Our heuristic presentation borrows from [318], in which a rigorous treatment is also given. See also [911] for an extension to non-i.i.d. random variables.

We assume that the pdf $P(X_i)$ can be represented as

$$P(X_i) = e^{-f(X_i)} , \tag{3.71}$$

where $f(X_i)$ tends to $+\infty$ with $|X_i|$ sufficiently fast to ensure the normalization of $P(X_i)$. We rule out the case where $f(X_i)$ becomes infinite at finite X_i, which corresponds to a distribution with compact support for which the extreme deviations would be trivial.

We get as usual the pdf of the sum of N random variables

$$P_N(X) = \int dX_1 \ldots \int dX_N \exp\left(-\sum_{j=1}^{N} f(X_j)\right) \delta\left(X - \sum_{i=1}^{N} X_i\right) . \tag{3.72}$$

We also make the assumption of convexity: $f''(X_i) > 0$, which is essential because it ensures that the minimum of $\sum_{j=1}^{N} f(X_j)$ is obtained for $X_1 = X_2 = \ldots = X_N = X/N$.

Stretched exponentials, for which $f(X_i) \simeq X_i^\alpha$ with $\alpha < 1$, are thus excluded since $f''(X_i) = \alpha(\alpha - 1)X_i^{\alpha-2}$ is negative. This reflects the fact that the tail of the sum S_N of N stretched exponentially distributed variables has the same order of magnitude as the tail of the maximum variable X_N^{max} among them:

$$\frac{\text{Probability}(S_N \geq x)}{\text{Probability}(X_N^{max} \geq x)} \xrightarrow{x \to +\infty} 1 . \tag{3.73}$$

This is rather remarkable considering that the typical values are very differ-ent: $S_N \approx N\langle X \rangle \sim N$ is much larger, for large N, than the typical value of the maximum $X_N^{\max} \sim (\ln N)^{1/\alpha}$. The proof goes as follows. We ask what are the set of positive X_i's such that $\sum_{i=1}^{N} X_i^\alpha$ is minimum so as to make maxi-mum the probability of this configuration, while the condition $\sum_{i=1}^{N} X_i > x$ is obeyed. We see immediately that, since $\alpha < 1$, $\sum_{i=1}^{N} X_i^\alpha \geq x^\alpha$ since $\sum_{i=1}^{N} X_i \geq x$. To see this intuitively, take for instance $X_i = x/N$ that realizes the condition that the sum adds up to x. Then, $\sum_{i=1}^{N} X_i^\alpha = x^\alpha N^{1-\alpha} > x^\alpha$ for $\alpha < 1$. The configurations that make $\sum_{i=1}^{N} X_i^\alpha = x^\alpha$ are those such that all X_i's are very small ($\rightarrow 0$) except one being almost equal to x. The correspond-ing probability is $\exp(-cx^\alpha)$ larger than any other configuration, where c is a constant depending on the configuration of the X_i's. We thus see how (3.73) emerges. This class of stretched exponential distribution is sometimes called "sub-exponentials" in the mathematical literature or Weibull laws in the engi-neering literature. The same phenomenon applies to Pareto distributions (any pdfs with a heavy-tail). Thus, heavy-tails and stretched-exponentials ($\alpha < 1$) both correspond to the situation where, with overwhelming probability, only one of the addends of the sum $X_1 + \ldots + X_n$ contribute significantly to this sum, while the contribution of most other addends is negligible.

We turn our attention to the exponential and "superexponential" cases where $f''(X_i) > 0$. This corresponds to the regular Cramer case discussed above (only in that case does the large deviation theorem holds). In this case, the "democratic" condition $X_1 = X_2 = \ldots = X_N = X/N$ realizes the minimum of $\sum_{j=1}^{N} f(X_j)$, in contrast to the sub-exponential case where the minimum is realized for one of the terms completely dominating the others. Indeed, let us set $\hat{X}_i = X_i - X/N$. Then, due to the convexity of $f(X_i)$, we have that

$$\sum_{j=1}^{N} f(X_j) = f\left(\frac{X}{N} + \hat{X}_1\right) + \ldots + f\left(\frac{X}{N} + \hat{X}_N\right) \geq N f\left(\frac{X}{N}\right), \quad (3.74)$$

under the constraint $\hat{X}_1 + \hat{X}_2 + \ldots + \hat{X}_N = 0$.

The large-X behavior of (3.72) can be obtained by Laplace's method. Basically, this consists in taking the Taylor expansion of the l.h.s. of (3.74) in powers of the $\hat{X}_i$'s up to second order, near its minimum at $\hat{X}_1 = \hat{X}_2 = \ldots = \hat{X}_N = 0$, to obtain

$$\sum_{j=1}^{N} f(X_j)$$
$$= N f\left(\frac{X}{N}\right) + \frac{1}{2}\left(\hat{X}_1^2 + \hat{X}_2^2 + \ldots + \hat{X}_N^2\right) f''\left(\frac{X}{N}\right) + \mathcal{O}(\hat{X}_i^3), \quad (3.75)$$

where $\mathcal{O}(\hat{X}_i^3)$ stands for higher order terms. When this is substituted into (3.72), the first term on the r.h.s. of (3.75) produces the leading-order contribution to $P_N(X)$, while the quadratic terms determine the widths $\Delta \sim [f''(X/N)]^{-1/2}$ of the Gaussian integrals. We want the X_i's to be well localized near the value X/N, that is, the width Δ should be small compared to X/N. Since we want this to hold for arbitrary values of N (and not only large N as for the large deviation regime), we strengthen the convexity condition into

$$f''(X_i) > 0 \quad \text{and} \quad X_i^2 f''(X_i) \to +\infty \quad \text{for} \quad |X_i| \to \infty . \tag{3.76}$$

This condition holds, for example, when for large $|X_i|$ one has $f(X_i) \simeq C|X_i|^\gamma$ with $C > 0$ and $\gamma > 1$, or $f(X_i) = AX_i - BX_i^m$ with $A > 0$, $B > 0$ and $0 < m < 1$. As we look for large deviations, the convexity condition needs only to hold for large positive or large negative values *separately*.

With the strengthened convexity assumption (3.76), we obtain the following leading-order behavior for the pdf of the sum of N variables:

$$\ln P_N(X) \simeq -Nf(X/N) = N \ln P_1(X/N) , \tag{3.77}$$

for $X \to \infty$ and N finite. The extreme tail behavior of the sum X of N random variables comes mostly from contributions where the individual variables in the sum are all close to X/N and the tail of the pdf is

$$\sim [P_1(X/N)]^N \sim e^{-Nf(X/N)} . \tag{3.78}$$

Comparison of (3.9) with (3.77) shows that the Cramér function $s(y)$ becomes equal to $-f(y)$ for large y. We can verify this statement by inserting the form $P_1(x) = e^{-f(x)}$ into (3.12) to get $Z(\beta) \sim \int_{-\infty}^{\infty} dx\, e^{-\beta x - f(x)}$. For large $|\beta|$, we can then approximate this integral by Laplace's method, yielding

$$Z(\beta) \approx \exp\left(-\min_x (\beta x + f(x))\right) . \tag{3.79}$$

Taking the logarithm and a Legendre transform, we recover the identification that $s(y) \to -f(y)$ for large y. Laplace's method is justified by the fact that $|y| \to \infty$ corresponds, in the Legendre transformation, to $|\beta| \to \infty$.

The large and extreme deviation régimes thus overlap when taking the two limits $N \to \infty$ and $X/N \to \infty$. Indeed, the large deviation theory usually takes $N \to \infty$ while keeping X/N finite, whereas the extreme deviation regime takes N finite with $X \to \infty$. The above analysis shows that, in the latter régime, Cramér's result already holds for finite N. The true small parameter of the large deviations theory is thus not $1/N$ but $\min(1/N, N/X)$.

3.4.2 Application to the Multiplication of Random Variables: a Mechanism for Stretched Exponentials

Consider the product

$$Y = m_1 m_2 \dots m_N \tag{3.80}$$

of N independent i.i.d. positive random variables with pdf $P(m)$. What follows can easily be extended to the case of signed m_i's with a symmetric distribution. Taking the logarithm of Y, it is clear that we recover the previous problem of the addition of N random variables with the correspondence $X_i \equiv \ln m_i$, $X \equiv \ln Y$ and $-f(x) = \ln P(e^x) + x$. Assuming again (3.76), we can apply the extreme deviations result (3.77) which translates into the following form for the pdf $P_N(Y)$:

$$P_N(Y) \sim [P(Y^{1/N})]^N, \qquad \text{for} \quad Y \to \infty \quad \text{and} \quad N \quad \text{finite}. \qquad (3.81)$$

This expression (3.81) has a very intuitive interpretation: the tail of $P_N(Y)$ is controlled by the realizations where all terms in the product are of the same order; therefore $P_N(Y)$ is, to leading order, just the product of the N pdfs, with each of their arguments being equal to the common value $Y^{1/N}$.

When $P(m)$ is an exponential, a Gaussian, or, more generally, $\propto \exp(-Cm^\gamma)$ with $\gamma > 1$, then (3.81) leads to stretched exponentials for large N. For example, when $P(m) \propto \exp(-Cm^2)$, then $P_N(Y)$ has a tail $\propto \exp(-CNY^{2/N})$.

Stretched exponentials are remarkably robust as they exhibit a pseudo-stability property: the pdf of the sum X_N of N random variables distributed according to a stretched exponential pdf with exponent c may be approxi-

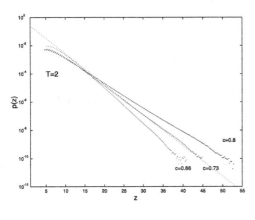

Fig. 3.6. Pdf P_2 of the sum of two stretched exponential variables with $c = 2/3$ and the choice $c_2 = 0.73$ as a function of $z \equiv X_2^{c_2}$, so that a stretched exponential is qualified as a straight line. (From [911])

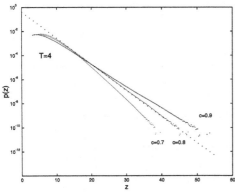

Fig. 3.7. Pdf P_4 of the sum of four stretched exponential variables with $c = 2/3$ and the choice $c_4 = 0.80$ as a function of $z \equiv X_4^{c_4}$, so that a stretched exponential is qualified as a straight line. (From [911])

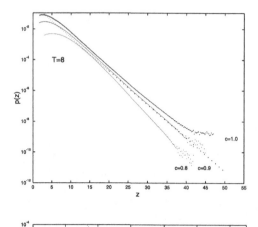

Fig. 3.8. Pdf P_8 of the sum of eight stretched exponential variables with $c = 2/3$ and the choice $c_4 = 0.90$ as a function of $z \equiv X_8^{c_8}$, so that a stretched exponential is qualified as a straight line. (From [911])

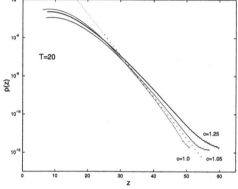

Fig. 3.9. Pdf P_{20} of the sum of twenty stretched exponential variables with $c = 2/3$ and the choice $c_{20} = 1.05$ as a function of $z \equiv X_{20}^{c_{20}}$, so that a stretched exponential is qualified as a straight line. The pdf P_{20} starts to deviate significantly from a stretched exponential form, as can be seen from the curvature. (From [911])

mated by a stretched exponential pdf with an apparent exponent c_N larger than the exponent c over a rather broad interval of X_N and for a rather large set of N values.

We test this idea by the following synthetic tests. Let us choose $c_1 = 2/3$ for the exponent of the stretched exponential pdf P_1 of the variables constituting the sum X_N. We construct the pdf P_N of the sum of N variables by taking the characteristic function of P_1 to the N-th power and then taking the inverse Fourier transform. Figures 3.6–3.9 plot the pdf's P_N as a function of $z \equiv X_N^{c_N}$ so that a stretched exponential is qualified as a straight line (dashed line on the plots). We show the cases $N = 2, 4, 8$ and 20 for which the best c_N are respectively $c_2 = 0.73, c_4 = 0.80, c_8 = 0.90$ and $c_{20} \approx 1.05$. The other curves allow one to estimate the sensitivity of the representation of P_N in terms of a stretched exponential pdf as a function of the choice of the exponent c_N. These simulations confirm convincingly that a stretched exponential distribution remains quasi-stable for a significant number of orders of convolutions, once the exponent c_N is correspondingly adjusted. We observe on Figs. 3.6–3.9 that the stretched exponential pdf representation is accurate over more than five orders of magnitude of the pdf P_N. Only for the largest number $N = 20$, do we observe significant departure from the stretched exponential representation.

3.4.3 Application to Turbulence and to Fragmentation

The result (3.81) has interesting applications in turbulence [316] and in fragmentation processes [167, 712, 765], where it is usually observed that the distribution of velocity increments at small scales and of fragment sizes in the small size limit have anomalously heavy tails of the stretched exponential or power law kind.

There is no generally accepted mechanism explaining the origin of these heavy tails and this is the subject of active investigations. The extreme deviations régime provides a very general and essentially *model-independent* mechanism, based on the extreme deviations of products of random variables [318].

Turbulence. In fully developed turbulence, random multiplicative models were introduced by the Russian school and have been studied extensively since. Indeed, their fractal and multifractal properties provide a possible interpretation of the intermittency phenomenon [316]. The pdf's of longitudinal and tranverse velocity increments clearly reveal a Gaussian-like shape at large separations and increasingly stretched exponential shapes at small separations as seen in Fig. 3.10.

Random multiplicative models cannot correctly account for *all* properties of increments. For example, they are inconsistent with the additivity of increments over adjacent intervals. Indeed, the pdf of velocity increments δv

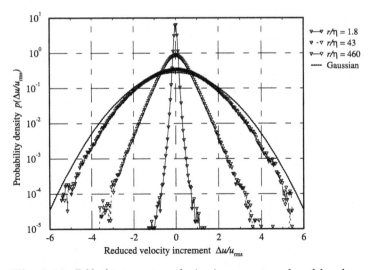

Fig. 3.10. Pdf of transverse velocity increments reduced by the r.m.s. velocity at various separations in units of the Kolmogorov dissipation scale η. (From [695])

cannot become much larger than the single-point pdf, as it would if the former were $\propto \exp\left(-|\delta v|^\beta\right)$ with $0 < \beta < 2$ while the latter would be close to Gaussian. Nevertheless, stretched exponentials could be working in an intermediate asymptotic range of not too large increments, where the controlling parameter of this intermediate asymptotic is the separation over which the increment is measured.

Fragmentation. This presentation is taken from [318]. Fragmentation occurs in a wide variety of physical phenomena from geophysics, material sciences to astrophysics and over a wide range of scales. The simplest (naive) model is to view fragmentation as a multiplicative process in which the sizes of children are fractions of the size of the parent. Neglecting the conservation of matter and assuming that the succession of breaking events are uncorrelated, this leads to a distribution of fragment size X, *conditioned by a given generation rank* N, which is log-normal in its center. Our above result (3.81) applies for $Y \to 0$, since the factors $m_1, m_2, \ldots, m_N$ are all less or equal to unity. This is easily checked by taking the logarithm and noting that the tail for $Y \to 0$ corresponds to the régime where the sum of logarithms goes to $-\infty$. Although $Y \to 0$, is not strictly speaking a "tail", we shall still keep this terminology.

In general, we expect a complicated functional dependence: for instance, if $P(m) \sim \exp\left(-cm^{-a}\right)$ for small m, we obtain $P_N(Y) \sim \exp\left(-cNY^{-a/N}\right)$. In contrast, most of the measured size distribution of fragments, not conditioned by generation rank, display a power-law behavior

$$P(Y) \sim \propto Y^{-\tau} \tag{3.82}$$

with exponents τ between 1.9 and 2.7 clustering around 2.4 [969]. Figure 3.11 illustrates the robustness of the power law description which also applies to the distribution of meteorite sizes and of debris sizes orbiting around the earth at two different altitudes.

The straight line is the power law $1.45 \times 10^{-6}/Y^{2.75}$. To quantify these figures in absolute numbers, there are about about 8000 man-made space objects, baseball-size (> 10 cm) and larger, orbiting Earth, each being individually tracked by the Space Surveillance Network of USSPACECOM [972]. About seven percent are operational satellites, 15 percent are rocket bodies, and about 78 percent are fragmented and inactive satellites.

Several models have been proposed to rationalize the observations of power law distributions [167, 712, 765] but there is no accepted theoretical description. The multiplicative model and extreme deviation régime offers a simple and general framework to understand these observations. Consider a fragmentation process in which, with a certain probability less than unity, a "hammer" repetitively strikes all fragments simultaneously. Then, the generation rank corresponds to the number of hammer hits. In real experiments, however, each fragment has suffered a specific number of effective hits which may vary greatly from one fragment to the other. The measurements of the

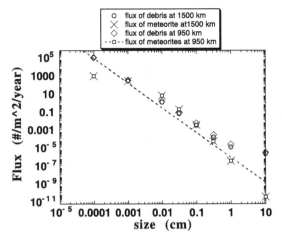

Fig. 3.11. Pdfs of meteorites and debris size Y orbiting around the earth at two different altitudes of 950 km and 1500 km above ground. The debris are man-made and are the remnents of rockets and satellites launched since the Soviets opened the space age with the launch of Sputnik I. The pdf's are given in terms of the number of objects of a given size in centimeter crossing one square meter per year. The straight line is the power law $1.45 \times 10^{-6}/Y^{2.75}$. The data has been provided by CNES, France

size distribution should thus correspond to a superposition of pdf's which are log-normal (see below) in their centers and of the form (3.81) in the tail $Y \to 0$.

We assume that the tail of the size distribution for a fixed generation rank N is given by (3.81) and that the mean number $\mathcal{N}(N)$ (per unit volume) of fragments of generation rank N grows exponentially: $\mathcal{N}(N) \propto e^{\lambda N}$ with $\lambda > 0$. It then follows that the tail of the unconditioned size distribution is given by

$$P_{\text{size}}(Y) \sim \sum_{N=0}^{\infty} [P(Y^{1/N})]^N e^{\lambda N} \approx \int_0^{\infty} dN \, e^{N \ln P(Y^{1/N}) + N\lambda} . \quad (3.83)$$

Laplace's method can now be applied to the integral since here N is assumed to be a continuous variable. A critical (saddle) point is found at

$$N_\star = -\frac{1}{\alpha} \ln Y , \quad (3.84)$$

where α is the solution of the transcendental equation

$$\lambda + \ln P(e^{-\alpha}) + \alpha e^{-\alpha} \frac{P'(e^{-\alpha})}{P(e^{-\alpha})} = 0 . \quad (3.85)$$

The leading-order tail behavior of the size distribution is then given by

$$P_{\text{size}}(Y) \sim Y^{-\tau} , \quad (3.86)$$

with an exponent

$$\tau = \frac{1}{\alpha} \left[\ln P \left(e^{-\alpha} \right) + \lambda \right] . \tag{3.87}$$

This solution (3.86) holds for λ below a threshold λ_c dependent on the specific structure of the pdf $P(m)$. For instance, if $P(m) \sim \exp\left(-Cm^\delta\right)$ for $m \to 0$, with $\delta > 0$, we get $\lambda_c = C$. Consequently, this power-law structure is very robust with respect to the choice of the distribution $P(m)$, while the exponent τ will vary. We notice that such a mechanism does not work if we replace $P_N(X)$ by the log-normal form valid in the center of the distribution.

What happens for $\lambda > C$? To find out, we return to the expression (3.83) giving the tail of the unconditioned size distribution and find that the exponential in the integral reads $e^{N(\lambda-CY^{\delta/N})}$. In the limit of small fragments $X \to 0$, the term $Y^{\delta/N} \leq 1$, where the upper bound 1 is reached in the limit $N \to \infty$. Thus, $\lambda - CY^{\delta/N} \geq \lambda - C$. Thus, for $\lambda > C$, the larger N is, the larger the exponential is. This is in constrast with the case $\lambda < C$ for which there is an optimal generation number $N_\star$, for a given size Y, given by (3.84). For $\lambda \geq C$, the critical value $N_\star$ moves to infinity.

Physically, this is the signature of a shattering transition occurring at $\lambda = C$: for $\lambda > C$, the number of fragments increases so fast with the generation number N (as $e^{\lambda N} > e^{CN}$) that the distribution of fragment sizes develops a finite measure at $Y = 0$. This result is in accordance with intuition: it is when the number of new fragments generated at each hammer hit is sufficiently large that a dust phase can appear. This *shattering transition* has first been obtained in the context of linear mean field rate equations [119, 606].

Consider another class of pdf $P(m) \propto \exp\left(-Cm^{-\delta}\right)$ for $m \to 0$, with $\delta > 0$. The pdf $P(m)$ goes to zero faster than any power law as $m \to 0$ (i.e. it has an essential singularity). The difference with the previous case is that, as the multiplicative factor $m \to 0$ occurs with very low probability in the present case, we do not expect a large number of small fragments to be generated. This should be reflected in a negative value of the exponent τ. This intuition is confirmed by an explicit calculation showing that τ becomes the opposite of the value previously calculated, i.e. $\tau/C\delta$ goes continuously from $-e \approx -2.718$ to -1 as λ goes from 0 to C.

In sum, we propose that the observed power-law distributions of fragment sizes could be the result of the natural mixing occurring in the number of generations of simple multiplicative processes exhibiting extreme deviations. This power-law structure is very robust with respect to the choice of the distribution $P(m)$ of fragmentation ratios, but the exponent τ is not universal. The proposed theory leads us to urge for experiments in which one can control the generation rank *of each* fragment. We then predict that the fragment distribution will not be (quasi-) universal anymore but on the contrary better characterizes the specific mechanism underlying the fragmentation process.

The result (3.86) only holds in the "tail" of the distribution for very small fragments. In the center, the distribution is still approximately log-normal.

We can thus expect a relationship between the characteristic size or peak fragment size and the tail structure of the distribution. It is in fact possible to show that the exponent τ given by (3.87) is a *decreasing* function of the peak fragment size: the smaller the peak fragment size, the larger will be the exponent (the detailed quantitative dependence is a specific function of the initial pdf). This prediction turns out to be verified by the measurements of particle size distributions in cataclastic (i.e. crushed and sheared rock resulting in the formation of powder) fault gouge [20]: the exponent τ of the finer fragments from three different faults (San Andreas, San Gabriel and Lopez Canyon) in Southern California was observed to be correlated with the peak fragment size, with finer gouges tending to have a larger exponent. Furthermore, the distributions were found to be a power law for the smaller fragments and log-normal in mass for sizes near and above the peak size.

3.5 Large Deviations in the Sum of Variables with Power Law Distributions

3.5.1 General Case with Exponent $\mu > 2$

The results of the previous sections on large deviations do not apply if the pdf $P(X_i)$ decays as a power law for large X_i with an exponent μ. Here, we consider the case where $\mu > 2$ for which the sum converges in probability towards the Gaussian law.

In this case, the cumulants of order larger than the exponent of the power law are infinite and the expansion (3.4) looses its meaning. As a result, the convergence to the Gaussian law is very slow. An interesting and concrete case is when a power law is truncated at some large value beyond which the pdf decays at least as fast as an exponential. Such laws obey the central limit theorem but their kurtosis is very large (and a function of the cut-off) and the criterion $N \gg \lambda_4$ becomes very drastic.

A very instructive example is a pdf with an asymptotic power law tail but of finite variance. Consider the following law:

$$P(x) = \frac{2a^3}{\pi(x^2 + a^2)^2} \sim \frac{1}{x^{1+\mu}} \quad \text{with } \mu = 3 \text{ for large } x , \qquad (3.88)$$

where a is a positive constant. This normalized pdf has a power law tail with $\mu = 3$ (see the definition (1.24)). Therefore, all its moments and cumulants of order larger or equal to 3 are infinite. However, its variance is finite and is equal to a^2. Its characteristic function is easy to evaluate exactly

$$\hat{P}(k) = (1 + a|k|)e^{-a|k|} . \qquad (3.89)$$

The first terms of its expansion around $k = 0$ read

$$\hat{P}(k) \simeq 1 - \frac{k^2 a^2}{2} + \frac{|k|^3 a^3}{3} + \mathcal{O}(k^4) . \qquad (3.90)$$

Notice that the third order derivative of $\hat{P}(k)$ is discontinuous at $k = 0$: the first singular term in the expansion of $\hat{P}(k)$ close to $k = 0$ is proportional to $|k|^3$. This is in agreement with the asymptotic behavior $P(x) \sim x^{-4}$ and the divergence of the moments of order larger and equal to three.

The pdf obtained by N convolutions has the following characteristic function:

$$\left[\hat{P}(k)\right]^N = (1 + a|k|)^N e^{-aN|k|} , \tag{3.91}$$

which can be expanded around $k = 0$ as

$$\left[\hat{P}(k)\right]^N \simeq 1 - \frac{Nk^2a^2}{2} + \frac{N|k|^3a^3}{3} + O(k^4) . \tag{3.92}$$

Notice that the singularity in $|k|^3$, which is the signature of the divergence of the moments m_l for $l \geq 3$, does not disappear upon convolution, even if the pdf converges to the Gaussian law. The explanation is, as usual, that the convergence to the Gaussian law occurs in the center of the law, while the behavior in the tails is still described by the power law $\sim |x|^{-4}$ which makes the higher moments diverge.

Since $P(x)$ defined in (3.88) has a finite variance, the sum of N variables is described by a Gaussian law of variance Na^2 and zero mean:

$$P_N(x) \simeq g\left(x/a\sqrt{N}\right) \equiv \frac{1}{\sqrt{2\pi Na}} \exp\left(-\frac{x^2}{2Na^2}\right) . \tag{3.93}$$

On the other hand, we have seen that the asymptotic power law behavior is conserved upon convolution and that the scale parameter C is additive (this will be elaborated further in Chap. 4). We can therefore write

$$P_N(x) \simeq \frac{2Na^3}{\pi x^4} \qquad \text{for } x \to \infty . \tag{3.94}$$

The equations (3.93) and (3.94) are not in contradiction. In fact, they describe two different regimes for the pdf $P_N(x)$. For N fixed, there exists a value $x_0(N)$ beyond which the Gaussian description becomes invalid and where the pdf is correctly described by the asymptotic power law. The value $x_0(N)$ is given approximately by matching the two regimes [115]:

$$\frac{1}{\sqrt{2\pi Na}} \exp\left(-\frac{x_0^2}{2Na^2}\right) \simeq \frac{2Na^3}{\pi x_0^4} . \tag{3.95}$$

We thus find

$$x_0 \simeq a\sqrt{N \ln N} , \tag{3.96}$$

neglecting subdominant terms for large N. In the rescaled variable $y = x/(a\sqrt{N})$, y tends to a normal variable with unit variance. However, this Gaussian description breaks down for $y \sim \sqrt{\ln N}$ or larger, and this value increases very slowly with N. Even for $N = 10^6$, the Gaussian description

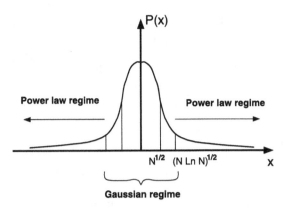

Fig. 3.12. Illustration of the cross-over from the Gaussian regime holding for $x < \sqrt{N \ln N}$ to the power law regime for $x \gg \sqrt{N \ln N}$

of $P_N(x)$ breaks down for fluctuations larger than about 3–4 standard deviations.

These results are not in contradiction with the central limit theorem, which only states that $P_N(x)$ converges to the Gaussian law in probability. This implies that the region where $P_N(x)$ is different from the Gaussian have a probability weight which decays to zero. In our example, the probability that X falls in the asymptotic region is given by

$$\mathcal{P}_<(-x_0) + \mathcal{P}_>(x_0) \simeq 2 \int_{a\sqrt{N \ln N}}^{\infty} \frac{2a^3 N}{\pi x^4} \mathrm{d}x = \frac{4}{3\pi \sqrt{N} \ln^{3/2} N} \ , \qquad (3.97)$$

which indeed goes to zero as $N \to \infty$.

The cross-over from the Gaussian description in the bulk of the pdf to the power law tail can actually be described precisely as follows:

$$\mathcal{P}_{>,N}\left(X_1 + \ldots + X_N > axN^{1/2}\right) \simeq 1 - G(x) + \frac{2}{3\pi \sqrt{N}} \frac{1}{x^3} \ , \qquad (3.98)$$

where $G(x) = \int_0^x g(u)\,\mathrm{d}u$ is the cumulative Gaussian pdf defined as the asymptotic regime in the bulk in (3.93). This formula (3.98) is valid uniformely for all $x \geq 1$ (not only for $x > N^{1/2+\epsilon}$ for arbitrary positive power ϵ), including all super-large values. The Gaussian term tends to zero very fast for $x > N^\epsilon$, but for, say, $x = 2$ it has a finite non-negligible contribution.

All these results apply to any law whose asymptotic behavior is a power law with $\mu > 2$, i.e. whose variance is finite. In this general case, one finds that the central limit theorem and the Gaussian description apply for $|X| \ll x_0 \simeq \sqrt{N \ln N}$ and the weight in probability of these tails is

$$\mathcal{P}_<(-x_0) + \mathcal{P}_>(x_0) \simeq \frac{1}{N^{\mu/2-1} \ln^{\mu/2} N}, \qquad (3.99)$$

which goes to zero as $N \to \infty$. Notice however that when μ approaches 2 from above, the weight of the tails is more and more important. For $\mu < 2$, everything breaks down! The weight in the tails *increases* with N and there is no convergence to the Gaussian law. However, it turns out that such a pdf does converge but to a different law, called a Lévy law.

The theorem by S. Nagaev generalizes the result (3.98). For i.i.d. random centered variables normalized to have a unit variance, then, under some (rather complicated) conditions that are fullfiled for the case of power-like tails (in particular, for the Pareto distribution), the following relation is true [660]:

$$\mathcal{P}_{<,N}(X_1 + \ldots + X_n > x) \simeq N\mathcal{P}_>(X_1 > x) , \tag{3.100}$$

for $N \to \infty$ and for $x > N^{1/2+\epsilon}$. The second term proportional to $1/x^3$ in the right-hand-side of expression (3.98) is a direct consequence of Nagaev's theorem (3.100). The Gaussian term $1 - G(x)$ does not appear in Nagaev's theorem because it tends to zero very fast and does not contribute to the asymptotic behavior.

3.5.2 Borderline Case with Exponent $\mu = 2$

The borderline case (1.24) with $\mu = 2$ is of special interest both from the point of view of the central limit theorem and for geophysical applications. We have seen in Chap. 2 that (2.72) is the general condition for the convergence of the pdf of the sum of N random variables to the Gaussian law. Ibragimov and Linnik [446] have reformulated (2.72) in a slightly different form

$$\lim_{X \to \infty} \frac{I(tX)}{I(X)} = 1 \qquad \text{for all } t , \tag{3.101}$$

where

$$I(X) = \int_{-X}^{+X} dx\, x^2 P(x) . \tag{3.102}$$

They also give an equation to determine the standard deviation σ_N of the pdf of the sum:

$$\lim_{X \to \infty} \frac{NI(\epsilon\sigma_N)}{\sigma_N^2} = 1 \qquad \text{for some } \epsilon > 0 . \tag{3.103}$$

Expression (3.103) assumes that the pdf P of the individual variables constituting the sum has zero mean, which is the case of interest here. That a formula like (3.103) is needed to get σ_N stems from the fact that the variance σ_N^2 is no longer equal to N times the variance of P, since the latter is infinite!

This problem is encountered in the problem of advection of passive and active tracers by two-dimensional systems of vortices, with applications to ozone transport in the stratosphere, pollutant dispersal in the atmosphere and the ocean, and plankton and salinity transport in the ocean. Consider an ensemble of many point vortices in two spatial dimensions, which capture many of the features of two-dimensional turbulent flows. Then, the pdf of vortex velocity has a power law tail with exponent $\mu = 2$ [1010]. This can be seen from the fact that the velocity u generated at a distance r by a single vortex of circulation Γ is $2\Gamma/r$. Choose the position of the velocity probe at random. The probability of the probe to be at a distance between r and $r + dr$ from the vortex is proportional to $2\pi r\, dr$, which gives

$$P(u) = P(r(u))2\pi r\frac{dr}{du} \sim \frac{1}{u^3} \ . \tag{3.104}$$

The power law is an example of the mechanism of a "power law change of variable close to the origin" discussed in Chap. 14. In the presence of an ensemble of N vortices, the velocity at the position of the probe is the sum over the velocities created by the N vortices. We thus have to estimate the pdf of the sum of N random variables distributed according to (3.104).

From (3.102), we find $I(X) \sim \ln X$. We verify that (3.101) holds and find the standard deviation

$$\sigma_N \sim \sqrt{N \ln N} \ , \tag{3.105}$$

from (3.103). Thus, the pdf of the velocity field created by N random vortices is a Gaussian of variance $\sigma_N^2 \sim N \ln N$ up to a scale $u \sim \sigma_N$. For larger velocities, the pdf deviates from a Gaussian and keeps the asymptotic $1/u^3$ power law tail. Note that this border case $\mu = 2$ is special in the sense that the standard deviation becomes equal to the domain $\sqrt{N \ln N}$ of validity of the Gaussian description shown in Fig. 3.12. This border case behavior has been studied in details in [163]. Chavanis has also shown that, in a statistical sense, the velocity created by a point vortex is shielded by the cooperative effects of the other vortices on a distance $\sim N^{-1/2}$, the inter-vortex separation [162]. For $R \gg N^{-1/2}$, the "effective" velocity decays as $1/r^2$ instead of the ordinary law $1/r$ recovered for $r \ll N^{-1/2}$. These results give further support to the observation that the statistics of velocity fluctuations are (marginally) dominated by the contribution of the nearest neighbor.

Finally, we note that this mechanism for the generation of power laws by an ensemble of sources with power law singularities of the field in their neighborhood has first been investigated in the problem of Holtsmark's gravitational force distribution created by an ensemble of stars in a random universe [293]. It has many other applications that we will cover in Chap. 17.

4. Power Law Distributions

4.1 Stable Laws: Gaussian and Lévy Laws

4.1.1 Definition

Summing N i.i.d. random variables with pdf $P_1(x)$, one obtains a random variable which is in general a different pdf $P_N(x)$ given by N convolutions of $P_1(x)$ with itself as given by (2.67).

Special pdf's have the remarkable property that $P_N(x)$ has the same form as $P_1(x)$. They are said to be "stable". More precisely, this similarity between $P_N(x)$ and $P_1(x)$ is required to hold, up to a translation and a dilation, which are often needed to accomodate the additivity property of the mean and of the measure of typical deviations from the mean:

$$P_N(x')\,\mathrm{d}x' = P_1(x)\,\mathrm{d}x \qquad \text{where} \qquad x' = a_N x + b_N \ , \tag{4.1}$$

for some constants a_N and b_N.

Within the formalism of the renormalization group (RG) presented in Chap. 2, a stable law corresponds to a fixed point of the RG process, which we have seen to involve both a decimation and rescaling. Fixed points of the RG usually play a very special role. Attractive fixed points, as in the present case, describe the macroscopic behavior observed in the large N limit. The introduction of correlations may lead to repulsive fixed points, which are the hallmark of a phase transition, i.e. the existence of a global change of regime at the macroscopic level under the variation of a control parameter quantifying the strength of the correlations. This will be discussed in more details in Chap. 11.

4.1.2 The Gaussian Probability Density Function

The best-known example of a stable law is the Gaussian law, also called the normal law or the Laplace–Gauss law and is, as we have seen, indeed a fixed point of the RG. For instance, the Binomial and Poisson distributions tend to the Gaussian law under the operation of the addition of a large number of random variables. It is often encountered and we have seen that a general explanation is provided by the central limit theorem.

The expression is

$$P_G(x) = \frac{1}{\sqrt{2\pi\sigma^2}} \exp\left(-\frac{(x-x_0)^2}{2\sigma^2}\right)$$
$$\text{defined for } -\infty < x < +\infty \ . \tag{4.2}$$

Deviations from the mean larger than a few standard deviations are rare for the Gaussian law. For instance, deviations larger than two standard deviations occur with probability 4.45% (2.2% for deviations above $x_0 + 2\sigma$ (resp. below $x_0 - 2\sigma$), while deviations larger than 3σ have a probability 0.3%. A deviation larger than 5 (resp. 10) σ has a probability 5.7×10^{-7} (resp. 1.5×10^{-23}), i.e. are never seen in practice.

4.1.3 The Log-Normal Law

Another pdf that is often encountered in natural sciences is the log-normal law. A variable X is distributed according to a log-normal pdf if $\ln X$ is distributed according to a Gaussian pdf. The log-normal distribution is stable, not under addition but under multiplication which is equivalent to the addition of logarithms. Therefore, the log-normal law is not a true stable law, but it is so intimately related to the Gaussian law that it is natural to discuss it here.

The change of variable $x \to \ln x$ in the Gaussian law gives the expression of the log-normal pdf:

$$P(x) = \frac{1}{\sqrt{2\pi\sigma^2}} \frac{1}{x} \exp\left(-\frac{\ln^2(x/x_0)}{2\sigma^2}\right) \ , \tag{4.3}$$

defined for $0 < x < +\infty$. The variable x_0 is a characteristic scale corresponding to the *median* $x_{1/2}$ as defined in (1.18). It has the property that

$$\langle \ln x \rangle \equiv \ln x_0 \ . \tag{4.4}$$

The variable σ is the standard deviation of the variable $\ln x$. The most probable value, defined in (1.19), is

$$x_{mp} = x_0 e^{-\sigma^2} \ . \tag{4.5}$$

The mean $\langle x \rangle$ is equal to

$$\langle x \rangle = x_0 e^{\sigma^2/2} \tag{4.6}$$

and can be much larger than x_0.

Notice that the knowledge of the moments does not allow us to determine the log-normal unambiguously. Indeed, the pdf

$$\frac{1}{\sqrt{2\pi}} x^{-1} e^{-(1/2)(\ln x)^2} [1 + a \sin(2\pi \ln x)] \ , \qquad \text{with } -1 < a < 1 \tag{4.7}$$

has exactly the same moments as (4.3) [293].

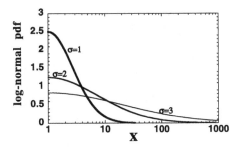

Fig. 4.1. Semi-log plot for $x \geq x_0$ of the log-normal pdf (4.3) for $x_0 = 1$ and three different values of $\sigma = 1, 2$ and 3. Notice how a relatively small change in σ expands considerably the range in x over which the log-normal pdf is non-negligible

It is interesting to notice that the log-normal pdf can be mistaken locally for a power law. This is an important remark for the analysis of data in which the standard procedure is to qualify the existence of a power law using double-logarithmic plots. Actually, as we are going to show, the log-normal distribution can mimic a power law very well over a relatively large interval. To see this, notice that

$$e^{a(\ln x)^2} = x^{a \ln x} \ . \tag{4.8}$$

Using this equality, we can rewrite (4.3) as

$$P(x) = \frac{1}{x_0 \sqrt{2\pi\sigma^2}} \left(x/x_0\right)^{-1-\mu(x)} \ , \tag{4.9}$$

with

$$\mu(x) = \frac{1}{2\sigma^2} \ln \frac{x}{x_0} \ . \tag{4.10}$$

Since $\mu(x)$ is a slowly varying function of x, this form shows that the log-normal distribution can be mistaken for an apparent power law with an exponent μ slowly varying with the range x. This was pointed out in [646], where it was noticed that for $x \ll x_0 e^{2\sigma^2}$, $\mu(x) \ll 1$ and the log-normal is undistinguishable from the $1/x$ distribution, providing a mechanism for $1/f$ noise. More generally, the larger σ is, the smaller μ is and the larger is the range in x over which the log-normal mimics a power law distribution with a small exponent.

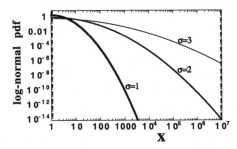

Fig. 4.2. log–log plot of the log-normal pdf (4.3) for $x_0 = 1$ and three different values of $\sigma = 1, 2$ and 3. Notice the considerable range over which the log-normal distribution extends for $\sigma = 3$. For the broad range of scales represented in this figure, a downward curvature is clearly apparent and distinguishes the log-normal pdf from a power law

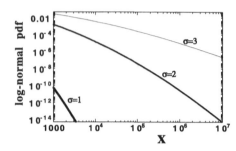

Fig. 4.3. Magnification of Fig. 4.2 showing that the log-normal pdf's with $\sigma = 2$ and 3 are close to perfectly linear over more than 4 decades both in abcissa and ordinate. With some additional noise, it would be difficult to distinguish them from pure power laws with constant exponent μ. Recall that a straight line qualifies a power law in such a log–log plot

The product of random variables is a generic mechanism for the creation of log-normal distributions (see Chap. 16 for a presentation of the general theorem by Kolmogorov on fragmentation processes and the log-normal pdf). There are interesting non-asymptotic deviations that have been exposed pedagogically by S. Redner [764]. Finite sums of log-normal random variables also exhibit a rich set of deviations from the asymptotic Gaussian pdf. Romeo et al. [790] have obtained approximate formulae for the pdf of the sum of lognormal variables valid for $\sigma \leq 4$. For larger σ, one may apply the theorems of Arous et al. [36] and of Bovier et al. [104]. The anomalous behaviour of the typical sums can be related to the broadness of lognormal distributions. For large enough shape parameter σ^2, the behavior of log-normal sums corresponds to that of broad distributions at small sample sizes and to properties of narrow distributions at large sample sizes, with a slow transition between the two regimes similar to that described for a truncated Lévy distribution in Fig. 4.6. There are some counter-intuitive effects, such as the decrease of the peak height of the sample mean distribution with the sample size and the fact that the typical sample mean and its inverse vary in the same way rather than in opposite ways as the sample size increases. These statistical effects arising from the broadness of log-normal distributions have observable consequences for physical systems of moderate size, such as in the electrical current flowing through small tunnel junctions [790].

4.1.4 The Lévy Laws

Properties. The stable laws have been studied and classified by Paul Lévy, who discovered that, in addition to the Gaussian law, there is a large number of other pdf's sharing the stability condition (4.1). One of their most interesting properties is their asymptotic power law behavior. We refer to [810] for a general presentation of their mathematical properties and to [456] for their stochastic representations and illustrative numerical simulations.

A symmetric Lévy law centered on zero is completely characterized by two parameters which can be extracted solely from its asymptotic dependence

$$P(x) \sim \frac{C}{|x|^{1+\mu}} \qquad \text{for } x \to \pm\infty \ . \tag{4.11}$$

C is a positive constant called the tail or scale parameter and the exponent μ is between 0 and 2 ($0 < \mu < 2$). Clearly, μ must be positive for the pdf to be normalizable. As for the other condition $\mu < 2$, we have seen that a pdf with a power law tail with $\mu > 2$ has a finite variance and thus converges (slowly) in probability to the Gaussian law. It is therefore not stable. Only its shrinking tail for $\mu > 2$ remains of the power law form. In contrast, the whole Lévy pdf remains stable for $\mu < 2$.

All symmetric Lévy laws with the same exponent μ can be obtained from the Lévy law $L_\mu(x)$ with exponent μ, centered on zero and with unit scale parameter $C = 1$, under the translation and rescaling transformations

$$P(x)\,\mathrm{d}x = L_\mu(x')\,\mathrm{d}x' \qquad \text{where } x' = C^{1/\mu}x + m \ , \tag{4.12}$$

m being the center parameter.

Lévy laws can be asymmetric and the parameter quantifying this asymmetry is

$$\beta = (C_+ - C_-)/(C_+ + C_-) \ , \tag{4.13}$$

where $C_\pm$ are the scale parameters for the asymptotic behavior of the Lévy law for $x \to \pm\infty$. When $\beta \neq 0$, one defines a unique scale parameter $C = (C_+ + C_-)/2$, which together with β allows one to describe the behavior at $x \to \pm\infty$. The completely antisymmetric case $\beta = +1$ (resp. -1) corresponds to the maximum asymmetry.

For $0 < \mu < 1$ and $\beta = \pm 1$, the random variables take only positive (resp. negative) values.

For $1 < \mu < 2$ and $\beta = +1$, the Lévy law is a power law for $x \to +\infty$ but goes to zero for $x \to -\infty$ as $P(x) \sim \exp(-|x|^{\mu/\mu-1})$. This decay is faster than the Gaussian law. The symmetric situation is found for $\beta = -1$.

An important consequence of (4.11) is that the variance of a Lévy law is infinite as the pdf does not decay sufficiently rapidly at $|x| \to \infty$ for the integral in (1.22) to converge. When $\mu \leq 1$, the Lévy law decays so slowly that even the mean (1.17) and the average of the absolute value of the spread (1.21) diverge. The median and the most probable value still exist and coincide, for symmetric pdf ($\beta = 0$), with center m. The characteristic scales of the fluctuations are determined by the scale parameter C, i.e. they are of the order of $C^{1/\mu}$.

There are no simple analytic expressions of the symmetric Lévy stable laws $L_\mu(x)$, except for a few special cases. The best known is $\mu = 1$, called the Cauchy (or Lorentz) law,

$$L_1(x) = \frac{1}{x^2 + \pi^2} \qquad \text{for } -\infty < x < +\infty \ . \tag{4.14}$$

The Lévy law for $\mu = 1/2$ is [649]

$$L_{1/2}(x) = \frac{2}{\sqrt{\pi}} \frac{\exp\left(-1/2x\right)}{(2x)^{3/2}} \qquad \text{for } x > 0 \ . \tag{4.15}$$

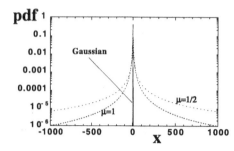

Fig. 4.4. Semi-log plot of the Lévy laws $L_1(x)$ and $L_{1/2}(x)$ (made symmetric) compared to the Gaussian with unit variance. These three distributions have similar half-widths but extraordinarily different tails

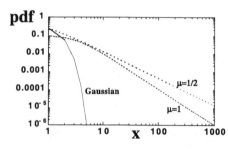

Fig. 4.5. Log–Log plot in the positive quadrant of the Lévy laws $L_1(x)$ and $L_{1/2}(x)$ compared to the Gaussian with unit variance. Notice the asymptotic linearity of the plots for the Lévy laws signaling their asymptotic power law dependence

This pdf $L_{1/2}(x)$ gives the distribution of first returns to the origin of an unbiased random walk.

Lévy laws are fully characterized by the expression of their characteristic functions. In the symmetric case ($\beta = 0$), the characteristic function reads

$$\hat{L}_\mu(k) = \exp\left(i\gamma k - a_\mu |k|^\mu\right) \ , \quad 1 < \mu < 2 \ , \tag{4.16}$$

where a_μ is a constant proportional to the scale parameter C:

$$a_\mu = \frac{\pi\, C}{\mu^2 \Gamma(\mu - 1) \sin(\pi\mu/2)} \qquad \text{for} \quad 1 < \mu < 2 \ . \tag{4.17}$$

A similar expression holds for $0 < \mu < 1$, while $\mu = 1$ and 2 requires a special form (see [350] for full details). For $\beta \neq 0$ (and omitting the contribution $i\gamma k$ that exists for $\mu > 1$ in the non-centered case), we have

$$\hat{L}_\mu^\beta(k) = \exp\left[-a_\mu |k|^\mu \left(1 + i\beta \tan(\mu\pi/2)\frac{k}{|k|}\right)\right] \ \text{for} \ \mu \neq 1 \ . \tag{4.18}$$

For $\mu = 1$, $\tan(\mu\pi/2)$ is replaced by $(2/\pi)\ln|k|$.

The behavior of Lévy laws $L_\mu(x)$ is also of interest close to the origin $x \to 0$, as this has been used to generate the stretched exponential relaxation for instance seen in dielectric experiments and in the dynamics of supercooled liquids [513]. Indeed, we can rewrite (4.16) using the Laplace transform instead of the Fourier transform and obtain the characteristic function as

$$\exp\left(-ct^\mu\right) = \int_0^\infty d\gamma\, L_{\mu,\beta=1}(\gamma) \exp\left(-\gamma t\right) \ . \tag{4.19}$$

The dual of the argument γ of the Lévy law through the Laplace transform is now interpreted as time t. The physical meaning of the right hand side of (4.19) is that the relaxation rate is the superposition of normal exponential relaxation rates with a weight given by the Lévy law $L_{\mu,\beta=1}(\gamma)$. From the definition of Lévy laws by their characteristic function (4.16), we see that this generates so-called anomalous Kohlrausch–Williams–Watts relaxation $\exp(-ct^\mu)$. This is a stretched exponential for $\mu < 1$. One is usually interested in the long-time behavior of the relaxation. We see from (4.19) that the long-time behavior is controlled by the contribution of exponentials in the right hand side of (4.19) with small γ. In other words, it is not the tail of the Lévy laws but their behavior close to zero that controls the Kohlrausch–Williams–Watts relaxation rate in this model. Assuming that $L_{\mu,\beta=1}(\gamma) \sim \exp(-\gamma^c)$, a saddle-node approximation of (4.19) gives

$$c = -\frac{\mu}{1-\mu} \quad < 0 \; . \tag{4.20}$$

Actually, the asymptotic expansion of the stable laws in the vicinity of zero is known [1066] and reads

$$L_{\mu<1,\beta=1}(\gamma) \approx \gamma^a \exp(-b\gamma^c) \; , \tag{4.21}$$

where the exponent c is indeed given by (4.20), $a = -(2-\mu)/2(1-\mu) < 0$ and $b = (1-\mu)/\mu^{\mu/(1-\mu)} > 0$. This shows that $L_{\mu<1,\beta=1}(\gamma)$ has an essential singularity close to the origin (all its derivatives are zero at the origin).

Scaling. Gaussian and Lévy laws present similar scaling properties with respect to the sum of random variables.

The pdf of the sum of N Gaussian i.i.d. random variables with mean $\langle x \rangle$ and variance σ^2 is also Gaussian with mean $N\langle x \rangle$ and variance $N\sigma^2$. Thus, the rescaled variable

$$\frac{X - N\langle x \rangle}{\sigma\sqrt{N}} \tag{4.22}$$

has exactly the same pdf as the initial variables, in other words, the pdf of the rescaled variable is independent of N. This remark is useful in practice because it provides a test for the Gaussian law: take different samples of different sizes $N_1, N_2, ...$ Construct the rescaled variables $(X - N_1\langle x\rangle)/\sigma\sqrt{N_1}$, $(X - N_2\langle x\rangle)/\sigma\sqrt{N_2}$, etc. Then, all these variables will exhibit the same pdf, in other words all the data will collapse onto the same Gaussian curve.

A similar property holds for Lévy laws and for pdfs with power law tails in their asymptotic regime.

- $1 < \mu < 2$. The pdf of the sum of N Lévy variables with the same exponent μ and distributed according to $L_{\mu,\beta}^C(x)$, where we now make the dependence on the three parameters μ, β and C explicit, can be rescaled by the change of variable

$$\frac{X - N\langle x \rangle}{N^{1/\mu}} \; , \tag{4.23}$$

which is the same as (4.22) with the replacement $2 \to \mu$. The pdf of the sum can thus be written

$$P_N(X) = \frac{1}{N^{1/\mu}} L^C_{\mu,\beta} \left(\frac{X - N\langle x \rangle}{N^{1/\mu}} \right) . \tag{4.24}$$

- $0 < \mu \leq 1$. A similar result holds:

$$P_N(X) = \frac{1}{N^{1/\mu}} L^C_{\mu,\beta} \left(\frac{X}{N^{1/\mu}} \right) . \tag{4.25}$$

Notice that $\langle x \rangle$ does not appear as it is no longer defined.

Table 4.1. The similarities between the Gaussian and Lévy pdfs for sums of N i.i.d. random variables are summarized

Property of each basin of attraction	Gaussian	Lévy
Tail decay	$\mu \geq 2$	$\mu < 2$
Characteristic fluctuation scale	$N^{1/2}$	$N^{1/\mu}$
Rescaled variable	$\dfrac{X - N\langle x \rangle}{\sqrt{N}}$	$\dfrac{X - N\langle x \rangle}{N^{1/\mu}}$ for $\mu > 1$
		$\dfrac{X}{N^{1/\mu}}$ for $\mu \leq 1$
Scale parameter	σ^2	C
Composition rule for the sum	$\Sigma_N^2 = N\sigma^2$	$C_N = NC$
of N variables		

These properties provide a general strategy for testing for the existence of Lévy laws and more generally for power laws. As for the Gaussian case, the idea is to compare different data sets, for instance obtained from different system sizes, and try to collapse the different pdfs onto a "universal" master curve by varying the exponent μ. This *finite size scaling* technique [147, 752] is one of the most powerful techniques to test and measure the exponents of power law distributions. It is widely used in all the disciplines where power laws are encountered.

When such a rescaling holds, one says that the quantity, here the pdf, exhibits scaling properties. In a given numerical procedure, a suitable translation and a change of unit scale will allow one to define a variable which remains described by the same law in different measurements. The concept of scaling has been found to be widely shared by many systems. It is observed in equilibrium systems at their critical points, as first discovered by B. Widom in the late sixties, as well as in out-of-equilibrium systems which are self-organizing in states with many differents scales.

Convergence to a Lévy Law. We first state the main results in simple terms. Upon N convolutions and as $N \to \infty$, the Gaussian pdf "attracts" all the pdf's decaying as or faster than $1/|x|^3$ at large $|x|$. Similarly, upon N convolutions, all pdfs with exponent $\mu < 2$ and scale parameter $C = C_+ = C_-$ are attracted to the symmetric Lévy law with exponent μ and scale parameter NC. If the initial pdf's have different scale parameters C_- and C_+, the convergence is to the asymmetric stable Lévy law with the same exponent, with the scale parameter NC and with the asymmetry parameter $\beta = (C_- + C_-)/(C_+ + C_-)$. If the exponents μ_- and μ_+ are different, the smallest one is the winner and the convergence is to a completely asymmetric stable Lévy law with exponent $\mu = \min(\mu_-, \mu_+)$ and asymmetry parameter $\beta = -1$ for $\mu_- < \mu_+$ or $\beta = 1$ for $\mu_- > \mu_+$.

This generalized limit theorem applies under the same restrictions (except for the finiteness of the variance) of independence and of large N. Roehner and Winiwarther give an explicit calculation of the aggregation of independent random variables distributed according to a power law [789].

Let us now state these results more rigorously. A pdf $P(x)$ belongs to the basin of attraction of the Lévy law $L_{\mu,\beta}$ if and only if

$$\lim_{x \to \infty} \frac{\mathcal{P}_<(-x)}{\mathcal{P}_>(x)} = \frac{1-\beta}{1+\beta} , \tag{4.26}$$

and for any r,

$$\lim_{x \to \infty} \frac{\mathcal{P}_<(-x) + \mathcal{P}_>(x)}{\mathcal{P}_<(-rx) + \mathcal{P}_>(rx)} = r^\mu . \tag{4.27}$$

This complicated expression essentially means that the asymptotic behavior is

$$\mathcal{P}_<(x) \simeq \frac{C_-}{\mu|x|^\mu} \qquad \text{for } x \to -\infty , \tag{4.28}$$

and

$$\mathcal{P}_>(x) \simeq \frac{C_+}{\mu x^\mu} \qquad \text{for } x \to \infty . \tag{4.29}$$

Such a law belongs to the basin of attraction of the Lévy law of exponent μ and asymmetry parameter $\beta = (C_+ - C_-)/(C_+ + C_-)$.

4.1.5 Truncated Lévy Laws

A Lévy or power law distribution measured in natural data sets or obtained from numerical simulations will not extend indefinitely due to obvious finite size limitations. For $\mu < 1$, there are even more compelling reasons to cause a cut-off in the power law behavior.

Consider, for instance, the case of earthquakes and their Gutenberg–Richter distribution. The Gutenberg–Richter magnitude–frequency law is often written as the cumulative distribution

$$\log_{10} N_> = a - bM_W , \tag{4.30}$$

where $N_>$ is the number of earthquakes whose magnitudes are equal to or greater than M_W. M_W is the moment magnitude, defined by

$$M_W = \frac{1}{\beta}[\log_{10}(m_0) - 9] . \tag{4.31}$$

m_0 is the seismic moment in N m. β is generally taken to be equal to 1.5. If we combine these two expressions, we get a power law distribution for the number of earthquakes having a given seismic moment, which is identical to (4.11) (with $x \equiv m_0$), characterized by the exponent $\mu = b/\beta$. For small and intermediate magnitude earthquakes, $b \approx 1.0$; thus $\mu \approx 2/3$.

Such a power law distribution, which holds for small earthquakes, cannot be extended to infinity because it would require that an infinite amount of energy be released from the Earth's interior [518]: taking the energy dissipated by an earthquake as proportional to the seismic moment, the average energy dissipated per unit time in a given region is then proportional to $\langle m_0 \rangle$, where the constant of proportionality contains the number per unit time of earthquakes of any magnitude belonging to the Gutenberg–Richter law. But, since the exponent $\mu < 1$, the average is mathematically infinite, which is clearly impossible since the earth is finite and cannot feed an infinite amount of energy. As we will discuss in details below, the solution of this paradox is that the average is essentially controlled by the largest event in the catalog. The finiteness of the energy available to trigger earthquakes must therefore correspond to a cut-off in the power law distribution, i.e. to the existence of a typical largest possible earthquake. There must be a crossover or a rollover to a second branch of the distribution. This second branch can for instance be a power law, but with an exponent larger than 1 such that the mean is defined [903], or an exponential or any other function decaying faster than m_0^{-2}. In Chap. 3, a solution has been proposed using a global constraint forcing the tail to bend down in the shape of an exponential.

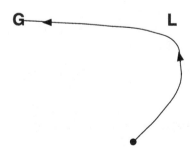

Fig. 4.6. Schematic view of the initial apparent convergence of the pdf of the sum to the Lévy fixed point L, leaving place for $N > N^*(\delta)$ to the asymptotic convergence to the Gaussian fixed point G

We now assume that the power law tail (4.11) holds up to a characteristic scale δ. What will be the consequence for the pdf of the sum of such N i.i.d. random variables? As long as all the variables are smaller than δ, the cut-off is not felt and the pdf of the sum progressively approaches the corresponding Lévy law, with the same exponent μ, as the number of terms N in the sum increases. However, as N increases further, the largest single value λ_N among the N variables increases according to $\lambda_N \sim (C_+ N)^{1/\mu}$ as found from (1.46). This maximum value λ_N is thus bound to reach δ as N increases. Equating λ_N with δ gives an estimate of the number of terms in the sum needed for the convergence to the Lévy law to be perturbed by the upper bound δ:

$$N^*(\delta) \simeq \frac{\delta^\mu}{C_+} . \tag{4.32}$$

Figure 4.6 provides a schematic view of what happens: first, the pdf of the sum starts to converge to the Lévy fixed point as N increases, but as N becomes larger than $N^*(\delta)$, the sum starts to "feel" the effect of the finite cut-off δ. In fact, due to the existence of this finite cut-off δ, the variance is finite and proportional to $\sim C_+ \delta^{2-\mu}$. Having a finite variance, the pdf of the sum converges eventually to the Gaussian stable law. The abstract trajectory taken by the pdf in the "space" of pdf's is shown in Fig. 4.6.

It is possible to be more precise by considering an explicit tractable form for the truncation, such as the so-called Gamma distribution, which in the tails is a power law multiplied by an exponential:

$$P(x) = C_\pm \frac{e^{-|x|/\delta}}{|x|^{1+\mu}} . \tag{4.33}$$

For $\delta \to +\infty$, the pure Lévy law is recovered. The characteristic function (Fourier transform) of the sum of N i.i.d. random variables whose pdf's are Lévy laws truncated by an exponential as in (4.33) can be calculated explicitely [527, 598]:

$$\ln \hat{P}(k) = -N a_\mu \left[\frac{(k^2 + \delta^{-2})^{\mu/2} \cos\left(\mu \tan^{-1}(\delta|k|)\right) - \delta^{-\mu}}{\cos(\pi\mu/2)} \right]$$
$$\left[1 + i\frac{k}{|k|} \beta \tan\left(\mu \tan^{-1}(\delta|k|)\right) \right] , \tag{4.34}$$

where a_μ is given by (4.17). We should warn that this explicit formula (4.34) is obtained by using an analytic continuation of (4.33) down to $|x| \to 0$ (at which $P(x)$ of the form (4.33) would be non-normalizable). What is important is that (4.34) recovers (4.18) for $\delta \to +\infty$. This expression (4.34) captures the cross-over from the Lévy form $\ln \hat{P}(k) \sim -|k|^\mu$ for $k \gg \delta^{-1}$ to the Gaussian form $\ln \hat{P}(k) \sim -k^2$ for $k \ll \delta^{-1}$.

When δ is large and k is not very small, $(k^2 + \delta^{-2})^{\mu/2} \approx k^\mu$ and $\hat{P}(k)$ has the form of a Lévy law. However, upon convolution (corresponding to increasing the number N of terms in the sum), the relevant values of k shrink down to $k = 0$ since larger and larger values of the sum are sampled which correspond to small wavenumbers. The cross-over from Lévy to Gaussian after N convolutions is obtained by setting

$$N[(k^2 + \delta^{-2})^{\mu/2} - \delta^{-\mu}] \simeq 1 , \tag{4.35}$$

to obtain the typical values of k representing the typical excursions of the sum. For $N \ll \delta^\mu$, one can safely neglect δ^{-2} compared to k^2 and obtain $k \simeq N^{-1/\mu}$. Translated in the sum domain, this means that the characteristic scales are of order $N^{1/\mu}$ for the sum of N terms, which characterize an ideal Lévy process. When on the contrary $N \gg \delta^\mu$, the relevant values of k become much smaller than δ^{-1}, and one finds

$$k \sim N^{-1/2}\delta^{\mu/2-1} , \tag{4.36}$$

and $\ln \hat{P}(k) \approx -(\mu/2)N\delta^{2-\mu}k^2$, which are the results corresponding to the usual central limit theorem for the convergence to a Gaussian. Hence, as expected, a truncated Lévy distribution is *not stable*: it "flows" to an ideal Lévy distribution for small N and then to a Gaussian distribution for large N, as schematically depicted in Fig. 4.6.

4.2 Power Laws

We follow Mandelbrot and contrast two broad classes of pdfs, the "mild" as opposed to the "wild" ones, which can be illustrated by the following questions.

- What is the probability that someone has twice your height? Essentially zero! The height, weight and many other variables are distributed with "mild" pdf's with a well-defined typical value and relatively small variations around it. The Gaussian law is the archetype of "mild" distributions.
- What is the probability that someone has twice your wealth? The answer of course depends somewhat on your wealth but in general, there is a nonvanishing fraction of the population twice, ten times or even one hundred times as wealthy as you are. This was noticed at the end of the last century by Pareto, after whom the Pareto law has been named, which describes the power law distribution of wealths [1043, 1046], a typical example of a "wild" distribution. The Lévy laws are also sometimes called Pareto–Lévy laws.

Wild pdf's have been found to quantify the size–frequency distribution of earthquakes (Gutenberg–Richter), hurricanes [686], volcanic eruptions, floods, meteorite sizes and so on. The distribution of seismic fault lengths is also documented to be a power law with exponent $\mu \simeq 1$. In the insurance business, recent studies has shown that the distribution of losses due to business interruption resulting from accidents [1044, 1045] is also a power law with $\mu \simeq 1$. These pdfs seem to characterize systems with very nonlinear dynamics or with stochastic multiplicative amplification effects. We have only a limited understanding of the physical origins of these behaviors. Several scenarii have been proposed, some of which are reviewed in Chaps. 14 and 15.

4.2.1 How Does One Tame "Wild" Distributions?

The behavior of the sum X of N random variables with a power law pdf with $\mu < 2$, for which the usual central limit theorem does not apply, can be qualitatively understood in simple terms. We follow [111] to recover in an intuitive way the scaling behavior of the sum and of the variance as a function of the number N of terms in the sum.

The main point to realize is that, even if the variance or the sum are mathematically not defined (infinite), one can make sense of their finite-size estimations to better understand and tame "wild" fluctuations occurring for power law pdfs with exponent $\mu < 2$. This comes about because, in a finite number N of terms, there is always a maximum value λ_N determined by the equation (1.37):

$$\lambda_N \sim (C_+ N)^{1/\mu} \; . \tag{4.37}$$

The probability to observe a value larger than λ_N in a typical series of N variables is small. This distribution of λ_N can be made precise using (1.38)–(1.42). The corresponding extreme value distribution is called the Fréchet distribution (1.59). The sum of N variables is thus insensitive to values larger than λ_N since this region is not sampled and this gives us the clue on how to control "wild" pdfs.

Using the result (1.56) of Sect. 1.8.1, we can express in closed form the expectation of the ratio of the sum $X_1 + ... + X_N$ divided by λ_N (also noted $X_{\max}$) in the case where the pdf of the i.i.d. variables is of the form $P(x) = \mu/x^{1+\mu}$ for $x \geq 1$. Then, $\mathcal{P}_<(x) = 1 - 1/x^\mu$ and expression (1.55) gives

$$T(y) = 1 - \frac{y^\mu}{1 - \mu} + \frac{\mu}{1 - \mu} \frac{1}{y} \; . \tag{4.38}$$

Using (1.56), we get

$$\mathrm{E}\left[\frac{X_1 + ... + X_N}{\max\{X_1, ..., X_N\}}\right] = \frac{1}{1 - \mu} \left[1 - NB(N, 1/\mu)\right] \; , \quad \mu \neq 1 \; , \tag{4.39}$$

$$= \sum_{k=1}^{n} \frac{1}{k} \; , \qquad \mu = 1 \; . \tag{4.40}$$

$B(a, b)$ is the beta function. This expression (4.40) provides the following asymptotic results (for $N \gg 1$):

$$E\left[\frac{X_1 + ... + X_N}{\max\{X_1, ..., X_N\}}\right] = \frac{1}{1 - \mu}, \quad \mu < 1, \tag{4.41}$$

$$= \frac{\Gamma(1/\mu)}{1 - \mu} N^{1-1/\mu}, \quad \mu > 1, \tag{4.42}$$

$$= 0.577... + \ln N, \quad \mu = 1. \tag{4.43}$$

In [737], V. Pisarenko extends these results to the case where N is itself a Poissonian random variable with parameter $\langle N \rangle$. Then, taking the additional expectation with respect to the Poisson variable N yields

$$E_N\left(E_X\left[\frac{X_1 + ... + X_N}{\max\{X_1, ..., X_N\}}\right]\right)$$
$$= \frac{1}{1 - \mu}\left(1 - \langle N \rangle^{1-1/\mu}\, \gamma(1/\mu, \langle N \rangle)\right), \tag{4.44}$$

for $\mu \neq 1$. $\gamma(a, x) = \int_0^x t^{a-1} e^{-t}\, dt$ is the incomplete Gamma-function.

$\mu \leq 1$. In this case, the divergence of the average of the sum X can be tamed by realizing that the integral in (1.17) must be truncated at λ_N:

$$X \sim N\langle x \rangle_N \sim N \int_{-\lambda_N}^{+\lambda_N} xP(x)\, dx. \tag{4.45}$$

This yields

$$X \sim N(N^{1/\mu})^{1-\mu} = N^{1/\mu} \quad \mu < 1, \tag{4.46}$$

$$= N \ln N \quad \mu = 1. \tag{4.47}$$

Expression (4.46) retrieves the characteristic scale $N^{1/\mu}$ quantifying the typical size of the sum of N random variables with $\mu < 1$. Note that, for $\mu < 1$, the sum X grows faster than linearly with N. This stems from the impact of increasingly larger extreme values that are sampled as the number N of terms increases.

Associated to this phenomenon, it is important to see that λ_N and X scale in the same way as a function of N and thus the expectation of the ratio λ_N/X

$$E\left[\frac{\lambda_N}{X}\right] \rightarrow 1 - \mu \tag{4.48}$$

goes to a constant for large N (see [293], p. 169 and (4.41)). The largest variable accounts for a finite fraction of the total sum! For instance, take $\mu = 2/3$ which is the exponent of the Gutenberg–Richter distribution for small and intermediate earthquakes. Then, the largest earthquake in any given sequence dissipates typically one-third of the total energy. When the next largest event occurs, it has a major impact on the sum, thus leading to the faster-than-linear growth (4.46). The results (4.41) and (4.48) provide the fundamental origin for the divergence of the average and contrasts from the behavior found for "mild" pdf's for which all terms in the sum contribute a similar amount. The variance of the ratio λ_N/X can be obtained as the second order derivative of the Laplace transform (called $\omega(\lambda)$ in Feller's exercise 20, Chap. XIII of [293]). The variance of the ratio equals

$$\frac{\mu(2-\mu)}{(1-\mu)^2} \ . \tag{4.49}$$

Thus, the variance of the ratio λ_N/X of the largest term over the sum of N random variables drawn from a heavy tailed pdf with exponent $\mu < 1$ does not tend to zero but converges to the finite constant (4.49). Typical realizations of λ_N/X oscillate around $1 - \mu$ indefinitely as N tends to infinity. Thus, one should be careful when speaking about an asymptotic relation between λ_N and X. This absence of convergence is called lack of self-averaging in Chap. 16 and is often encountered in the physics of quenched random media.

Another interesting example has been studied by Pisarenko [737] who finds that the cumulative sum $X(t)$ of losses caused by floods worldwide has been increasing with time approximately as $t^{1.3}$ since 1964. Here, losses are evaluated by the number of homeless caused by floods, since these data are the most systematically reported. This $t^{1.3}$ law is faster than the linear growth expected for a stationary process and signals an increasing loss rate.

At the same time, the rate of floods and the distribution of losses appear to be stationary over the period of observation. A possible explanation of this paradox comes from the fact that the distribution of losses is a power law with exponent $\mu \approx 0.75 \pm 0.1$ which is less than 1. Then from (4.46), we expect the sum $X(t)$ to behave as $t^{1/\mu} \sim t^{4/3}$, which is indeed very close to the observed time-dependence. As for the earthquakes, the cumulative loss over a time interval $[0,t]$ is determined in large part by a single major event λ_t occurring during this period. This is confirmed from the analysis of the database of losses caused by floods which indicate that the largest loss amounts approximately to 45% of the total cumulative loss over the 28 year period since 1964. This number is larger than the most probable amplitude $(1 - \mu)X(t)$ [293] of the largest event, equal in the present case to 25% of the total cumulative loss. The discrepancy is however well within the typical error bar equal to a factor of two found for the largest event in a set distributed with a power law pdf [903], as described in Chap. 6.

Another application is for earthquakes. Figures 1.2 and 3.5 illustrate the well-documented fact that the pdf of the seismic moments or equivalently the energies released by earthquakes is a power law with exponent $\mu \approx 2/3$. By the mechanism just discussed above for floods, the cumulative energy released by earthquakes in a given region should thus increase as $\sim t^{1/\mu} = t^{3/2}$, that is, much faster than linear. This is actually observed. A good example of the non-linear growth of the cumulative seismic energy can be found in [497], which cites a paper by the seismologist J. Brune (1968) who studied the earthquakes in the region of Imperial Valley in California in the time window from 1934 to 1963. For this catalog, the ratio (4.41) is found equal to 2.6, giving the estimation $\mu \approx 0.61$ in good agreement with the exponent of the pdf measured independently. One can thus conclude that, for this catalog, the Gutenberg–Richter pdf is a pure power law with no discernable truncation, implying that the recurrence time for the largest earthquakes in Imperial Valley is more than 29 years. Considering the same problem worldwide, a different picture emerges. The growth of the cumulative seismic energy on a global scale and on large time intervals, typically 1900–2000, is approximately linear in time. This can be rationalized by the bend of the Gutenberg–Richter law observed for the largest events, as discussed in Sect. 3.3.5. This provides an indirect confirmation that one cannot model the distribution of seismic moments only as a pure power law distribution with $\mu \approx 2/3$. Since the non-linearity of the cumulative sum depends heavily on the very end of the tail of the pdf, this provides a sensitive test of the existence of a truncation or bend in the Gutenberg–Richter distribution as demonstrated from a similar argument in Sect. 3.3.5. One can thus conclude that the recurrence time of the largest typical event in the world is less than 100 years. Thus, the two strongest earthquakes in the twentieth century (Chile, 22.05.1960, energy released $\simeq 1.1 \times 10^{26}$ erg), (Alaska, 28.03.1964, energy released $\simeq 4 \times 10^{25}$ erg) are already typical of the largest possible events compatible with the physics of earthquakes and the finite length of the plate boundaries on which they occur. This is notwithstanding the fact that these two earthquakes released more energy by themselves that the average yearly seismic energy by all other earthquakes, which oscillates around 3×10^{24} erg. For comparison, the largest nuclear explosion till present (1961, Novaya Zemlya, former USSR) had an estimated TNT equivalent of 58 Mt, corresponding to the energy 2.4×10^{24} erg. It should be noted that, for some small regions as in the above Imperial Valley, the corresponding recurrence time can be much larger than for the whole Earth.

In their statistical analysis of earthquake losses and casualties [786], Rodkin and Pisarenko have also observed a non-linear growth of cumulative casualties and losses, which cannot be explained neither by changes in the occurrence of disasters nor by an incomplete reporting of casualties for earlier periods. The non-linear growth in the cumulative numbers of casualties and losses can mainly be explained by the presence of a heavy tail in the respective distributions, as for the distribution of losses due to floods [737]. Non-linearities in the growth of earthquake-caused casualties and losses are observed over time intervals less than 20–30 years. For longer intervals, the size of the maximum disaster stops increasing owing to natural restrictions on the amount of the maximum possible losses. As a consequence, the total cumulative loss increases approximately linearly in time over time intervals of 40–50 years or longer, due to the same mechanism as for the cumulative energy released by earthquakes.

Another famous domain of application of power laws with exponent $\mu \leq 1$ is provided by the St. Petersburg paradox [811]: for a fee, a player is given the right to play a coin game in which heads in the first throw gives him \$2 while tails makes him loose the fee. Having won in the first throw, he can continue to play by putting his gain at risk and will win a total of \$4 for heads while tails make him loose everything. The game can continue indefinitely at the player's will such that, conditioned on having thrown n heads in a row, the next throw gives him a total of \$$2^n$ for heads while tails make him loose everything. What should be the fee requested by the casino and the strategy of the player? A good starting point is to calculate the expected gain of the player:

$$\frac{1}{2} \times 2 + \frac{1}{4} \times 4 + ... + \frac{1}{2^n} \times 2^n + ... = 1 + 1 + ... = \infty . \tag{4.50}$$

The reason for this divergence is that the distribution of gains is a power law with exponent $\mu = 1$. Indeed, the probability to gain a sum larger than or equal to $g = \$2^n$ is $2/2^n = 2/g$. Therefore, a fair game condition would require in principle that the casino should charge an infinite fee to cover its potential loses. This makes the game unplayable. In practice, the casino can make the game playable by limiting the maximum number of throws in a given game to some finite value $n_{\max}$. This limits the largest potential loss of the casino to $2^{n_{\max}}$ and the mean becomes finite and equal to $n_{\max}$. The results discussed above on the slow convergence of a truncated power law to a Gaussian law teach us that the casino would need a number of customers much larger than $n_{\max}$ in order for the large fluctuations from one player to the next to cross-over from the power law regime to the Gaussian regime by aggregation over many games. In other words, from the perspective of the casino, this game is potentially profitable only if the expected number of players is much larger than $n_{\max}$.

$1 < \mu \leq 2$. We follow the same strategy as for the mean in the previous case $\mu \leq 1$ and obtain the typical value of the variance $(X - \langle X \rangle)^2$ as

$$(X - \langle X \rangle)^2 \sim N \langle (x - \langle x \rangle)^2 \rangle_N \sim N \int_{-\lambda_N}^{+\lambda_N} (x - \langle x \rangle)^2 P(x) \, dx . \tag{4.51}$$

This yields

$$(X - \langle X \rangle)^2 \sim N(N^{1/\mu})^{2-\mu} = N^{2/\mu} \qquad \mu < 2, \tag{4.52}$$
$$= N \ln N \qquad \mu = 2 . \tag{4.53}$$

This recovers the result that, for $1 < \mu < 2$, the characteristic scale of the fluctuation around the mean is of order $N^{1/\mu}$, i.e. larger than the Gaussian case $N^{1/2}$. The faster than $\sqrt{N}$ rate of growth of the standard deviation is the hallmark of the breakdown of the usual central limit theorem and explains the non-convergence to the Gaussian law. Figure 4.7 presents a "Lévy walk" compared to a standard random walk to illustrate the fast growth of typical excursions in this regime.

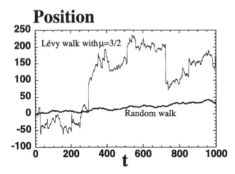

Fig. 4.7. Comparison between a Lévy walk and a random walk. The Lévy walk is constructed by taking the running sum up to time t of random steps with random signs and amplitudes distributed according to $\mu/x^{1+\mu}$ for $x > 1$ with $\mu = 1.5$. The typical size 100–200 of the excursion after 1000 time steps is in good agreement with the prediction (4.52) giving $1000^{1/\mu} = 100$. In comparison, the random walk is constructed by taking the running sum up to time t of random steps uniformly distributed in $[-1, +1]$. The typical excursion is in good agreement with the prediction of the standard deviation of the random walk after 1000 steps equal to $\sqrt{1000/3} \approx 18$

$2 < \mu.$ The mean and variance converge when $\lambda_N \to \infty$, and one recovers a purely linear dependence on N for the mean and a $\sqrt{N}$ dependence of the standard deviation, which are characteristic of the convergence to the Gaussian law for the sum of N variables.

Moments of order higher than or equal to μ will however diverge mathematically and their divergence can be tamed exactly in the same fashion by truncating the integrals at the largest value λ_N sampled among the N terms.

4.2.2 Multifractal Approach

As shown graphically in Fig. 4.7 and as seen from the scaling properties (4.46, 4.52), the sum of N random variables with a power law pdf has a *self-similar* structure. This remark can be used quantitatively to recover the anomalous (with respect to the usual random walk) scaling behavior (4.46, 4.52) for $\mu < 2$ from a multifractal view point. This view allows us to introduce in a simple way the concept of multifractality [316] (see also Chap. 5) and to link it to power law distributions.

We start from the remark that the variables constituting the sum are of very different amplitude. The idea is then to class them into groups of similar amplitude, for instance class 1 contains the variables between x_1 and x_2, class 2 contains the variables between x_2 and x_3, etc. What is the best choice of $x_1, x_2, x_3, \ldots$ for the classification and what is the corresponding cardinal of each class in this optimal classification?

A clue comes from the remark that sums of power law variables are mainly controlled by the largest variables, as we have seen in the derivation of (4.46,

4.52). It is thus natural to start the classification from the larger values, which are of order $N^{1/\mu}$. The cardinal of the set of variables among N of amplitude $N^{1/\mu}$ is very low, typically only a few variables are of this magnitude. On the other hand, there are many variable of order $N^0 = 1$.

This suggests the following classification in sets which contain variables of typical amplitude of the order of N^α with α continuously spanning the interval from $1/\mu$ to 0. Each family is thus defined by its exponent α and is the set of all the variables of size of the order of N^α. The family with $\alpha = 1/\mu$ corresponds to the few largest events in the sum and its cardinal is of order 1 as we already pointed out. At the other extreme, the family $\alpha = 0$ contains the crowd of small variables whose cardinal is proportional to but less than N.

From the scale invariance explicit in the power law distribution, we can guess that the cardinal of a family α is of order $N^{f(\alpha)}$, i.e. also a power of N with an exponent which is a function of α. We already know that $f(\alpha = 1/\mu) = 0$ and $f(\alpha = 0) = 1$. To determine $f(\alpha)$ for intermediate values, we ask how many variables have a size of order N^α. By definition of the pdf, the answer is

$$N^{f(\alpha)} \simeq N P(x \sim N^\alpha) N^\alpha \ , \tag{4.54}$$

where the last factor N^α accounts for the width Δx of the interval of the variables of order N^α in the definition of the probability $P(x)\,\mathrm{d}x$ (the width is simply proportional to the size, in order to conserve the scaling). We thus get

$$f(\alpha) = 1 - \alpha\mu \ . \tag{4.55}$$

$f(\alpha)$ is often called the multifractal spectrum of exponents α. It provides a natural classification of the self-similar structure of the variables in the sum in terms of a hierarchical set of families of variables of the same order. A power law pdf thus produces naturally a multifractal spectrum. Notice however its linear dependence, which is a special case of the more general convex form discussed in Chap. 5.

This classification allows us to avoid completely the complication brought by the power law pdf: in each family α, the variables are comparable and in each family the central limit theorem can be applied! For instance, the mean is obtained by summing the contributions over all families, such that each contribute by their cardinal $N^{f(\alpha)}$ times the typical size of their variables N^α:

$$X \sim \sum_{\alpha=0}^{1/\mu} N^\alpha N^{f(\alpha)} = \sum_{\alpha=0}^{1/\mu} N^{1+\alpha(1-\mu)}. \tag{4.56}$$

- For $\mu < 1$, the sum is controlled by the largest value $\alpha = 1/\mu$ and goes as $N^{1/\mu}$.

- For $\mu > 1$, the sum is controlled by $\alpha = 0$ and we recover the linear dependence with N. In this case, the speed of the convergence is obtained by applying the central limit theorem to each family individually: the error in the sum of $N^{f(\alpha)}$ random variables is of order $[N^{f(\alpha)}]^{1/2}$. If each term in the sum is of size $\sim N^\alpha$, the error on the mean due to these $N^{f(\alpha)}$ terms is

$$\frac{N^\alpha N^{f(\alpha)/2}}{N} \sim N^{-1/2+(\alpha/2)(2-\mu)} . \tag{4.57}$$

1. For $1 < \mu \leq 2$, the error on the mean (X/N) is dominated by the effect of the largest exponent $\alpha = 1/\mu$, corresponding to the largest events. This leads to a slow convergence of the mean with fluctuations decaying as $N^{-(\mu-1)/\mu}$.
2. For $\mu > 2$, the error in the mean is dominated by the family of the weakest variables with $\alpha = 0$ and we retrieve the standard rate of convergence $N^{-1/2}$ as $N \to \infty$ given by the central limit theorem.

A similar reasoning holds for the variance which can be expressed similarly to (4.56) under the form

$$V_N \sim \sum_{\alpha=0}^{1/\mu} N^{2\alpha} N^{f(\alpha)} = \sum_{\alpha=0}^{1/\mu} N^{1+\alpha(2-\mu)}. \tag{4.58}$$

- For $\mu < 2$, V_N/N is controlled by $\alpha = 1/\mu$ and the reduced variance diverges with N as $N^{2/\mu-1}$.
- For $\mu > 2$, V_N/N is controlled by $\alpha = 0$ and we recover the standard convergence to a finite value. In order to study the convergence rate of the variance in this regime, we can again apply the central limit theorem within each family α: each family provides a contribution to the error on the variance of order

$$\frac{N^{2\alpha} N^{f(\alpha)/2}}{N} \sim N^{-1/2+(\alpha/4)(4-\mu)} . \tag{4.59}$$

We thus obtain

1. for $2 < \mu \leq 4$, the error on the variance is controlled by the largest exponent $\alpha = 1/\mu$, corresponding to the largest fluctuations and leading to a slow convergence of the variance as $N^{2/\mu-1}$. This result shows that, even for $\mu > 2$, the variance can be difficult to estimate in practice.
2. for $\mu > 4$, we recover the usual convergence of the variance as $N^{-1/2}$.

4.3 Anomalous Diffusion of Contaminants in the Earth's Crust and the Atmosphere

Chemical transport in geological formations of the earth's subsurface and in the earth's atmosphere is often observed to be anomalous, i.e. non-Gaussian.

For example, in a large-scale field study carried out in a heterogeneous alluvial aquifer at the Columbus Air Force Base (Mississipi), bromide was injected as a pulse and traced over a 20 month period by sampling from an extensive 3D well network (see [81, 82] for a quantitative analysis along the lines presented here). The tracer plume was found to be remarkably asymmetric with a sharp peak close to the injection point and a long tail, which cannot be explained by classical Gaussian diffusion models.

The understanding and quantification of flow and contaminant transport in fractured and heterogeneous structures is of considerable practical importance for the exploitation and preservation of aquifers, and for the assessment of potential underground repository sites for the storage of radioactive and industrial wastes. We now show that the anomalous diffusion of elements in geological formations stems from the distribution of trapping times which is generically a power law distribution [81, 82], with exponent μ less than one. The resulting extremely broad range of time scales explains the anomalous behavior. It is bad news because there is a significant amount of tracers that can remain trapped for extraordinary long time scales. Thus, a pollutant may reappear at the surface tens of years or centuries later than would have been expected from the available Gaussian models.

4.3.1 General Intuitive Derivation

Suppose the distribution of trapping times is

$$P(\tau) \sim \frac{C}{\tau^{1+\mu}} \, , \qquad \text{for large } \tau \text{ with } \mu < 1 \, . \tag{4.60}$$

If N different trapping sites are encountered over the time t along the diffusion path, the typical maximum trapping time is $\tau_{\max}(N) \sim N^{1/\mu}$. As a consequence, the average trapping time is $\langle \tau \rangle_N \sim N^{1/\mu - 1}$. The interval of time t is the sum (neglecting the time needed to jump from one trap to the next)

$$t = \tau_1 + \dots + \tau_N \sim N \langle \tau \rangle_N \sim N^{1/\mu} \, , \tag{4.61}$$

as shown in (4.45). Thus, the number N of traps encountered is proportional to t^μ. This retrieves the scaling behavior of the typical mean position and standard deviation of the diffusing plume found in [81, 82] using the formalism of continuous time random walks.

4.3.2 More Detailed Model of Tracer Diffusion in the Crust

Consider a source of contaminants at some spatial origin. To make the discussion simple, we model the space from the source to a typical measurement point placed at a distance L away by a regular cubic network of tubes with characteristic length a. At each node of this network, there is a crack, joint or fault, in which the contaminant may be trapped such as in a dead end.

It is well-documented that the distribution of fault lengths in the crust is approximately a power law [892]

$$P(l) \sim l^{-(1+\mu_l)} \, dl , \quad \text{with } \mu_l \approx 1 . \tag{4.62}$$

The typical time of residence of a diffusing chemical in a fault of length l with diffusion coefficient D is

$$\tau_l \sim \frac{l^2}{D} . \tag{4.63}$$

As a consequence of the diffusion in a the dead end of length l, a contaminant may be trapped for a typical time τ_l before being released to diffuse to the next site. From the distribution of fault lengths, we get the distribution of trapping times as

$$P(\tau) \, d\tau = P(l) \, dl \sim \tau^{-(1+\mu_\tau)} \, d\tau , \quad \text{with } \mu_\tau = \frac{\mu_l}{2} \approx \frac{1}{2} . \tag{4.64}$$

The time needed to diffuse from the source to the detection point a distance L away involves two contributions:

1. the diffusion time

$$t_d \sim \frac{L^2}{D} \tag{4.65}$$

to cover the distance in a random walk fashion;
2. the trapping time t_t which is the sum of all τ_l's spent in the dead ends at each visited site.

In order to estimate t_t, we need to determine the number of sites which have been visited by this random walk. It is of order $\mathcal{N} \sim L^2/a$, from the definition of the fractal dimension of the random walk in Sect. 2.1.4. The total time spent in the traps can therefore be written as

$$t_t = \tau_1 + \dots + \tau_{\mathcal{N}} , \tag{4.66}$$

where the trapping times τ_i are distributed according to a power law with exponent $\mu_\tau \approx 1/2 < 1$. We are thus in a regime described in Sect. 4.2.1 where the average is mathematically infinite, therefore, $t_t \gg \mathcal{N} \sim L^2$ and thus $t_t \gg t_d$, implying that the trapping process is the dominating factor which controls the time to diffusion to the receptor. In order to estimate t_t, we use the results above and find

$$t_t \sim \mathcal{N}^{1/\mu_\tau} \sim L^{2/\mu_\tau} \approx L^4 . \tag{4.67}$$

This result corresponds to an extremely slow diffusion (compare with the usual diffusion law (4.65)). The complete space–time distribution of the contaminants can also be obtained [111]:

$$P(x,t) \sim_{t \to \infty} a^{-d} \left(\frac{\tau_0}{t} \right)^{\nu d} f \left(\frac{\tau_0^\nu}{a} \frac{|x|}{t^\nu} \right) , \tag{4.68}$$

where $\nu = \mu_\tau/2 \approx 1/4$ and τ_0 is a microscopic time scale. The function $f(u)$ scales as

$$f(u) \sim_{u \gg 1} e^{-u^{1/(1-\nu)}} = e^{-u^{4/3}} \, . \tag{4.69}$$

These results imply long anomalous tails: after the first significant signal occurring a time $t_d \sim L^2$ after the injection of a tracer at the origin, the contaminants will continue to be detected at a distance L from the source over an exceedingly long time. The large deviations explored in the tail of the power law pdf of the trapping times are reflected in this anomalous diffusion behavior. Practically, the conclusion is not encouraging: a storage deposit may pollute the environment over time scales that are much larger than naively extrapolated from standard diffusion models.

4.3.3 Anomalous Diffusion in a Fluid

Anomalous diffusion processes can occur even in absence of a frozen distribution of traps [382]. Some interesting examples are treated in [111, 449, 802].

Let us briefly discuss the situation where a fluid is partitionned into convective cells (think of a steady state of the atmosphere). Close to the ground, the velocity of the wind has to vanish ("stick" boundary condition). For weak wind conditions, we neglect the influence of turbulence and the velocity vanishes as d^β, where d is the distance to the ground. The exponent $\beta = 1$ for laminar flow with "stick" boundary condition. Other values for β can be found to represent more complicated situations.

Consider a particle transported by the convective wind. Most of the time, it is convected and thus remains trapped in a convective cell for some time. But, when it approaches the boundary between two cells, it can cross this boundary diffusively and then be carried away to the other cell. The motion of such a particle is thus characterized by two types of motions and two different time scales:

- the fast convective motion within a cell and
- the random walk-like behavior characterizing the cell crossing, leading to diffusion behavior at large scales.

This picture must in fact be corrected due to the presence of large fluctuations in the residence time in a given cell. This occurs due to the fact that the particle can approach, due to diffusion, arbitrarily close to the ground and thus take a very large time to be carried away by the flow, since the velocity of the carrying flow vanishes close to the ground. The calculation can be done explicitly (see [745]) to obtain the distribution of waiting times in a given cell due to this effect. It is characterized by a power law with exponent

$$\mu = \frac{1+\beta}{2+\beta} = \frac{2}{3} \quad \text{for } \beta = 1 \, . \tag{4.70}$$

The previous derivation for diffusion in the crust in the presence of a large distribution of trapping sites can be applied to the present problem, by replacing $\mu_\tau = 1/2$ by $2/3$. As a consequence, a particle will exhibit an anamalous diffusion accross the set of convective cells with a very large distribution of residence times. For further readings on this problem, consult [525, 526].

Finally, let us mention that Ferrari et al. [298] have called attention to the fact that the tails of the distribution of particles in many dispersive processes may not be described by the same functional law that applies to the central part of the distribution. It is indeed in general true that, at long times, the core of the concentration relaxes to a self-similar profile, while the tails, consisting of particles which have experienced exceptional displacements, may not be self-similar. This absence of self-similarity may lead to complicated cross-over in the scaling of the moments of the concentration field. Ferrari et al. [298] have illustrated these ideas on a stochastic model, that they call the generalized telegraph model, and on a deterministic area-preserving map, the kicked Harper map. For these models, moments of arbitrary orders can be obtained from the Laplace–Fourier representation of the concentration.

4.4 Intuitive Calculation Tools for Power Law Distributions

We conclude this chapter by providing a set of simple derivations to manipulate sums and products of variables distributed according to a power law. We follow [115] and call μ-variable a variable with power law pdf with exponent μ:

$$P(w) \simeq \frac{C_\pm}{|w|^{1+\mu_\pm}} \qquad w \longrightarrow \pm\infty . \tag{4.71}$$

Let us study the positive part of the distribution, that we write

$$P(w) = \frac{C}{w^{1+\mu}} \qquad \text{for } 1 \le w < +\infty . \tag{4.72}$$

A useful tool is the Laplace transform. The Laplace transform plays the same role for functions bounded on one side as the Fourier transform for functions unbounded on both sides. The computation rules of the Fourier transform thus extend straightforwardly to the Laplace transform. The Laplace transform

$$\hat{P}(\beta) \equiv \int_0^\infty dw P(w) e^{-\beta w} \tag{4.73}$$

of (4.72) is

$$\hat{P}(\beta) = C \int_1^\infty dw \frac{e^{-\beta w}}{w^{1+\mu}} = \mu \beta^\mu \int_\beta^\infty dx \frac{e^{-x}}{x^{1+\mu}} . \tag{4.74}$$

Denote l the integer part of μ ($l < \mu < l+1$). Integrating by parts l times, we get (for $C = \mu$)

$$\hat{P}(\beta) = \mathrm{e}^{-\beta}\left(1 - \frac{\beta}{\mu - 1} + \dots + \frac{(-1)^l \beta^l}{(\mu - 1)(\mu - 2)\dots(\mu - l)}\right)$$
$$+ \frac{(-1)^l \beta^\mu}{(\mu - 1)(\mu - 2)\dots(\mu - l)} \int_\beta^\infty \mathrm{d}x\, \mathrm{e}^{-x} x^{l-\mu} \, . \tag{4.75}$$

This last integral is equal to

$$\beta^\mu \int_\beta^\infty \mathrm{d}x\, \mathrm{e}^{-x} x^{l-\mu} = \Gamma(l+1-\mu)[\beta^\mu + \beta^{l+1}\gamma^*(l+1-\mu, \beta)] \, , \tag{4.76}$$

where Γ is the Gamma function ($\Gamma(n+1) = n!$) and

$$\gamma^*(l+1-\mu, \beta) = \mathrm{e}^{-\beta} \sum_{n=0}^{+\infty} \frac{\beta^n}{\Gamma(l+2-\mu+n)} \tag{4.77}$$

is the incomplete Gamma function [1]. We see that $\hat{P}(\beta)$ presents a regular Taylor expansion in powers of β up to the order l, followed by a term of the form β^μ. We can thus write

$$\hat{P}(\beta) = 1 + r_1\beta + \dots + r_l\beta^l + r_\mu\beta^\mu + \mathcal{O}(\beta^{l+1}), \tag{4.78}$$

with $r_1 = -\langle w \rangle$, $r_2 = \langle w^2 \rangle/2, \dots$ and, reintroduce C, where r_μ is proportional to the scale parameter C. For small β, we rewrite $\hat{P}(\beta)$ under the form

$$\hat{P}(\beta) = \exp\left[\sum_{k=1}^{l} d_k\beta^k + d_\mu\beta^\mu\right] \, , \tag{4.79}$$

where the coefficient d_k can be simply expressed in terms of the r_k's in a way similar to the transformation from the moments to the cumulants. The expression (4.79) generalizes the canonical form (4.18) of the characteristic function of the stable Lévy laws, for arbitrary values of μ, and not solely for $\mu \leq 2$ over which Lévy laws are defined. The canonical form is recovered for $\mu \leq 2$ for which the coefficient d_2 is not defined (the variance does not exist) and the only analytical term is $\langle w \rangle \beta$ (for $\mu > 1$).

This expression (4.79) allows us to obtain interesting results on combinations of power law variables in an economical way:

1. if w_i and w_j are two independent μ-variables characterized by the scale factors $C_i^\pm$ and $C_j^\pm$, then $w_i + w_j$ is also a μ-variable with $C^\pm$ given by $C_i^\pm + C_j^\pm$.
2. If w is a μ-variable with scale factor C, then $p \times w$ (where p is a real number) is a μ-variable with scale factor $p^\mu C$.
3. If w is a μ-variable, then w^q is a (μ/q)-variable.
4. If w_i and w_j are two independent μ-variables, then the product $x = w_i w_j$ is also a μ-variable up to logarithmic corrections.

Consider the sum of two μ-variables. The pdf of the sum is the convolution of the two pdfs and the Laplace transform of the sum is the product of the two Laplace transforms. Therefore, the term proportional to β^μ is additive and the scale coefficient for the tail is the sum of the two scale coefficients C (property (1)).

Consider the variable $w_p \equiv p \times w$ where p is some real number. Writing $P(w_p)\,dw_p = P(w)\,dw$ (invariance of the probability, a pure number, with respect to a change of representation, i.e. *variable*), we get for large w_p

$$P(w_p) \simeq \frac{Cp^\mu}{|w_p|^{1+\mu}} \text{ for } w_p \longrightarrow \infty , \tag{4.80}$$

yielding the property (2). The same argument applied to w^q yields

$$P(w_q) \simeq \frac{C}{q|w_q|^{1+\mu/q}} \text{ for } w_p \longrightarrow \infty , \tag{4.81}$$

corresponding to the property (3).

Intuitively, the property (4) expresses the fact that $x = w_i w_j$ is large when either one of the factors is large and the other one takes a typical (small) value in the central region of its pdf. This is the opposite to the "democratic" rule found to hold in Sect. 3.4.1 for distributions decaying faster than an exponential. The contributions where the two variables are simultaneously large are negligible in probability. The pdf of $x = w_i w_j$ is

$$P(x) \equiv \int^\infty \int^\infty dw_i\,dw_j\,P_i(w_i)P_j(w_j)\delta(x - w_i w_j)$$
$$= \int^\infty \frac{dw_j}{w_j} P_j(w_j) P_i \left(\frac{x}{w_j} \right) . \tag{4.82}$$

For large x, we find

$$P(x) \propto \frac{C_i}{x^{1+\mu}} \int^x dw_j\,w_j^\mu P_j(w_j) \simeq \frac{C_i C_j \ln x}{x^{1+\mu}} \tag{4.83}$$

which is the announced result (4).

4.5 Fox Function, Mittag–Leffler Function and Lévy Distributions

This exposition is a synthesis of [57, 607, 619] and borrows significantly from [619].

The Fox function is also referred to as the Fox's H-function, the H-function, the generalised Mellin–Barnes function, or the generalised Meijer's G-function. The importance of the Fox function lies in the fact that it includes nearly all special functions occurring in applied mathematics and statistics, as its special cases. Even sophisticated functions like Wright's generalised

Bessel functions, Meijer's G-function or Maitland's generalized hypergeometric function are embraced by the class of Fox functions. The Fox function was introduced in physics in the 1980s [84, 823, 824, 1030] as analytic representations for Lévy distributions in direct-space, and as solutions of fractional equations. Bernasconi et al. [84] introduced them in the study of conductivity in disordered systems. Schneider [823] demonstrated that Lévy stable densities can be expressed analytically in terms of Fox functions. Wyss [1030] and Schneider and Wyss [824] uses Fox functions for the solution of the fractional diffusion equation.

In 1961, Fox defined the H-function in his studies of symmetrical Fourier kernels as the Mellin–Barnes type path integral [307, 607, 920, 921]:

$$
\begin{aligned}
H_{m,n}^{p,q}(z) &= H_{m,n}^{p,q}\left[z \,\middle|\, \begin{matrix} (a_p, A_p) \\ (b_q, B_q) \end{matrix}\right] \\
&= H_{m,n}^{p,q}\left[z \,\middle|\, \begin{matrix} (a_1, A_1), (a_2, A_2), ..., (a_p, A_p) \\ (b_1, B_1), (b_2, B_2), ..., (b_q, B_q) \end{matrix}\right] \\
&= \frac{1}{2\pi} \int_L \mathrm{d}s\, \chi(s) z^s ,
\end{aligned}
\tag{4.84}
$$

with

$$
\chi(s) = \frac{\prod_1^m \Gamma(b_j - B_j s) \prod_1^n \Gamma(1 - a_j + A_j s)}{\prod_{m+1}^q \Gamma(1 - b_j + B_j s) \prod_{n+1}^p \Gamma(a_j - A_j s)} .
\tag{4.85}
$$

The last equality of (4.84) shows that $\chi(s)$ is nothing but the Mellin transform of the Fox function $H_{m,n}^{p,q}(z)$ since the path integral in this last equality represents just the inverse Mellin transform of $\chi(s)$. According to [921], the Fox functions have been known since at least 1868, and Fox rediscovered them in his studies.

Due to the structure of the defining integral kernel $\chi(s)$ from (4.84), the Fox functions fulfil several convenient properties.

$$
H_{m,n}^{p,q}\left[z \,\middle|\, \begin{matrix} (a_p, A_p) \\ (b_q, B_q) \end{matrix}\right] = k H_{m,n}^{p,q}\left[z \,\middle|\, \begin{matrix} (a_p, k A_p) \\ (b_q, k B_q) \end{matrix}\right] .
\tag{4.86}
$$

$$
x^\sigma H_{m,n}^{p,q}\left[z \,\middle|\, \begin{matrix} (a_p, A_p) \\ (b_q, B_q) \end{matrix}\right] = H_{m,n}^{p,q}\left[z \,\middle|\, \begin{matrix} (a_p + \sigma A_p, A_p) \\ (b_q + \sigma B_q, B_q) \end{matrix}\right] .
\tag{4.87}
$$

The fractional differential and integral of the Fox function is a map into the Fox function class:

$$
{}_0 D_z^\nu \left(z^\alpha H_{m,n}^{p,q}\left[(az)^\beta \,\middle|\, \begin{matrix} (a_p, A_p) \\ (b_q, B_q) \end{matrix}\right] \right)
$$

$$
= z^{\alpha - \nu} H_{p+1,q+1}^{m,n+1}\left[(az)^\beta \,\middle|\, \begin{matrix} (-\alpha, \beta), (a_p, A_p) \\ (b_q, B_q), (\nu - \alpha, \beta) \end{matrix}\right] .
$$

${}_0 D_z^\nu$ is the Riemann–Liouville differential operator of fractional order ν, which is defined in Sect. 8.4.3 with (8.49).

An H-function can be expressed as a computable series in the form [607, 920]:

$$H_{m,n}^{p,q}(z) = \sum_{h=1}^{m} \sum_{\nu=0}^{+\infty} \frac{\prod_{j=1,j\neq h}^{m} \Gamma(b_j - B_j(b_h + \nu)/B_h)}{\prod_{j=m+1}^{q} \Gamma(1 - b_j + B_j(b_h + \nu)/B_h)}$$
$$\times \frac{\prod_{j=1}^{n} \Gamma(1 - a_j + A_j(b_h + \nu)/B_h)}{\prod_{j=n+1}^{p} \Gamma(a_j - A_j(b_h + \nu)/B_h)} \times \frac{(-1)^{\nu} \, z^{(b_h+\nu)/B_h}}{\nu! B_h} , \qquad (4.88)$$

which is an alternating series with slow convergence. For large argument $|z| \to \infty$, Fox functions can be expanded as a series over the residues

$$H_{m,n}^{p,q}(z) \sim \sum_{\nu=0}^{\infty} \mathrm{res}(\chi(s) z^s) \qquad (4.89)$$

to be taken at the poles $s = (a_j - 1 - \nu)/A_j$, for $j = 1, ..., n$.

The Mittag–Leffler function [276, 632, 633] is the natural generalisation of the exponential function and is a special Fox function, often encountered in anomalous diffusion problems. It is defined for $\alpha > 0$ by

$$E_\alpha(z) = \sum_{k=0}^{+\infty} \frac{z^k}{\Gamma(\alpha k + 1)} . \qquad (4.90)$$

Note that $E_1(z) = \mathrm{e}^z$. Other specials cases are $E_0(z) = 1/(1 - z)$, $E_2(z) = \cosh(\sqrt{z})$, $E_{1/2}(z) = \exp(z^2) \, [1 + \mathrm{erf}(z)]$, where $\mathrm{erf}(z)$ is the error function. The general Mittag–Leffler function is defined as

$$E_{\alpha,\beta}(z) = \sum_{k=0}^{+\infty} \frac{z^k}{\Gamma(\alpha k + \beta)} , \qquad (4.91)$$

for $\alpha > 0, \beta > 0$. Note that $E_{\alpha,\beta=1}(z) = E_\alpha(z)$. $E_{\alpha,\beta}(z)$ satisfies

$$\int_0^{+\infty} \mathrm{d}t \, \mathrm{e}^{-t} t^{\beta-1} E_{\alpha,\beta}(t^\alpha z) = \frac{1}{z-1} , \qquad (4.92)$$

and

$$\int_0^{+\infty} \mathrm{d}t \, \mathrm{e}^{pt} t^{\beta-1} E_{\alpha,\beta}(at^\alpha) = \frac{1}{p^\beta} \frac{1}{1 - ap^{-\alpha}} , \qquad (4.93)$$

with the conditions that the real part of p should be larger than $|a|^{1/\alpha}$ and the real part of β should be positive. This last relationship gives the Laplace transform of the general Mittag–Leffler function $E_{\alpha,\beta}(z)$. Specifically, in terms of the Laplace transform, the Mittag–Leffler function $E_\alpha(z)$ is defined by

$$E_\alpha(-(t/\tau)^\alpha) = \mathcal{L}^{-1}\left[\frac{1}{p + \tau^{-\alpha} u^{1-\alpha}}\right] , \qquad (4.94)$$

where $\mathcal{L}^{-1}$ denotes the inverse Laplace transform.

For $0 < \alpha < 1$, $t^{\alpha-1}E_\alpha((t/\tau)^\alpha)$ interpolates between $1/t^{1-\alpha}$ for $t < \tau$ to the exponential $\sim \exp(t/\tau)$ for $t > \tau$. The expression $t^{\alpha-1}E_\alpha(-(t/\tau)^\alpha)$ interpolates between $1/t^{1-\alpha}$ for $t < \tau$ to $1/t^{1+\alpha}$ for $t > \tau$ [411]. Such a crossover describes the behavior of aftershocks of earthquakes according to a so-called modified Omori's law $1/t^{1-\theta}$ for $t < t^*$ to $1/t^{1+\theta}$ for $t > t^*$ found in [411, 864], where the cross-over time t^* is controlled by the distance of the model to a critical branching point. The occurrence of the Mittag–Leffler function in this context reflects the anomalous diffusion associated with cascades of earthquake triggering [412]. Such a behavior was also studied by Scher and Montroll [817] in a continuous-time random walk (CTRW) with absorbing boundary condition to model photoconductivity in amorphous semi-conductors As_2Se_3 and an organic compound TNF-PVK finding $\theta \approx 0.5$ and $\theta = 0.8$ respectively. In a semiconductor experiment, electric holes are injected near a positive electrode and then transported to a negative electrode where they are absorbed. The transient current follows exactly the transition $1/t^{1-\theta}$ for $t < t^*$ to $1/t^{1+\theta}$ for $t > t^*$ mentioned above for the Omori law for earthquake aftershocks. In the semiconductor context, the finiteness of t^* results from the existence of a force applied to the holes. When the force goes to zero, $t^* \to +\infty$. A similar transition has been recently proposed to model long-term time series measurements of chloride, a natural passive tracer, in rainfall and runoff in catchments [816]. The quantity analogous to the function describing the decay of seismic aftershocks is the effective travel time distribution $h(t)$ which governs the global lag time between injection of the tracer through rainfall and outflow to the stream. $h(t)$ has been shown to have a power-law form $h(t) \sim 1/t^{1-m}$ with m between -0.3 and 0.2 for different time series [507]. This variability may be due to the transition between an exponent $1 - \theta$ at short times to $1 + \theta$ at long times [816], where θ is the exponent of the distribution of individual transition times. Let us also mention a deep connection between the seismic relaxation in aftershock sequences due to cascades of earthquake triggering and continuous-time random walks [412].

The general Mittag–Leffler function $E_{\alpha,\beta}(z)$ can be expressed as follows in terms of the Fox function:

$$E_{\alpha,\beta}(z) = H_{1,2}^{1,1}\left[z \left| \begin{matrix} (0,1) \\ (0,1), (1-\beta, \alpha) \end{matrix} \right. \right] . \tag{4.95}$$

Finally, the stable Lévy law defined through its characteristic function (4.18) is given explicitly in terms of a Fox function for $\mu > 1$ as [823]

$$L_\mu^\beta(x) = \frac{1}{\mu} x^2 H_{2,2}^{1,1}\left[x \left| \begin{matrix} (1 - 1/\mu, 1/\mu), (1 - (\mu - \beta)/2, (\mu - \beta)/2) \\ (0,1), (1 - (\mu - \beta)/2, (\mu - \beta)/2)) \end{matrix} \right. \right] . \tag{4.96}$$

This expression (4.96) allows one to use the various known power expansions of the Fox function for the Lévy law [607].

5. Fractals and Multifractals

5.1 Fractals

Clouds are not spheres, mountains are not cones, coastlines are not circles, and bark is not smooth, nor does lightning travel in a straight line.
B.B. Mandelbrot [592]

5.1.1 Introduction

During the third century BC, Euclid and his students introduced the concept of space dimension which can take positive integer values equal to the number of independent directions. A smooth line has dimension 1, a smooth surface has dimension 2 and our space (seen at large scales) has dimension 3.

In the second half of the nineteen century, several mathematicians started to study the generalization of dimensions to fractional values. The concept of non-integer dimensions was fully developed in the first half of the twentieth century. The book [263] compiles some of the most important mathematical works. To capture this novel concept, the word "fractal" was coined by Mandelbrot [592], from the Latin root fractus to capture the rough, broken and irregular characteristics of the objects he intended to describe. In fact, this roughness can be present at all scales, which distinguishes fractals from Euclidean shapes. Mandelbrot worked actively to demonstrate that this concept is not just a mathematical curiosity but has strong relevance to the real world. The remarkable fact is that this generalization, from integer dimensions to fractional dimensions, has a profound and intuitive interpretation: non-integer dimensions describe irregular sets consisting of parts similar to the whole. This generalization of the notion of a dimension from integers to real numbers reflects the conceptual jump from translational invariance to continuous scale invariance.

According to Mandelbrot [592], a fractal is a rough or fragmented geometric shape that can be subdivided into parts, each of which is (at least approximately) a reduced-size copy of the whole. Mathematically, a fractal is a set of points whose fractal dimension exceeds its topological dimension.

There are many examples of (approximate) fractals in Nature, such as the distribution of galaxies at large scales, certain mountain ranges, fault networks and earthquake locations, rocks, lightning bolts, snowflakes, river networks, coastlines, patterns of climate change, clouds, ferns and trees, mammalian blood vessels, etc. Nowadays, the notion of fractality has so much permeated all the scientific disciplines that there is even an excess in its attribution to systems which may not be fractal. Maybe the most useful achievement provided by the introduction of fractals is the epistemological breakthrough of considering seriously and quantitatively complex irregular structures.

It is useful to state at the outset some important limitations of fractals. First, many systems present only an apparent but not a genuine fractality, which results from measurement and quantification artifacts [393, 711] as we will discuss below. Second, most physical laboratory experiments, which found evidence for fractal structures, have documented their claim over a rather narrow range of typically 1.3 decade, i.e. over a range of scale from 1 to $10^{1.3} \approx 20$ [88, 89, 583, 593]. We will also come back to this question related to the limited range of scales over which fractals appear: there is always a lower length scale below which and an upper size beyond which the fractal description breaks down.

Experience has shown that it is often difficult to be precise when defining fractals. The definition based solely on dimension is too narrow and it is better to view a fractal set as possessing a fine structure, too much irregularity to be described in traditional geometric language, both locally and globally, some form of self-similarity, perhaps approximate or statistical, a "fractal dimension" (somehow defined) which is greater than its topological dimension, and a simple definition, usually recursive.

5.1.2 A First Canonical Example: the Triadic Cantor Set

Figure 5.1 shows the first five iterations of the construction of the so-called triadic Cantor set. At the zeroth level, the construction of the Cantor set begins with the unit interval, that is, all points on the line between 0 and 1. This unit interval is depicted by the filled bar at the top of the figure. The first level is obtained from the zeroth level by deleting all points that lie in the middle third, that is, all points between 1/3 and 2/3. The second level is obtained from the first level by deleting the middle third of each remaining interval at the first level, that is, all points from 1/9 to 2/9, and 7/9 to 8/9. In general, the next level is obtained from the previous level by deleting the middle third of all intervals obtained from the previous level. This process continues ad infinitum, and the result is a collection of points that are tenuously cut out from the unit interval. At the n-th level, the set consists of $N_n = 2^n$ segments, each of which has length $\ell_n = 1/3^n$, so that the total length (i.e. measure in a mathematical sense) over all segments of the Cantor set is $(2/3)^n$. This result is characteristic of a fractal set: as $n \to +\infty$, the number of details (here the segments) grows exponentially to infinity while the total mass goes to zero also exponentially fast. In the limit

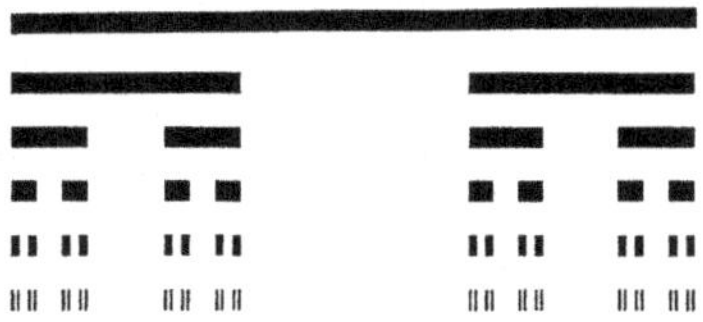

Fig. 5.1. The initial unit interval and the first five iterations of the construction of the triadic Cantor set are shown from the *top* to *bottom*

of an infinite number of recursions, we find the Cantor set made of an infinite number of dots of zero size.

The topological dimension of the Cantor set is $d_t = 0$ since its total measure (length) is zero. We see that this notion of dimension is not very useful since it does not distinguish between this rather complex set of elements and a single point, which also has a vanishing topological dimension. To cope with this degeneracy, mathematicians have introduced alternative concepts of dimensions that give useful information for quantifying such sets. The simplest non-trivial dimension that generalize the topological dimension is the so-called capacity dimension, which in this context can be simply defined as follows:

$$D_c = \lim_{n \to +\infty} \frac{\ln N_n}{\ln(1/\ell_n)} = \frac{\ln 2}{\ln 3} \approx 0.63 . \tag{5.1}$$

The fractal dimension D_c quantifies the rate at which the number N_n of observable elements proliferate as the resolution $1/\ell_n$ increases. There are many other measures of dimension, for example the Hausdorff dimension and the multifractal dimensions discussed below.

5.1.3 How Long Is the Coast of Britain?

In his founding paper [591], Mandelbrot revisited and extended the investigation launched by Richardson [778], concerning the regularity between the length of national boundaries and scale size. He dramatically summarized the problem by the question written in the title of this subsection [591]. This question is at the core of the introduction of "fractal" geometry. Figure 5.2 shows a synthetically generated coastline that has a corrugated structure reminiscent of the coastline of Brittany in France.

Such a coastline is irregular, so a measure with a straight ruler, as in Fig. 5.3, provides only an estimate. The estimated length $L(\epsilon)$ equals the length of the ruler ϵ multiplied by the number $N(\epsilon)$ of such rulers needed to cover the measured object. In Fig. 5.3, the length of the coastline is measured twice with two rulers ϵ_1 and $\epsilon_2 \approx \epsilon_1/2$. It is clear that the estimate of the length $L(\epsilon_2)$ using the smaller ruler ϵ_2 is significantly larger than the length $L(\epsilon_1)$ using the larger ruler ϵ_1. For very corrugated coastlines exhibiting roughness at all length scales, as the ruler becomes very small, the length grows without bound. The concept of (intrinsic) length begins to make little sense and has to be replaced by the notion of (relative) length measured

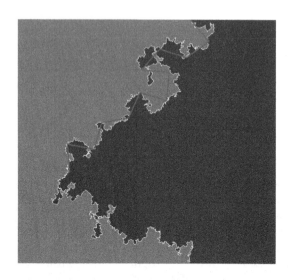

Fig. 5.2. Synthetic fractal coastline (courtesy of P. Trunfio)

at a given resolution. Then, the fractal dimension quantifies precisely how the relative length changes with the resolution. The general relationship for fractal coastlines is

$$L(\epsilon) = C\epsilon^{1-D} , \tag{5.2}$$

where C is a constant. For Great Britain, $D \approx 1.24$ which is a fractional value. In constrast, the coastline of South Africa is very smooth, virtually an arc of a circle and $D \approx 1$. In general, the "rougher" the line, the larger the fractal dimension.

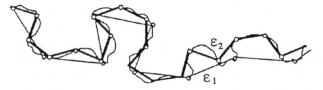

Fig. 5.3. Implementation of the ruler method consisting in covering the rough line by segments of fixed size. As the ruler length decreases, finer details are captured and the total length of the line increases

This expression (5.2) is a special case of the more general relationship

$$M(\epsilon) = C\epsilon^{d_t} \left(\frac{L}{\epsilon}\right)^{D} , \tag{5.3}$$

where $M(\epsilon)$ is the mass of the object measured at the resolution ϵ, d_t is the topological dimension (equal to 1 for a line) and L is the linear size for the object from one extremity to the other. This formula (5.3) shows that the measure ("mass") of the object is dimensionally proportional to the resolution raised to the power equal to the topological dimension. The multiplying factor is the typical number L/ϵ of balls of linear size ϵ needed to cover the object, that is raised to the power D, thus defining the fractal dimension. Notice that if $D = d_t$, the dependence of the mass M on the resolution ϵ disappears as it should for regular objects. This illustrates the striking and essential feature of fractal sets, namely that their measures and properties become functions of the scale at which they are observed.

As an illustration, let us consider the regular construction of a fractal coastline, called the Koch curve. The six first iterations of its construction are shown in Fig. 5.4 from bottom to top. At the n-th iteration, there are 4^n segments, each of size $1/3^n$. The total length is thus $L(n) = (4/3)^n$ which diverges as $n \to +\infty$. The total length $L(n) = (4/3)^n$ can also be written $L(n) = (1/3)^{1-D}$ which defines the fractal dimension of the Koch curve

$$D = \frac{\ln 4}{\ln 3} \ . \tag{5.4}$$

5.1.4 The Hausdorff Dimension

In the 1920s, the mathematician Hausdorff developed another way to "measure" the size of a set [286]. He suggested that we should examine the number $N(\epsilon)$ of small intervals, needed to "cover" the set at a scale ϵ. This method is illustrated in Fig. 5.5 for the case of a line that is covered by disks of radius ϵ.

The Hausdorff dimension is defined by considering the quantity

$$M \equiv \lim_{\epsilon \to 0} \mathrm{Inf}_{\epsilon_m < \epsilon} \sum_m \epsilon_m^d \ , \tag{5.5}$$

which is constructed by summing the volumes ϵ_m^d of balls of radii ϵ_m not exceeding ϵ that cover the fractal set. The "Inf" means that all partitions of balls of radius less than ϵ that cover the set are considered and the one which gives the smallest possible value for the sum is kept. Hausdorff showed that there is a special value D_H for the exponent d, called the Hausdorff dimension, such that

- for $d < D_H$, $M = \infty$ because too much weight is given to the balls that proliferate in number as $\epsilon \to 0$;
- for $d > D_H$, $M = 0$ as too little weight is given to the balls that proliferate in number as $\epsilon \to 0$;

This shows that the Hausdorff dimension D_H is a metric concept.

5.1.5 Examples of Natural Fractals

In sum, a fractal is an irregular geometric object with an infinite nesting of structure at all scales. In practice, the work "infinite" has to be replaced

Fig. 5.4. Koch curve: starting from the unit interval, we divide it in three equal segments of length 1/3 and replace the central one by two segments forming a tent-like structure. This gives the first iteration shown at the *bottom* of the figure. The second iteration is obtained by performing the same step on each of the four segments of the first iteration, namely divide each of them in three sub-segments of size 1/9 and replace the central one of each by two sub-segments to form a tent-like structure. This is iterated an infinite number of times. The first six generations are shown

Fig. 5.5. "Sausage" covering method illustrating how a metric measure can be defined by covering a rough line by disks of different radii

by "large," so that fractals are objects that exhibit a self-similar nesting of structures over a large interval of scales, bounded from below by a microscopic cut-off (for instance a building block) and from above by a macroscopic size (for instance the linear size of the system).

Have you ever wondered why careful geologists always include a scale or reference when taking a picture of geologic interest? The reason is that if they did not, the actual size or scale of the object pictured could not be determined. This is because most geoforms are self-similar, i.e., a fold 1 cm long looks quite the same as if it were 10 m or 10 km long. This is illustrated

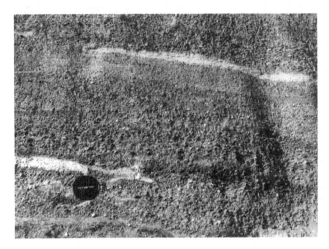

Fig. 5.6. Two typical geological sedimentary structures sedimentary layers with lens caps serving as scale reference

by Figs. 5.6 and 5.7, which provide probably one of the best pedagogical demonstrations of self-similarity in the real world.

The same scale-free characteristics are observed for most fault systems, layering, foliations, coastlines, topographic features, drainage patterns, etc. Self-similarity characterizes many physical systems and seems to result generically from complex internal dynamics [251].

Fractal Fault Systems. The self-similar nature of fracture and faulting is widely documented from the analysis of data from both field observations [60, 61, 826] and experiments [416, 862]. Figure 5.8 shows successive magnification of a fracture network from the scale of 1 m to the scale of 400 km. Fractal geometry provides a first-order description of the complexity of fault systems. In particular, it is well-adapted to the large range of scales, from microns (microcracks) to thousands of kilometers (major mature faults). These fault networks shown in Fig. 5.8 have been mapped in the

Fig. 5.7. Same as Fig. 5.6: the finger at the *top* and the scientist at the *bottom* examplify the self-similarity of sedimentary layers: without a genuine scale of reference, the dummy lens cap in the bottom picture fools the observer. Courtesy of S.W. Wheatcraft, Professor of Hydrogeology and S.W. Tyler, Professor of Soil Physics, University of Nevada, Reno, California

Arabian plate, which was deformed during the Pan-African orogeny and is covered by Phanerozoic platform deposits including sedimentary and volcanic rocks. One of the most important features developed during the closing stages of the Pan-African orogeny was a broad zone of strike-slip faults, called the Nadj Fault System. These faults, striking WNW–ESE through the basement, strongly influenced the fracture pattern of the platform. The plate is bounded by the Taurus–Zagros collision zone to the north-northeast, the Gulf of Aden and the Arabian sea to the south-southeast, and the Red sea spreading center, Gulf of Aqaba and Dead Sea transform to the west-southwest. The plate is estimated to have been displaced 105 km to the north along the sinistral Dead sea transform fault since the Miocene.

A careful study of these different fault maps shows that different fractal geometrical structures hold separately in distinct limited ranges, separated by well-defined characteristic scales. This is shown by using a combination of tools, from box-counting multifractal (see below) improved to correct for

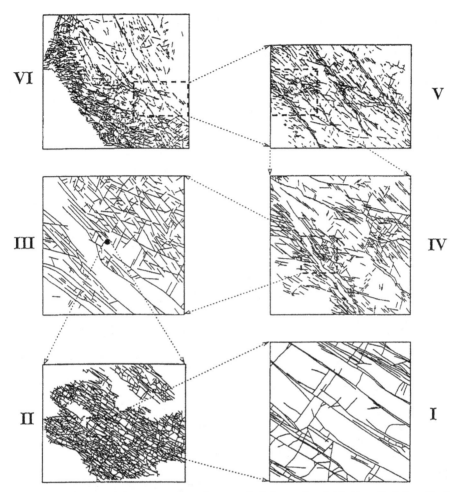

Fig. 5.8. Series of fracture networks sampled from the field (1 : 1) (plate I of size about one meter across) to the continental scale (1 : 1,000,000) (plate VI of size about 400 km accross), based on field mapping, interpretation (checked on the ground) and digitization of photographs taken from a helicopter, classical aerial photography and satellite images on the western Arabian plate. Reproduced from [711]

irregular geometry of mapped domains and finite size effects and from local wavelet analysis (adapted to the fault anisotropic) [711]. Fracture and faulting can thus be better modeled as a hierarchical process, controlled in large part by geometry and preexisting heterogeneity (vertical and/or horizontal). Microscopic analysis of faults confirm a complex interplay of cracks acting with different mechanisms and at different scales [985]. Thus, the pure fractal description is too naive and more sophisticated quantifiers must be introduced that reconcile the existence of structures at many scales giving an impression of fractality and the presence of characteristic scales that break down the pure scaling.

Another example of a complex fractal fault system is shown in Fig. 5.9 showing one of the most spectacular cases of continental collisions, namely the India–Asia collision occurring in the last fifty million years. One can see on the figure the complex network of faults developing in Asia and extending

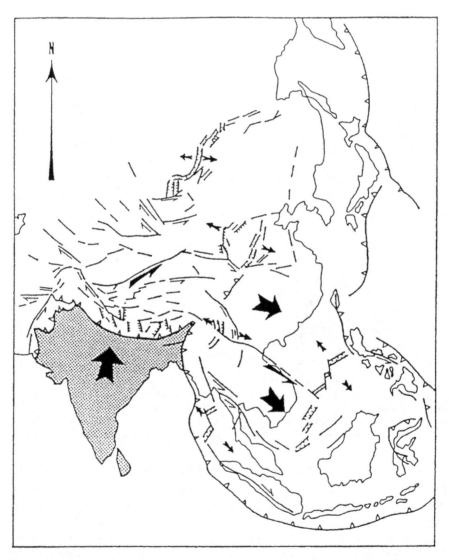

Fig. 5.9. Map of the main faults resulting from the collision between the India and the Asia tectonic plates. The *fat arrows* represent the mean motion of India penetrating within Asia and the corresponding extrusion of Asia to the East

over more than 5000 km from the collision. Analog laboratory models of such continental collisions using layers of granular media and silicon to represent respectively the upper brittle crust and the lower more ductile crust and mantle have shown that the fault network is fractal [863]. Different models with varying rheologies have been analyzed which attempt to respect the brittle–ductile stratification corresponding to the crust and mantle structure in the Earth. Each experiment has been quantified by studying the strain field and the fault pattern, as a function of the position of the penetrating wedge within the system. The fractal dimension of the fault network was found to be almost constant within experimental uncertainty (1.7 ± 0.1) and thus appears rather insensitive to the particular chosen rheology [201, 862]. There exists a correlation between the generalized multifractal dimensions and two exponents, the fractal dimension of the set of fault barycenters and the exponent of the fault length distribution [202], that indicates that the different scaling laws are not independent: the multifractality results from the interplay between the fractal geometry of the positions of the faults and the power law distribution of the fault lengths. This later distribution provides a kind of self-similar weight decorating the fractal set of fault barycenter, in agreement with a known mechanism for multifractality [391].

Fractal Systems of Earthquake Epicenters and Seismograms. A complex spatial distribution of epicenters is shown in Figs. 5.10 (data from http://quake.geo.berkeley.edu/cnss/maps/cnss-map.html) and 5.11 (data from http://quake.geo.berkeley.edu/cnss/maps/cnss-map.html). To the naked eye, the fractality corresponds heuristically to the existence of "holes", i.e. domains with no events, of many scales from the largest one equal to a finite fraction of the total map to the smallest size given by the resolution of the seismic network. The qualitative notion of fractals attributed to earthquake epicenters is related to the ubiquitous observation that earthquakes are clustered in space: they tend to aggregate on or close to major fault structures. The fractal structure of earthquake epicenters is thus in part controlled by the fractal nature of fault networks. The complex spatio-temporal dynamics of earthquakes can engender additional clustering. Studying the interplay between such pre-existing fault patterns and the earthquake dynamics is an active field of research to understand the organization of earthquakes and their predictability [666].

The first quantitative measurement of the fractal (correlation) dimension of earthquake epicenters was performed by Kagan and Knopoff [492]. The first extension to a multifractal analysis of the distribution of earthquake epicenters is by Geilikman et al. [335].

An interesting practical application of the concept of fractals to seismology has been developed by Tosi et al. [961], who propose a detection algorithm with improved local and regional seismic signal recognition. The method consists in calculating the fractal dimension of seismic records (vibration amplitude as a function of time), where the fractal dimension is related

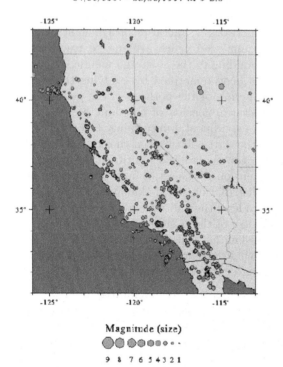

CNSS Earthquake Catalog
04/01/1997 - 06/30/1997 M > 2.0

Magnitude (size)

9 8 7 6 5 4 3 2 1

Fig. 5.10. Map of seismicity for all earthquakes of magnitude larger than 2 in California from 1st April to 30th June 1997. This map has been prepared by the CNSS, Council of the National Seismic System. One sees a complex and irregular structure of epicenters, documented [492] to be of fractal dimension close to 1.2 (corresponding to a dimension 2.2 in 3D)

to the exponent describing how the variogram of the signal scales with the time interval separating pairs of points along the seismic signal. The method of recognition of earthquakes is based on the difference between the fractal dimensions of genuine seismic signals and of background noise. Results from the comparison with standard methods show that the new method recognizes the seismic phases already detected by existing procedures and, in addition, it presents a greater sensitivity to smaller signals, without an increase in the number of false alarms. The efficiency of the method relies on its multiscale nature, that is, on the fact that the fractal signature implies to search for a relationship between the spectral amplitude of different frequencies.

Fractal Structure of Craters. The fractal landscapes of craters on the moon and other bodies in the solar system are the signature of the power law distribution (3.82) of impactor sizes, whose small range limit distributions (from the micron scale to the ten centimeter scale) are depicted in Fig. 3.11. Figure 5.12 shows the self-similar distribution of craters on Callisto, Jupiter's second largest moon. This portion of the surface of Callisto contains an immensely varied crater landscape. A large, degraded crater dom-

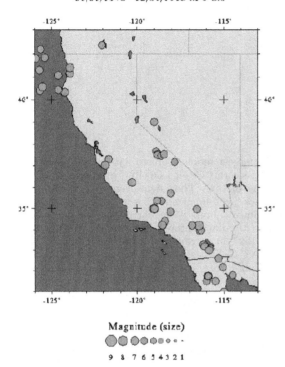

Magnitude (size)

9 8 7 6 5 4 3 2 1

Fig. 5.11. Map of seismicity for all earthquakes of magnitude larger than 6 in California from 1st January 1946 to 31st December 1996. This map has been prepared by the CNSS, Council of the National Seismic System

inates the southern (bottom) portion of the image. There are fresh to highly degraded craters at all sizes, but a relatively low number of small, fresh craters.

River Networks and Fractal Topography. The transportation networks for precipitated water are known to obey a series of approximately universal scale-free distributions [788], which have mainly been obtained through analyses of soil-height maps of the drainage basins of rivers. Figure 5.13 shows such a typical river network. These networks are thought to result from self-organizing mechanisms and a large number of models based on nonlinear dynamical laws (with and without heterogeneity) or on energy minimization principles have been proposed to account for their complex shapes and their universal properties [237].

Observing the existence of self-similarity accross many scales, theories of scaling can then be applied to unravel the origin of universal properties that arise when the qualitative character of a system is sufficient to quantitatively predict its essential features, such as the exponents that characterize scaling laws. In the case of river networks, the empirical scaling laws follow from three simple assumptions that (1) river networks are structurally self-similar,

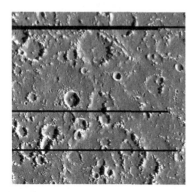

Fig. 5.12. Image of Jupiter's second largest moon, Callisto with its craters from Catalog PIA01630, mission Galileo, Spacecraft Galileo Orbiter, produced by JPL, 1998-10-13. North is to the *top* of the picture. The image, centered at 13.4 degrees north latitude and 141.8 degrees west longitude, covers an area approximately 61 km by 60 km. The resolution is about 85 m per picture element. The *horizontal black lines* indicate gaps in the data received for this image. The image was taken on September 17th, 1997 at a range of 8400 km by the Solid State Imaging (SSI) system on NASA's Galileo spacecraft during its tenth orbit of Jupiter. From the Jet Propulsion Laboratory, (http://www.jpl.nasa.gov/galileo and http://www.jpl.nasa.gov/galileo/sepo)

Fig. 5.13. Typical river network

(2) single channels are self-affine, and (3) overland flow into channels occur over a characteristic distance (drainage density is uniform) [238].

The power spectrum of linear transects of the earth's topography is often observed to be a power-law function of wave number k with exponent close to -2, thus corresponding to an approximate self-affine random walk. An aerial view of such a complex topography is shown in Fig. 5.14 (see [427] for a collection of pictures). The simplest mechanism to obtain such structures is to couple a simple diffusive erosion process, according to which the local flux is proportional to the slope (Culing's law), to a nonlinear process with stochastic driving (embodying the many noise sources such as rain and rock heterogeneity) [918]. The resulting fluctuations of landscape and of mountain topography are found to be self-affine with an exponent in agreement with observations. Many other models have been developed more recently that capture more precisely the relationships between the fractal tree structure of river networks and the self-affine nature of the topography.

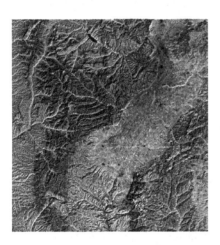

Fig. 5.14. Fractal topography of mountain range. From the internet document http://daac.gsfc.nasa.gov/DAAC_DOCS/ geomorphology/GEO_HOME_PAGE.html in the chapter on Fluvial Processes. Courtesy of J. Pelletier

"River networks" can also be obtained as mathematical solution of global optimization problems. An example is provided by Fig. 5.15. It shows a series of magnifications of so-called random directed polymers (RDP) at zero temperature [388]. The model is defined as follows. Consider a square lattice oriented at $45°$ with respect to the x axis and such that each bond carries a random number, interpreted as an energy. An arbitrary directed path (a condition of no backwards turn) along the x-direction and of length W (in this direction) corresponds to the configuration of a RDP of W bonds. In the present version of the model, a given RDP minimizes the sum of the W bond energies along it, while having its two end-points fixed at the same ordinate $y(x = 0) = y(x = W)$. The set of lines shown in Fig. 5.15 corresponds to all the optimal conformations with minimal energy and fixed end-points spanning all possible values verifying $y(x = 0) = y(x = W)$. We allow in this construction the superposition of conformations. The tree-like structures with branches that proliferate close to the borders result from the fact that RDP's find optimal paths with minimum energy. Such paths are organized in families with approximately $W^{2/3}$ elements per family [475]. A family is defined by the common ancestor, i.e. main trunk far from the borders, from which all conformations branch out. This simple model, with its much varied behavior, has become a valuable tool in the study of self-similar surface growths [495], interface fluctuations and depinning [442], the random stirred Burgers equation in fluid dynamics [496] and the physics of spin glasses [91, 622], as well as complex fault networks self-organized by earthquakes [627, 907]. The complexity of random directed polymers result from the competition between the quenched energy that the RDP attempts to optimize globally and the entropy quantifying the reduction of the natural fluctuation of random walks. As a result of this competition, a hierarchical structure of quasi-degenerate random directed polymers is found that can be quantified by several universal scaling laws [388, 475, 495].

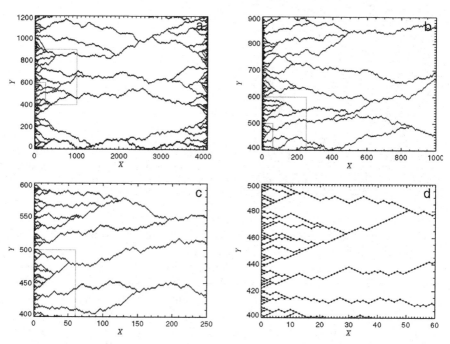

Fig. 5.15. A typical set of optimal configurations for a random directed polymer of length $W = 4096$ whose end-points take all possible values $0 \leq y(x = 0) = y(x = W) \leq 1200$. (**a**) global system; (**b**) magnification of the largest box in (a); (**c**) magnification of the largest box in (b); (**d**) magnification of the box in (c). The magnifications illustrate the self-affine structure of the RDPs and the self-similar hierarchical pattern of the local branching structure. Reproduced from [475]

Clouds. Figure 5.16 shows a picture of a cloud configuration taken from high altitude. This picture shows a qualitative fractal coverage of earth by clouds, as a large range of hole scales can be observed. This impression can be quantified precisely [727] by measuring the cumulative frequency-area distribution of tropical cumulus clouds as observed from satellite and space shuttle images from scales of 0.1 to 1000 km. The cumulative distribution is found to be a power-law function of area with exponent $\mu = 0.8$. This result and the fractal dimension of cloud perimeters can be interpreted from the fact that the top of the convective boundary layer is a self-affine interface [727].

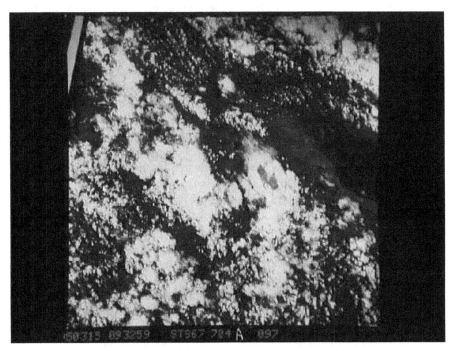

Fig. 5.16. Image of clouds illustrating their fractal nature. From the space shuttle STS-67 series images. Courtesy of J. Pelletier

DLA and Fractal Laplacian Growth. The processes underlying the formation of lightning, frost, coral, polymers, mineral crystallinity in igneous rocks, dendrites in many chemical reactions and post-percolation coffee ground clumps are now understood to fall under a general class related to the simple model known as Diffusion Limited Aggregation (DLA) invented in 1981 by Witten and Sander [1024]. Simply put, Diffusion Limited Aggregation is an algorithm which simulates the formation of an aggregate. The algorithm starts with some initial structure of "particles" and then adds to this structure by sending a particle on a random walk far from the aggregate (this is the diffusion part of DLA), until the particle approaches sufficiently close from the perimeter of the aggregate that some kind of binding condition is satisfied. Once this binding condition is satisfied, the random walker is added to the aggregate and the process is repeated with a new particle. The result of an off-lattice simulation is shown in Fig. 5.17. DLA clusters are thought to be fractal with a dimension approximately equal to 1.7 in two dimensions and 2.5 in three dimensions.

Intuitively, the complex fractal structure of DLA aggregates results from two competing mechanisms. The tendancy for branches to branch out due to the sticking of particles to the side is compensated by the screening from large branches that grow faster at the expense of smaller branches [615]. This results from the so-called Mullins–Sekerka instability, which is nothing but the "lighting-rod-effect" in electrostatics (i.e. in the presence of a Laplacian field, here the concentration of diffusive particles). The random behavior of the random walking particles will cause small irregularities to develop on the surface of the cluster. Particles diffusing randomly toward the cluster will be more likely to stick to external perimeter sites than to those in the cluster's interior. The outer sites effectively screen particles from the interior. As a result, small distortions in the cluster become large distortions, an effect known as growth instability. The competition between branching and screening in the presence of these fluctuations leads to a hierarchy of branching and screening instabilities resulting in a fractal structure.

There are many parameters in Diffusion Limited Aggregation which can be varied to suit the need of modeling. For instance, the binding conditions can be modified to simulate an aggregate composed of a type of molecule with a limited number of binding sites. More than one species of particle can be introduced. The random walk each particle undergoes can be made less random with the introduction of repulsive and/or attractive forces between particles. The introduction of new particles might be biased to favor a particular initial direction of incidence. Some sort of time dependence can be added. In general, however, the resultant aggregate is typically a self-similar structure which can be described mathematically as a fractal.

Fig. 5.17. Off-lattice DLA Cluster of 10^6 particles

5.2 Multifractals

5.2.1 Definition

Many fractals arising in Nature have a far more complex scaling relation than simple fractals, usually involving a range of scales that can depend on their location within the set. Such fractals are called multifractals.

The multifractal formalism is a statistical description that provides global information on the self-similarity properties of fractal objects [391]. A practical implementation of the method consists first in covering the system of linear dimension L under study by a regular square array of some given mesh size l. One then defines the measure or weight p_n of a given box n: it is defined as the sum of the measure of interest within the box. A simple fractal of dimension α is defined by the relation

$$p_n \sim l^\alpha . \tag{5.6}$$

Simply put, a multifractal is a generalization in which α may change from point to point and is a local quantity. The standard method to test for multifractal properties consists in calculating the so-called moments of order q of the measure p_n defined by

$$M_q(l) = \sum_{n=1}^{n(l)} p_n^q, \tag{5.7}$$

where $n(l)$ is the total number of non-empty boxes. Varying q allows one to characterize the inhomogeneity of the pattern, for instance the moments with large q being controlled by the densest boxes.

If scaling holds, then one has

$$M_q(l) \sim l^{(q-1)D_q} , \tag{5.8}$$

which defines the generalized dimensions D_q. For instance, D_0 (respectively D_1 and D_2) corresponds to the so-called capacity (respectively information and correlation) dimensions.

A monofractal has the same fractal dimension α in each box, which is expressed by $p_n = Cl^\alpha$ for all n's. By normalization, we have

$$1 = \sum_{n=1}^{n(l)} p_n = \sum_{n=1}^{n(l)} Cl^\alpha = \left(\frac{L}{l}\right)^d Cl^\alpha . \tag{5.9}$$

In the monofractal case, expression (5.7) leads to

$$M_q(l) = \left(\frac{L}{l}\right)^d Cl^{\alpha q} = l^{(q-1)\alpha} . \tag{5.10}$$

The last equality is obtained by using (5.9). Comparing (5.10) with (5.8), this shows that a monofractal has, as expected, all its dimensions identical $D_q = \alpha$.

In multifractal analysis, one can also determine the number $N(l)$ of boxes having similar local scaling characterized by the same exponent α. Assuming self-similar scaling, the expression

$$N_\alpha(l) \sim (L/l)^{f(\alpha)}, \tag{5.11}$$

defines $f(\alpha)$, called the multifractal singularity spectrum, as the fractal dimension of the set of singularities of strength α. The sum (5.7) can then be rewritten

$$M_q(l) = \sum_\alpha J p_n^q N_\alpha(l) = \sum_\alpha J l^{\alpha q}(L/l)^{f(\alpha)} , \tag{5.12}$$

where J is the Jacobian of the transformation from the index of a given box to the exponent α characterizing that box. To obtain $f(\alpha)$ as a function of the D_q's, we use a saddle-node estimation of the sum over α in (5.12). First, we note that the Jacobian J does not exhibit a singular behavior for small l and thus does not contribute to the leading behavior of the sum. The standard saddle-node argument is that the sum over α is correctly estimated, with respect to its dependence on l, by its leading term. This leader is such that the exponent $\alpha q - f(\alpha)$ of l is extremum with respect to α. By taking the derivative of $\alpha q - f(\alpha)$ with respect to α and putting to zero, this yields

$$q = \left. \frac{\mathrm{d}f}{\mathrm{d}\alpha} \right|_{\alpha^*} . \tag{5.13}$$

For a given moment order q, this is an equation in the variable α whose solution $\alpha^*(q)$ is the saddle node solution. The moment $M_q(l)$ in (5.12) can thus be estimated by

$$M_q(l) \sim l^{\alpha^*(q)q - f(\alpha^*(q))} . \tag{5.14}$$

Comparing (5.14) with the definition (5.8), we find the general relationship between the set of dimensions D_q and the multifractal spectrum. This relationship is expressed mathematically as a Legendre transformation:

$$f(\alpha) = q\alpha - (q-1)D_q , \tag{5.15}$$

where the * has been dropped. Physically, expression (5.15) relies on (5.13), which says in essence that one set of boxes characterized by the same singularity α provides the leading contribution to a given moment of order q. Plotting $\ln M_q$ as a function of $\ln l$ yields the exponent $(q-1)D_q$. It can also be shown [792] that the quantity $L_q(l) = \sum_{n=1}^{n(l)} p_n^q \ln p_n$ is proportional to $M_q(l)\alpha(q)\ln l$. Thus plotting $L_q(l)/M_q(l)$ as a function of $\ln l$ yield α. $f(\alpha)$ can then be obtained using the above Legendre transformation.

The multifractal spectrum $f(\alpha)$ given by (5.15) has several interesting properties which have been studied in depth in the literature [391, 722, 792]. First, standard properties of the Legendre transform (5.15) with (5.13) imply that $f(\alpha)$ is concave. In addition, expression (5.15) shows that $\mathrm{d}f(\alpha)/\mathrm{d}\alpha = q$ (which is nothing but (5.13)): the slope of $f(\alpha)$ is thus given by the order

q of the moment related to α via (5.13). The maximum of $f(\alpha)$ thus occurs for $\alpha(q = 0)$ and is thus equal to the capacity dimension $f(\alpha(q = 0)) = D_0$, using (5.15). Taking the derivative of (5.15) gives

$$\alpha = D_q + (q - 1)\frac{dD_q}{dq} . \qquad (5.16)$$

This expression relates the abscissa of the function $f(\alpha)$ to its slope q.

It is interesting to realize that multifractality imposes the existence of a physical scale. Indeed, the multifractal scaling of moments is valid for the resolution scale l going either to 0 or to infinity. In the former (resp. later) case, the multifractal spectrum is concave (resp. convex). These two cases are mutually exclusive. This implies that multifractality describes a self-similar property starting either from a large length scale down to a vanishing length scale or from a small finite length scale up to infinity.

Schertzer and Lovejoy [818, 819] have introduced the notion of universal multifractals obtained by mixing of identical independent multiplicative processes. This classification allows one a parsimonious description of different multifractal processes in terms of very few exponents. When the mean flux is conserved (corresponding to an average branching ratio 1, see Chap. 13), they are described by only two exponents μ and C_1. The first exponent $0 < \mu < 2$ is the Lévy exponent of the generator of the multiplicative process and quantifies the deviation from monofractality: the monofractal case corresponds to $\mu = 0$ since, in this case, only the strongest singularity is seen as it completely dominates all other singularities; the lognormal model is obtained for the other bound $\mu = 2$ as it should since it corresponds to a multiplicative process with Gaussian variables. The second exponent C_1 is the codimension of the mean inhomogeneity of the hierarchy of singularities constituting the multifractal: the radius of curvature of $f(\alpha)$ at $\alpha = C_1$ is $2^{2/3}\mu$.

5.2.2 Correction Method for Finite Size Effects and Irregular Geometries

The previous section described the standard method used widely in pratical applications. However, one must realize that the exponents D_q and the $f(\alpha)$ spectrum are very sensitive to the finite size and shape of the sampled sets. For instance, in geological settings where one is interested in measuring the fractal properties of faults, sediments cover some parts of the studied areas and correspond to holes in the sampling which could wrongly be interpreted as reflecting a genuine multifractality, whereas this only reflects the sampling bias. An irregular border geometry may lead to similar effects. There is an efficient method that can be used to correct for these biases [711]. One first considers the natural measure and computes with the above method the various sums $M_q(l)$ and $L_q(l)$. We then generate a synthetic system by taking each element of the measure of the initial system and position it at random (without changing its geometrical structure) in the sampling domain with its specific geometry. We then compute the sums $M'_q(l)$ and $L'_q(l)$ for this system. The artificial system has been constructed so that the "true" unbiased values of the exponents are known: the random and uniform sampling procedure ensures that all D_q and α are equal to the space dimension 2 (if we deal with 2D problems) in the absence of finite size effects. The corrected exponents of the initial system are then given by representing $M_q(l)$ as a function of $M'_q(l)$ in a logarithmic plot. This generally yields a straight line of slope $\beta(q)$. The correct exponents of the initial system are then given by $D_q = 2\beta(q)$ and similarly for the α's. Tests on well-controlled synthetic sets show that the accuracy of the determined exponents is within 0.05. The gist of this method is that the same distortions apply to the two sets. Therefore, the correction needed to correct the calculation on the real data set is determined from the known synthetic data set.

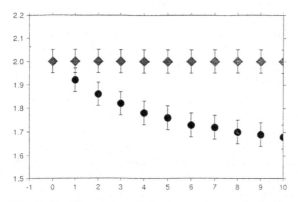

Fig. 5.18. Illustration of the correction method presented in the text on the generalized set of multifractal dimensions D_q. The multifractal analysis is performed on the fault system shown in panel I in Fig. 5.8. The direct application of the box-counting method and the formula (5.7) gives a non-trivial monotonically decaying dependence of D_q as a function of q (*black circles*), which would qualify this fault system as a genuine multifractal. The application of the correction method described in the text shows that all D_q's are in fact equal to 2 and that the multifractality is spurious and stems from finite size and boundary effects. Reproduced from [710, 711]

A test of this method is shown in Fig. 5.18 on the fault system shown in panel I in Fig. 5.8. In this case, the application of the correction method shows that the apparent multifractality is spurious. Figure 5.19 shows the multifractal spectrum $f(\alpha)$ for the two fault systems V and IV shown in Fig. 5.8. In this case, the two fault networks exhibit a genuine multifractality. As a bonus, the multifractal spectra that appeared different before correction are found to be essentially identical after correction, showing a universal behavior at these two sampling scales.

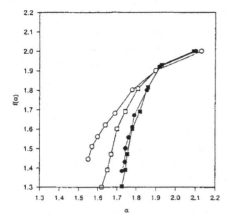

Fig. 5.19. Illustration of the correction method presented in the text on the generalized set of multifractal dimensions D_q. The multifractal analysis is performed on the fault systems shown in panels V and VI in Fig. 5.8. The *open circles* and *squares* are the spectra $f(\alpha)$ calculated before correction for finite size and boundary effects, while the *black circles* and *squares* represent the spectra $f(\alpha)$ after correction. Taken from [710, 711]

5.2.3 Origin of Multifractality and Some Exact Results

Heuristically, multifractality results when a "fractal measure" is defined on a "fractal" geometrical object. In other words, multifractality stems from the interplay between two sets of singularities [391, 722]. A particularly important and general class is the classical potential, for instance the electrostatic or diffusion field, near random fractal boundaries, whose self-similarity is reflected in a multifractal behavior of the potential. Generalizations include the stress field in fractal networks of faults and its feedback on the fault organization [862]. The paradigm of this class of problems is Diffusion Limited Aggregation (DLA), which produces fractal clusters, as shown in Fig. 5.17, immersed in a concentration field (that of the incomming particles) which has multifractal properties. The concentration field, also called the potential, is termed the "harmonic measure" in mathematical terms. It actually determines the growth process and its scaling properties are intimately related to those of the cluster itself [615].

The multifractal spectra describing the singularities of the potential, or, equivalently, the distribution of wedge angles along the boundary, are amenable to exact analytical treatments in two dimensions. A first exact example contains the whole universality class of random or self-avoiding walks, and percolation clusters, which all possess the same harmonic multifractal spectrum [137, 253, 254]. Another exactly soluble case is the potential distribution near any conformal fractal in two dimensions, using the classical method of conformal transformations to solve two-dimensional electrostatics on Euclidean domains. The multifractal spectra of the potential singularities on the perimeter of such fractal objects are found to depend only on the so-called "central charge c", a parameter which labels the universality class of the underlying conformal field theory [255]. These results apply directly to well-recognized universal fractals, like $O(N)$ loops or Potts clusters. The dimension $\hat{f}(\theta)$ of the boundary set with local wedge angle θ is found to be given by $\hat{f}(\theta) = \pi/\theta - [(25 - c)/12]\{(\pi - \theta)^2/[\theta(2\pi - \theta)]\}$, with c the central charge of the model. As a corollary, the dimensions D_{EP} of the external perimeter and D_H of the hull of a Potts cluster obey the duality equation $(D_{EP} - 1)(D_H - 1) = 1/4$.

5.2.4 Generalization of Multifractality: Infinitely Divisible Cascades

Infinitely divisible cascades (IDC) have been introduced in turbulence [144, 145] and applied in other fields such as in finance [34] and in network traffic data on the Internet [982]. IDC generalize multifractals in a very interesting way and rationalize the empirically observed extended self-similarity and generalized scale-invariance (see the contributions in [251]).

Consider a decomposition of a signal X across several scales. The signal can be a time series $X(t)$, a spatial function or a space–time recording. A very efficient scale decomposition scheme is provided by the wavelet transform [306, 586]. In a nutshell, a wavelet transform is a convolution of the signal with a "mother" wavelet Ψ which can be translated and dilated at will. It thus offers the flexibility of a genuine mathematical microscope that can focus on details on the signal at arbitrary positions and scales. In the case of a time signal $X(t)$ for instance, the wavelet transform defines wavelet coefficients given by

$$d_X(b,a) \equiv \int_{-\infty}^{+\infty} dt \, X(t)\Psi\left(\frac{t - b}{a}\right) , \qquad (5.17)$$

where a and b are the dilation and translation parameters. The knowledge of all the wavelet coefficients is more than equivalent to the information contained in $X(t)$ and is in fact redondant. It is enough to estimate the wavelet coefficients on a subset of scales, ordered for instance in powers 2^j of two, where j spans a set of integer values, and on a subset of the positions in order to have an information that is equivalent to the initial data $X(t)$ [306, 586]. Let us thus consider wavelet coefficients $d_X(j, k)$, where 2^j is the scale and k is the time or position at which the wavelet is centered and the coefficient is calculated.

A simple fractal corresponds to the case where

$$\langle |d_X(b,a)|^q \rangle \propto \exp\left(qH \ln(2^j)\right) , \qquad (5.18)$$

for any real q. The self-similarity is expressed in (5.18) which indicates that all moments $\langle |d_X(b,a)|^q \rangle$ of the wavelet coefficients behaves as power laws of the scale $a = 2^j$, controlled by a single exponent, the self-similarity parameter H. This simplicity contains its own limitation and this leads, as we have seen above, to generalize (5.18) by allowing that the exponent H becomes a function $H(q)$ of the moment order q. This multi-scaling behavior is thus defined by

$$\langle |d_X(b,a)|^q \rangle \propto \exp\left(H(q) \ln(2^j)\right) , \qquad (5.19)$$

where $H(q)$ is nothing but the exponent $(q-1)D_q$ in the multifractal formalism discussed above. The main limitation of this model is that the moments must still exhibit power law dependences as a function of the scale a. Another level of generalization is thus still possible which consists in writing

$$\langle |d_X(b,a)|^q \rangle \propto \exp\left(H(q)n(2^j)\right) , \qquad (5.20)$$

where the function $n(a)$ does not necessarily reduce to the logarithmic function. This expression (5.20) defines the infinitely divisible cascade (IDC) model. The moments are not required to be power laws of the scale a; however, IDC maintains a fundamental feature common to exact self-similarity (5.18) and multiscaling (5.19), namely the separability of the moment structure in the moment order q and the scale $a = 2^j$.

IDC have been introduced in a different way [144, 145] through the relationship between probability density functions (pdf) $P(d)$ of wavelet coefficients d measured at different scales:

$$P_a(d) = \int G_{a,a'}(\ln \alpha) P_{a'}(d/\alpha) \, d\ln \alpha . \qquad (5.21)$$

The function $G_{a,a'}$ is called the kernel or propagator of the cascade. If $G_{a,a'}$ is a Dirac function $G_{a,a'}(\ln \alpha) = \delta\left(\ln \alpha - H \ln(a/a')\right)$, IDC reduces to exact self-similarity with exponent H. This definition (5.21) shows that the pdf of the log-coefficients $\ln |d|$ are related by a convolution:

$$P_a(\ln |d|) = \int G_{a,a'}(\ln \alpha) P_{a'}(\ln |d| - \ln \alpha) \, d\ln \alpha$$

$$= (G_{a,a'} \otimes P_{a'})(\ln |d|) . \qquad (5.22)$$

Starting from the definition of a cascade between scales a'' and a' with propagator $G_{a'',a'}$ and between scale a and a'' with propagator $G_{a,a''}$, the transitivity of the definition (5.22) shows that there is a cascade between scales a and a' with propagator $G_{a,a'} = G_{a,a''} \otimes G_{a'',a'}$. Infinite divisibility means that any scale a'' can be used between a and a' as no scale plays any special role. Infinite divisibility therefore implies that the propagator consists of an elementary function G_0 convolved with itself as

$$G_{a,a'}(\ln \alpha) = [G_0(\ln \alpha)]^{\otimes(n(a)-n(a'))} . \qquad (5.23)$$

Using the Laplace transform $\hat{G}_{a,a'}(q)$ of $G_{a,a'}$, (5.23) becomes

$$\hat{G}_{a,a'}(q) = \exp[H(q)(n(a) - n(a'))] , \qquad (5.24)$$

with $H(q) = \ln \hat{G}_0(q)$ and $a = 2^j$. This result allows us to recover (5.20) which rationalizes the empirical observation of extended self-similarity: a moment $\langle |d_X(b,a)|^q \rangle$ plotted as a function of another moment $\langle |d_X(b,a)|^{q'} \rangle$ gives a power law relationship

$$\langle |d_X(b,a)|^q \rangle \propto \langle |d_X(b,a)|^{q'} \rangle^{H(q)/H(q')} , \qquad (5.25)$$

and therefore a linear dependence in a log–log representation. This property documented in many systems such as turbulence [74], diffusion-limited-aggregation [756] and kinetic surface roughening [539] results from the separability of the moment structure in the moment order q and the scale $a = 2^j$.

5.3 Scale Invariance

5.3.1 Definition

The symmetry of "scale invariance" generalizes the geometrical concept of fractals. Speaking about a (material or mathematical) object, scale invariance refers to its invariance over changes of scales of observation (see [251] for a general introduction). In a nutshell, scale invariance simply means reproducing itself on different time or space scales. But what object are we speaking about?

1. some people are talking about the invariance of their equations;
2. some are talking about one physical quantity;
3. Some people consider invariance in a statistical sense:

    - geometrically, they are discussing fractal shapes;
    - analytically, they are discussing the invariance of a probability distribution function or a correlation function, or sometimes only the shape of this function.

Some people use the word "fractal" in this broad sense, and not only for refering to the self-similar geometrical properties discussed above in this chapter.

Once the observable one wishes to study has been clearly identified, the question is how this observable changes with the scale of observation. Precisely, an observable $\mathcal{O}$ which depends on a "control" parameter x is scale invariant under the arbitrary change $x \to \lambda x$ if there is a number $\mu(\lambda)$ such that

$$\mathcal{O}(x) = \mu \mathcal{O}(\lambda x) . \tag{5.26}$$

Here, we implicitly assume that a change of scale leads to a change of control parameter as in the renormalization group formalism discussed in Chap. 11. More directly, x can itself be a scale.

Equation (5.26) defines a homogeneous function and is encountered in the theory of critical phenomena, in turbulence, etc. Its solution is simply a power law

$$\mathcal{O}(x) = Cx^\alpha , \qquad \text{with} \quad \alpha = -\frac{\ln \mu}{\ln \lambda} , \tag{5.27}$$

which can be verified directly by insertion. Power laws are the hallmark of scale invariance as the ratio

$$\frac{\mathcal{O}(\lambda x)}{\mathcal{O}(x)} = \lambda^\alpha \tag{5.28}$$

does not depend on x, i.e. the relative value of the observable at two different scales only depend on the *ratio* of the two scales. This is only true for a function of a single parameter. Homogeneous functions of several variables take a more complex form than (5.26).

The equation (5.26) is the fundamental property that associates power laws to scale invariance, self-similarity and criticality. Self-similarity is the same notion as scale invariance but is expressed in the geometrical domain, with application to fractals. As discussed in Chaps. 8 and 9, criticality refers to the state of a system which has scale invariant properties. The critical state is usually reached by tuning a control parameter as in liquid–gas and paramagnetic–ferromagnetic phase transitions. Many driven extended out-of-equilibrium systems also seem to exhibit a kind of dynamical criticality. This has been coined "self-organized criticality" [44] (see Chap. 15).

The influence of scale symmetry on objects is reflected in the appearance of power laws. Extending the concept of scale symmetry to physical laws generalizes the concept of scale *invariance* to that of scale *covariance*. This concept applies to laws (as opposed to observables or geometrical objects that we considered up until now) and requires the independence of the *shape* of the laws with respect to changes of scales. This notion is more general and actually encompasses the notion of scale invariance: solutions of scale covariant laws can sometimes be scale *invariant*, i.e. power laws. In general, however, boundary conditions, forcing or dissipation spoil this invariance, and the solutions are no longer power laws. The requirement of scale *covariance* is then very useful to study the breaking of scale invariance [251].

Historically, one of the first uses of the notion of scale covariance was made in the framework of critical systems, via the Renormalization Group, as explained in Chap. 11: the coarse-graining rescaling procedure alters the hamiltonian (its coupling constants are different), but its shape remains the same (up to irrelevant terms). This is an example of scale-covariance, i.e. invariance of a law under a scale transformation. We refer to [251] and references therein for other examples in open dissipative nonlinear systems.

Symmetries constitute the organizing principle for shaping our theories of nature. In modern physics, fundamental symmetry principles dictate the basic laws of physics, control the structure of matter, and define the fundamental forces in nature. Symmetry is also an essential concept in mathematics, chemistry, geology and biology. In 1905, Emmy Amalie Noether proved the following theorem [126, 424, 708]: "for every continuous symmetry of the laws of physics, there must exist a conservation law. For every conservation law, there must exist a continuous symmetry." This gives the following amazing correspondences:

1. The conservation law corresponding to space translational symmetry is the Law of Conservation of Momentum.
2. The conservation law corresponding to time translational symmetry is the Law of Conservation of Energy.
3. The conservation law corresponding to rotational symmetry is the Law of Conservation of Angular Momentum.
4. The conservation law corresponding to the symmetry operation of a global phase shift in quantum mechanics is the Law of conservation of electric charge.

It is a natural question to ask what is the conservation law corresponding to the scale invariance symmetry? To address this question, we notice the following: since $x \rightarrow \lambda x$ and $\mathcal{O}(x) \rightarrow \mu \mathcal{O}(\lambda x)$ given by (5.26) are equivalent to $T \equiv \ln x \rightarrow T + \ln \lambda$ and $X(T) \equiv \ln \mathcal{O}(T) \rightarrow X(T + \ln \lambda) + \ln \mu$, an arbitrary scale transformation is simply a translation of "time" T associated to a translation of "position" X. Continuous scale invariance is thus the same as a coupled continuous translational invariance expressed on the logarithms of the variables. This kind of symmetry links a translation in "time" to a translation in "space": this is not the same as the first or second symmetry enumerated in the list above. It rather states that the system is seen to move at a constant velocity α. Noether's theorem does not apply to this special case.

However, if we request that the system is invariant with respect to arbitrary translation of X, i.e. a change of units of $\mathcal{O}$, Noether's theorem leads to *dimensional analysis* which we discuss below. Extending works of L. Nottale [689–694], B. Dubrulle and F. Graner have investigated the correspondence between the three symmetries of translation in X and in T as well as the Galilean invariance in the framework of scale invariance. They have shown a deep analogy between scale symmetry and relativistic mechanics. In this context, the change of frame leads to a change of moments quantifying the fluctuations of a given statistical field (for instance turbulent velocity) and the generalized Lorenz transformations relate different moments [248–250, 363].

5.3.2 Relation with Dimensional Analysis

A basic postulate in physics is the independence of the laws with respect to changes of units: the physics should be the same for a French experimentator recording in meters and kilograms, or for his English colleague thinking in terms of inches and pounds. The invariance of a quantity under change of units necessarily involves power laws. As a result, any physical quantity can then be expressed in terms of monomials of a few "basic units" such as meters, seconds or kilograms. The exponents appearing in these monomials describe how the quantity varies with respect to unit changes. They define the "dimension" of the quantity. A quantity which remains invariant under changes of units is called "dimensionless".

Laws of physics must be such as to be expressed using only dimensionless variables. This is a powerful statement, mathematically described by the famous Π theorem [54, 55]. It helps reduce the number of physically pertinent parameters and shows the equivalence between similar systems of different sizes. Fluid mechanics is notorious for its extensive use of similitude ratios, such as the Reynolds number. A famous example is the prediction due to Kolmogorov in 1941, who proposed that, in a turbulent fluid, there exist a range of scales called "inertial range" where the velocity u_ℓ of an eddy of size ℓ is only determined by the energy transfer rate $\epsilon = \mathrm{d}(u^2)/\mathrm{d}t$. Dimensional analysis then tells you that $u_\ell = C(\epsilon\ell)^{1/3}$. This simple reasoning predicts that the turbulent spectral energy spectrum $E(k) = u(k)u(-k)$ scales like $k^{-5/3}$ in the inertial range. This prediction has been checked in a variety of turbulent flows and is extremely well satisfied [316].

Scale invariant quantities are power laws, and power laws are naturally generated by dimensional analysis. Does it mean that any scale invariant system can be studied by dimensional analysis? The answer is negative. This is strictly true when you have only one scale, i.e. when you are dealing with an isotropic problem or with a time scale, because power laws are the only homogeneous functions of one variable. As soon as you have more than one variable, you must examine the homogeneous function describing your problem more carefully. There are conspiracies, first discussed by Barenblatt [55], which can drive a system away from a "dimensional" scaling (one then speaks about "anomalous" scaling). These conspiracies involve additional non-dimensional variables, which couple to the initial one. To understand this, let us take again the case of turbulence. Consider the scaling law on the second order longitudinal structure function which provides a quantification of the complexity of turbulent flows:

$$\langle (v_r)^2 \rangle = C_\mathrm{K} (\bar{\epsilon} r)^{2/3} \ . \tag{5.29}$$

The equality is a prediction of Kolmogorov (1941). The variable r, which lies in the inertial range, is the scale at which velocity differences are measured, and $\bar{\epsilon}$ is the mean rate of energy dissipation per unit mass. Dimensional analysis shows that

$$\langle (v_r)^2 \rangle = (\bar{\epsilon} r)^{2/3} F(\Re, r/L) \ , \tag{5.30}$$

where $F(x, y)$ is a universal function to be determined, L is the external or integral scale. Kolmogorov's assumption is that, for the Reynolds number $\Re \to \infty$ and $r/L \to 0$, $F(x, y)$ goes to a constant C_K. This is the so-called complete similarity of the first kind [55] with respect to the variables $\Re$ and r/L.

The existence of the limit of $F(\Re, r/L \to 0)$ has first been questioned by L.D. Landau and A.M. Obukhov, on the basis of the existence of intermittency – fluctuations of the energy dissipation rate about its mean value

$\bar{\epsilon}$. Indeed, Barenblatt's classification leads to the possibility of an *incomplete similarity* in the variable r/L. This would require the absence of a finite limit for $F(\Re, r/L)$ as $r/L \to 0$, and leads in the simplest case to the form

$$\langle (v_r)^2 \rangle = C_{\mathrm{K}} (\bar{\epsilon} r)^{2/3} \left(\frac{r}{L} \right)^\alpha , \tag{5.31}$$

where α is the so-called intermittency exponent, believed to be small and positive. If α is real, this corresponds to a similarity of the second kind [55]. In critical phenomena, this anomalous scaling is understood as the consequence of fluctuations of the order parameter at all scales between the microscopic cut-off up to the macroscopic scale as discussed in Chaps. 8, 9 and 11. The renormalization group has provided the framework to systematically calculate the anomalous corrections to the mean field exponents. The renormalization group has been shown [351] to explain the "similarity of second kind" classified by Barenblatt and found in nonlinear partial differential equations.

Incomplete self-similarity [56, 247] may stem from a possible $\Re$-dependence of the exponents. The case where α is complex, leading to

$$\langle (v_r)^2 \rangle = C_{\mathrm{K}} (\bar{\epsilon} r)^{2/3} \left(\frac{r}{L} \right)^{\alpha_{\mathrm{R}}} \cos[\alpha_{\mathrm{I}} \ln(r/L)] , \tag{5.32}$$

could be termed a similarity of the third kind, characterized by the absence of limit for $F(\Re, r/L)$ and accelerated (log-periodic) oscillations [881], discussed more thoroughly in the next section. To our knowledge, Novikov has been the first to point out in 1966 that structure functions in turbulence could contain log-periodic oscillations [696]. His argument was that, if an unstable eddy in a turbulent flow typically breaks up into two or three smaller eddies, but not into 10 or 20 eddies, then one can suspect the existence of a prefered scale factor, hence the log-periodic oscillations. They have been repeatedly observed in hydrodynamic experiments but do not seem to be stable and depend on the nature of the global geometry of the flow and recirculation [30, 316] as well as the analyzing procedure. Demonstrating unambiguously the presence of log-periodicity and thus of discrete scale invariance (see the next section) in turbulent time-series would provide an important step towards a direct demonstration of the Kolmogorov cascade or at least of its hierarchical imprint. For partial indications of log-periodicity in turbulent data, we refer the reader to Fig. 5.1 p. 58 and Fig. 8.6 p. 128 of [316], Fig. 3.16 p. 76 of [32], Fig. 1b of [948] and Fig. 2b of [146]. Freely decaying 2-d turbulence is a good candidate for a demonstration of discrete scale invariance and log-periodic signatures in the time-evolution of the merging of vortices, in particular in the number of vortices, their radius and separation as a function of time [483]. The log-periodicity could embody the punctuated dynamics of vortices merging in the average scale-free time decay of 2D-turbulence.

5.4 The Multifractal Random Walk

5.4.1 A First Step: the Fractional Brownian Motion

Starting with Hurst's study of 690 time series records of 75 geophysical phenomena, in particular river flow statistics, documenting the so-called "Hurst effect" of long term persistence [441], many studies in the last decades have investigated the existence of long memory effects in a large variety of systems, including meteorology (wind velocity, moisture transfer in the atmosphere, precipitation), oceanography (for instance wave-height), plasma turbulence, solar activity, stratosphere chemistry, seismic activity, internet traffic, financial price volatility, cardiac activity, immune response, and so on.

Mandelbrot and Van Ness [594] introduced the fractional Brownian motion (fBm) as the unique possible extension of the memoryless continuous time random walk which has an exact self-similarity with an arbitrary exponent H which can be different from the value $1/2$ for the standard random walk. The standard memoryless continuous time random walk has a unique specification described in Sect. 2.1.3 and, in mathematics, it is called the Wiener process. The motivation for the fBm was to account for the Hurst effect. From an initial value $B_H(0)$, the fBm is defined by

$$B_H(t) - B_H(0) = \frac{1}{\Gamma(H + (1/2))} \int_{-\infty}^{t} d\tau\, \eta(\tau) K(t - \tau) , \qquad (5.33)$$

where $d\tau\, \eta(\tau) = dW_\tau$ is usually taken as the increment of the standard random walk with white noise spectrum and Gaussian distribution with variance $E[dW_\tau] = d\tau$. The memory kernel $K(t - \tau)$ is given by

$$K(t - \tau) = (t - \tau)^{H-1/2} , \qquad \text{for } 0 \le \tau \le t \qquad (5.34)$$

$$= (t - \tau)^{H-1/2} - (-\tau)^{H-1/2} , \qquad \text{for } \tau < 0 . \qquad (5.35)$$

For $H > 1/2$, the fBm $B_H(t)$ exhibits long term persistence and memory, since the effect of past innovations of dW_τ is felt in the future with a slowly decaying power law weight $K(t-\tau)$. For $H < 1/2$, the fBm is anti-persistent, meaning that the fBm tends to come back more often to its past tracks than would a memoryless random walk.

Fractional noise motion (fNm), which is defined as the time derivative of $B_H(t)$, possesses the property of statistical stationarity. A fNm is defined by

$$A(t) = \int_{-\infty}^{t} d\tau\, \eta(\tau) K(t - \tau) , \qquad (5.36)$$

with

$$K_{\text{fNm}}(t - \tau) = \frac{1}{(t - \tau)^{3/2-H}} = \frac{1}{(t - \tau)^{1-\theta}} , \qquad (5.37)$$

for $H = 1/2 + \theta$. Persistence $1/2 < H < 1$ (respectively antipersistence $0 < H < 1/2$) corresponds to $0 < \theta < 1/2$ (respectively $-1/2 < \theta < 0$). Such a memory kernel describes also the renormalized Omori's law for earthquake aftershocks [411, 414, 864] and many other processes and has important implications allowing to distinguish exogenous versus endogenous shocks [897, 906].

Both the fBm and fNm exhibit simple scaling and are thus "monofractals."

5.4.2 Definition and Properties of the Multifractal Random Walk

Motivated by the structure (5.21) at different scales of probability density functions and by the dependence properties of velocity increments in hydro-dynamic turbulence and of financial returns, the multifractal random walk (MRW) has been introduced by Bacry and Muzy as the unique continuous time random walk that generalizes the fBm to exhibit exact multifractal properties [41, 658]. The MRW is the continuous time limit of a random walk with a stochastic variance such that the correlation in time of the logarithm of the variance decays logarithmically. It possesses a nice stability property related to its scale invariance property. For each time scale $\Delta t \leq T$, the increment at time t of the MRW at scale Δt, $r_{\Delta t}(t)$, can be described as follows:

$$r_{\Delta t}(t) = \epsilon(t)\sigma_{\Delta t}(t) = \epsilon(t)e^{\omega_{\Delta t}(t)} \,, \tag{5.38}$$

where $\epsilon(t)$ is a standardized Gaussian white noise independent of $\omega_{\Delta t}(t)$ and $\omega_{\Delta t}(t)$ is a nearly Gaussian process with mean and covariance:

$$\mu_{\Delta t} = \frac{1}{2}\ln(\sigma^2 \Delta t) - C_{\Delta t}(0) \tag{5.39}$$

$$C_{\Delta t}(\tau) = \text{Cov}[\omega_{\Delta t}(t), \omega_{\Delta t}(t+\tau)] = \lambda^2 \ln\left(\frac{T}{|\tau| + e^{-3/2}\Delta t}\right) \,, \tag{5.40}$$

where $\sigma^2 \Delta t$ is the variance of the increment at scale Δt and T represents an "integral" (correlation) time scale. Such logarithmic decay of the log-variance covariance at different time scales as been documented empirically for financial returns in [34, 658, 659]. Typical values for T and λ^2 are respectively 1 year and 0.04. According to the MRW model, the variance correlation exponent ν is equal to λ^2 according to (5.48) below.

It is important to stress that it is only for this specification (5.39) and (5.40) that the process has a bona fide continuous limit. Rigorously, the process (5.38) is defined such that the t-th realization of $r_{\Delta t}$, defined as the increment of the MRW between t and $t + \Delta t$, is obtained as

$$r_{\Delta t} = \lim_{\delta t \to 0} \sum_{k=1}^{\Delta t / \delta t} \epsilon_{\delta t}(k) e^{\omega_{\delta t}(k)} \, , \tag{5.41}$$

where $\epsilon_{\delta t}$ and $\omega_{\delta t}$ have the properties (5.39) and (5.40) (with Δt replaced by δt). For $\lambda^2 = 0$, $\omega_{\delta t} = 0$ and expression (5.41) recovers the standard construction of the Wiener process. In this later case, the stability property is expressed by the fact that both the infinitesimal increments $\epsilon_{\delta t}(k)$ and the increment $r_{\Delta t}$ at arbitrary finite time scales are distributed according to Gaussian laws with variances in ratio $\Delta t / \delta t$.

The MRW model can be expressed in a different form, in which the log-variance $\omega_{\Delta t}(t)$ obeys an auto-regressive equation whose solution reads

$$\omega_{\Delta t}(t) = \mu_{\Delta t} + \int_{-\infty}^{t} d\tau \, \eta(\tau) K_{\Delta t}(t - \tau) \, . \tag{5.42}$$

Here, $\eta(t)$ denotes again a standardized Gaussian white noise and the memory kernel $K_{\Delta t}(\cdot)$ is a causal function, ensuring that the system is not anticipative. The process $\eta(t)$ can be seen as the information flow. Thus $\omega(t)$ represents the response of the system to incoming information up to the date t. At time t, the distribution of $\omega_{\Delta t}(t)$ is Gaussian with mean $\mu_{\Delta t}$ and variance $V_{\Delta t} = \int_0^\infty d\tau \, K_{\Delta t}^2(\tau) = \lambda^2 \ln(T e^{3/2} / \Delta t)$. Its covariance, which entirely specifies the random process, is given by

$$C_{\Delta t}(\tau) = \int_0^\infty dt \, K_{\Delta t}(t) K_{\Delta t}(t + |\tau|) \, . \tag{5.43}$$

Performing a Fourier transform, we obtain

$$\hat{K}_{\Delta t}(f)^2 = \hat{C}_{\Delta t}(f)$$
$$= 2\lambda^2 f^{-1} \left[\int_0^{T f} \frac{\sin(t)}{t} dt + O\left(f \Delta t \ln(f \Delta t)\right) \right] \, , \tag{5.44}$$

where we have used (5.40). This shows that for τ small enough

$$K_{\Delta t}(\tau) \sim K_0 \sqrt{\frac{\lambda^2 T}{\tau}} \qquad \text{for} \ \ \Delta t \ll \tau \ll T \, . \tag{5.45}$$

This slow power law decay (5.45) of the memory kernel in (5.42) ensures the long-range dependence and multifractality of the stochastic variance process (5.38). Note that (5.42) for the log-volatility $\omega_{\Delta t}(t)$ takes a form similar to but simpler than the ARFIMA models usually defined on the (linear) variance σ [43, 75].

The MRW has three parameters:

- σ^2: variance of increments;
- λ^2: intermittency coefficient;
- T: integral correlation time.

For any $\Delta t \leq T$, the MRW is characterized by an exact continuous scale invariance (multifractality) which takes the following form for q even:

$$\langle |r_{\Delta t}|^q \rangle \sim (\Delta t)^{\zeta_q} , \tag{5.46}$$

with

$$\zeta_q = \left(\frac{1}{2} + \lambda^2 \right) q - \frac{\lambda^2}{2} q^2 . \tag{5.47}$$

Here, the scale l of (5.8) is the time scale Δt. Note that the definition of multifractaly differs slightly from that in (5.8) and the correspondence is $\zeta_q = qD_q$ rather than $(q - 1)D_q$ due to the absence of a condition of normalization in (5.46). For $\Delta t \gg T$, $\zeta_q = q/2$ which expresses the fact that the MRW recovers a standard random walk structure. For $q > q^* = (1 + 2\lambda^2)/\lambda^2$, ζ_q becomes negative, implying that $\langle |r_{\Delta t}|^q$ diverges for $\Delta t \to 0$. The divergence of the moments of order larger than q^* implies that the probability density function of $r_{\Delta t}$ has a heavy (power law-like) tail with exponent μ equal to q^*. For turbulence or financial applications, $\lambda^2 \approx 0.04$ which gives $q^* \approx 27$. Such large power law exponent would be very hard if not impossible to qualify in practice. Lastly, the MRW predicts a very slow decay of the correlation functions of the powers of the variances of the increments:

$$\langle |r_{\Delta t}(t)|^q \ |r_{\Delta t}(t + \tau)|^q \rangle \sim \left(\frac{\tau}{T} \right)^{-\lambda^2 q^2} , \tag{5.48}$$

for q even.

5.5 Complex Fractal Dimensions and Discrete Scale Invariance

Fractals have dimensions that are in general real numbers. As already pointed out, the generalization from the set of integers to the set of real numbers embodies the transition from the symmetry of translational invariance to the symmetry of scale invariance. It is possible to generalize further and ask what could be the properties of a set whose fractal dimension belongs to the set of complex numbers. It turns out that this generalization captures the interesting and rich phenomenology of systems exhibiting discrete scale invariance, a weaker form of scale invariance symmetry, associated with log-periodic corrections to scaling [878].

5.5.1 Definition of Discrete Scale Invariance

Let us start from the concept of (continuous) scale invariance defined by (5.26). Discrete scale invariance (DSI) is a weaker kind of scale invariance according to which the system or the observable obeys scale invariance as defined above only for specific choices of λ (and therefore μ), which form in general an infinite but countable set of values $\lambda_1, \lambda_2, \ldots$ that can be written as $\lambda_n = \lambda^n$. λ is the fundamental scaling ratio. This property can be qualitatively seen to encode a *lacunarity* of the fractal structure [592].

As we already mentioned, continuous scale invariance is the same as continuous translational invariance expressed on the logarithms of the variables. DSI is then seen as the restriction of the continuous translational invariance to a *discrete* translational invariance: $\ln \mathcal{O}$ is simply translated when translating y by a multiple of a fundamental "unit" size $\ln \lambda$. Going from continuous scale invariance to DSI can thus be compared with (in logarithmic scales) going from the fluid state to the solid state in condensed matter physics! In other words, the symmetry group is no longer the full set of translations but only those which are multiples of a fundamental discrete generator.

5.5.2 Log-Periodicity and Complex Exponents

We have seen that the hallmark of scale invariance is the existence of power laws. The signature of DSI is the presence of power laws with *complex* exponents α which manifests itself in data by log-periodic corrections to scaling. To see this, consider the triadic Cantor set shown in Fig. 5.1. It is usually stated that this triadic Cantor set has the fractal (capacity) dimension $D_0 = \ln 2/\ln 3$, as the number of intervals grows as 2^n while their length shrinks as 3^{-n} at the n-th iteration.

It is obvious to see that, by construction, this triadic Cantor set is geometrically identical to itself *only* under magnification or coarse-graining by factors $\lambda_p = 3^p$ which are arbitrary powers of 3. If you take another magnification factor, say 1.5, you will not be able to superimpose the magnified part on the initial Cantor set. We must thus conclude that the triadic Cantor set does not possess the property of continuous scale invariance but only that of DSI under the fundamental scaling ratio 3.

This can be quantified as follows. Call $N_x(n)$ the number of intervals found at the n-th iteration of the construction. Call x the magnification factor. The original unit interval corresponds to magnification 1 by definition. Obviously, when the magnification increases by a factor 3, the number $N_x(n)$ increases by a factor 2 independent of the particular index of the iteration. The fractal dimension is defined as

$$D = \lim_{x \to \infty} \frac{\ln N_x(n)}{\ln x} = \lim_{x \to 0} \frac{\ln N_x(n)}{\ln x} = \frac{\ln 2}{\ln 3} \approx 0.63 . \qquad (5.49)$$

However, the calculation of a fractal dimension usually makes use of arbitrary values of the magnification and not only those equal to $x = 3^p$. If we increase the magnification continuously from say $x = 3^p$ to $x = 3^{p+1}$, the numbers of intervals in all classes jump by a factor of 2 at $x = 3^p$, but then remains unchanged until $x = 3^{p+1}$, at which point they jump again by an additional factor of 2. For $3^p < x < 3^{p+1}$, $N_x(n)$ does not change while x increases, so the measured fractal dimension $D(x) = (\ln N_x(n))/\ln x$ decreases. The value

$D = 0.63$ is obtained only when x is a positive or negative power of three. For continuous values of x one has

$$N_x(n) = N_1(n)x^D P\left(\frac{\ln x}{\ln 3}\right), \tag{5.50}$$

where P is a function of period unity. Now, since P is a periodic function, we can expand it as a Fourier series

$$P\left(\frac{\ln x}{\ln 3}\right) = \sum_{n=-\infty}^{\infty} c_n \exp\left(2n\pi i \frac{\ln x}{\ln 3}\right). \tag{5.51}$$

Plugging this expansion back into (5.50), it appears that D is replaced by an infinity of complex values

$$D_n = D + ni\frac{2\pi}{\ln 3}. \tag{5.52}$$

We now see that a proper characterization of the fractal is given by this set of *complex dimensions* which quantifies not only the asymptotic behaviour of the number of fragments at a given magnification, but also its modulations at intermediate magnifications. The imaginary part of the complex dimension is directly controlled by the prefered ratio 3 under which the triadic Cantor set is exactly self-similar. Let us emphasize that DSI refers to discreteness in terms of scales, rather than discreteness in space (like discreteness of a cubic lattice approximation to a continuous medium).

If we keep only the first term in the Fourier series in (5.51) and insert in (5.50), we get

$$N_x(n) = N_1(n)x^D\left(1 + 2\frac{c_1}{c_0}\cos\left(2n\pi\frac{\ln x}{\ln 3}\right)\right), \tag{5.53}$$

where we have used $c_{-1} = c_1$ to ensure that $N_x(n)$ is real. Expression (5.53) shows that the imaginary part of the fractal dimension translates itself into a log-periodic modulation decorating the leading power law behavior. Notice that the period of the log-periodic modulation is simply given by the logarithm of the prefered scaling ratio. The higher harmonics are related to the higher dimensions D_n defined in (5.52) for $n > 1$.

It is in fact possible to directly obtain all these results from (5.26). Indeed, let us look for a solution of the form $\mathcal{O}(x) = Cx^\alpha$. Putting in (5.26), we get the equation $1 = \mu\lambda^\alpha$. But 1 is nothing but $e^{i2\pi n}$, where n is an arbitrary integer. This leads to

$$\alpha = -\frac{\ln\mu}{\ln\lambda} + i\frac{2\pi n}{\ln\lambda}, \tag{5.54}$$

which has exactly the same structure as (5.52). The special case $n = 0$ gives the usual real power law solution corresponding to fully continuous scale invariance. In contrast, the *more general* complex solution corresponds to a possible DSI with the prefered scaling factor λ. The reason why (5.26) has solutions in terms of complex exponents stems from the fact that a finite rescaling has been done by the finite factor λ. In critical phenomena presenting continuous scale invariance, (5.26) corresponds to the linearization, close to the fixed point, of a renormalization group equation describing the behavior of the observable under a rescaling by an arbitrary factor λ. The power law solution and its exponent α must then not depend on the specific choice of λ, especially if the rescaling is taken infinitesimal, i.e. $\lambda \to 1^+$. In the usual notation, $\lambda = e^{a_x \ell}$ which implies that $\mu = e^{a_\phi l}$ and $\alpha = -a_\phi/a_x$ is independent of the rescaling factor ℓ as a_x and a_ϕ are independent of ℓ. In this case, the imaginary part in (5.54) drops out.

As we have seen, going from integer dimensions to fractional dimensions corresponds to a generalization of the translational symmetry to the scaling symmetry. It may come as a surprise to observe that generalizing further the concept of dimensions to the set of complex numbers is in constrast reducing the scale symmetry into a sub-group, the discrete scale symmetry. This results from the fact that the imaginary part is actually introducing an additional constraint that the symmetry must obey. Chapter 11 expands on the role played by complex fractal dimensions as being the critical exponents emerging from renormalization group equations in hierarchical systems.

5.5.3 Importance and Usefulness of Discrete Scale Invariance

Existence of Relevant Length Scales. Suppose that a given analysis of some data shows log-periodic structures. What can we get out of it? First, as we have seen, the period in log-scale of the log-periodicity is directly related to the existence of a prefered scaling ratio. Thus, logperiodicity must immediatly be seen and interpreted as the existence of a set of prefered characteristic scales forming all together a geometrical series $..., \lambda^{-p}, \lambda^{-p+1}, ..., \lambda, \lambda^2, ..., \lambda^n, ...$ The existence of such prefered scales appears in contradiction with the notion that a critical system exhibiting scale invariance has an infinite correlation length, hence only the microscopic ultraviolet cut-off and the large scale infra-red cut-off (for instance the size of the system) appear as distinguishable length scales. This recovers the fact that DSI is a property different from continuous scale invariance. Examples where complex exponents can be found are random systems, out-of-equilibrium situations and irreversible growth problems. In addition to the existence of a single prefered scaling ratio and its associated log-periodicity discussed above, there can be several prefered ratios corresponding to several log-periodicities that are superimposed. This can lead to a richer behavior such as log-quasiperiodicity, which has been suggested to describe the scaling properties of diffusion-limited-aggregation clusters [899].

Log-periodic structures in the data indicate that the system and/or the underlying physical mechanisms have characteristic length scales. This is extremely interesting as this provides important constraints on the underlying physics. Indeed, simple power law behaviors are found everywhere, as seen from the explosion of the concepts of fractals, criticality and self-organized-criticality [44]. For instance, the power law distribution of earthquake energies known as the Gutenberg–Richter law can be obtained by many different mechanisms and a variety of models and is thus extremely limited in constraining the underlying physics. Its usefulness as a modelling constraint is even doubtful, in contradiction with the common belief held by physicists on the importance of this power law. In contrast, the presence of log-periodic features would teach us that important physical structures, hidden in the fully scale invariant description, existed.

Prediction. It is important to stress the practical consequence of log-periodic structures. For prediction purposes, it is much more constrained and thus reliable to fit a part of an oscillating data than a simple power law which can be quite degenerate especially in the presence of noise. This remark has been used and is vigorously investigated in several applied domains, such as earthquakes [477, 808, 809, 910], rupture prediction [29] and financial crashes [300, 479, 898, 900].

5.5.4 Scenarii Leading to Discrete Scale Invariance

After the rather abstract description of DSI given above, let us briefly discuss the physical mechanisms that may be found at its origin. It turns out that there is not a unique cause but several mechanisms which may lead to DSI. Since DSI is a partial breaking of a continuous symmetry, this is hardly surprising as there are many ways to break a symmetry. Some mechanisms have already been unravelled while others are still under investigation. The list of mechanisms and poorly understood example is by no mean exhaustive and others will certainly be found in the future (see [878] and references therein):

1. Built-in geometrical hierarchy,
2. Programming and number theory,
3. Newcomb–Benford law of first digits [425] and the arithmetic system,
4. Eigenfunctions of the Laplace transform,
5. Diffusion in anisotropic quenched random lattices,
6. Cascade of ultra-violet instabilities: growth processes and rupture,
7. Deterministic dynamical systems

   - Cascades of sub-harmonic bifurcations in the transition to chaos,
   - Two-coupled anharmonic oscillators,
   - Near-separatrix Hamiltonian chaotic dynamics,
   - Kicked charged particle moving in a double-well potential: physical realization of Mandelbrot and Julia sets,
   - Log-periodic topology in chaotic scattering,

8. Animals (configurations of percolation clusters),
9. Quenched disordered systems,

10. Turbulence [483, 1058, 1059],
11. The bronchial tree of mammals,
12. Titius–Bode law,
13. Gravitational collapse and black hole formation,
14. Spinodal decomposition of binary mixtures in uniform shear flow,
15. Cosmic lacunarity,
16. Rate of escape from stable attractors,
17. Life evolution...

It is essential to notice that all the mechanisms involve the existence of a characteristic scale (an upper and/or lower cut-off) from which the DSI can develop and cascade. In fact, for characteristic length scales forming a geometrical series to be present, it is unavoidable that they "nucleate" from either a large size or a small mesh. This remark has the following important consequences: even if the mathematical solution of a given problem contains in principle complex exponents, if there are no such cut-off scales to which the solution can "couple" to, then the log-periodicity will be absent in the physical realization of the problem. An example of this phenomenon is provided by the interface-crack stress singularity at the tip of a crack at the interface between two different elastic media [775, 776].

6. Rank-Ordering Statistics and Heavy Tails

6.1 Probability Distributions

Many physical and natural systems exhibit probability density functions (pdf's) that are different from the most commonly used Gaussian or log-normal distributions. Many phenomena are in fact characterized by "heavy tails", i.e. larger probabilities for large event sizes compared to the prediction given by a Gaussian or log-normal fit using the variance estimated from the data. Even if less frequent, the large events often play an important if not leading role. The largest earthquakes account for the majority of the tectonic deformation in seismic regions as well as cause the largest damages, the largest floods and largest storms provide major erosion boosts and so on. It is thus of utmost interest and importance to characterize these large events and their distributions. However, by definition, the statistics of rare events is limited and the distributions are thus difficult to constrain.

There is a large literature on the empirical determination of distributions that we barely skim here. One school advocates the use of cumulative distribution which presents the advantage of not requiring binning. The problem with binning is that bins become undersampled in the tail. On the other hand, cumulative distributions suffer from systematic biais due to their cumulative nature which introduces correlations in their fluctuations. Diverse regularization methods can be introduced to deal with pdf's (see for instance [28] with an application on the distribution of velocity increments in turbulence), using for instance "line" and "curvature energy" to smooth out the pdf, optimized by guessing the functional form of the tail of the distribution. A simpler but efficient approach consists in computing histograms with different bin sizes, evaluate the pdf from each histogram and then combine those values to form a completed pdf [563]. From a more theoretical point of view, recent approaches have developed a field-theoretical point of view incorporating Bayesian principles [87, 430].

It is important to stress that distributions are statistical objects and hence should be discussed by using the language of the statistical science rather than "deterministic" methods consisting in a more or less naive visual inspection and fit of log–log plots. The method of least-square fitting of the log–log plots grossly underestimates the standard errors of the slope (exponent) parameter and has other deficiencies in evaluating the parameters of statistical distributions. Vere-Jones ([984], p. 287) remarks in regard to the estimation of the b-value in the magnitude–frequency Gutenberg–Richter relation for earthquakes: "Unweighted least squares should not be used; weighted least squares, with weights proportional to the expected cell entries, should give comparable values for the slope (although there is a problem with empty cells) but still leads to spurious estimates of the standard error – in particular a perfect fit does not mean zero standard error estimate!"

The determination of power-law exponents of cumulative distributions from a finite number of data may be extremely sensitive to the values of the largest sampled values. A proper method to estimate the exponent is to use a logarithmic transformation of the data, which converts the power-law into an exponential distribution. The corresponding maximum-likelihood technique [9, 422] allows one to evaluate the truncation point at large values [209, 536] and the possible consequences of incomplete data [699]. Let us also mention Adler et al. [3] who review modern methods of statistical analysis for the power-law distributions. The problem of truncation and censoring effects has been addressed by many studies. In the geophysical context, let us mention the abundant literature dealing with fracture distributions [174, 198, 551, 736]. In this case, the problem is made worse by that fact that large fractures are only partially sampled, leading to a bad estimation of both the density and cumulative distributions. Corrections have been developed for instance in studies dealing with scaling properties of channel network [188, 544] and faulting [710].

In the following, we touch on several of these issues by studying the rank ordering technique [1065]. This approach presents the advantage of constraining the fit of empirical distributions by the large events. This is in contrast to usual fits by density or cumulative distributions that are usually constrained by the large majority of small and intermediate event sizes. A second advantage is that it allows a simple and quantitative assessment of the statistical significance of the fits. For instance, it turns out that it is possible to get a reasonable estimation of the exponent of a power law from the measurement of a few tens of data points [903]. We show how the rank-ordering technique applies to the exponential, power law, Gamma law and stretched exponential distributions. Motivated by the versatility of stretched exponential distributions [547] as intermediate between "thin tail" (Gaussian, exponential, ...) and very "fat tail" distributions, we end this chapter by a review of calibration methods relevant to this class.

6.2 Definition of Rank Ordering Statistics

Consider N observations and let us reorder them by decreasing values

$$v_1 \geq v_2 \geq ... \geq v_n \geq ... \geq v_N \ . \tag{6.1}$$

v_1 is the largest observed value, v_2 is the second largest one and so on. The rank-ordering method consists in quantifying the dependence of the nth value v_n as a function of the rank n. This method is well-suited when the data is naturally ordered. For instance, eonomic and business statistics are often presented in this format (sales of the largest firms, incomes, etc.).

The rank-ordering method is very close to the construction of the cumulative distribution $P_>(v)$. Recall that $P_>(v)$ is defined as the probability to find a value larger or equal to v:

$$P_>(v) = \int_v^\infty p(x)\,\mathrm{d}x \; , \tag{6.2}$$

where $p(x)$ is the pdf. The integer part of $NP_>(v_n)$ is thus the expected number of values larger or equal to v_n. Equating it to n simply expresses that v_n is indeed the nth largest observed value:

$$NP_>(v_n) = n \; . \tag{6.3}$$

The expression (6.3) provides a determination of v_n knowing $p(v)$ or $P_>(v)$. Inversely, measuring v_n, one can determine $p(v)$. This reasoning ignores fluctuations which appear in real data.

For any random variable v with continuous cumulative distribution $P_<(v)$, the random value $P_<(v)$ has a uniform distribution on the interval $[0, 1]$. Thus, for ordered sample $v_1, ..., v_N$, the random variables $P_<(v_1), ..., P_<(v_N)$ are distributed as N uniform ordered random variables. The mean values and variances of these random variables are easily calculated:

$$\mathrm{E}[P_<(v_k)] = \frac{k}{N+1} \; ; \quad \mathrm{Var}[P_<(v_k)] = \frac{k(N-k+1)}{(N+1)^2(N+2)} \; . \tag{6.4}$$

The expression (6.3) has a simple geometric interpretation: the rank-ordering method simply consists in interchanging the $(x = v, \; y = P(v))$ axis into $(x = NP(v), \; y = v)$. However, this interchange is not as innocent as it seems *a priori*. In the first representation $(x = v, \; y = P(v))$ used for the cumulative distribution, the inherent statistical fluctuations always present in finite data sets are decorating the ordinate $P_>(v)$ at fixed v. In the second representation $(x = NP(v), \; y = v)$ used in the rank-ordering method, the fluctuations occur on the variable v itself at fixed rank n.

For the large events where the data is sparse, the fluctuations of the estimated $P_>(v) = n/N$ are controlled by integers added or subtracted from small integers. This introduces a significant bias that is no present in the fluctuations of the values at fixed rank. We will illustrate this point in what follows.

Let us now give the exact formula making (6.3) more precise. The probability $F(v_n) \, dv_n$ that the nth value is equal to v_n to within dv_n reads [381]:

$$F(v_n) \, dv_n = (N - n + 1) \binom{N}{n} \left(1 - \int_{v_n}^{+\infty} p(v) \, dv \right)^{N-n}$$

$$\times p(v_n) \, dv_n \left(\int_{v_n}^{+\infty} p(v) \, dv \right)^{n-1} . \tag{6.5}$$

The term $\binom{N}{n}$ is the number of combinations of n elements among N. The term $(1 - \int_{v_n}^{+\infty} p(v) \, dv)^{N-n}$ is the probability that $N - n$ values are smaller than v_n. The term $p(v_n) \, dv_n$ is the probability for finding one value between v_n and $v_n + dv_n$. The last term $[\int_{v_n}^{+\infty} p(v) \, dv]^{n-1}$ is the probability that the $n - 1$ remaining values are larger than v_n. These different terms contribute multiplicatively to $F(v_n) \, dv_n$ due to the assumed statistical independence of the events. This expression (6.5) is valid for arbitrary pdf's $p(v)$. From (6.5), we can get an estimate of the "typical" value v_n: it is simply the value that maximizes $F(v_n)$. Let us now illustrate this method by discussing a few parametric examples.

6.3 Normal and Log-Normal Distributions

The log-normal distribution is defined by

$$p(v) \, dv = \frac{1}{\sqrt{2\pi}\sigma} e^{[\ln(v/v_0)]^2/2\sigma^2} \frac{dv}{v} , \tag{6.6}$$

where σ is the standard deviation of the variable $\ln v$.

Maximizing the expression (6.5) provides, as we said, the most probable nth rank v_n, which is a solution of

$$\frac{e^{-x_n^2}}{\sqrt{\pi} x_n} = \frac{n}{N} , \tag{6.7}$$

where

$$x_n = \frac{\ln v_n}{\sqrt{2}\sigma} . \tag{6.8}$$

This expression also applies to the determination of a normal (or Gaussian) distribution since the variable x_n is distributed according to a Gaussian pdf. The leading term for large v's is

$$x_n \simeq \sqrt{\ln \frac{N}{n\sqrt{2\pi}}} , \tag{6.9}$$

which is correct up to terms of order $\ln \ln v$. Expanding (6.5) up to second order around the maximum, we obtain the typical uncertainty Δx_n in the determination of x_n:

$$\Delta x_n = \frac{1}{\sqrt{2n}} \frac{1}{x_n} \; . \tag{6.10}$$

Translating these results to the log-normal case by inverting the formula (6.8), we obtain v_n as a function of x_n. The main point to note is the large amplitude of the fluctuations for the first ranks:

$$\frac{\Delta v_n}{v_n} = \frac{1}{\sqrt{\ln N/(n\sqrt{2\pi})}} \frac{\sqrt{2}\sigma^2}{\sqrt{n}} \approx \frac{\sigma^2}{\sqrt{n}} \; . \tag{6.11}$$

For instance, take $N = 4000$ and $\sigma = 2.7$. This yields $\Delta v_1/v_1 = 3.7$ and $\Delta v_{10}/v_{10} = 1.4$. These numbers of the order or larger than unity illustrate the fact that the very largest number in a sample exhibits very large fluctuations.

6.4 The Exponential Distribution

It is defined by

$$p(v) = \mu e^{-\mu v} \; , \qquad \text{for } 0 \le v < \infty \; . \tag{6.12}$$

Applying the formula (6.3) yields the typical value

$$v_n = -\frac{1}{\mu} \ln \frac{n}{N} \; . \tag{6.13}$$

Thus, a straight line obtained by plotting v_n as a function of $\ln n$ qualifies an exponential pdf. The slope of the straight line gives $1/\mu$.

Expression (6.5) shows that the tail of the pdf of the largest value v_1 is

$$F(v_1)\,dv_1 \simeq N\mu e^{-\mu v_1} e^{-(N-1)e^{-\mu v_1}} \; , \tag{6.14}$$

in the limit of large N. As we have seen in Chap. 1, this law is known as the Gumbel distribution for extremes [381]. The most probable value of the largest value is given by

$$v_1^{\mathrm{mp}} = \frac{1}{\mu} \ln \frac{N}{\mu} \; , \tag{6.15}$$

which is consistent with (6.13) but not exactly the same.[1] The asymptotic tail of (6.14) is an exponential $N\mu e^{-\mu v_1}$ for $v_1 \gg v_1^{\mathrm{mp}}$. The fluctuations of v_1

---

[1] They are equivalent to within a factor μ in the logarithm: indeed, the typical and most probable values are close but not exactly identical.

for one realization of the N events to another are thus correctly estimated by the standard deviation Δv_1, which is well-defined mathematically:

$$\Delta v_1 = \frac{1}{\mu} \; . \tag{6.16}$$

The relative error $\Delta v_1/v_1^{\mathrm{mp}} = 1/\ln(N/\mu)$ decays very slowly to zero as the size N of the data set increases.

The statistical literature is replete with estimators for μ. Before recalling the main one, it is useful to stress how difficult it is in practice to estimate the reliability of an estimator. It is recommended to use the graphical representation of the rank-ordering plots in order to get a qualitative insight on the quality of the fit. A simple least-square fit of a rank-ordering plot provides a fast and often reliable estimation of μ.[2]

The best known estimator for μ is obtained by the maximum-likelihood approach [267]. The problem is to obtain the best estimate of μ from the knowledge of the first n ranks $v_1 > v_2 > ... > v_n$. Note that, as we have already mentioned, proceeding in this way ensures that the estimation of μ is controlled by the largest values. In order to simplify, consider v_n fixed. The probability to observe a value v conditioned on being larger than or equal to v_n is [3]

$$p_{\mathrm{c}}(v) = \mu e^{-\mu(v-v_n)} \; . \tag{6.17}$$

The probability to observe n values $v_1, v_2, ..., v_n$ conditioned to be all larger than or equal to v_n is the product of n terms of the form (6.17)

$$p_{\mathrm{c}}(v_1, v_2, ..., v_n) = \mu^n e^{-\mu n(\langle x \rangle_n - x_n)} \; , \tag{6.18}$$

where $\langle x \rangle_n = (1/n) \sum_{i=1}^{n} x_i$. $p_{\mathrm{c}}(v_1, v_2, ..., v_n)$ is also called the likelihood. The most probable value of μ maximizes $p_{\mathrm{c}}(v_1, v_2, ..., v_n)$, and is thus the solution of $dp_{\mathrm{c}}(v_1, v_2, ..., v_n)/d\mu = 0$. The solution is

$$\mu = \frac{1}{(1/n) \sum_{i=1}^{n} (x_i - x_n)} \; . \tag{6.19}$$

This is the Hill estimator [422, 423] obtained by maximizing the likelihood function (6.18), hence the name "maximum likelihood estimator." The most correct estimator is obtained by relaxing the constraint on x_n. Then, $p_{\mathrm{c}}(v_1, v_2, ..., v_n)$ reads

$$p_{\mathrm{c}}(v_1, v_2, ..., v_n) = \mu^n e^{-\mu(1/n) \sum_{i=1}^{n} v_i} \left(1 - e^{-\mu v_n}\right)^{N-n} \; , \tag{6.20}$$

---

[2] In principle, a least-squares fit is not the best method as the fluctuations are not expected to be independent and Gaussian, due to the cumulative nature of the plot. In practice however, these effects are not strong and do not lead to significant deviations from other methods.

[3] We make use of the identity that the probability of an event A conditioned on an event B is the ratio of the two probabilities.

where the additional factor is the probability that the $N - n$ other variations are smaller than v_n. This equation (6.20) is nothing but another form of the expression (6.5). Maximizing $p_c(v_1, v_2, ..., v_n)$ with respect to μ yields

$$\frac{n}{\mu} - \sum_{i=1}^{n} \ln v_i + (N - n)\frac{v_n^{-\mu} \ln v_n}{1 - v_n^{-\mu}} = 0 . \tag{6.21}$$

In principle, this estimator is better than the first (6.19). In practice, it turns out to be worse! The reason is simply that it is quite rare that the whole set of data is available with the same quality. In addition, if the tail is a power law, the smaller events may have a different pdf shape. Taking them into account as in (6.21) leads to a severe distortion. It is thus better to use Hill's estimator using data from the first to the nth rank and test its stability as n is progressively increased.

A rigorous justification for Hill's estimator can be obtained from the general expression (6.5). Let us condition the observation v_n as being the nth largest one. By virtue of Bayes' theorem, the pdf of v_n which is conditionned in this way reads

$$F(v_n)\, \mathrm{d}v_n = \binom{n}{1}\ p(v_n)\, \mathrm{d}v_n\ \left(\int_{v_n}^{+\infty} p(v)\, \mathrm{d}v\right)^{n-1} , \tag{6.22}$$

where the factor $\binom{N}{n}\left(1 - \int_{v_n}^{+\infty} p(v)\, \mathrm{d}v\right)^{N-n}$ stems from the condition. This expression retrieves Hill's estimator (6.19) by maximization.

The quality of Hill's estimator is obtained from the typical amplitude of the fluctuations

$$\Delta\mu = \left(\frac{\mathrm{d}^2 p_c(v_1, v_2, ..., v_n)}{\mathrm{d}\mu^2}\right)^{-1/2} . \tag{6.23}$$

In the limit of large n, one gets

$$\frac{\Delta\mu}{\mu} = \frac{1}{n^{1/2}} . \tag{6.24}$$

These expressions (6.23) and (6.24) are justified through the Fisher Information, which states that the asymptotic variance of the estimator of μ is the inverse of the Fisher information $I = -\mathrm{E}[\mathrm{d}^2 \ln L/\mathrm{d}\mu^2]$, where L stands for the likelihood defined in (6.18) or (6.20). In the case where several parameters exist, I becomes a hessian matrix.

One can do better and get the full distribution of μ: asymptotically, it takes the shape of a Gaussian distribution with variance $(\Delta\mu)^2$. In the statistical literature, the determination of non-asymptotic corrections to the Gaussian law is still open. In practice, this is a very important question since most concrete applications involve typically a few tens to a few hundred data points. By definition, the tails of distributions have few events.

In [903], the leading non-asymptotic correction to the Gaussian law is given in the case where only the first n ranks are used. If μ_0 is the true value, one finds that $1/\mu$ has a Gaussian pdf with mean $1/\mu_0$ and variance $1/n\mu_0^2$:

$$P\left(\frac{1}{\mu}\right) = \left(\frac{n}{2\pi}\right)^{1/2} \mu_0 e^{-(1/\mu-1/\mu_0)^2 n\mu_0^2/2} , \tag{6.25}$$

leading to an asymmetric distribution $P(\mu)$:

$$P(\mu) = \left(\frac{n}{2\pi}\right)^{1/2} \frac{\mu_0}{\mu^2} e^{-[n(\mu-\mu_0)^2]/2\mu^2} . \tag{6.26}$$

The most probable value is

$$\mu^{\mathrm{mp}} = \mu_0 \frac{2}{1 + (1 + [8/n])^{1/2}} , \tag{6.27}$$

which converges to μ_0 from below, as $n \to +\infty$. Thus, there is a systematic bias since the most probable estimator is shifted from the true value. This bias is not negligible in practice: for $n = 25$, $\mu^{\mathrm{mp}} = 0.93\mu_0$, corresponding to an error of 7%.

6.5 Power Law Distributions

6.5.1 Maximum Likelihood Estimation

There is a huge literature on the identification of power laws [216, 252, 422, 423, 458, 573, 634]. As already discussed in Chap. 5, a power law distribution is endowed with a remarkable property, known as "self-similarity" or "scale invariance". Indeed, the ratio of the probabilities of two values is only a function of the ratio of these variations:

$$\frac{p(v_a)}{p(v_b)} = \left(\frac{v_a}{v_b}\right)^{-\mu} . \tag{6.28}$$

In other words, the relative occurrence of the two values is invariant with respect to a homothetic transformation on the two values. This self-similarity is observed to hold approximately in many natural phenomena.

We now use expression (6.5) for this case parameterized by

$$p(v) \simeq_{v \to +\infty} \frac{C}{|v|^{1+\mu}} , \tag{6.29}$$

where C is the scale factor. To simplify, we assume that (6.29) holds for the interval $1 \le v \le \infty$. The lower bound can always be rescaled to the value 1 by dividing v by the minimum value. For concrete applications, it may be useful to reintroduce the minimum value v_{min}. Then, the scale factor C becomes $C v_{\mathrm{min}}^{\mu}$.

The maximization of $F(v_n)$ provides the most probable v_n:

$$v_n^{\mathrm{mp}} = \left[\frac{(\mu N + 1)C}{\mu n + 1}\right]^{1/\mu}. \tag{6.30}$$

Thus, $v_n \simeq n^{-1/\mu}$ for $1 \ll n \leq N$. A plot of $\ln v_n$ as a function of $\ln n$ gives a *straight line* with slope $-1/\mu$. Figure 6.1 shows a rank-ordering plot of the distribution of earthquake moments worldwide documented in the Harvard catalog [258]. The main body of the distribution is well-described by a pure power law distribution with an exponent close to $\mu = 0.7$ while the tail for the largest earthquakes exhibits a significant departure that can be described convincingly by an exponential tail [491, 912]. The rank-ordering representation emphasizes the tail region.

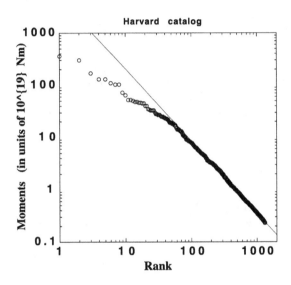

Fig. 6.1. Log–log plot of the rank-ordered seismic moments M (in units of 10^{19} N m) of the largest shallow earthquakes (depth < 70 km) in the Harvard catalog versus its rank. The Harvard catalog we use for this figure [258] spans the time interval from 1977 to 1992 for earthquakes worldwide. The straight line shows the best fit for ranks $n \geq 100$ by a power law $M \propto n^{-1/\mu}$, giving $1/\mu = 1.38$, i.e. $\mu = 0.72$

Note that for small μ, typically less than one, a distortion to the pure power law dependence $n^{-1/\mu}$ occurs that must be taken into account. This introduces a curvature in the log–log plot that has nothing to do with a change of regime but is intrinsic to the tail of the power law distribution. In the cumulative plot, this is reflected in a downward sloping curve. It is important to recognize this effect before concluding a change of regime. See [712] for an illustration of this effect.

We now characterize the fluctuations of v_n for n small. For $\mu < 2$, the variance is not mathematically defined (it diverges) and thus cannot characterize the amplitude of the fluctuations. The expression (6.5) shows that the distribution of v_1 for $v_1 > (NC)^{1/\mu}$ is also a power law

$$p(v_1)\,\mathrm{d}v_1 = \frac{NC}{v_1^{1+\mu}}\mathrm{d}v_1 = \left(\frac{v_1}{(NC)^{1/\mu}}\right)^{-(1+\mu)} \mathrm{d}\left(\frac{v_1}{(NC)^{1/\mu}}\right). \tag{6.31}$$

The last expression in the r.h.s. shows that the natural variable is $v/(NC)^{1/\mu}$. Thus, the amplitude of fluctuations of v_1 are of order $(NC)^{1/\mu}$, i.e. of the small order as v_1 itself. We can make this result more precise by expanding $F(v_n)$ in (6.5) up to second order around its maximum:

$$F(v_n) \simeq F(v_n^{\mathrm{pp}}) - \frac{1}{2} \frac{\mathrm{d}^2 F(v_n)}{\mathrm{d}v_n^2}(v_n - v_n^{\mathrm{pp}})^2 + \dots \tag{6.32}$$

The fluctuation $|v_n - v_n^{\mathrm{pp}}|$ which makes $F(v_n)$ equal to half its maximum value is

$$\frac{|v_n - v_n^{\mathrm{pp}}|}{v_n^{\mathrm{pp}}} = \left(\frac{1}{\mu(n\mu + 1)}\right)^{1/2}. \tag{6.33}$$

The relative fluctuation amplitude is independent of N for the largest ranks and decreases when the rank n increases, i.e. when one goes to less extreme values.

The maximum-likelihood estimation of μ as well as its full distribution is easily obtained by the same procedure described for the exponential case. A more direct approach is based on the remark that the change of variable $v = \mathrm{e}^x$ transforms an exponential distribution $p(x) = \mu\mathrm{e}^{-\mu x}$ into a power law distribution $p(v) = \mu/v^{1+\mu}$. The most probable value for μ in the exponential case is thus the most probable value for the exponent of the corresponding power law pdf. We thus get [9, 422]

$$\mu = \frac{1}{(1/n)\sum_{i=1}^{n} \ln(v_i/v_n)}. \tag{6.34}$$

The amplitude of the fluctuations and the full distribution of μ are given by the formulas (6.24) and (6.26) that we now derive explicitly. Let μ_0 be the true value of the exponent. In order to derive the distribution of μ from a finite data sample, we note that

$$\langle \ln E \rangle|_{E > E^*} = \int_{E^*}^{\infty} \ln E P(E)\,\mathrm{d}E = \ln E^* + \frac{1}{\mu_0}. \tag{6.35}$$

The quantity $\langle x \rangle|_{E > E^*}$ is the average of x with respect to its distribution, for those samples with E having values larger than E^*. With (6.34), which can also be written

$$\left\langle \frac{1}{\mu} \right\rangle|_{E > E_n} = \langle \ln E \rangle|_{E > E_n} - \ln E_n, \tag{6.36}$$

this yields

$$\left\langle \frac{1}{\mu} \right\rangle|_{E > E_n} = \frac{1}{\mu_0}. \tag{6.37}$$

Next, we note that

$$\langle (\ln E)^2 \rangle|_{E > E^*} = (\ln E^*)^2 + \frac{2\ln E^*}{\mu_0} + \frac{2}{\mu_0^2}, \tag{6.38}$$

which yields

$$\text{Var}[\ln E] = \langle (\ln E)^2 \rangle |_{E>E^*} - \langle \ln E \rangle^2 |_{E>E^*} = \frac{1}{\mu_0^2} \ . \tag{6.39}$$

With (6.34), we obtain

$$\text{Var}\left(\frac{1}{\mu}\right) = \text{Var}\left(\frac{\sum_{i=1}^n \ln E_i - n \ln E_n}{n}\right) = \frac{1}{n\mu_0^2} \ . \tag{6.40}$$

The central limit theorem states that the distribution of $\sum_{i=1}^n \ln E_i$ will have an approximate normal distribution for sufficiently large n. Thus, for known μ_0, $1/\mu$ has an approximately normal distribution of mean $1/\mu_0$ and variance $1/n\mu_0^2$ and the distribution $P(1/\mu)$ is given by the expression (6.25), leading to $P(\mu)$ given by (6.26). We note again that the distribution $P(\mu)$ is skewed owing to the prefactor μ_0/μ^2 in front of the exponential term.

6.5.2 Quantiles of Large Events

We now turn to the estimation of quantiles for the large events. A typical question of interest for the purpose of risk assessment is the following: what is the largest typical event size that is expected over a given period of time T? T can be ten years (the decadal storm), a century (the centenial flood), a thousand years (the millenium wave), etc. The quantile v_T over the period T is defined by

$$\int_{v_T}^{+\infty} p(v)\,\mathrm{d}v = \epsilon \ , \tag{6.41}$$

where $\epsilon = 1/T$ and $p(v)$ has been normalized with respect to a suitable unit time scale (for instance, the daily time scale). This equation (6.41) is nothing but (6.3) with $n = 1$, which determines the typical value v_T that appears only once in the series over the time T. For a power law pdf with exponent μ, we obtain

$$v_T = v_0 \epsilon^{-1/\mu} \ , \tag{6.42}$$

where we have reintroduced the scale factor v_0.

Let us now quantify the effect brought by the uncertainty on the determination of the exponent μ on the estimation of the quantile. If μ is distributed according to (6.26), we find that the quantile is distributed according to a log-normal distribution:

$$p(v_T) = \frac{1}{v_T}\left(\frac{n}{2\pi}\right)^{1/2}\frac{1}{\ln(1/\epsilon)}e^{-\{n\mu_0^2/[2(\ln\epsilon)^2]\}[\ln(v_T/v_0)\epsilon^{1/\mu_0}]^2} \ . \tag{6.43}$$

The average of the logarithm of the quantile and its variance are given by

$$\left\langle \ln \frac{v_T}{v_0} \right\rangle = \ln(\epsilon^{-1/\mu}) \ , \tag{6.44}$$

and

$$\mathrm{Var}\left(\ln\frac{v_T}{v_0}\right) = \sigma^2 = \frac{(\ln\epsilon)^2}{n\mu_0^2} \ . \tag{6.45}$$

The quantile corresponding to m standard deviations σ is:

$$v_T^{\pm m\sigma} = v_0\epsilon^{-(1/\mu)(1\pm m/\sqrt{n})} \ . \tag{6.46}$$

Table 6.1 gives the decadal quantile for daily measurements for two different values of the exponent μ.

Table 6.1. Fluctuations of decadal quantiles in units of v_0 at plus or minus two standard deviations, for two values of the exponent μ. For $\mu = 3.7$, which is such that the mean and variance are defined while the fourth and higher order moments are not defined, the largest daily event typically expected once every ten years is estimated to lie between 6.2 and 11.1 times the characteristic scale v_0. For a much smaller exponent $\mu = 1.5$ such that the mean is defined but the variance is not defined, the largest daily event typically expected once every ten years is estimated to lie between 89 and 382 times the characteristic scale v_0. This shows that smaller exponents μ gives wilder fluctuations for the largest possible events

μ	$\dfrac{v_{10}^{-2\sigma}}{v_0}$	$\dfrac{v_{10}^{-\sigma}}{v_0}$	$\dfrac{v_{10}}{v_0}$	$\dfrac{v_{10}^{+\sigma}}{v_0}$	$\dfrac{v_{10}^{+2\sigma}}{v_0}$
3.7	6.2	7.1	8.3	9.6	11.1
1.5	89	128	184	265	382

The estimations given in the table describe the fraction of the uncertainty on v_T stemming only from the imperfect determination of the exponent. There is another important contribution coming from the fact that, conditioned on a fixed exponent, the distribution of v_T is itself a power law (6.31).

6.5.3 Power Laws with a Global Constraint: "Fractal Plate Tectonics"

Plate Tectonics. Plate tectonics is a relatively new theory (1968) that has revolutionized the way geologists think about the Earth. According to the theory, the surface of the Earth is broken into large plates. The size and position of these plates change over time. The edges of these plates, where they move against each other, are sites of intense geologic activity, such as earthquakes, volcanoes, and mountain building. Plate tectonics is a combination of two earlier ideas, continental drift and sea-floor spreading. Continental drift is the movement of continents over the Earth's surface and in their change in position relative to each other. Sea-floor spreading is the creation of new oceanic crust at mid-ocean ridges and movement of the crust away from the mid-ocean ridges.

According to the theory of plate tectonics, the outermost 100 km of the solid Earth is composed of a relatively small number of internally rigid plates (lithosphere) that move over a weak substrate (asthenosphere). The plate sizes and positions change over time and are driven by the internal thermal engine of the Earth that also drives mantle convection. DeMets et al. [210] performed a global inversion to determine the relative rotation rates of the 12 largest plates (the NUVEL-1 model) later refined into the NUVEL-1A solution [211]. These plate edges are not sharp narrow boundaries but are often constituted of complex systems of competing faults and other geological structures of width extending over several hundreds of kilometers for transform and subduction boundaries and up to thousands kilometers for continental collisions. It is now common lore to view such tectonic deformation as possessing some kind of self-similarity or fractal properties [827], or better a hierarchical structure [711]. In addition, several researchers have repeatedly proposed to describe this multi-scale organization of faulting by a hierarchy of blocks of multiple sizes [23, 336, 355]. In these models, blocks slide against each other, rotate, lock at nodes or triple junctions which may represent the loci of major earthquakes, in a way similar to (but at a reduced scale compared to) the relative motion of the major tectonic plates. An example of a complex network of faults is observed in the broad San Andreas transform boundary between the Pacific and the North American plates in California, for which Bird and Rosenstock [93] have suggested 21 possible microplates within southern California alone.

Keeping in mind these ingredients of a few major plates at large scales on one hand and a hierarchical self-similar organization of blocks at the boundary scale on the other hand, the recent reassessment of present plate boundaries on the Earth by P. Bird [92] is particularly interesting: taking into account relative plate velocities from magnetic anomalies, moment tensor solutions, and/or geodesy, to the 14 large plates whose motion was described by the NUVEL-1A poles, Bird's model in PB2001 included 28 additional small plates, for a total of 42 plates. In his latest revision, he has added ten more even smaller plates leading to a total of 52 reported plates in [92]. Bird suggests that the cumulative-number/area distribution for his model follows a power-law for plates of less than 1 steradian of area [92].

Power Law Distribution of Plate Sizes. We summarize the results of the analysis of the 42 plates of the PB2001 model, obtained in [909]. The latest 10 additional plates are the smallest ones and P. Bird argues that these population is not complete [92]. The statistical analysis of such a data set is very difficult due to its small size $N = 42$ but is not impossible. Figure 6.2 shows the complementary cumulative number $N(A)$ of plates as a function of area A in steradians, i.e., the number of plates with an area equal to or larger than A.

Most of the data except for a few largest plates follow a linear dependence in the double log-scale of the figure. The slope of this straight line is close

Cumulative number

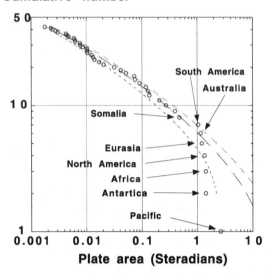

Fig. 6.2. Complementary cumulative distribution of the areas of tectonic plates in the PB2001 model of P. Bird (*open circles*) compared to the fit with the formula (6.49) for a power law (*central long-dashed line*) with exponent $\mu = 0.25$ and $a = 0.002$. The *small-dashed line* and *medium-dashed line* provide the 90% confidence interval. Reproduced from [909]

to 0.25. We first compare this sample with a pure power law (the Pareto distribution). For this purpose we use the rank-ordering method previously discussed. We put the sample into descending order $A_1 \geq A_2 \geq ... \geq A_N$. The probability density function (PDF) of the n-th rank, denoted $\phi_{n,N}(x)$, is well-known and we rewrite (6.5) here for convenience

$$\phi_{n,N}(x) = (N - n + 1)\binom{N}{n}F^{N-n}(x)(1 - F(x))^{n-1}f(x) , \tag{6.47}$$

where $F(x), f(x)$ are the distribution function (DF) and PDF of the random values in question. Putting the Pareto law $F(x) = 1 - (a/x)^\mu$, $x \geq a$, into (6.47), we get

$$\phi_{n,N}(x) \propto (1 - (a/x)^\mu)^{n-1} x^{\mu(N-n+1)-1} . \tag{6.48}$$

The mode $M_{n,N}$ of the PDF (6.48) (i.e., the maximum of the PDF) is the most probable value of the random variable A_n:

$$M_{n,N} = a \left(\frac{N\mu + 1}{n\mu + 1}\right)^{1/\mu} , \tag{6.49}$$

which is the same as (6.30). Besides, an interval around the mode containing some prescribed probability (say, 90%) can be derived from the density $\phi_{n,N}(x)$. The dependence (6.49) with $\mu = 1/4$ is shown as the long-dashed line in Fig. 6.2. The two short-dashed lines represent the upper/lower limits of the 90%-confidence interval. The data are well accounted for by the power law, except, perhaps, for the three smallest ranks, i.e. the three largest plates, the Pacific, the Antarctica, and the Africa plates, which fall outside the confidence interval.

"Finite-Size" Constraint on the Distribution of Plate Sizes. From a visual inspection of Fig. 6.2, it might be argued that the deviation from the power law prediction (6.49) occurs somewhat earlier, say, at rank $n = 7$, i.e., the seven largest plates with area more than 1 steradian belong to a different population than the rest of the plates. Following this hypothesis that is refuted in [909] (see below), one could argue that the largest continent and ocean plates are commensurable with the size of the convection cells in the upper mantle while the smaller plates appeared as a result of the interaction and collision between the larger plates. The study of [909] provides a test of the hypothesis [24] that a superficial self-organizing process is sufficient for understanding the mosaic of plates, in contrast with the idea that the lithosphere necessarily mirrors the planform of mantle convection.

The following analysis shows that the full distribution of the 42 plate sizes can be fully accounted by assuming a power law distribution constrained by the "finite-size" condition that the sum of areas over all plates must sum up to 4π.

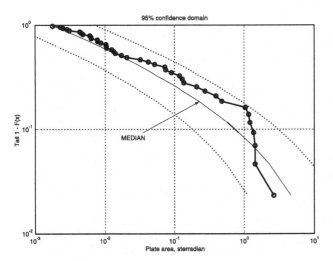

Fig. 6.3. Medians $M_1, ..., M_{42}$ (*continuous line*) and their corridor at the 95% confidence level delimited by the two *dotted lines* of the conditional Pareto distribution $\phi(x_1, ..., x_{42}|\mu)$ given by (6.51) compared with the empirical cumulative distribution of the 42 plate areas (*circles* linked by straight segments), documented in the PB2001 model of P. Bird. Reproduced from [909]

The distribution $g_{N,C}(x)$ of sample values $x_1, ..., x_N$ conditioned by the constraint

$$S_N = x_1 + ... + x_N = C , \tag{6.50}$$

where C is a constant (4π for the plates) is modified from its unconditional Pareto density expression $f(x) = \mu a^\mu / x^{1+\mu}$ for $x \geq a$ and $f(x) = 0$ for

$x < a$. The lower threshold (the minimum value) for the plate data is $a = 0.00178$ steradian and corresponds to the smallest documented plate in the PB2001 model. Denoting the unconditional density of the sum $S_k = x_1 + \ldots + x_k, k = N - 1, N$ by $s_k(x)$, we have $g_{N,C}(x) = s_{N-1}(C - x)f(x)/s_N(C)$, for $a \leq x \leq C$. Thus, the constraint (6.50) decreases the unconditional Pareto density $f(x)$ by a factor $s_{N-1}(C - x)/s_N(C)$ which acts as a "taper."

In order to use the Maximum Likelihood method for the estimation of the exponent μ, we need the vectorial distribution of conditional sample to take into account the interdependence between the different variables (areas of the plates) $x_1, \ldots, x_{42}$ induced by the constraint (6.50). The corresponding likelihood function is therefore

$$\phi(x_1, \ldots, x_{42}|\mu) = \frac{\delta(x_1 + \ldots + x_{42} - 4\pi)}{s_{42}(4\pi)} f(x_1|\mu) \ldots f(x_{42}|\mu) \ . \qquad (6.51)$$

The resulting ML estimate is $\mu = 0.25 \pm 0.05$. With this value, we generate an artificial sample of 1045 conditional 42-dimensional (42D) vectors with the condition (6.50) with $C = 4\pi$. Rank-ordering each of these 1045 vectors, we determine their sample medians $M_1, \ldots, M_{42}$, where M_j is the median of the j-th rank. These conditional medians are slightly smaller than given by (6.49) for the unconditional Pareto distribution. The conditional distribution (6.51) allows us to construct a confidence domain for the 42D random vectors, defined as a "corridor" of the form $[cM_j; (1/c)M_j]$, the constant $c = 0.244$ being chosen such that 95% of vectors fall within this corridor. The medians $M_1, \ldots, M_{42}$ and their corridor are shown in Fig. 6.3. All samples of the tectonic plates falls within the 95% confidence corridor, showing that (in contrast with the "pure" Pareto used in Fig. 6.2) the Pareto model together with the total area constraint (6.50) accounts satisfactorily for all the data, including the largest plates [909].

Given this strength of the validity of the Pareto distribution for the 42 plates documented here, one can expect that it can be extrapolated beyond this range to smaller yet undetected plates. Using the complementary Pareto distribution $(a/x)^\mu$ with $a = 0.00178$ steradian and $\mu = 0.25$, this extrapolation predicts 100 plates larger than 1.35×10^{-4} steradian (5,500 km^2 or 74×74 km^2) [92]. The total area needed to define another 58 plates ranging from this size up to the smallest size in PB2001 would be only about 0.028 steradians, which could be taken from large plates like EU and NA without materially affecting their areas. As discussed in [92], the places where additional small plates are most likely to be recognized are within the zones of distributed deformation identified in [92], which have total area of 0.838 steradians (6.7% of Earth).

This analysis [909] on a universal fractal character of plate fragmentation does not prevent the existence of some underlying distinctions. For instance, P. Bird (private communication and [92]) proposes three main tectonic origins for plates: plume-assisted rifting for the larger plates (ranks 1–8 in Figs. 6.2 and 6.3), continental collision stress for the intermediate plate sizes (ranks 8–20), and back-arc spreading for the smallest plates (ranks 21–42). In Figs. 6.2 and 6.3, one can discern slightly different patterns of the tail behavior of these three subgroups. However, any formal statistical analysis of these possible distinctions would be, to our opinion, excessive due to the extreme smallness of the data set.

This study suggests that the plate mosaic may be the result of a self-organized process of fragmentation and aggregation (see Sect. 16.3.2 and [712, 909]), readily reorganized by stress changes. Thus, in contrast with the idea that the lithosphere necessarily mirrors the planform of mantle convection, the texture of tectonic plates may have simple and surficial explanations. This suggests to revise the model of plate tectonics in terms of a dynamical model of plates with creation, fragmentation and destruction that may occur at all scales. At any given time, the plate sizes are somewhat arbitrary and changing according to an ever evolving dynamics. Acknowledging the robustness of self-similar structures in general and of power laws in particular, we conjecture that the present observation of a fractal structure of plates is a robust property of plate tectonics that should be used to further our understanding of plate self-organization through the super-continent cycle.

6.6 The Gamma Law

The Gamma law is defined by

$$p(v)\, dv = \frac{v_0^\mu}{v^{1+\mu}} \exp\left(-\frac{v}{v_g}\right) dv \ . \tag{6.52}$$

For $v < v_g$, the Gamma law reduces to a pure power law which becomes progressively an exponential for $v > v_g$. The Gamma distribution is found in critical phenomena in the presence of a finite size effect or at a finite distance from the critical point. It is also expected more generally as a result of constraints in the maximum-likelihood estimation of power law distributions. The Gamma law has been advocated for the description of the Gutenberg–Richter distribution of earthquake sizes [491]. The self-similarity holds up to the value v_g and is broken for larger values due to the appearance of this characteristic scale.

The rank-ordering method allows us to determine this distribution (6.52). In this goal, we express the cumulative distribution $P_>(v) = \int_v^\infty dx\, p(x)\, dx$ under a simple approximate form

$$P_{\mathrm{app}}(v) \simeq \frac{v_0^\mu}{v^\mu} \exp\left(-\frac{v}{v_g}\right) \left(\frac{v}{v_g} + (1+\mu)\left(1 - \frac{v_g}{v + (1+\mu)v_g}\right)\right)^{-1} \ . \tag{6.53}$$

The difference $[P_{\mathrm{app}}(v) - P(v)]/P(v)$, for $v_g = 1$ and $\mu = 1.7$ and for v between 0 and 10, is no more than 4.6%. If a better precision is needed, a more complex formula can be written. Once the cumulative distribution function (6.53) is obtained, this gives the rank n in the rank ordering plot.

6.7 The Stretched Exponential Distribution

It is useful to extend the toolbox of "fat tail" distributions that can be used to describe the distributions of natural phenomena. The power-exponential distribution is defined by

$$p(v)\,dv = \alpha\left(\frac{v}{v_0}\right)^{(\alpha-1)} e^{-(v/v_0)^\alpha} \frac{dv}{v_0} \qquad (v > 0) . \tag{6.54}$$

This law introduces a characteristic scale v_0 and is parameterized by the exponent α. The exponential is the special case $\alpha = 1$. The power law with $\mu \to 0$ is recovered for the special case $\alpha \to 0$. The smaller α is, the larger is the probability to observe large values. The case where $\alpha < 1$ is called the stretched exponential. The stretched exponential provides an interesting alternative for the description of fat tails, which is intermediate between the exponential and the power laws. We have shown in Chap. 3 [318] that it naturally occurs in the extreme regime of multiplicative processes. It has been found to provide a reasonable and parsimonious description of many distributions in Nature and in the Economy [547]. In the next section, we describe its properties and some generalizations and give a review of maximum likelihood methods to estimate its parameters.

As we already said, the law (6.54) has all its moments finite. The first two read

$$\left\langle \frac{v}{v_0} \right\rangle = \frac{1}{\alpha} \Gamma\left(\frac{1}{\alpha}\right) \tag{6.55}$$

and

$$\left\langle \left(\frac{v}{v_0}\right)^2 \right\rangle = \frac{2}{\alpha} \Gamma\left(\frac{2}{\alpha}\right) , \tag{6.56}$$

where $\Gamma(x)$ is the Gamma function $\Gamma(x) = \int_0^\infty t^{x-1} e^{-t}\,dt$ reducing to $(x-1)!$ for integer x.

Proceeding as before for the other distributions, we obtain the expression v_n of the nth rank:

$$v_n^\alpha \simeq -v_0^\alpha \ln n + B , \tag{6.57}$$

where B is a constant. Thus, a plot of v_n^α as a function of $\ln n$ gives a straight line (up to logarithmic corrections) which qualifies the stretched exponential. The slope gives an estimation of v_0.

This is illustrated in Fig. 6.4 for the distribution of price variations: it plots the n-th price variation taken to the power 0.7 for the French Franc expressed in German marks (in the period from 1989 to the end of 1994) as a function of the decimal logarithm of the rank. The positive variations are represented with square symbols and the negative variations represented with diamond symbols. We observe an excellent description with straight lines over the full range of quotation variations. We note that the exponent α is smaller than 1, which corresponds to a "fatter" tail than an exponential, i.e. the existence of larger variations. Note that the characteristic scales v_0 of the positive and negative exchange rate variations are different, characterizing a clear asymmetry with larger negative variations of the Franc expressed in Marks. This asymmetry corresponds to a progressive depreciation of the Franc with respect to the Mark occurring in bursts rather than as a continuous drift. One could have imagined that such a depreciation would correspond to a steady drift on which are superimposed symmetric variations. We find something else: the depreciation is making its imprints at all scales of price variations and is simply quantified, not by a drift, but by a different reference scale v_0. For positive variations, we have $\langle v \rangle = 1.4 v_0 = 0.17\%$ and $v_{95\%} = 5.6 v_0 = 0.7\%$. The difference between the mean $\langle v \rangle$ and the 95%-confidence extreme $v_{95\%}$ clearly illustrates the wild character of the fat tail of the Franc–Mark exchange rate variations. For negative variations, we have $\langle v \rangle = 0.2\%$ and $v_{95\%} = 4.6 v_0 = 0.7\%$.

The maximum-likelihood method provides the following estimates for v_0 and α

$$v_0^\alpha = \frac{1}{n} \sum_{i=1}^{n} v_i^\alpha - v_n^\alpha \; , \tag{6.58}$$

and

$$\frac{1}{\alpha} = \frac{\left(\sum_{i=1}^{n} v_i^\alpha \ln v_i - v_n^\alpha \ln v_n \right)}{\sum_{i=1}^{n} v_i^\alpha - v_n^\alpha} - \frac{1}{n} \sum_{i=1}^{n} \ln v_i \; . \tag{6.59}$$

The probability that the quantile v_T is larger than or equal to v^* is given by

$$P(v_T \geq v^*) = 1 - [1 - e^{-(v^*/v_0)^\alpha}]^N \simeq 1 - e^{-N e^{-(v^*/v_0)^\alpha}} \; , \tag{6.60}$$

where N is the number of measurements during the period T. If p is the confidence level for the quantile v_T, i.e. $P(v_T \geq v^*) = p$, we find

$$v^* = v_0 \left(\ln \frac{N}{\ln(1/1-p)} \right)^{1/\alpha} \; . \tag{6.61}$$

6.8 Maximum Likelihood and Other Estimators of Stretched Exponential Distributions

In this section, we survey a large body of the literature to provide a synthesis of useful methods for estimating the parameters of stretched expo-

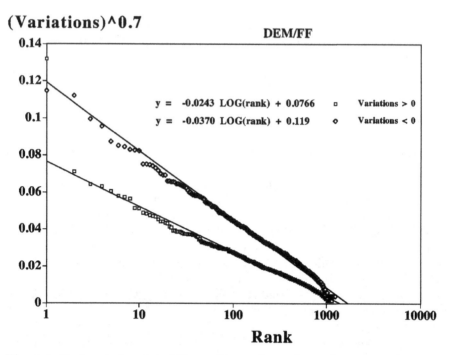

Fig. 6.4. Rank-ordering plot of the positive and negative exchange rates between the French Franc and the German mark in the period from 1989 to the end of 1994. Taken from [547]

nentials from a finite set of observations. This literature is scattered across three essentially non-overlapping fields, mathematics concerned with "sub-exponentials", physics with stretched relaxation exponentials, and the largest and most relevant body of literature found in the engineering and reliability disciplines concerned with Weibull distributions.

6.8.1 Introduction

Definition and Applications. The two-parameter stretched exponential (SE) complementary cumulative distribution is

$$P_>(x) = e^{-(x/\chi)^c} , \qquad \text{for } x \geq 0 . \tag{6.62}$$

Its pdf is $p(x) = -dP_>(x)/dx$:

$$p(x)\, dx = c \left(\frac{x}{\chi}\right)^{c-1} e^{-(x/\chi)^c} \frac{dx}{\chi} \qquad (x \geq 0) . \tag{6.63}$$

Stretched exponentials are characterized by an exponent $c < 1$. The borderline $c = 1$ corresponds to the usual exponential distribution. For $c < 1$, the distribution (6.62) presents a clear downward curvature in a log–log plot while exhibiting a relatively large apparent linear behavior (that can thus be mistaken for a limited power law regime) all the more so, the smaller the shape parameter c is. The distribution (6.62) can thus be used to account both for a limited scaling regime and a cross-over to non-scaling. When using the stretched exponential pdf, the rational is that the deviations from a power law description is fundamental and not only a finite-size correction.

In a recent empirical study [547], we find that the stretched exponential (6.62) provides an parsimonious description of the distribution of radio and light emissions from galaxies, of US GOM OCS oilfield reserve sizes, of World, US and French agglomeration sizes, of the United Nation 1996 country sizes, of daily Forex US–Mark and Franc–Mark price variations, and of the Raup–Sepkoskis kill curve in evolution. Even the distribution of biological extinction events is much better accounted for by a stretched exponential than by a power law. We also find that the distribution of the largest 1300 earthquakes in the world from 1977 to 1992 and the distribution of fault displacements can be well-described by a stretched exponential. Similar results are obtained for the temperature variations over the last 420 000 years obtained for ice core isotope measurements and for the distribution of citations of the most cited physicists in the world.

In condensed-matter physics, stretched exponential laws are familiar. They are found to describe the rate dependence of anomalous relaxations in glasses and in the Ising ferromagnet in two dimensions. The term "stretched" reflects the fact that the relaxation is slower than exponential with a time dependence t^c in the exponential which is slower than t and thus corresponds to an effective exponential time $t_{eff} = t^c$ such that $t = t_{eff}^{1/c}$ is effectively "stretched".

There are several mechanisms that produce stretched exponentials. One recently discovered is based on a generalization of the central limit theorem of the tail of product of random variables [318]. In this scenario, the exponent c is the inverse of the number of generations (or products) in a multiplicative process.

Sub-Exponentials. The stretched exponentials are known in the literature under other names. In mathematics, they belong to the class of so-called "subexponentials" [274]. Technically, subexponential distributions $P_>(x)$ are defined by the following property:

$$\lim_{x \to +\infty} \frac{1 - P_>^{*2}(x)}{1 - P_>(x)} = 2 , \tag{6.64}$$

where $*$ denotes the convolution product. $P_>^{*2}$ is thus the cumulative distribution for the sum of two i.i.d. variables with cdf $P_>(x)$. In practice, this means that the tail of the distribution of the sum has the same form as the initial distribution. This property is thus similar to the stability condition for Gaussian and Lévy laws, except that it survives only in the tail. The stretched exponential with $c < 1$ belongs to this class (6.64). However, there are many other functions in this subexponential class formally defined by (6.64).

For the stretched exponentials, we can express (6.64) as follows: the tail of the sum S_N of N stretched exponential distributed variables has the same order of magnitude as the tail of the maximum variable $X_N^{\max}$ among the N ones:

$$\frac{\text{Probability}(S_N \geq x)}{\text{Probability}(X_N^{\max} \geq x)} \xrightarrow{x \to +\infty} 1 \ . \tag{6.65}$$

This is rather remarkable considering that the typical values are very different since $S_N \approx N\langle X \rangle \sim N$ is much larger, for large N, than the typical value of the maximum $X_N^{\max} \sim (\ln N)^{1/c}$. The proof goes as follows. We ask what are the set of positive X_i's such that $\sum_{i=1}^{N} X_i^c$ is minimum so as to make maximum the probability of this configuration, while the condition $\sum_{i=1}^{N} X_i > x$ is obeyed. In this goal, we rewrite the constraint as $\sum_{i=1}^{N} x_i > 1$ where $x_i \equiv X_i/x$. The minimum of $\sum_{i=1}^{N} X_i^c = x^c \sum_{i=1}^{N} x_i^c$ is thus minimum for the minimum of $\sum_{i=1}^{N} x_i^c$. We see immediately that, since $c < 1$, $\sum_{i=1}^{N} x_i^c \geq 1$ derives from $\sum_{i=1}^{N} x_i \geq 1$. The configuration that makes the inequality as the minimum one, i.e. transforms it into an equality are those such that all x_i's are all very small ($\to 0$) except one almost equal to 1. The corresponding probability is $\exp[ax^c]$ larger than any other configuration, where a is a constant depending on the configuration of the X_i's. We thus see how (6.65) emerges.

Note that this property (6.65) on the tails is not in contradiction with the fact that the *expectation* of ratio S_N/X_N tends to infinity when the sample size $N \to \infty$. This must be contrasted to the power law distribution with exponent $\alpha < 1$ which is such that the ratio $S_N/X_N^{\max}$ goes to a constant $1/(1 - \alpha)$ ([293] in Chap. XIII). Notwithstanding this property that the tail of the maximum is equivalent to the tail of the sum for subexponential distributions, all the moments of stretched exponentials at all orders are finite. This distinguishes the stretched exponential from power laws. The law of large numbers and the central limit theorem are valid for the sum of stretched exponential distributions.

The references [37, 175, 271–273, 275, 353, 515, 516, 656, 740, 951, 952, 1020] discuss mathematical properties of general subexponential distributions.

Weibull Distributions. In the field of reliability, the distribution (6.62) is known as the Weibull distribution. It is widely used to fit distributions of strengths of materials and of failure times. It has a long history starting with Fisher and Tippet [305] who introduced a reduced from of the Weibull distribution for extreme values, followed by [1006] to describe distribution of material failure strengths. We refer to [386] for an historical review of Weibull's distribution.

For failure applications, the variable x in (6.62) can be the stress at failure or the lifetime of the structure. c is called the Weibull shape parameter and χ is the Weibull scale parameter.

In these applications, an additional parameter is often added in the form

$$P_>(x) = e^{-(x-b/\chi)^c} , \qquad \text{for } x \geq b . \tag{6.66}$$

$b > 0$ is the minimum stress or lifetime. Introducing the shift parameter b may be of interest to describe the fact that the stretched exponential may not describe accurately the very small values.

In reliability analysis, it is customary to use "Weibull" paper, similar to the log–log paper, such that

$$\ln[-\ln P_>(x)] = c \ln \frac{x - b}{\chi} \tag{6.67}$$

is linear in x. The slope gives the exponent c directly.

6.8.2 Two-Parameter Stretched Exponential Distribution

Maximum Likelihood Estimation. Suppose we observe $0 < x_1 \leq x_2 \leq \dots \leq x_n$. The likelihood function is

$$L \equiv \prod_{i=1}^{n} p(x_i) . \tag{6.68}$$

The log-likelihood is

$$\ln L = n \ln \frac{c}{\chi} + (c - 1) \sum_{i=1}^{n} \ln \frac{x_i}{\chi} - \sum_{i=1}^{n} \left(\frac{x_i}{\chi} \right)^c . \tag{6.69}$$

Maximizing $\ln L$ with respect to χ gives

$$0 = \frac{\partial \ln L}{\partial \chi} \qquad \to \qquad \chi^c = \frac{1}{n} \sum_{i=1}^{n} x_i^c . \tag{6.70}$$

Maximizing $\ln L$ with respect to c gives

$$0 = \frac{\partial \ln L}{\partial c} \qquad \to \qquad \frac{1}{c} = h(c) , \tag{6.71}$$

where

$$h(c) \equiv \frac{\sum_{i=1}^{n} x_i^c \ln x_i}{\sum_{j=1}^{n} x_j^c} - \frac{1}{n} \sum_{i=1}^{n} \ln x_i . \tag{6.72}$$

Note that the equation (6.71) with (6.72) involves c alone and can be solved by iterative methods. Then, one substitutes the resulting value for c in (6.70) to get χ.

Farnum and Booth [289] have shown that

1. $h(c)$ is an increasing function of c for $c \geq 0$.
2. $\lim_{c \to +\infty} h(c) = \ln x_n - \frac{1}{n} \sum_{i=1}^{n} \ln x_i \equiv V .$ $\tag{6.73}$

3. $h(0) = 0$.
4. V is always positive or zero and vanishes only when all x_i's are equal.
5. For $V > 0$ (interesting case), the solutions $\hat{\chi}$ and $\hat{c}$ of (6.70) and (6.71) are unique.
6. From the data, it is easy to calculate V. Then, we have the following lower bounds

$$\hat{c} > \frac{1}{V} \qquad \text{and} \qquad \hat{\chi} > \left[\frac{1}{n} \sum_{i=1}^{n} x_i^{1/V} \right]^V . \qquad (6.74)$$

7. Thus, $1/V$ provides a possible initial value of the iterative solution of (6.71) with (6.72).
8. Empirical tests [289] give a better starting point equal to $2/V$ for c in the iterative solution of (6.71) with (6.72).

The computation may be based on Newton–Raphson or quasi-Newton iterative method [853]. The Newton–Raphson method depends on evaluation of the first- and second-order derivatives of the log likelihood in each iteration, whereas the quasi-Newton methods require only the first-order derivatives to be evaluated explicitly and use an approximation to the second-order derivatives. In both case, it is advisable to incorporate checks to make sure the log likelihood increases at each iteration (if not, use a shorter step length or switch to the direction of steepest descent). It is usually recommented to construct the probability plots to assess goodness of fit.

Qiao and Tsokos [757] have introduced an iterative precedure for estimating the parameters c and χ that always converges and does not depend on the initial point of the iteration. This method converges faster than the popular Newton–Raphson method (which in addition depends on the initial point and which does not always converge). Defining

$$s_1(c) = \sum_{i=1}^{n} \ln x_i , \qquad s_2(c) = \sum_{i=1}^{n} x_i^c , \qquad (6.75)$$

$$s_3(c) = \sum_{i=1}^{n} x_i^c \ln x_i , \qquad s_2(c) = \sum_{i=1}^{n} x_i^c (\ln x_i)^2 , \qquad (6.76)$$

the "simple iterative procedure" (SIP) invented by Qiao and Tsokos [757] corresponds to iterate

$$c_{k+1} = \frac{c_k + q(c_k)}{2} , \qquad (6.77)$$

where

$$q(c) = \frac{n s_2(c)}{n s_3(c) - s_1(c) s_2(c)} . \qquad (6.78)$$

The convergence is at least at a geometrical rate of $1/2$.

Thomas et al. [958] have developed a genetic algorithm for Weibull parameter estimation. The method uses a simultaneous random search function by integrating the principles of the genetic algorithm and the methods of maximum likelihood estimation. The results indicate that GA is superior to the Newton–Raphson method. Its major advantage is being able to find optimal values withoug requiring good initial estimates of the Newton–Raphson method. This result is attributed to the GA's ability to search multiple points simultaneously over a wider search space. Use of GA methods may be recommended for generalizations discussed below of the Weibull distribution with a larger number of parameters.

Thoman et al. [956] have calculated the asymptotic covariance matrix of the estimated parameters $(\hat{\chi}, \hat{c})$. We give here the square root of the elements of the covariance matrix:

$$\frac{\sigma_\chi}{\chi} = \frac{1.053}{c} \frac{1}{\sqrt{n}} , \qquad \frac{\sigma_c}{c} = 0.780 \frac{1}{\sqrt{n}} , \qquad \sigma_{\chi,c} = 0.507 \sqrt{\frac{\chi}{n}} . \quad (6.79)$$

Cohen [176] gives the expressions of the second-order derivatives of the logarithm of the likelihood function. They allow us to obtain the confidence level on the values of the parameters since

$$\ln L(\theta, c) = \ln L(\hat{\theta}, \hat{c}) + \frac{1}{2}(\theta - \hat{\theta})^2 \left. \frac{\partial^2 \ln L}{\partial \theta^2} \right|_{\hat{\theta}, \hat{c}}$$

$$+ (\theta - \hat{\theta})(c - \hat{c}) \left. \frac{\partial^2 \ln L}{\partial \theta \partial c} \right|_{\hat{\theta}, \hat{c}}$$

$$+ \frac{1}{2}(c - \hat{c})^2 \left. \frac{\partial^2 \ln L}{\partial c^2} \right|_{\hat{\theta}, \hat{c}} + \text{h.o.t.} , \quad (6.80)$$

where h.o.t. means higher order terms. θ is defined by

$$\theta = \chi^c . \quad (6.81)$$

We have [176]

$$-\left. \frac{\partial^2 \ln L}{\partial c^2} \right|_{\hat{\theta}, \hat{c}} = \frac{n}{\hat{c}^2} + \frac{1}{\hat{\theta}} \sum_{i=1}^{n} x_i^c (\ln x_i)^2 . \quad (6.82)$$

$$\left. \frac{\partial^2 \ln L}{\partial \theta \partial c} \right|_{\hat{\theta}, \hat{c}} = \frac{1}{\hat{\theta}^2} \sum_{i=1}^{n} x_i^c \ln x_i . \quad (6.83)$$

$$-\left. \frac{\partial^2 \ln L}{\partial \theta^2} \right|_{\hat{\theta}, \hat{c}} = -\frac{n}{\hat{\theta}^2} + \frac{2}{\hat{\theta}^3} \sum_{i=1}^{n} x_i^c . \quad (6.84)$$

Asymptotic Variance-Covariance of Maximum Likelihood Estimators of the SE Parameters. Malevergne et al. [584] provide the asymptotic variance-covariance of maximum likelihood estimators of the parameters of the stretched exponential distribution (6.62). Actually, they consider a slight generalization of the stretched-exponential (SE) or Weibull parametric family with complementary distribution function

$$\bar{F} = 1 - F(x) = \exp\left[-\left(\frac{x}{\chi}\right)^c + \left(\frac{u}{\chi}\right)^c\right] \quad x \geq u \,, \tag{6.85}$$

where c, χ are unknown parameters and u is a known lower threshold.

It is convenient to take a new parameterization of the distribution (6.85), more appropriate for the derivation of asymptotic variances. It should be noted that this reparameterization does not affect the asymptotic variance of the form parameter c. In the new parameterization, the complementary distribution function has the form:

$$\bar{F}(x) = \exp\left[-v\left(\left(\frac{x}{u}\right)^c - 1\right)\right], \quad x \geq u \,. \tag{6.86}$$

Here, the parameter v involves both unknown parameters c, d and the known threshold u:

$$v = \left(\frac{u}{\chi}\right)^c \,. \tag{6.87}$$

The log-likelihood L for sample $(x_1 \ldots x_N)$ has the form:

$$L = N \ln v + N \ln c + (c-1) \sum_{i=1}^{N} \ln \frac{x_i}{u} - v \sum_{i=1}^{N}\left[\left(\frac{x_i}{u}\right)^c - 1\right]. \tag{6.88}$$

The Fisher matrix Φ reads:

$$\Phi = \begin{pmatrix} E\left[-\partial_v^2 L\right] & E\left[-\partial_{v,c}^2 L\right] \\ E\left[-\partial_{c,v}^2 L\right] & E\left[-\partial_c^2 L\right] \end{pmatrix} \,. \tag{6.89}$$

We find:

$$\frac{\partial^2 L}{\partial v^2} = -\frac{N}{v^2} \,,$$

$$\frac{\partial^2 L}{\partial v \, \partial c} = -N \frac{1}{N} \sum_{i=1}^{N} \left(\frac{x_i}{u}\right)^c \ln \frac{x_i}{u} \xrightarrow{N \to \infty} -N E\left[\left(\frac{x}{u}\right)^c \ln \frac{x}{u}\right] \,,$$

$$\frac{\partial^2 L}{\partial c^2} = -\frac{N}{c^2} - N v \frac{1}{N} \sum_{i=1}^{N} \left(\frac{x_i}{u}\right)^c \ln^2 \frac{x_i}{u}$$

$$\xrightarrow{N \to \infty} -\frac{N}{c^2} - N v E\left[\left(\frac{x}{u}\right)^c \ln^2 \frac{x}{u}\right] \,.$$

After some calculations we find:

$$E\left[\left(\frac{x}{u}\right)^c \ln \left(\frac{x}{u}\right)\right] = \frac{1 + E_1(v)}{cv} \,, \tag{6.90}$$

where $E_1(v)$ is the integral exponential function:

$$E_1(v) = \int_v^\infty \frac{e^{-t}}{t}\, dt \; . \tag{6.91}$$

Similarly we find:

$$E\left[\left(\frac{x}{u}\right)^c \ln^2 \frac{x}{u}\right] = \frac{2e^v}{vc^2}\left[E_1(v) + E_2(v) - \ln(v)E_1(v)\right] , \tag{6.92}$$

where $E_2(v)$ is the partial derivative of the incomplete Gamma function:

$$E_2(v) = \int_v^\infty \frac{\ln(t)}{t} e^{-t}\, dt = \left.\frac{\partial}{\partial a} \int_v^\infty t^{a-1}e^{-t}\, dt\right|_{a=0}$$

$$= \left.\frac{\partial}{\partial a}\Gamma(a,x)\right|_{a=0} \; . \tag{6.93}$$

The Fisher matrix (multiplied by N) is given by

$$N\Phi = \begin{pmatrix} \dfrac{1}{v^2} & \dfrac{1 + e^v E_1(v)}{cv} \\ \dfrac{1 + e^v E_1(v)}{cv} & \dfrac{1}{c^2}(1 + 2e^v\left[E_1(v) + E_2(v) - \ln(v)E_1(v)\right]) \end{pmatrix} \tag{6.94}$$

The covariance matrix B of ML-estimates $(\tilde{v}, \tilde{c})$ is equal to the inverse of the Fisher matrix. Thus, inverting the Fisher matrix Φ in equation (6.94) we find:

$$B = \begin{pmatrix} \dfrac{v^2}{NH(v)}[1 + 2e^v E_1(v) + 2e^v E_2(v) - \ln(v)e^v E_1(v)] & -\dfrac{cv}{NH(v)}[1 + e^v E_1(v)] \\ -\dfrac{cv}{NH(v)}[1 + e^v E_1(v)] & \dfrac{c^2}{NH(v)} \end{pmatrix} \tag{6.95}$$

where $H(v)$ has the form:

$$H(v) = 2e^v E_2(v) - 2\ln(v)e^v E_1(v) - (e^v E_1(v))^2 \; . \tag{6.96}$$

Thus, the matrix (6.95) provides the desired covariance matrix.

We present here as well the covariance matrix of the limit distribution of ML-estimates for the SE distribution on the whole semi-axis $(0, \infty)$:

$$1 - F(x) = \exp(-gx^c), \quad x \geq 0 \; . \tag{6.97}$$

After some calculations by the same scheme as above we find the covariance matrix B of the limit Gaussian distribution of ML-estimates $(\tilde{g}, \tilde{c})$:

$$B = \frac{6}{N\pi^2}\begin{pmatrix} g^2\left[\dfrac{\pi^2}{6} + (\gamma + \ln(g) - 1)^2\right] & gc[\gamma + \ln(g) - 1] \\ gc[\gamma + \ln(g) - 1] & c^2 \end{pmatrix} \tag{6.98}$$

where γ is the Euler number: $\gamma \simeq 0.577\,215\ldots$

Testing the Pareto Model Versus the (SE) Model Using Wilks' Test. Malevergne et al. [584] have discovered that the Pareto power law distribution is a special limit of the Weibull or stretched exponential (SE) family. Consider the parameterization (6.85) and the limit $c \to 0$ and $u > 0$. In this limit, and provided that

$$c \left(\frac{u}{\chi} \right)^c \to \beta, \quad \text{as } c \to 0 , \tag{6.99}$$

the (SE) model goes to the Pareto model. Indeed, we can write

$$\frac{c}{\chi^c} x^{c-1} \exp \left(-\frac{x^c - u^c}{\chi^c} \right)$$

$$= c \left(\frac{u}{\chi} \right)^c \frac{x^{c-1}}{u^c} \exp \left[-\left(\frac{u}{\chi} \right)^c \left(\left(\frac{x}{u} \right)^c - 1 \right) \right] ,$$

$$\simeq \beta x^{-1} \exp \left[-c \left(\frac{u}{\chi} \right)^c \ln \frac{x}{u} \right] , \quad \text{as } c \to 0$$

$$\simeq \beta x^{-1} \exp \left[-\beta \ln \frac{x}{u} \right] ,$$

$$\simeq \beta \frac{u^\beta}{x^{\beta+1}} , \tag{6.100}$$

which is the pdf of the Pareto power law model with tail index β. This implies that, as $c \to 0$, the characteristic scale χ of the (SE) model must also go to zero with c to ensure the convergence of the (SE) model towards the (PD) model.

This shows that the Pareto model can be approximated with any desired accuracy on an arbitrary interval $(u > 0, U)$ by the (SE) model with parameters (c, χ) satisfying equation (6.99) where the arrow is replaced by an equality. Although the value $c = 0$ does not give strickly speaking a Stretched-Exponential distribution, the limit $c \to 0$ provides any desired approximation to the Pareto distribution, uniformly on any finite interval (u, U). This deep relationship between the SE and power law models allows us to understand why it can be very difficult to decide, on a statistical basis, which of these models fits the data best [547, 584].

We use this insight to show how to develop a formal statistical test of the (SE) hypothesis $f_1(x|c, b)$ versus the Pareto hypothesis $f_0(x|b)$ on a semi-infinite interval (u, ∞), $u > 0$. Here, we use the parameterization

$$f_1(x|c, b) = b u^c x^{c-1} \exp \left[-\frac{b}{c} \left(\left(\frac{x}{u} \right)^c - 1 \right) \right]; \quad x \geq u \tag{6.101}$$

for the stretched-exponential distribution and

$$f_0(x|b) = b \frac{u^b}{x^{1+b}}; \quad x \geq u \tag{6.102}$$

for the Pareto distribution.

Theorem [585]. Assuming that the sample $x_1 \ldots x_N$ is generated from the Pareto distribution (6.102), and taking the supremums of the log-likelihoods L_0 and L_1 of the Pareto and (SE) models respectively over the domains $(b > 0)$ for L_0 and $(b > 0, c > 0)$ for L_1, then Wilks' log-likelihood ratio W [761]:

$$W = 2 \left[\sup_{b,c} L_1 - \sup_b L_0 \right], \qquad (6.103)$$

is distributed according to the χ^2-distribution with one degree of freedom, in the limit $N \to \infty$.

This theorem is very interesting because it allows us to use the full apparatus of statistical testing of nested-hypotheses. Without this theorem, the comparison of the Stretched-Exponential versus the Pareto distribution should in principle require that we use the methods for testing non-nested hypotheses [359], such as the Wald encompassing test or the Bayes factors [498], which are much weaker. Indeed, the Pareto model and the (SE) model are not, strictly speaking, nested. However, the Theorem shows that the Pareto distribution is a limit case of the Stretched-Exponential distribution, as the fractional exponent c goes to zero. In addition, it shows that we can use the standard Wilks' log-likelihood ratio [761] and corresponding statistics. We refer to [584] for the proof of the theorem and for the empirical tests using the Wilk procedure on empirical financial data.

Small Data Sets and Unbiasing Methods. General results on the principles of likelihood imply that the maximum likelihood estimators of the shape c and scale χ parameters based on independent and identically distributed sample of size n should be asymptotically unbiased, with variances obtaining the appropriate Cramer–Rao bound: expressions for the asymptotic bias in $\hat{c}$ and $\hat{\chi}$ are both of order $1/n$. Thus, for large samples, it is impossible to outperform maximum likelihood estimators in terms of estimator accuracy in any systematic way. However, we emphasize that these results are asymptotic. It is known that the bias observed for estimates calculated in small to moderate samples need not be insignificant. This small-sample behavior has prompted research into alternative methods for estimating the parameters of the Weibull distribution.

The estimators of χ and c discussed above usually assume that the sample size is large and/or the observations include no outliers. However, real data often contradict these assumptions; these estimators may then not always give desirable results. One answer is to get rid of outliers, but for small sample sizes, elimination of data is difficult and not always a good idea. The alternative is to find a robust estimation method when outliers are present. A suitable solution is the bootstrap method [264]. In this method, the determined distribution $\hat{P}_>(x)$ in terms of the parameters $\hat{\chi}$ and $\hat{c}$ is used to generate many sets of n i.i.d. random variables obtained from the initial data set by replacing about $1/e \approx 37\%$ of the points by duplicated data drawn from the distribution $\hat{P}_>(x)$. On each of these n data points sets, the estimation procedures for the parameters χ and c are applied. This allows one to calculate the interval of uncertainty on the parameters. This bootstrap method has been applied to the Weibull distribution in [831].

For small data sets ($n < 20$), unbiasing methods have thus been proposed for the estimation of the Weibull parameters. In [129, 791], comparisons between several unbiasing methods have been performed. The following unbiasing methods have been found to exhibit satisfactory performance, independently of the value of the shape parameter c and the sample size n. The improved Ross method, valid for c, gives the unbiased $\hat{c}_U$ as

$$\hat{c}_U = \hat{c}_{ML} \, \frac{n - 1.93}{n - 0.6} \, , \tag{6.104}$$

where $\hat{c}_{ML}$ is the maximum likelihood estimate given above. The modified Harter and Moore algorithm, valid for χ, is applied to the expected value of the sampling distribution of χ and reads

$$\hat{\chi}_U = \hat{\chi}_{ML} \, \frac{n^{1/\hat{c}} \Gamma(n)}{\Gamma(n + 1/\hat{c})} \, , \tag{6.105}$$

where $\Gamma(x)$ is the Gamma function equal to $(x - 1)!$ for integer x. The estimates given by these two formulas do not differ significantly from the true values.

Dubey [246] presents results of a Monte Carlo study on the problem of the small sample properties of the moment and the maximum likelihood estimators of Weibull parameters. He concludes that the moment and maximum likelihood estimators of the Weibull parameters may not be found satisfactory when the sample size is small.

Weighted Least-Square Linear Regression Method. Bergman [79] has shown that the conventional least-square linear regression method can be improved by using an appropriate weight function

$$W_i = \left(P_>(x_i) \ln P_>(x_i) \right)^2 . \tag{6.106}$$

This weight function is to be incorporated into the sum of squares of the deviations of the data for the straight line fit. The weight function varies strongly with $P_>(x_i)$ and has a peak at $P_>(x_i) = 1/e \approx 0.37$, where x is equal to the characteristic value χ. This approach has been implemented by Chandrasekhar [157] for the estimation of the Weibull parameters.

Jacquelin [454] gives the least square method estimations

$$\hat{c} = \frac{\sum\limits_{i=1}^{n} \ln x_i \sum\limits_{i=1}^{n} \ln\ln(1/P_>(x_i)) - n \sum\limits_{i=1}^{n} \ln x_i \, \ln\ln(1/P_>(x_i))}{\left(\sum\limits_{i=1}^{n} \ln x_i \right)^2 - n \sum\limits_{i=1}^{n} (\ln x_i)^2} \, ,$$

$$\tag{6.107}$$

$$\hat{\chi} = \exp\left(\frac{1}{n}\sum_{i=1}^{n}\ln x_i - \frac{1}{n\hat{c}}\sum_{i=1}^{n}\ln\ln\frac{1}{P_>(x_i)}\right). \tag{6.108}$$

$P_>(x_i)$ can be estimated from the rank ordering of the x_i's: noting $x_1 \geq x_2 \geq \ldots \geq x_n$, we have $P_>(x_i) \approx i/n$. Jacquelin [454] also gives extended tables to evaluate the confidence bounds for the Weibull parameter estimates in the cases of the mean square method, maximum likelihood and generalized maximum likelihood methods. The range of sample sizes investigated is from $n = 3$ to 100 and the computed confidence intervals range from 0.005 to 0.995.

Reliability. In the engineering literature, $P_>(x)$ is called the reliability. It measures the probability that x will be exceeded. For finance application, x is called a VaR (value-at-risk) which can be fixed at a given level, say a 5% dayly loss. Then, $P_>(x)$ is the probability that this loss is exceeded in any one day. Suppose we find $P_>(x) = 10^{-2}$. This means that we will see typically one loss of this magnitude or larger every one hundred days.

There is an extensive literature to obtain accurate efficient estimators for $P_>(x)$ for the Weibull distribution. One has indeed to realize that the quality of the estimator of probability level $P_>(x)$ depends on the underlying distribution. It is important to have reliable estimators tailored to the relevant distribution (here the Weibull). The difficulty stems from the fact that the parameters c and χ of the Weibull distribution are unknown a priori and must be estimated from a finite sample. The question is then to find efficient confidence bounds for the reliability or confidence level $P_>(x)$.

From (6.62), we obtain the $100p$-th percentile of the Weibull distribution

$$x_p = \chi\left(-\ln(1-p)\right)^{1/c}. \tag{6.109}$$

Then the $100(1 - e^{-1}) = 63.2$-th percentile is $x_{0.632} = \chi$ for any Weibull distribution. Therefore, we can obtain the estimate of the shape parameter c from (6.109) by

$$\hat{c} = \frac{\ln[-\ln(1-p)]}{\ln(x_p/x_{0.632})}, \tag{6.110}$$

where $0 < x_p < x_{0.632}$. From hundred of thousands of numerical simulations, Wang and Keats [1005] find that a single approximately optimal value is $p = 0.15$. The x_p value is then found using the linear interpolation equation

$$x_p = x_r + (x_{r+1} - x_r)\frac{p - (1 - P_>(x_r))}{P_>(x_r) - P_>(x_{r+1})}, \tag{6.111}$$

where x_r is the r-th ordered (by increasing) value and $1 - P_>(x_r)$ is the proportion of the data less than x_r.

A problem concerns the reliability of the estimation of large risks from the extrapolation of the stretched exponential using the estimated (and therefore uncertain) parameters $\hat{\chi}$ and $\hat{c}$. Let us call $\hat{P}_>(x)$ the SE distribution expressed in terms of the estimated parameters $(\hat{\chi}, \hat{c})$ from a sample of n data points. Thoman et al. [957] have shown that $\hat{P}_>(x)$ is nearly a minimum variance unbiased estimator of the true $P_>(x)$. In particular, one can easily show that $\hat{P}_>(x)$ depends on the true values of the parameters χ and c only through $P_>(x)$. This property makes it possible to obtain confidence intervals for $P_>(x)$ based on $\hat{P}_>(x)$. This has important applications for the estimation of the VaR (Value at Risk in financial applications).

The asymptotic variance of $\hat{P}_>(x)$ is given by

$$\sigma_P^2 = \sigma_\chi^2 \left(\frac{\partial P_>}{\partial \chi}\right)^2 + 2\sigma_{\chi,c}^2 \frac{\partial P_>}{\partial \chi} \frac{\partial P_>}{\partial c} + \sigma_c^2 \left(\frac{\partial P_>}{\partial c}\right)^2 . \tag{6.112}$$

This allows one to get confidence intervals for $P_>(x)$ by assuming that $P_>(x)$ is normally distributed with mean $P_>(x)$ and variance σ_P^2.

In 1952, Halperin has shown that

$$\frac{\sqrt{n} \ [\hat{P}_>(x) - P_>(x)]}{\sqrt{V[\hat{P}_>(x)]}} \tag{6.113}$$

is asymptotically normally distributed with zero mean and unit variance. The variance $V[\hat{P}_>(x)]$ of $\sqrt{n} \ [\hat{P}_>(x) - P_>(x)]$ is given by

$$V[\hat{P}_>(x)]$$
$$= \left(P_>(x) \ \ln P_>(x)\right)^2 \left(1 + \frac{1}{\gamma_2 - \gamma_1^2}\left[1 + \gamma_1 - \ln \ln \frac{1}{P_>(x)}\right]^2\right), \tag{6.114}$$

where $\gamma_1 = \int_0^{+\infty} dy \ \ln y \, e^{-y} = -C \approx -0.577...$ is the Euler constant and $\gamma_2 = \int_0^{+\infty} dy \ (\ln y)^2 e^{-y} = \pi^2/6 + C^2$.

Improved estimators that are exact even for small sample sizes, i.e. attain the desired confidence level precisely, are described in rather cumbersome tables in [484].

6.8.3 Three-Parameter Weibull Distribution

Haan and Beer [383] first developed the equations that be solved to obtain the maximum likelihood estimators for the three parameters of the Weibull probability density function and constructed a numerical solution for these equations.

We consider the distribution (6.66). The maximum likelihood approach generalizes that of Sect. 6.8.2. The log-likelihood is

$$\ln L = n \ln \frac{c}{\chi^c} - \frac{1}{\chi^c} \sum_{i=1}^n (x_i - b)^c + (c - 1) \sum_{i=1}^n \ln(x_i - b) . \tag{6.115}$$

The three parameters are determined by the solution of the following likelihood equations [717]:

$$0 = \frac{\partial \ln L}{\partial b} = \frac{c}{\chi^c} \sum_{i=1}^{n} (x_i - b)^{c-1} - (c-1) \sum_{i=1}^{n} (x_i - b)^{-1} . \tag{6.116}$$

$$0 = \frac{\partial \ln L}{\partial \chi^c} = -\frac{n}{\chi^c} + \frac{1}{\chi^{2c}} \sum_{i=1}^{n} (x_i - b)^c . \tag{6.117}$$

$$0 = \frac{\partial \ln L}{\partial c} = \frac{n}{c} + \sum_{i=1}^{n} \ln(x_i - b) - \frac{1}{\chi^c} \sum_{i=1}^{n} (x_i - b)^c \ln(x_i - b) . \tag{6.118}$$

From (6.117), we get

$$\chi^c = \frac{1}{n} \sum_{i=1}^{n} (x_i - b)^c \tag{6.119}$$

which may be substituted into (6.116) and (6.118) to yield two equations in terms of the parameters c and b.

The solution of this system of nonlinear equations is difficult and several iterative numerical schemes have been proposed. Panchang and Gupta [717] state "Our experience with some of these algorithms has frequently been frustrating. Thus a practitioner, who is anxious to obtain the appropriate solutions rapidly, has no assurance of doing so even after investing considerable time and effort in setting up and running the program for a chosen algorithm. Frequently, it is found that a particular method does not work for a given sample". This provides a good warning. However, the situation is not as bad as there is now a simple method that will definitely yield parameter estimates that maximize the likelihood function for the sample [717]. This method always works!

The idea is first to make use of the constraint that $c > 0$ and that all data points x_i's are larger than b. Let us call $x_{\min}$ the smallest value: $b < x_{\min}$.

1. The domain of b, which is $0 \le b \le x_{\min} - \epsilon$ where ϵ is a small number, is divided into I intervals of size Δb and b is sampled with the values $b_i = (i-1)\Delta b$.
2. For a given value $b = b_i$, (6.117) and (6.118) are solved to get $\chi = \chi_i$ and $c = c_i$. This is equivalent to solving the two parameter Weibull maximum likelihood problem, since $x_j - b_j$ may be considered as another sample point.
3. Instead of solving (6.116), $\ln L(b_i, \chi_i, c_i)$ is next computed.
4. The above process is repeated for all b_i's and $\ln L$ is obtained as a function of b. It is a simple matter to scan all the values of $\ln L$ to determine the maximum.

Note that this method does not depend, like some other methods do, on the choice of good initial estimates for its success. The computational effort required is less than that of most other schemes. We refer to [717] for details, implementation and examples of applications.

6.8.4 Generalized Weibull Distributions

Bradley and Price [120] showed that generalized Weibull distributions numerically fitted by nonlinear least-square or maximum-likelihood procedures are comparable in performance to the fits of more traditional functions, such as Johnson and Pearsonian function families. The generalized Weibull distributions do not require the sometimes ambiguous evaluation procedures for their selection. In addition, numerical methods provide a quick means of determining the parameters and the inverse function has a closed form for easy generation of random variates and for the problem of iterative treatment of missing data.

Four-Parameter Weibull Distribution. The four-parameter generalization of Weibull distribution is

$$
P_>(x) = 1 - \left[1 - e^{-((x-b)/\chi)^c} \right]^\gamma .
\tag{6.120}
$$

For $\gamma = 1$ and $b = 0$, this expression retrieves the simple definition (6.62), while, for $\gamma = 1$ and $b \neq 0$, one retrieves (6.66). Bradley and Price [120] introduces a further generalization of the Weibull function in terms of five parameters. To fit these functions, numerical methods are used to determine the parameters. Levenberg–Marquardt's method for nonlinear least-square estimates and pattern-search techniques for the maximum likelihood measures. The least-square estimates, under commonly used assumptions, are consistent but not efficient. Maximum likelihood estimators are consistent and efficient. Neither is unbiased asymptotically.

Generalized Gamma Distribution. The generalized Gamma distribution is defined by

$$
p(x)\,dx = dx\, \frac{c x^{c\alpha-1}}{\chi^{c\alpha}\Gamma(\alpha)}\, e^{-(x/\chi)^c} .
\tag{6.121}
$$

The case $\alpha = 1$ recovers the two-parameter Weibull distribution (6.62), since $\int_x^{+\infty} p(x)\,dx = e^{-(x/\chi)^c}$. The likelihood function of the generalized Gamma distribution is

$$
L = \frac{c^n \prod_{i=1}^n x_i^{c\alpha-1}}{\chi^{n c\alpha}[\Gamma(\alpha)]^n}\, e^{-\sum_{i=1}^n (x/\chi)^c} .
\tag{6.122}
$$

Equating to zero the partial derivatives of $\ln L$ with respect to the three parameters χ, α and c yields the following three equations

$$
\sum_{i=1}^n \left(\frac{x_i}{\chi} \right)^c = n\alpha ,
\tag{6.123}
$$

$$
c \sum_{i=1}^n \ln \frac{x_i}{\chi} = n\psi(\alpha) ,
\tag{6.124}
$$

where $\psi(\alpha) = \partial(\ln \Gamma(\alpha))/\partial\alpha$ is the digamma function, and

$$\frac{n}{c} - n\alpha \ln \chi + \alpha \sum_{i=1}^{n} \ln x_i - \sum_{i=1}^{n} \left(\frac{x_i}{\chi}\right)^c \ln \frac{x_i}{\chi} = 0 . \tag{6.125}$$

They can be manipulated to yield the equivalent and simpler equations

$$\chi = \left(\frac{\langle x_c \rangle}{\alpha}\right)^{1/c} , \tag{6.126}$$

$$\psi(\alpha) - \ln \alpha = \frac{1}{n} \sum_{i=1}^{n} \ln \frac{x_i^c}{\langle x_c \rangle} , \tag{6.127}$$

$$\frac{1}{\alpha} = \frac{1}{n} \sum_{i=1}^{n} \left(\frac{x_i^c}{\langle x_c \rangle} - 1\right) \ln \frac{x_i^c}{n\langle x_c \rangle} , \tag{6.128}$$

where

$$\langle x_c \rangle = \frac{1}{n} \sum_{i=1}^{n} x_i^c . \tag{6.129}$$

Wong [1025] implements a computer program (and gives its listing) to solve this equation, using a Lanczos series to evaluate the Gamma function.

7. Statistical Mechanics: Probabilistic Point of View and the Concept of "Temperature"

The concept of *temperature* is intuitive and constitutes an everyday experience. From a fundamental point of view, the notion of temperature is rigorously defined within the thermodynamic theory of systems at equilibrium, as the inverse of the variation ΔS of the entropy with respect to a variation of the energy ΔE, at fixed system volume. If the variation of entropy ΔS is large for a given variation of energy ΔE, the temperature is small: the system has few excited states and is quite ordered; a change of energy leads to the activation of many states and thus to a large change of the entropy. If the variation of entropy ΔS is small for a given variation of ΔE of energy, the temperature is large: the system is sufficiently excited and disordered that the entropy (which roughly quantifies the number of excited states) is almost insensitive to the energy.

The extension of the notion of temperature to statistical mechanics allows one to quantify the relative probability, given by the exponential Boltzmann factor, that the system possesses a given energy. In a nutshell, the temperature measures the amplitude of the noise or of the fluctuations of the physical variables around their average or most probable values. This can be seen for instance in the writing of the dynamical evolution of a field Φ in a system with a partial differential equation with a stochastic noise source [651, 1003]:

$$D_t\Phi = D_x\Phi + F(\Phi) - f\Phi + \eta(x,t) , \tag{7.1}$$

where D_t is a time differential operator like $\partial/\partial dt$, D_x is a space differential operator like ∇^2, f is a friction coefficient ensuring that Φ does not blow up due to the noise source term $\eta(x,t)$. $\eta(x,t)$ is often assumed to be white noise:

$$\langle \eta(x,t)\eta(x',t') \rangle = A\, T\delta(t-t')\, \delta(x-x') , \tag{7.2}$$

which defines the temperature T as proportional to the variance of the noise. The fluctuation–dissipation theorem ensures the proportionality between the strength f of the dissipation and the variance T of the noise. Stochastic partial differential equations such as (7.1) offer a convenient and powerful formulation for a large class of problems involving in particular the interplay between nonlinearity (in the function $F(\Phi)$) and stochasticity (in the noise source $\eta(x,t)$). Noisy hydrodynamics, stochastic surface growth processes and stochastic diffusion-reactions are three among many examples of applications.

The concept of temperature has been repeatedly applied to systems that are out-of-equilibrium, in the hope that their variability and fluctuations could also be captured by the simple Boltzmann factor. However, hidden behind this thermodynamic description of systems at equilibrium is an important and non-trivial assumption which is examined in the next section. This assumption is often not warranted quantitatively for out-of-equilibrium systems. Notwithstanding these restrictions, attempts are still being made to apply the concept of a temperature to out-of-equilibrium chaotic or turbulent systems, using clever choices of variables and subtle analogies, as we will summarize below. This is motivated by the hope that the thermodynamic formalism which has been so successful in equilibrium systems could help understand and simplify the complexity of chaotic and turbulent systems [190, 191]. Here, we attempt to present the concepts and their derivations in the simplest and most intuitive way, rather than emphasize mathematical rigor.

7.1 Statistical Derivation of the Concept of Temperature

Consider a (small) system at equilibrium which can exchange energy with another (large) system called a reservoir. The distinctive feature of this so-called "canonical ensemble" is described by the probability theory of independent events. We need two concepts, probability and energy. Let us consider two subparts I and II of the system and call $P(E_{Ia})$ $(P(E_{IIb}))$ the probability to find the system I (II) in a given state a (b) of energy E_{Ia} (respectively E_{IIb}). By the assumption that these two subsystems are independent, the joint probability $P(E)$ for the systems to be simultaneously in the states a and b respectively is equal to the product $P(E_{Ia})$ $P(E_{IIb})$:

$$P(E) = P(E_{Ia})P(E_{IIb}) , \tag{7.3}$$

where

$$E = E_{Ia} + E_{IIb} \tag{7.4}$$

is the sum of the two energies of the subsystems. We look for an expression of the probability $P(E)$ which is independent of the specific partition into two subsets. The only functional form that satisfies (7.3) together with (7.4) gives the so-called Boltzmann function

$$P(E) = Ce^{\alpha E} , \tag{7.5}$$

where C and α are constants. In this derivation, and in the equation (7.3), we neglect the correlations and also the energy provided by the boundary between the two systems. Indeed, in a system of volume L^d (for a cube of linear size L in d dimensions), the boundary effect is of order L^{d-1} and is thus negligible in relative size for large L, provided the influence of the boundary is short-ranged.

The Boltzmann function (7.5) describes the probability to find a system in any given state of energy E. This is not the same as the probability that the system has an energy E. This latter probability is the sum of (7.5) over all configurations that have this energy E. This leads to the multiplication of (7.5) by the so-called density of state $g(E)$ defined below, which is nothing but the number of configurations that the system can take at a given energy E.

Let us now connect this description with the mechanical description of the world in which systems obey extremal principles corresponding for instance to the minimization of the energy. This leads us to expect that the probability $P(E)$ is a decreasing function of E, hence $\alpha \leq 0$ and we note it $\alpha = -\beta = -1/k_B T$, which defines the temperature T. k_B is called the Boltzmann constant (introduced so that $k_B T$ has the units of an energy). Defining

$$Z(\beta) = \sum_E e^{-\beta E} , \tag{7.6}$$

which provides the normalization of the probability $P(E)$, we can write

$$P(E) = \frac{e^{-\beta E}}{Z} . \tag{7.7}$$

We have just derived the fundamental quantities of statistical physics! The expression (7.7) gives the probability that the system be in a configuration with the energy E while Z, called the partition function (because it is a sum over the partitionning of the system into different energy states), is simply the normalizing constant.

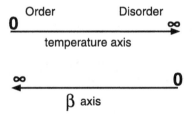

Fig. 7.1. Schematics of the temperature and β scales

The "temperature" appears as a parameter quantifying the degree of stochasticity of the system. If $T \to \infty$, $\beta \to 0$, and all states have the same probability, independently of their energy. In this limit of infinite disorder, no interaction is capable of ordering the system which takes all possible configurations that are available with the same probability.

In the other limit $T \to 0$, $\beta \to +\infty$, the system takes the only configurations characterized by the minimum energy. Any other state with a larger energy is infinitely less probable. This limit thus corresponds to the classical non-disordered case where the system minimizes its energy.

In between $0 < T < +\infty$, the system tends to have a larger probability of being in states with less energy but can nevertheless explore configurations with a higher energy, more so with larger temperature. We see clearly the competition between the disordering effect of the stochasticity (measured by the temperature) and the interactions determining the energies accessible to the system.

The expression (7.7) gives the probability of finding the system in *one* particular (macroscopic state) of energy E. The probability $p(E)$ that the system has an energy in a small *range* between E and $E + dE$ is simply obtained by summing the probabilities for all states whose energy lies in this range. Since all these states are characterized by essentially the same exponential factor $e^{-\beta E}$, one thus simply needs to multiply this common exponential factor $e^{-\beta E}$ by the number $g(E) dE$ of states in this energy range dE to get

$$p(E)\, dE = g(E) e^{-\beta E}\, dE \ , \tag{7.8}$$

$g(E)$ is called the density of states [770]. Usually, it is a rapidly increasing function of E as states proliferate at higher energies; the combination with the fast decreasing exponential results in a maximum of $g(E) e^{-\beta E}$, which is all the sharper for larger systems as more states become available and the law of large number starts to apply. For most situations, this product converges to a Gaussian distribution as a consequence of the central limit theorem. However, there are certain cases, that we will examine below, for which long range correlations appear and lead to distributions which are not Gaussian: they correspond to critical phase transitions. Other anomalous situations also arise from the competition between frozen disorder and ordering interactions [622].

7.2 Statistical Thermodynamics

All the usual thermodynamic variables can be derived from these two fundamental expressions (7.6) and (7.7) if we add the definition of the entropy

$$S = -k_B \sum_k P(E_k) \ln P(E_k) \ , \tag{7.9}$$

where the sum is carried out over all possible configurations, with each configuration k characterized by its energy E_k as well as by other defining variables Q_k. This expression can be justified from purely statistical considerations, using the Shannon definition of information. The constant k_B is again introduced for dimensionality reasons in order to connect this approach with the classical thermodynamic approach.

Expression (7.9) has an intuitive meaning:

$$\exp\left(-\frac{NS}{k_B}\right) = \prod_k [P(E_k)]^{NP(E_k)} \ , \tag{7.10}$$

where N is the total number of configurations that the system can take and $NP(E_k)$ is the number of configurations with energy E_k and defining variables Q_k. The r.h.s. of (7.10) is the product of independent probabilities over all possible states and thus quantifies the probability for the system to exist (in the corresponding so-called "macro-state").

The average energy $\langle E \rangle$ of the system is defined by

$$\langle E \rangle = \sum_E E \, \frac{e^{-\beta E}}{Z} \, , \tag{7.11}$$

which can also be written as

$$\langle E \rangle = -\frac{\partial \ln Z}{\partial \beta} \, . \tag{7.12}$$

The free energy is defined by

$$F = \langle E \rangle - TS \, , \tag{7.13}$$

which leads to

$$F = \sum_E \left(EP(E) - \frac{1}{\beta} P(E) \ln P(E) \right) \tag{7.14}$$

It is straightforward to verify that

$$F = -k_B T \ln Z \, . \tag{7.15}$$

From these expressions, we can recover all the quantities usually measured in thermodynamics: specific heat, susceptibility, enthalpy, etc. [131].

To summarize, if the dynamics of a system is sufficiently stochastic, either due to its intrinsic nature or due to external factors, so that it is possible to consider different parts of the system as independent, then the fundamental principles of thermal statistical physics can be derived in a simple and general way from probabilistic considerations. The properties of a system are completely embodied in the partition function Z, which is the fundamental quantity to calculate and from which everything else derives. After having defined a model for a given problem at equilibrium, the main task is thus to calculate its partition function.

7.3 Statistical Mechanics as Probability Theory with Constraints

7.3.1 General Formulation

We use the formalism of the previous section to establish a general relationship between the formulation of statistical mechanics and the theory of constrained fluctuations described in Chap. 3. Our presentation is inspired from [762]. In Chap. 3, we have seen the crucial role played by the entropy H_p:

the algorithm which yields the best estimate for an unknown frequency distribution is found from the maximization of H_p. Furthermore, we have seen that hypotheses can be tested with the help of (14.101), i.e., by comparing the experimental and theoretical values of H_p.

For complex systems with many degrees of freedom (like a fluid, a plasma or a solid), the exact configuration of all degrees of freedom is usually not known. It is therefore impossible to assign to the system a unique point in phase space (classical) or a unique wave function (quantum). Instead one must resort to a statistical description in which each point in phase space characterizes a different state that the system may take. The system is described by a classical *phase space distribution* $\rho(\pi)$ or an incoherent *mixture* of mutually orthogonal quantum microstates $\{|i\rangle\}$ in the quantum regime. For our purpose, the distinction between classical and quantum mechanics does not matter and we shall use the generic symbol ρ. Probabilities must be real, non-negative, and normalized to one. This implies the respective properties

$$\rho(\pi)^* = \rho(\pi) \ , \quad \rho(\pi) \geq 0 \ , \quad \int \mathrm{d}\pi \, \rho(\pi) = 1 \ . \tag{7.16}$$

In this statistical description, every observable A is assigned an *expectation value*

$$\langle A \rangle_\rho = \int \mathrm{d}\pi \, \rho(\pi) A(\pi) \ . \tag{7.17}$$

Typically, not even the distribution ρ is a priori known. Rather, the state of a complex physical system is characterized by very few macroscopic data. These data may come in different forms:

- as *data given with certainty*, such as the type of particles that make up the system, or the shape and volume of the box in which they are enclosed. We take into account these exact data through the definition of the phase space in which we are working;
- as *prescribed expectation values*

$$\langle G_a \rangle_\rho = g_a \quad , \quad a = 1, \ldots, m \tag{7.18}$$

of some set $\{G_a\}$ of selected macroscopic observables. Examples might be the average total energy, average angular momentum, or average magnetization. Such data, which are of a statistical nature, impose constraints of the type (3.29) on the distribution ρ; or

- as additional *control parameters* on which the selected observables $\{G_a\}$ may explicitly depend, such as an external electric or magnetic field.

According to our general considerations of Chap. 3, the best estimate for the macrostate, which takes into account these constraints, is a distribution of the form (3.45)

$$\rho(\pi) = \frac{1}{Z} \exp\left(\ln \sigma(\pi) - \langle \ln \sigma \rangle_\sigma - \sum_a \lambda^a G_a(\pi) \right) \tag{7.19}$$

with

$$Z = \int d\pi \, \exp \left(\ln \sigma(\pi) - \langle \ln \sigma \rangle_\sigma - \sum_a \lambda^a G_a(\pi) \right) . \tag{7.20}$$

Here, σ denotes the a priori distribution. As we have seen, the auxiliary quantity Z is referred to as the *partition function*. Readers already familiar with statistical mechanics might be disturbed by the appearance of σ in the definitions of ρ and Z. Yet, this is essential for a consistent formulation of the theory, as seen for instance from the possibility of iterating the frequency estimation algorithm described in Sect. 3.3.2. In most practical applications, σ is uniform and hence $\ln \sigma - \langle \ln \sigma \rangle_\sigma = 0$. The definitions of ρ and Z then reduce to the conventional expressions.

The phase space integral for Z depends on the specific choice of the phase space or Hilbert space; hence they may depend on parameters like the volume or particle number. Furthermore, there may be an explicit dependence of the observables $\{G_a\}$ or of the a priori distribution σ on additional control parameters. Therefore, the partition function generally depends not just on the Lagrange multipliers $\{\lambda^a\}$ but also on some other parameters $\{h^b\}$. One then defines new variables

$$\gamma_b \equiv \frac{\partial \ln Z}{\partial h^b} . \tag{7.21}$$

The $\{g_a\}$, $\{\lambda^a\}$, $\{h^b\}$ and $\{\gamma_b\}$ are called the *thermodynamic variables* of the system; together they specify the system's macrostate. The thermodynamic variables are not all independent: Rather, they are related by (7.21), that is via partial derivatives of $\ln Z$. One says that h^b and γ_b, or g_a and λ^a, are *conjugate* to each other.

Some combinations of thermodynamic variables are of particular importance, which is why the associated distributions go by special names. If the observables that characterize the macrostate – in the form of precisely determined values given with certainty, or in the form of expectation values – are all constants of the motion then the system is said to be in *equilibrium* to which there is an associated *equilibrium distribution* of the form (7.19), with all $\{G_a\}$ being constants of the motion. Such an equilibrium distribution is itself constant in time, and so are all expectation values calculated from it, assuming that there is no time-dependence of the a priori distribution σ. The set of constants of the motion always includes the Hamilton function, provided it is not explicitly time-dependent. If its value for a specific system, the *internal energy*, and the other macroscopic data are all given with certainty, then the resulting equilibrium distribution is called *microcanonical*; if just the energy is given on average, while all other data are given with certainty, it is called *canonical*; and if both energy and total particle number are given on average, while all other data are given with certainty, it is called *grand canonical*.

Every description of the macrostate in terms of thermodynamic variables represents a hypothesis: namely that the sets $\{G_a\}$ and $\{h^b\}$ are actually complete. This is analogous to Jaynes' model for Wolf's die discussed in Sect. 3.3.4, which assumes that just two imperfections (associated with two observables G_1, G_2) suffice to characterize the experimental data. Such a hypothesis may well be rejected by experiment. If so, this does not mean that our rationale for constructing ρ by maximizing H_σ under given constraints was wrong. Rather, it means that important macroscopic observables or control parameters (such as "hidden" constants of the motion, or further imperfections of Wolf's die) have been overlooked, and that the correct description of the macrostate requires additional thermodynamic variables.

7.3.2 First Law of Thermodynamics

Changing the values of the thermodynamic variables alters the distribution ρ and with it the associated

$$H_\sigma^{\max} \equiv H_\sigma(\rho) = \langle \ln \sigma \rangle_\sigma + \ln Z + \sum_a \lambda^a \, g_a \; . \tag{7.22}$$

By virtue of (7.21), its infinitesimal variation is given by

$$dH_\sigma^{\max} = d\langle \ln \sigma \rangle_\sigma + \sum_a \lambda^a \, dg_a + \sum_b \gamma_b \, dh^b \; . \tag{7.23}$$

As the set of constants of the motion always contains the Hamiltonian, its value for the given system, the internal energy U, and the associated conjugate parameter, which we denote by β, play a particularly important role. Depending on whether the energy is given with certainty or on average, the pair (U, β) corresponds to a pair (h, γ) or (g, λ) respectively. For all remaining variables, one then defines new conjugate parameters

$$l^a \equiv \lambda^a/\beta \quad , \quad m_a \equiv \gamma_a/\beta \tag{7.24}$$

such that in terms of these new parameters the energy differential reads

$$dU = \beta^{-1} \, d(H_\sigma^{\max} - \langle \ln \sigma \rangle_\sigma) - \sum_a l^a \, dg_a - \sum_b m_b \, dh^b \; . \tag{7.25}$$

A change in internal energy that is effected solely by a variation of the parameters $\{g_a\}$ or $\{h^b\}$ is defined as *work*

$$\delta W \equiv - \sum_a l^a \, dg_a - \sum_b m_b \, dh^b \; . \tag{7.26}$$

If, on the other hand, these parameters are held fixed ($dg_a = dh^b = 0$), then the internal energy can still change through the addition or subtraction of *heat*

$$\delta Q \equiv \frac{1}{k_B \beta} \, k_B \, d(H_\sigma^{\max} - \langle \ln \sigma \rangle_\sigma) \; , \tag{7.27}$$

where k_B is the Boltzmann constant, equal to $k_B = 1.381 \times 10^{-23} \, \text{J/K}$.

We retrieve the *temperature*

$$T \equiv \frac{1}{k_{\mathrm{B}}\beta} \tag{7.28}$$

and the *entropy*

$$S \equiv k_{\mathrm{B}}(H_\sigma^{\mathrm{max}} - \langle \ln \sigma \rangle_\sigma) \tag{7.29}$$

and obtain δQ in the more familiar form

$$\delta Q = T \, \mathrm{d}S . \tag{7.30}$$

The entropy is related to the other thermodynamic variables via (7.22), i.e.

$$S = k \ln Z + k \sum_a \lambda^a \, g_a . \tag{7.31}$$

Even though the entropy is related to measurable quantities, it is essentially an auxiliary concept and does not itself constitute a physical observable. This is also true for the partition function.

The relation

$$\mathrm{d}U = \delta Q + \delta W , \tag{7.32}$$

which reflects nothing but energy conservation, is known as the *first law of thermodynamics*.

7.3.3 Thermodynamic Potentials

Like the partition function, thermodynamic potentials are auxiliary quantities used to facilitate calculations. One example is the (generalized) *grand potential*

$$\Omega(T, l^a, h^b) \equiv -\frac{1}{\beta} \ln Z , \tag{7.33}$$

related to the internal energy U via

$$\Omega = U - TS + \sum_a l^a \, g_a . \tag{7.34}$$

Its differential

$$\mathrm{d}\Omega = -S \, \mathrm{d}T + \sum_a g_a \, \mathrm{d}l^a - \sum_b m_b \, \mathrm{d}h^b \tag{7.35}$$

shows that S, g_a and m_b can be obtained from the grand potential by partial differentiation; e.g.,

$$S = -\left(\frac{\partial \Omega}{\partial T}\right)_{l^a, h^b} , \tag{7.36}$$

where the subscript means that the partial derivative is to be taken at fixed l^a, h^b. In addition to the grand potential, there are many other thermodynamic potentials. In a given physical situation, it is most convenient to work with that potential which depends on the variables being controlled or measured in the experiment. For example, if a chemical reaction takes place at constant temperature and pressure (controlled variables T, $\{m_b\} = \{p\}$), and the observables of interest are the particle numbers of the various reactants (measured variables $\{g_a\} = \{N_i\}$), then the reaction is most conveniently described by the free enthalpy $G(T, N_i, p)$ well-known in thermodynamics [131].

When a large system is physically divided into several subsystems, then in these subsystems, the thermodynamic variables generally take values that differ from those of the total system. In the special case of a *homogeneous system*, all variables of interest can be classified either as extensive – varying proportionally to the volume of the respective subsystem – or intensive – remaining invariant under the subdivision of the system. Examples for the former are the volume itself, the internal energy or the number of particles; whereas amongst the latter are the pressure, the temperature or the chemical potential. In general, if a thermodynamic variable is extensive then its conjugate is intensive, and vice versa. If we assume that the temperature and the $\{l^a\}$ are intensive, while the $\{h^b\}$ and the grand potential are extensive, then

$$\Omega_{\text{hom}}(T, l^a, \tau h^b) = \tau \Omega_{\text{hom}}(T, l^a, h^b) \tag{7.37}$$

for all $\tau > 0$ and hence

$$\Omega_{\text{hom}} = -\sum_b m_b h^b . \tag{7.38}$$

This implies the *Gibbs–Duhem relation*

$$S \, dT - \sum_a g_a \, dl^a - \sum_b h^b \, dm_b = 0 . \tag{7.39}$$

The Gibbs–Duhem relation is nothing but the Euler property for an homogeneous function. For an ideal gas in the grand canonical ensemble, for instance, we have the temperature T and the chemical potential $\{l^a\} = \{-\mu\}$ intensive, whereas the volume $\{h^b\} = \{V\}$ and the grand potential Ω are extensive; hence

$$\Omega_{\text{i.gas}}(T, \mu, V) = -p(T, \mu) \, V . \tag{7.40}$$

7.4 Does the Concept of Temperature Apply to Non-thermal Systems?

7.4.1 Formulation of the Problem

How general is the applicability of this modeling strategy? Our understanding of out-of-equilibrium spatially extended systems with many degrees of freedom is far from approaching that of systems at the thermodynamic equilibrium for which the concept of a temperature is well-established. In this latter case, a remarkable simplification occurs as we have just seen in that the fluctuations can be embodied by a single parameter, the temperature ($\propto \beta^{-1}$), thus leading to the possible use of powerful statistical tools relying on the Boltzmann distribution $P(E) \sim e^{-\beta E}$, based on the multiplicative dependence of the number of microstates in the number of degrees of freedom while the energy is additive.

Claims have been made in the past about the relevance of the Boltzmann distribution and the analogy with a temperature in many systems: neural networks [399], spatio-temporal chaos [134], turbulence [145] and granular media [616, 815]. This quest to find equipartition of energy and Boltzmann statistics from nonlinear dynamics was first initiated by Fermi, Pasta and Ulam [296], who failed. More recent works (see for instance [569]) have shown the subtlety of this problem. In their pioneering work, Fermi, Pasta, and Ulam revealed that even in strongly nonlinear one-dimensional classical lattices, recurrences of the initial state prevented the equipartition of energy and consequent thermalization. The questions following from this study involve the interrelations between equipartition of energy (Is there equipartition? In which modes?), local thermal equilibrium (Does the system reach a well-defined temperature locally? If so, what is it?), and transport of energy/heat (Does the system obey Fourier's heat law? If not, what is the nature of the abnormal transport?). In sorting through these questions, it is important to recall that the study of heat conduction (Fourier's heat law), stating that the heat flux is proportional to minus the gradient of the temperature, is the search for a non-equilibrium steady state in which heat flows across the system. However, this situation is usually analyzed, using the Green–Kubo formalism of linear response [534], in terms of the correlation functions in the thermal equilibrium (grand canonical) state. The surprising result of Fermi, Pasta and Ulam has now been understood: under general conditions for classical many-body lattice Hamiltonians in one dimension, it has been shown that total momentum conservation implies anomalous transport in the sense of the divergence of the Kubo expression for the coefficient of thermal conductivity [753]. The absence of equipartition is thus an anomalous feature of one-dimensional systems.

We must also mention the emerging application of the theory of smooth dynamical systems to nonequilibrium statistical mechanics, based on the chaotic character of the microscopic time evolution, as reviewed in [328, 796]. In this dynamical approach, the emphasis is on non-equilibrium steady states rather than on the traditional approach to equilibrium point of view of Boltzmann. The nonequilibrium steady states, in presence of a Gaussian thermostat, are described by "statistics of the motion" measures, in terms of which one can prove the Gallavotti–Cohen fluctuation theorem and prove a general linear response formula not restricted to near-equilibrium situations. At equilibrium, this recovers in particular the Onsager reciprocity relations. This formalism allows a quantitative definition of intermittency in statistical mechanics and in fluid mechanics.

7.4.2 A General Modeling Strategy

Kurchan [543] has shown that, in the limit of a large number of degrees of freedom, slowly flowing granular media driven by both tapping and shearing can still be described by thermodynamic concepts for the low frequency motion. The derivation uses the framework of glassy systems with driving and friction that are generic and do not correspond to a thermal bath (whose microscopic "fast" motion is hence not thermal). There is thus a well-defined macroscopic temperature associated with the slow degrees of freedom.

We follow [543] for this presentation. For gently driven glasses, a picture has emerged [1036] that involves multiple thermalisations at widely separated timescales. In the simplest description and for any two observables A and B belonging to the system, the correlation function is defined as

$$\langle A(t)B(t')\rangle = C_{AB}(t,t') \tag{7.41}$$

and the response of A to a field conjugate to B is

$$\frac{\delta}{\delta h_B(t')}\langle A(t)\rangle = R_{AB}(t,t') . \tag{7.42}$$

For a pure relaxational (undriven) glass, the correlation breaks up into two parts:

$$C_{AB}(t,t') = C_{AB}^{\mathrm{F}}(t-t') + \tilde{C}_{AB}\left(\frac{H(t')}{H(t)}\right) \tag{7.43}$$

where H is the same growing function for all observables A and B. If the glass is gently driven (with driving forces proportional to, say, ϵ), we have:

$$C_{AB}(t,t') = C_{AB}^{\mathrm{F}}(t-t') + \tilde{C}_{AB}\left(\frac{t-t'}{\tau_o}\right) \tag{7.44}$$

where τ_o is a time scale that diverges as the driving forces proportional to ϵ go to zero.

In the long time limit and in the small drive limit, the time scales become very different. When this happens, the responses behave as:

$$R_{AB}(t,t') = \beta\frac{\partial}{\partial t'}C_{AB}^{\mathrm{F}} + \beta^*\frac{\partial}{\partial t'}\tilde{C}_{AB} . \tag{7.45}$$

The fast degrees of freedom behave as if thermalised at the bath inverse temperature β. On the other hand, the effective, system-dependent temperature $T^* = 1/\beta^*$ indeed deserves its name: it can be shown that it is what a "slow" thermometer measures, and it controls the heat flow and the thermalisation of the slow degrees of freedom. It is *the same* for any two observables at a given timescale, whether the system is aging or gently driven.

Furthermore, it is macroscopic in the sense it remains non-zero in the limit in which the bath temperature is zero.

If the system is not coupled to a true thermal bath but instead energy is supplied by shaking and shearing while it is dissipated by some linear or nonlinear friction, there is no bath inverse temperature β. What has been argued [543] is that even so, the "slow" inverse temperature β^* survives despite the fact that the fast motion is not thermal. Indeed, if we have correlations having fast and slow components:

$$C_{AB}(t,t') = C_{AB}^{\mathrm{F}}(t,t') + \hat{C}_{AB}^{\mathrm{S}}(t,t') \, , \tag{7.46}$$

the response is of the form:

$$R_{AB}(t,t') = R_{AB}^{\mathrm{F}}(t,t') + \beta^* \frac{\partial}{\partial t'} \hat{C}_{AB}^{\mathrm{S}}(t,t') \tag{7.47}$$

with the fast response $R_{AB}^{\mathrm{F}}(t,t')$ bearing no general relation with the fast correlation $C_{AB}^{\mathrm{F}}(t,t')$.

7.4.3 Discriminating Tests

In order to qualify an analogy with thermal equilibrium, one should be careful. Not only must the probability of an energy of the microscopic level be of the form (7.7), but this holds true for the total system. Indeed, Boltzmann statistics should also describe the *bulk* energy distribution of the system. We need to be more precise and express the probability to find a system in a state of energy E as $P_\beta(E) = g(E)e^{-\beta E}$, where $g(E)$ is interpreted as the density of microstates (number of configurations of the system having the same energy) defined in Sect. 7.1.

To qualify Boltzmann statistics, the standard method is to take the ratio

$$\frac{P_{\beta_1}(E)}{P_{\beta_2}(E)} \, , \tag{7.48}$$

which is independent of the unknown $\rho(E)$ and should be a pure exponential. To provide an additional test of Boltzmann statistics, one should also verify the validity of the fluctuation–dissipation theorem according to which the derivative of the average bulk energy with respect to the temperature should be proportional to the variance of the energy fluctuations [934]. Another version of the fluctuation–dissipation theorem describes the probability ratio of observing trajectories that satisfy or violate the second law of thermodynamics which in a nutshell states that the entropy of closed systems is a non-decreasing function of transformations. Recent works show that the fluctuation-dissipation theorem does not rely on the reversibility or the determinism of the underlying dynamics and applies to thermostated deterministic nonequilibrium as well as for a class of stochastic nonequilibrium systems [829].

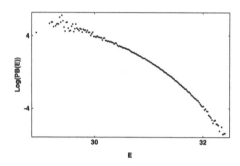

Fig. 7.2. $P_{\beta_1}(E)/P_{\beta_2}(E)$ for the same system size and parameters as in [798] with uniform breaking strength. When the force on a site becomes larger than 1, the stress drop in this site is reshuffled as $0.8 + (1/\beta)$ran, where ran is a random number drawn uniformly in the interval $[0, 1]$ and β is the inverse temperature. See [1031] for details

For systems self-organizing through spanning "avalanches" such as in models of self-organized criticality (which we will discuss later), we do not expect Boltzmann statistics to hold, because the avalanches are the vehicle of the correlations and of the self-organization. As a consequence, a system cannot be partitionned without changing its organization: due to the effective long-range coupling induced by large avalanches, extensivity is no longer present and the concept of temperature, if any, needs much more care to be defined. The concept of a temperature and of Boltzmann statistics suggested in [798, 799] for dynamical systems with threshold dynamics (sandpile models) have been tested and the results are negative [1031], as shown in Fig. 7.2. Boltzmann statistics should give a straight line in this representation. However, Boltzmann statistics may be retrieved in the limit of infinite range interactions [799].

The generalization of the entropy and of the temperature by the so-called non-extensive statistics [964] may provide a better description of the energy distribution in these sandpile models (C. Tsallis, private communication). To see this, let us use expressions (14.50, 14.51, 14.52) in Chap. 14 for the Tsallis distribution for arbitrary q. In terms of energies E, we rewrite (14.50, 14.51, 14.52) as $\propto [1 - (1-q)\beta E]^{1/(1-q)}$. Thus, according to the Tsallis statistics,

$$P_\beta(E) \propto g(E) \, [1 - (1-q)\beta E]^{1/1-q}$$

$$\propto g(E) \exp\left(\frac{1}{1-q} \ln[1 - (1-q)\beta E] \right)$$

$$\sim g(E) \exp\left(-\beta E - \frac{1-q}{2}\beta^2 E^2 \right) , \qquad (7.49)$$

where $g(E)$ is the density of states. Then

$$\ln\left(P_{\beta_1}(E)/P_{\beta_2}(E) \right)$$

$$= -(\beta_1 - \beta_2)E - (1-q)\frac{\beta_1^2 - \beta_2^2}{2} \, E^2 + \mathcal{O}[(1-q)^2] . \qquad (7.50)$$

Expression (7.50) predicts a downward quadratic correction, as observed in Fig. 7.2 (for $\beta_1 > \beta_2$ since the overall slope is negative), only if we take $q < 1$, for which the distribution of energies has a compact support as seen in expression (14.52).

Simulations of the Bak–Tang–Wiesenfeld sandpile model [46] (see Chap. 15) with non-vanishing driving rate h (incoming sand flux) and dissipation rate [59] have allowed the calculation of the susceptibility $\chi(t)$ and the correlation function $C(t)$ of the sandpile. They are defined as

$$\Delta\rho_a(r,t) = \int \chi(t-t')\,\Delta h(r,t')\,\mathrm{d}r'\,\mathrm{d}t' \qquad (7.51)$$

and

$$C(t) \equiv \langle \rho_a(r,t)\rho_a(r,0)\rangle - \langle\rho_a\rangle^2 \;, \qquad (7.52)$$

where $\rho_a(r,t)$ is the density of active site (i.e. above threshold) at point r and at time t. A version of the fluctuation dissipation theorem would predict that the response function $\chi(t)$ is proportional to the time derivative of the correlation function $C(t)$. Barrat et al. [59] find that this is not born out by the data: both $\chi(t)$ and $C(t)$ exhibit exponential relaxation behaviors but with *different* time scales such that their ratio does not depend on driving and dissipation rates. Simulations of a random neighbor sandpile model, described by mean-field theory as in [798, 799], show that the fluctuation–dissipation relation is not satisfied either.

7.4.4 Stationary Distribution with External Noise

In the next two paragraphes, we follow [51] and consider the motion of a particle of unit mass moving in potential $V(x)$ and subjected to random forces of both internal $f(t)$ and external origin $e(t)$ according to the Langevin equation

$$\ddot{x} + \int_0^t \gamma(t-\tau)\dot{x}(\tau)\,\mathrm{d}\tau + V'(x) = f(t) + e(t) \;, \qquad (7.53)$$

where the friction kernel $\gamma(t)$ is connected to internal noise $f(t)$ by the fluctuation-dissipation relationship

$$\langle f(t)f(t')\rangle = k_{\mathrm{B}}T\gamma(t-t') \;. \qquad (7.54)$$

Both the noises $f(t)$ and $e(t)$ are stationary and Gaussian, their correlation times may be of arbitrary decaying type and the external noise is independent of the memory kernel and there is no corresponding fluctuation–dissipation relation. In addition, $f(t)$ is independent of $e(t)$:

$$\langle f(t)e(t)\rangle = 0 \;. \qquad (7.55)$$

The external noise modifies the dynamics such that the equilibrium distribution of the particle is disturbed and no longer given by the standard Boltzmann distribution. This new stationary distribution must be a solution of the generalized Fokker–Planck equation.

Linearizing the dynamics around a minimum of the potential up to quadratic order, corresponding to a linear restoring force, the stationary probability distribution which is the solution of the Fokker–Planck equation of the problem is given by [51]

$$p_{st}^0(x, v) = \frac{1}{Z} \exp \left[-\frac{v^2}{2D_0} - \frac{\tilde{V}(x)}{D_0 + \psi_0} \right] \tag{7.56}$$

where x and v are the position and velocity of the particle and D_0 and ψ_0 are functions of the memory kernel γ and the correlation functions of the internal and external noises and Z is the normalization constant. $\tilde{V}(x)$ is a renormalized linearized potential with a renormalization in its frequency [51]. The important point behind this result is that the distribution (7.56) is not an equilibrium Botzmann distribution even if there are some similarities. This stationary distribution for the open system plays the role of an equilibrium distribution for the closed system which may be however recovered in the absence of external noise terms. In this simple example, the steady state is unique and the question of multiple steady states does not arise because the potential $\tilde{V}(x)$ has been chosen quadratic.

7.4.5 Effective Temperature Generated by Chaotic Dynamics

In addition to recovering distributions of the Bolzmann form, some simple, far-from-equilibrium, dissipative, extensively chaotic systems can recover the equilibrium properties of ergodicity, detailed balance, partition functions, and renormalization group flow at coarse-grained scales with the underlying chaotic dynamics serving as a temperature bath. A well-studied case consists in a coupled map lattice made of a set of scalar variables $u(\mathbf{r}, t)$ defined at discrete integer times on a square two-dimensional spatially periodic $L \times L$ grid with positions indicated by $\mathbf{r}$. The rule for updating the variables from time t to $t+1$ is

$$u(\mathbf{r}, t+1) = \phi(u(\mathbf{r}, t)) + g \sum_{\mathbf{r}_n(\mathbf{r})} [\phi(\mathbf{r}_n(\mathbf{r}), t)) - \phi(u(\mathbf{r}, t))] , \tag{7.57}$$

where g indicates the strength of the spatial coupling and $\mathbf{r}_n(\mathbf{r})$ denotes nearest neighbors of site $\mathbf{r}$. The local map $\phi(u)$ is chaotic and can be of several forms. The case studied in [265, 626] corresponds to a zig-zag curve, made of three linear segments. This coupled map lattice exhibits chaotic, spatially disordered dynamics for values of g within the range $[0, 0.25]$. Miller and Huse [626] reported that at $g_c \approx 0.2054$, this system undergoes a kind of paramagnetic-to-ferromagnetic transition exhibiting a number of features in common with the equilibrium transition in the Ising ferromagnet. Egolf [265] has extended this observation and has shown that the long-wavelength behavior of this far-from-equilibrium system can be understood by using the powerful tools of equilibrium statistical mechanics. This system possesses however some important differences from true equilibrium systems, the more important one being that the effective noise strength (or temperature) is internally generated and dependent on the state of the system, rather than imposed by an external temperature bath. This difference poses a challenge for explorations of the second law of thermodynamics (entropy growth) in these systems.

A natural extension for stationary out-of-equilibrium systems of the classical ergodic hypothesis introduced for systems at equilibrium is called the chaotic hypothesis [332], which amounts to stating that the system is chaotic enough so that typical initial data generate a unique stationary distribution with several features of the so-called SRB (Sinai, Ruelle, Bowen) distributions [329, 795, 796]. In this case, the concept of entropy can be replaced by the definition of a Lyapunov function defined as minus the divergence of the equations of motion, which measures the phase space contraction rate. The role of the Lyapunov function is to indicate which will be the final equilibrium state of an initial datum in phase space. Gallavotti [331] proposes to define the "temperature" T of a mechanical thermostat in contact with an out-of-equilibrium system by remarking that the Lyapunov function is proportional to the work per unit time that the mechanical forces perform, the proportionality constant being in general a function of the point in phase space. Therefore, Gallavotti calls $1/T$ the time average of the proportionality constant between the Lyapunov function and the work per unit time that the mechanical thermostatting forces perform. The utility of such a notion of temperature stems from the possible equivalence between a vast classes of thermostats, deterministic or stochastic, in the sense that they produce motions which although very different when compared at equal initial conditions and at each time have, nevertheless, the same statistical properties [329, 330]. One among the most striking examples of such equivalence is the equivalence between the dissipative Navier Stokes fluid and the Euler fluid in which the energy content of each shell of wave numbers is fixed (via Gauss' least constraint principle) to be equal to the value that Kolmogorov's theory predicts to be the energy content of each shell at a given (large) Reynolds number [836].

In Chap. 13, we alluded to a simple 2D dynamical model of earthquake in a tectonic plate with long range elastic forces and quenched disorder, which captures both the spontaneous formation of fractal fault structures by repeated earthquakes, the short-time spatio-temporal chaotic dynamics of earthquakes, and the Gutenberg–Richter power law of earthquake size distribution [907]. This model is a deterministic self-organized critical model (see Chap. 15) controlled by two parameters, the relative stress drop $\delta\sigma$ and the relative amplitude of the disorder $\Delta\sigma$. Simulations have shown that the stress drop parameter is similar to a temperature, in the sense that the larger it is, the larger are the fluctuations of the various variables of the problems, such as the elastic stress field, the strain field and the total elastic energy stored in the whole system. As a function of time, the elastic energy

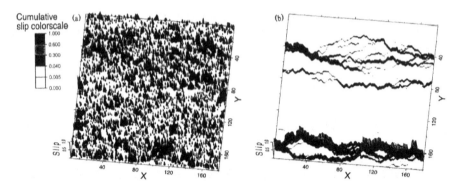

Fig. 7.3. Evolution of the cumulative earthquake slip, represented along the vertical axis in perspective in the white to black color code shown on the left of the picture, at two different times: (**a**) early time and (**b**) long time. The system has a size of 90×90. The width of the distribution of stress thresholds at rupture is $\Delta\sigma = 1.9$ which is must larger than the stress drop associated with each earthquake rupture equal to $\delta\sigma = 0.05$ in reduced units. Reproduced from [627]

exhibits small fluctuations for small stress drop and larger and larger fluctuations as the stress drop parameter increases. This is compatible with the idea that the stress drop controls the degree of chaoticity of the model (the Lyapunov exponent increases with the stress drop) and can thus be viewed as a kind of effective temperature. Similarly to the coupled map lattice model discussed above [265, 626], the temperature is generated dynamically by the chaotic nature of the deterministic threshold dynamics. In this class of models with thresholds, it is not possible to use the standard method involving the linearized dynamics for the computation of the Lyapunov exponents, since the time evolution cannot be written in terms of an explicit map. However, one can follow the divergence of nearby trajectories, and thus determine the maximum Lyapunov exponent. Following the procedure proposed in [189] for a cellular automaton model of earthquakes, one finds [907] that, for a very large range of disorder and stress drop parameters, the maximum Lyapunov exponent is positive, indicating that deterministic chaos is a generic feature of the model.

For stress drop comparable to but smaller than the disorder in stress threshold for rupture, the earthquakes occur on several competing faults as shown in Fig. 7.3. One fault system is typically active for very long periods of time until the locus of ruptures spontaneously switches to a different fault and the previously active fault may become completely silent for a long time. In some cases, the switch to another structure is brief. This feature of alternate fault activity over long periods of time has been described in several geological contexts and its origin is still a mystery. In this model [907], these observations find a natural explanation within an analogy developed with random directed polymers (see Chap. 5 and Fig. 5.15) by viewing the

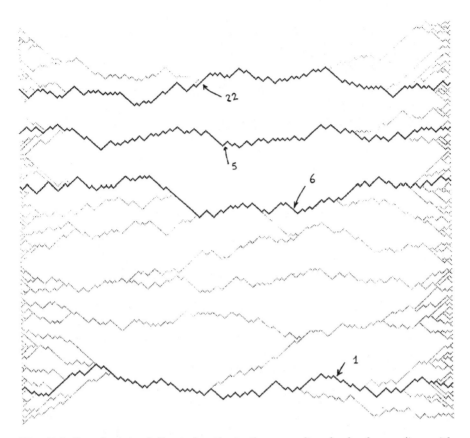

Fig. 7.4. Set of minimal directed paths in the same disorder landscape (i.e., with exactly the same realization of the random numbers $\{\sigma(x,y)\}$ characterizing the rupture threshold) as Fig. 7.3. The paths shown here solve the following problem: find the directed path connecting the left side to the right side at a same arbitrary elevation such that the sum of the quenched random numbers $\{\sigma(x,y)\}$ characterizing each fault among the 90×90 fault elements of the network along the path is minimum. This problem is also known as the random directed polymer problem at zero temperature (see Chap. 5 and Fig. 5.15). The remarkable result shown here is that the faults self-organize dynamically on these optimal paths which compete as a function of time. The rupture stress drop $\delta\sigma$ plays the role of a temperature activating the earthquake dynamics that may jump from one optimal path to the next. These optimal paths are in general widely separated. The *numbers* and *arrows* indicate the ranks of some the main optimal paths (rank 1 corresponds to the path with the absolute minimum of the sum of the random numbers; rank 2 is the second minimum, and so on). Comparing with Fig. 7.3, one can see that the optimal path with rank 1 is indeed the geometrical structure accumulating the largest amount of slip over long times, and is thus activated more often than all the other faults. Reproduced from [627]

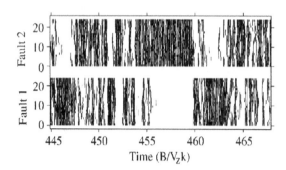

Fig. 7.5. Illustration of the competition in time between the seismic activity of two faults. The horizontal axis is the time. The vertical axis gives the coordinate along each competing faults. Observe the large distribution of time spans over which each fault is quiescent. Reproduced from [554]

stress drop as an effective temperature. Qualitatively, when several faults have almost the same "energy" and are thus almost equally optimal, the competition between them results from a slow exploration of these "energy" minima in the presence of a non-zero "temperature". The configurations of the fault with minimum "energy" are shown in Fig. 7.4. The underlying chaotic "earthquake" dynamics is similar to the thermal activation of the motion of a polymer due to the coupling to a thermal bath at a non-zero temperature T. If $T = 0$, the unique optimal path configuration is recovered (a single fault is active). If $T > 0$, the polymer explores slowly a number of other configurations with a probability governed by the Boltzmann factor. It is a general property that when a random medium is subjected to a small noise or a slow drift, the best path can undergo long period of stability and be suddenly drastically alterned [124, 1056], due to the family-tree structure of local optimal paths. In the present model, the small noise can be envisonned to be due to the fluctuating nature of the release by earthquakes at short time scales of the elastic energy stored in the system. The resulting competition between two faults is illustrated in Fig. 7.5. Reference [554] extends this model to focus on fault competition and shows that the distribution of time intervals between shifts of fault activation is a power law with an exponent which is function of the coupling strength of the faults. The coupling strength is itself determined by their distance and by the heterogeneity of the rupture thresholds.

7.4.6 Principle of Least Action for Out-Of-Equilibrium Systems

In non-equilibrium statistical mechanics, Onsager pointed out that the mean histories (or mean trajectories) of a system satisfy a "principle of least action" [701, 702]. The so-called Onsager–Machlup action determines the probability of fluctuations away from the most probable state. Close to thermal equilibrium, there is a standard fluctuation–dissipation relation and the action has the physical interpretation of a "dissipation function". Onsager's variational principle reduces then to a principle of least dissipation. This principle of least dissipation has been revived by modern interest in the self-organizing properties of out-of-equilibrium systems, such as the formation of complex river networks [447, 641, 779, 787, 788, 845, 939].

In its original form, Onsager's principle was restricted to weakly noisy systems and could not be applied to turbulence for instance, where fluctuations are large. This concept has been generalized to deal with strongly noisy systems [15, 282–284]: a similar "effective action" $\Gamma[z]$ exists in strongly stochastic out-of-equilibrium systems, such as in turbulent flows, for any random variable $z(t)$. This action functional has the following properties.

1. it is non-negative, $\Gamma[z] \geq 0$,
2. it has the ensemble mean $\bar{z}(t)$ as its unique minimum $\Gamma[\bar{z}] = 0$, and
3. it is convex,

$$\Gamma[\lambda z_1 + (1 - \lambda)z_2] \geq \lambda\Gamma[z_1] + (1 - \lambda)\Gamma[z_2], \ 0 < \lambda < 1 \ . \qquad (7.58)$$

These are realizability conditions which arise from positivity of the underlying statistical distributions. As a consequence, the ensemble-mean value $\bar{z}$ is characterized by a "principle of least effective action".

Just as Onsager's action, this functional is related to fluctuations. For example, in statistically stationary turbulence, the time-extensive limit of the effective action, called the *effective potential*,

$$V[z] \equiv \lim_{T \to \infty} \frac{1}{T} \Gamma[\{z(t) = z : 0 < t < T\}] \ , \qquad (7.59)$$

determines the probability of fluctuations in the empirical time-average

$$\bar{z}_T = \frac{1}{T} \int_0^T dt \, z(t) \qquad (7.60)$$

over a finite time interval T away from the (time-independent) ensemble-mean value $\bar{z}$. According to (7.59), $V[z]$ is thus an average of the action functional over the realizations for which the variable $z(t)$ visits the value z. More precisely, the probability for any value z of the time-average $\bar{z}_T$ to occur is given by

$$\text{Prob}\left(\{\bar{z}_T \approx z\}\right) \sim \exp\left(-TV[z]\right). \qquad (7.61)$$

This agrees with the standard ergodic hypothesis, according to which, as $T \to \infty$, the empirical time-average must converge to the ensemble-mean, $\bar{z}_T \to \bar{z}$, with probability one in every flow realization. Equation (7.61) refines that hypothesis by giving an exponentially small estimate of the probability at a large (but finite) T to observe fluctuations away from the ensemble-mean. This result generalizes the large deviation theorem discussed in Chap. 3.

7.4.7 Superstatistics

We conclude this chapter by pointing out a novel direction for generalizing statistical mechanics for out-of-equilibrium systems. A particular class of more general statistics relevant for nonequilibrium systems, containing Tsallis statistics (see Chap. 14) as a special case, has been termed 'super-statistics' [63–65, 68, 965]. A superstatistics arises out of the superposition of two statistics, namely one described by ordinary Boltzmann factors $e^{-\beta E}$ and another one given by the probability distribution of β. This means that the inverse temperature parameter β is assumed not to be constant but to be fluctuating on a relatively large time scale or spatial scale. This kind of approach is physically relevant for driven nonequilibrium systems with fluctuations, rather than for equilibrium systems. Recent applications of the superstatistics concept include fully developed hydrodynamic turbulence, pattern formation in thermal convection states and the statistics of cosmic rays [67].

Depending on the probability distribution of β, there are infinitely many superstatistics. It has been shown that Tsallis statistics is a particular superstatistics obtained under the assumption that β is χ^2-distributed [62]. Various other examples of superstatistics have been studied [68], among them superstatistics of the log-normal type. A main result of [68] was that for small E all superstatistics behave in a universal way, i.e., they generate probability distributions close to Tsallis distributions. But for large E, the various superstatistics can have quite different properties.

The idea behind the "superstatistics" concept is applicable to many systems, including turbulent systems. Consider a driven nonequilibrium systems with spatio-temporal fluctuations of an intensive parameter β. This can e.g. be the inverse temperature, or a chemical potential, or a function of the fluctuating energy dissipation in the flow (for the turbulence application). Locally, i.e., in cells where β is approximately constant, the system is described by ordinary statistical mechanics, i.e., ordinary Boltzmann factors $e^{-\beta E}$, where E is an effective energy in each cell. To describe the system in the long-term run, one has to do a spatio-temporal average over the fluctuating β. One obtains a superposition of two statistics (that of β and that of $e^{-\beta E}$), hence the name 'superstatistics'. One may define an effective Boltzmann factor $B(E)$ given by

$$B(E) = \int_0^\infty d\beta \, f(\beta) e^{-\beta E} , \qquad (7.62)$$

where $f(\beta)$ is the probability distribution of β. For type-A superstatistics, one normalizes this effective Boltzmann factor, obtaining the stationary probability distribution

$$p(E) = \frac{1}{Z} B(E) , \qquad (7.63)$$

where

$$Z = \int_0^\infty B(E) \, dE . \qquad (7.64)$$

For type-B superstatistics, the β-dependent normalization constant is included into the averaging process, obtaining

$$p(E) = \int_0^\infty f(\beta) \frac{1}{Z(\beta)} e^{-\beta E} \, d\beta \; , \tag{7.65}$$

where $Z(\beta)$ is the normalization constant of $e^{-\beta E}$ for a given β. Both approaches can be easily mapped into each other, by defining a new probability density $\tilde{f}(\beta) \sim f(\beta)/Z(\beta)$. Type-B superstatistics with f is equivalent to type- A superstatistics with $\tilde{f}$.

A simple dynamical realization of a superstatistics can be constructed by considering stochastic differential equations with spatio-temporally fluctuating parameters [62]. Consider the Langevin equation

$$\dot{u} = \gamma F(u) + \sigma L(t) \; , \tag{7.66}$$

where $L(t)$ is Gaussian white noise, $\gamma > 0$ is a friction constant, σ describes the strength of the noise, and $F(u) = -(\partial/\partial u)V(u)$ is a drift force. If γ and σ are constant, then the stationary probability density of u is proportional to $e^{-\beta V(u)}$, where $\beta \equiv \gamma/\sigma^2$ can be identified with the inverse temperature of ordinary statistical mechanics. Most generally, however, we may let the parameters γ and σ fluctuate so that $\beta = \gamma/\sigma^2$ has probability density $f(\beta)$. These fluctuations are assumed to occur on a long time scale so that the system can temporarily reach local equilibrium. In this case, one obtains for the conditional probability $p(u|\beta)$ (the probability of u given some value of β)

$$p(u|\beta) = \frac{1}{Z(\beta)} \exp\left\{-\beta V(u)\right\} \; , \tag{7.67}$$

for the joint probability $p(u, \beta)$ (the probability to observe both a certain value of u and a certain value of β)

$$p(u, \beta) = p(u|\beta) f(\beta) \; , \tag{7.68}$$

and for the marginal probability $p(u)$ (the probability to observe a certain value of u no matter what β is)

$$p(u) = \int_0^\infty p(u|\beta) f(\beta) \, d\beta \; . \tag{7.69}$$

This marginal distribution is the generalized canonical distribution of the superstatistics considered.

Beck has applied this formalism to Eulerian and Lagrangian turbulence experiments and has found them well described by simple superstatistics models [66]. Significant differences between different superstatistics only arise for very large velocity differences (and large accelerations), where Tsallis statistics predicts a power law decay of probability density functions, whereas log-normal superstatistics yields tails that decay in a more complicated way. It is indeed the tails that contain the information on the most appropriate superstatistics for turbulent flows.

8. Long-Range Correlations

This chapter extends the results obtained in Chap. 2 by considering the existence of correlations in the sum of random variables. This situation is obviously the general situation and we will encounter it in different guises when dealing with the collective behavior of systems made up of many interacting elements. Studying the correlations and their consequences is an essential part of the analysis of a system since the correlations are the signatures that inform us about the underlying mechanisms.

In the presence of correlations, how is the central limit theorem discussed in Chap. 2 modified? The answer to this question cannot be developed as generally and precisely as in Chap. 2 because we do not have yet a full classification and understanding of all possible scenarii that can occur in the sum of correlated random variables. Correlations can modify the results previously presented in Chap. 2. The most obvious case is the extreme where all the variables contributing to the sum are perfectly correlated and thus identical. On the other hand, we can hope that sufficiently short-range correlations will not modify the central limit theorem. What is thus needed is a criterion for the relevance of the correlations. The presentation of this chapter is inspired by [111].

8.1 Criterion for the Relevance of Correlations

Let us come back to the study of the sum of random variables l_i for which we have defined the correlation function C_{ij} by

$$C_{ij} = \langle l_i l_j \rangle - \langle l_i \rangle \langle l_j \rangle \ . \tag{8.1}$$

We consider a stationary process, i.e. C_{ij} only depends on the difference $n = j - i$. We thus write

$$C(n) = \langle l_i l_{i+n} \rangle - \langle l_i \rangle \langle l_{i+n} \rangle \ , \tag{8.2}$$

In practice, $C(n)$ can be estimated by taking the sum

$$C(n) = \frac{1}{N-n} \sum_{i=1}^{N-n} l_i l_{i+n} - \left(\frac{1}{N} \sum_{i=1}^{N} l_i \right)^2 \ . \tag{8.3}$$

The conditions for $C(n)$ to exist are the same as those ensuring the convergence of the variance: the pdfs of the l_i must decay faster than $l_i^{-(1+\mu)}$ with $\mu > 2$ at large l_i. For concrete applications, if $2 < \mu < 4$, large fluctuations may still spoil a correct estimation of $C(n)$ from finite samples, whose convergence to its asymtotic value for large N remains very slow, as discussed in Chap. 4.

To simplify notations, we assume $\langle l_i \rangle = 0$ and rewrite the variance of the sum $x(t)$ defined by

$$x(t) = l(t - \tau) + l(t - 2\tau) + ... + l(\tau) + l(0) \ , \tag{8.4}$$

as

$$\langle (x(t))^2 \rangle = tC(0) + 2\sum_{k=1}^{t-1}(t-k)C(k)$$

$$= t\left(C(0) + 2\sum_{k=1}^{t-1}C(k)\right) - 2\sum_{kw=1}^{t-1}kC(k) \ . \tag{8.5}$$

We have cut the double sum occurring in the calculation of $\langle (x(t))^2 \rangle$ into three groups, (i) $i = j$, (ii) $i < j$ and (iii) $i > j$. The first one gives t times $C(0)$. The second and third group are identical as i and j are dummy indices, hence the factor 2 multiplying the second term.

It is clear that if $\sum_{k=1}^{t}C(k)$ converges for large t to a finite number, then $\langle (x(t))^2 \rangle \propto t$, which recovers the diffusive result obtained for a sum of independent variables with finite variance presented in Chap. 2. In this case, the existence of a correlation changes the numerical factor in front of the t-dependence, i.e. it modifies the value of the diffusion coefficient so it will no longer be given by $D \equiv \sigma^2/2\tau$ in general (where σ is the standard deviation of the i.i.d. variables constituting the sum).

The condition for this regime to exist is, as we said, that $\sum_{k}^{t}C(k)$ converges. For large t, the discrete sum can be approximated, using the Poisson formula, by the integral $\int_1^t dk\, C(k)$. This integral converges absolutely (i.e. the integral of $|C(k)|$ converges) if $C(k) \ll 1/k$ for large k. Thus, in order for the standard diffusion regime to hold, the condition on the decay of the correlation function is rather mild as it need only to decay faster than $1/k$. We refer to Ibragimov and Linnik [446] for a complete exposition of relevant results and, in particular, on the importance of so-called mixing properties that quantify the tendency for the random variables generated by the stationary process to take different values. It is important to stress that the proof of validity of the Central Limit Theorem for a stationary process requires restriction on its mixing properties [446].

Let us now examine the opposite situation where $C(k)$ decays slower than $1/k$, for instance as

$$C(k) \sim \frac{1}{k^y} \qquad \text{with } 0 \le y \le 1 \ . \tag{8.6}$$

In this case, $\sum_k^t C(k) \sim t^{1-y}$ for $y < 1$ and $\sim \ln t$ for the border case $y = 1$. We thus see that the second term $\sum_{k=1}^t (t - k)C(k)$ in the r.h.s. of (8.5) behaves as t^{2-y} for $y < 1$ and $\sim t \ln t$ for the border case $y = 1$. This dominates the first term in the r.h.s. of (8.5) and we thus get a *super-diffusive* behavior[1]

$$\langle (x(t))^2 \rangle \sim t^{2-y} \qquad \text{for } y < 1 \ , \tag{8.7}$$

$$\sim t \ln t \qquad \text{for } y = 1 \ . \tag{8.8}$$

Diffusion is thus *enhanced* by the long-range correlations and the typical value of $x(t)$ is of order $t^{1-y/2} \gg t^{1/2}$ at large times. Notice that we recover standard diffusion (up to logarithmic corrections for the border case $y = 1$) while perfect correlations ($y = 0$) lead to an average drift with a finite velocity ($x(t) \propto t$).

As an interesting example of random walks with long-range correlations with exponent $y < 1$, let us mention the structure of DNA. Analysis of non-coding parts of DNA sequences have used the random walk analogy to test for long-range correlations. The first step is to group the four constituant bases of DNA into two families A and B characterized by similar chemical structure with approximately the same fraction $1/2$. The second step is to imagine that time t corresponds to the length covered when counting the bases in one direction along a DNA chain. When one encounters the family A (B), the random walk is assumed to make a $+1$ (-1) step along a 1D-line. If the correlations in the position of the bases decay faster than $1/k$, the r.m.s. (root mean square) of the displacement of this walker should be $\propto t^{1/2}$ while, if long-range correlations exist, the r.m.s. will exhibit super-diffusive behavior with an exponent larger than $1/2$ (but less than 1). This analysis has indeed shown [31, 601, 1000] that the non-coding parts of DNA seem to have long-range correlations with an exponent $y < 1$. In contrast, the coding regions seem to have short-range or no correlations. This might be interpreted by the fact that a coding region must appear random since all bases contain useful information. If there were some correlation, it would mean that it is possible to encode the information in fewer bases and the coding regions would not be optimal. In contrast, non-coding regions contain few or no information and can thus be highly correlated. This paradox, that a message with a lot of information should be uncorrelated while a message with no information is highly correlated, is at the basis of the notion of random sequences. A truly random sequence of numbers or of symbols are those that contain the maximum possible information, in other words it is not possible to define a shorter algorithm which contains the same information [156]. The condition for this is that the sequence be completely uncorrelated so that each new term carries new information.

---

[1] There can exist cases where the integral of $|C(k)|$ does not converge but $\sum_{k=1}^t (t - k)C(k)$ remains of order t due to subtle cancellations that might occur if $C(k)$ is oscillating.

8.2 Statistical Interpretation

There is an intuitive explanation for the effect of long-range correlations on the behavior of the sum. The idea is that if the variables are strongly correlated, many of them are similar to a given one and it is possible to rearrange the sum in families of similar variables. If these families saturate their growth as the number t of terms in the sum increases, we are left with of the order of t different independent families with no correlation between them, on which we can apply the central limit theorem and recover the previous results of Chap. 2. On the other hand, if the size of the families grow with t, the properties of the sum will be modified. Let us give a quantitative foundation to this heuristic explanation.

The sum

$$N_{\mathrm{id}}(t) = \sum_{n=1}^{t} \frac{C(n)}{\langle l^2 \rangle - \langle l \rangle^2} \; , \tag{8.9}$$

of the correlation function gives us, by its definition, a good estimation of the number of variables in the sum that are similar to a given one. This is because

$$\frac{\langle l_0 l_n \rangle - \langle l_0 \rangle \langle l_n \rangle}{\langle l^2 \rangle - \langle l \rangle^2} \tag{8.10}$$

measures the probability for l_n to be "close" to l_0.

Intuitively, this ratio gives the relative width of the joint probability to find l_0 and l_n. We can be more specific in the special case where the correlations have a Gaussian structure with pdf

$$P(l_0, l_n) = (\det C)^{-1/2} \exp\left(-[C_{00-1} l_0^2 + C_{00}^{-1} l_n^2 + 2C_{0n}^{-1} l_0 l_n]\right) \; , \tag{8.11}$$

where we have assumed symmetry: $C_{00} = C_{nn}$ and $C_{0n} = C_{n0}$. The coefficients C_{00}^{-1} and C_{0n}^{-1} are the elements of the inverse matrix of C. In this case,

$$\frac{\langle l_0 l_n \rangle - \langle l_0 \rangle \langle l_n \rangle}{\langle l^2 \rangle - \langle l \rangle^2} = \frac{C_{0n}}{C_{00}} \; , \tag{8.12}$$

which can be compared to an explicit calculation of the probability that l_0 and l_n be equal to within a fraction $\eta \ll 1$ of their standard deviation $\sqrt{C_{00}}$

Probability of $l_0 = l_n$ to within $\eta\sqrt{C_{00}} \approx$

$$\eta\sqrt{C_{00}} \int \mathrm{d}l_0 \int \mathrm{d}l_n \, P(l_0, l_n)\delta(l_0 - l_n) \approx \eta\sqrt{\frac{C_{00}}{C_{00} - C_{0n}}} \; . \tag{8.13}$$

If there are no correlations ($C_{0n} = 0$), this probability is a constant that represents the probability that one variable comes close to the other solely by chance. Since this contribution is already taken into account in the sum, this constant must be subtracted to obtain the effect of the correlations. We thus obtain that the *increase* in probability that the two variables be close to each other due to their correlation is proportional to

$$\sqrt{\frac{C_{00}}{C_{00} - C_{0n}}} - 1 \simeq \frac{C_{0n}}{C_{00}} \ , \quad \text{for small } C_{0n} \ , \tag{8.14}$$

thus confirming that (8.10) and (8.12) are proportional to the probability that the two variables are similar due to their mutual correlations.

Building on this understanding, we use (8.9) to regroup the set $\{l_k, 0 \leq k \leq t\}$ into families of $N_{\mathrm{id}}(t)$ variables each. There are $N_{\mathrm{fam}}(t) = t/N_{\mathrm{id}}(t)$ such families which can be considered as statistically independent *with respect* to the observable we are studied, namely the sum $x(t)$. Each family can now be seen as an effective variable L_k of scale $N_{\mathrm{id}}(t)$. Their sum obeys the central limit theorem since they are essentially uncorrelated and we get

$$x(t) \approx \sum_{k=1}^{N_{\mathrm{fam}}(t)} L_k \ . \tag{8.15}$$

The standard deviation of this sum is thus, according to Chap. 2 and using $\langle L_k \rangle \simeq N_{\mathrm{id}}(t)$,

$$x(t) \sim N_{\mathrm{id}}(t)[N_{\mathrm{fam}}]^{1/2} \sim (tN_{\mathrm{id}}(t))^{1/2} \ , \tag{8.16}$$

which gives back (8.7) and (8.8) since $N_{\mathrm{id}}(t) \sim$ constant if $y > 1$, $\sim \ln t$ if $y = 1$ and $\sim t^{1-y}$ if $y < 1$. Note that, in defining the variables L_k, we have replaced the long-range dependence into short-range or zero correlation between variables which have a rather special property: their sizes depend on the global length t of the series.

• We stress that this approach contains the basic idea of the renormalization group [351, 1021] that we will discuss in Chap. 11, which deals with correlated variables by grouping them in such a way as to simplify the correlations. In the present case, the grouping can be done in one step by introducing the families of N_{id} variables. In this interpretation, a family is a *renormalized* variable. Indeed, going from the initial variables to the families has increased the "strength" or amplitude (up to $\sim N_{\mathrm{id}}$) of the variables that are summed up, while their number has decreased accordingly.

• The grouping of the correlated variables into families (or renormalized variables) must not be taken literally. It is true only in a statistical sense. However, from a correlation function decaying like $C_{ij} \sim |i - j|^{-y}$, it is possible to construct a time series where literally the probability to get each variable l_i equal to the first one decays as $|i - j|^{-y}$. We encounter here again the dual interpretation of probabilities (and thus correlations) in terms of frequency versus scenarii.

8.3 An Application: Super-Diffusion in a Layered Fluid with Random Velocities

Consider the simple model shown in Fig. 8.1 made of layers, each having a specific fluid velocity taken from a distribution with zero average [111, 112]. In other words, the velocity field has a "quenched" randomness. This might represent an idealization of a porous medium in which permeability fluctuations induce a random distribution of local flow velocities. Suppose that a chemical molecule or a contaminant is diffusing with a diffusion coefficient D. There are two components in this diffusion, one parallel to the layers and the other perpendicular to them. In the parallel direction, we assume that the diffusion can be neglected compared to the velocity drift carrying the molecule with the flow. The relative strength of diffusion versus drift is measured by the Péclet number $Pe = Rv/D$, which is the ratio of the diffusion time R^2/D to diffuse a distance R to the convection time R/v to cover the same distance ballistically at the spead v. We thus assume a large Péclet number in the direction parallel to the layers.

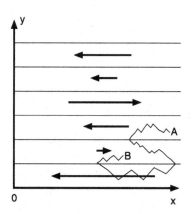

Fig. 8.1. Two-dimensional medium layered along the y-direction with frozen velocity fields as shown by the large *arrows* within each layer. The *thin curve* from A to B represents a possible particle trajectory which is diffusive in the y-direction and ballistic in the direction of the arrow of each layer in the x-direction

Since there is no drift in the direction perpendicular to the layers, the diffusive motion is the sole dynamics of the process (in the limit of vanishing Peclet number) and makes the particle cross from one layer to another. When present in a given layer, the particle is carried away horizontally in the direction and at the velocity characterizing this layer. The vertical motion, being purely diffusive, we can use the results of Chap. 2 and, in particular, obtain that the probability that a given layer be revisited at time t after having been visited at time 0 is $\sim (Dt)^{-(1/2)}$. This is seen by putting $v = 0$ and $y = 0$ in the expression of the probability

$$P_G(y,t) = \frac{1}{\sqrt{2\pi}} \frac{1}{\sqrt{2Dt}} e^{-(y-vt)^2/4Dt} , \qquad (8.17)$$

given in Chap. 2 for the particle to be at position y at time t after starting at 0 at $t = 0$. When the particle comes back at time t to a layer already visited, the horizontal velocity that it encounters is thus exactly the same as at time 0, since we have assumed the velocity field to be frozen. This implies that the correlation $\langle V(0)V(t)\rangle$ of the horizontal velocities is proportional to $(Dt)^{-(1/2)}$, i.e. decays with t with an exponent $y = 1/2 < 1$. The total horizontal displacement of the particle is the sum of the displacements travelled while in each layer, each single displacement being the product of the velocity in this layer by the time of sojourn in this layer. This time of sojourn is about the same and approximately equal to the time necessary to diffuse across its width. Different sojourn durations in each layer correspond to different times taken by the particle to diffuse across them.

We thus obtain that the total horizontal displacement is proportional to the sum of the velocities of the explored layers. We can thus put this problem in the previous framework of a sum of random variables with long-range correlations decaying as $t^{-1/2}$ at long times t. Using (8.7), we get

$$x_{//}(t) \sim D^{-1/2}\, t^{3/4} \,, \tag{8.18}$$

with an exponent $3/4$ larger than $1/2$, that characterizes a super-diffusion process.

This mechanism, in terms of long-range correlations to generate super-diffusion, is distinct from the mechanism discussed in Chap. 4 in which super-diffusion results from the divergence of the variance of the *uncorrelated* variables contributing to the sum. In the present example, the variance is finite and the super-diffusion stems from the existence of the long-range correlations. Hence, when confronted with a super-diffusion process, one must bear in mind that these two mechanisms can be present, possibly co-existing.

8.4 Advanced Results on Correlations

8.4.1 Correlation and Dependence

Absence of correlation is not synonymous of independence! More precisely, the vanishing of the two-point correlation function C_{ij} defined by (8.1) does not necessarily imply that the two variables are independent. For this to be true, all correlation functions that one can construct with arbitrary functions of the two variables must be zero.

Consider, for instance, a random variable x with symmetric pdf $P(x) = P(-x)$. We construct the random variable $y = x^2$. Obviously, y is not independent of x. In fact, y is completely dependent on x, since the knowledge of x fully determines that of y. However, the correlation between x and y (noted $\mathrm{cov}\{x, y\}$) is zero:

$$\mathrm{cov}\{x, y\} = \langle xy\rangle - \langle x\rangle\langle y\rangle = \langle x^3\rangle - \langle x\rangle\langle x^2\rangle = 0 \,, \tag{8.19}$$

because $\langle x^3\rangle = \langle x\rangle = 0$ due to the symmetry of $P(x)$.

To generalize this result, we now construct a (non-stationary) random process $x(t)$ defined by

$$x(t) = f(x_0)g(t) , \tag{8.20}$$

where f is an even function, for instance $f(x_0) = x_0^2$ and $g(t)$ is arbitrary and independent of x_0. We also assume that, for a given scenario, the initial value x_0 is taken randomly for a centered distribution (such that $\langle x_0 \rangle = 0$). Then, the correlation between $x(t)$ and the initial value x_0 calculated over a statistical ensemble of possible scenarii is

$$C_{0,t} = \langle x_0 x(t) \rangle - \langle x_0 \rangle \langle x(t) \rangle = \langle x_0^3 \rangle \langle g(t) \rangle - \langle x_0 \rangle \langle x_0^2 \rangle \langle g(t) \rangle = 0 , \tag{8.21}$$

because $\langle x_0^3 \rangle = \langle x_0 \rangle = 0$ for a symmetric distribution of initial values x_0.

A different class of systems exhibiting the same property is given by

$$x(t + 1) = f(x(t)) , \tag{8.22}$$

where $f(x(t))$ is decomposed onto the complete set of Hermite polynomials[2] of order larger than or equal to 2. This series exhibits zero correlation, while the process is completely predictable (deterministic) and $x(t)$ depends only on x_0. One can show [260] that $\langle x(t)x(0) \rangle - \langle x(t) \rangle \langle x(0) \rangle = 0$ for any $t \neq 0$, where the average is performed in a statistical sense, i.e. over all possible initial values x_0 distributed according to a *symmetric* distribution.

The fact that correlation and dependence are two distinct concepts is further illustrated with the following nonlinear model with zero correlation but non-vanishing dependence leading to significant predictability. Let us construct a stochastic process $x(t)$ according to the following rule:

$$x(t) = \epsilon(t) + b\epsilon(t - 1)\epsilon(t - 2) , \tag{8.23}$$

where $\epsilon(t)$ is a white noise process with zero mean and unit variance and b is a fixed parameter. For instance, $\epsilon(t)$ is either $+1$ or -1 with probability $1/2$. The definition (8.23) means that the value of the variable $x(t)$ at time t is controlled by three random coin tosses, one at time t, one at time $t - 1$ and one of time $t - 2$. It is easy to check that the average $E(x(t))$ as well as the two-point correlation $E(x(t)x(t'))$ for $t \neq t'$ are zero and $x(t)$ is thus also a white noise process. Intuitively, this stems from the fact that an odd number of coin tosses ϵ's enter in these diagnostics whose average is zero $((1/2) \times (+1) + (1/2) \times (-1) = 0)$. However, the three-point correlation function $E(x(t-2)x(t-1)x(t))$ is non-zero and equal to b and the expectation of $x(t)$ given the knowledge of the *two* previous realizations $x(t-2)$ and $x(t-1)$ is non-zero and reads

$$E(x(t)|x(t - 2), x(t - 1)) = bx(t - 2)x(t - 1) . \tag{8.24}$$

---

[2] The Hermite polynomials form an orthogonal set of functions with respect to the norm defined with the Gaussian distribution and are themselves directly obtained by successive differentiation of the Gaussian law.

This means that it is possible to predict the variable $x(t)$ with better success than 50% chance, knowing the two past values $x(t-2)$ and $x(t-1)$. Thus, absence of correlation does not imply absence of predictability.

8.4.2 Statistical Time Reversal Symmetry

Description of the Problem. Assume that we are studying a time ordered data sequence,

$$\cdots, X_{-2}, X_{-1}, X_0, X_1, X_2, X_3 , \cdots \tag{8.25}$$

which is the realization of some stationary stochastic process. Is it possible to discern a preferred direction of time such that the appropriate algorithm indicates the presence of such a direction when (8.25) is reversed as in

$$\cdots, X_3, X_2, X_1, X_0, X_{-1}, X_{-2}, \cdots \tag{8.26}$$

Pomeau [743] has introduced such an idea of testing for the time-reversal symmetry in a *statistical* sense. If the fluctuations of a random variable $X(t)$ are stationary on average, then the usual two-point correlation function $C(\tau) = \langle (X(t) - \langle X \rangle)(X(t+\tau) - \langle X \rangle) \rangle$ is an even function of τ: $C(\tau) = C(-\tau)$. If $X(t)$ is recorded on a magnetic tape, its examination does not allow to distinguish between the two directions of reading the tape. Limiting the analysis of the time series to the calculation of $C(\tau)$ thus corresponds to a loss of information since the signal $X(t)$ can be symmetric or asymmetric statistically under time reversal but yet absent in the information contained in $C(\tau)$.

It is in fact possible to construct correlation functions that test for the statistical symmetry under time reversal. The simplest one is

$$C_3(\tau) = \langle X(t)[X(t+2\tau) - X(t+\tau)]X(t+3\tau) \rangle . \tag{8.27}$$

To understand the meaning of $C_3(\tau)$, notice that it is equal to the difference of $\langle X(t)X(t+2\tau)X(t+3\tau) \rangle$ and $\langle X(t)X(t+\tau)X(t+3\tau) \rangle$. The first term $\langle X(t)X(t+2\tau)X(t+3\tau) \rangle$ corresponds to three signals shifted with respect to t as $0, 2\tau, 3\tau$, while the second term $\langle X(t)X(t+\tau)X(t+3\tau) \rangle$ corresponds to $0, \tau, 3\tau$. Upon time reversal, they respectively change into time shifts $0, \tau, 3\tau$ and $0, 2\tau, 3\tau$, i.e. they exchange their role. If the signal is time reversal invariant, $C_3(\tau)$ will thus be zero. If there is some statistical assymmetry under time reversal, $C_3(\tau)$ will be non-zero.

It can be shown that fluctuations of many-body systems at the thermodynamic equilibrium are reversible in this sense while they are not time symmetric if the many-body system is out-of-equilibrium in a steady state. This last case corresponds to chaotic phenomena, turbulence and self-organized (critical or not) systems. An application of these ideas to turbulence can be found [754] showing the importance of viscosity to break down the statistical time symmetry.

Examples and Qualifiers. We consider a simple analytically workable example, the Kesten multiplicative process defined by

$$X_{n+1} = a_n X_n + b_n, \quad \text{given} \quad X_0 , \tag{8.28}$$

where the stochastic variables a_n, b_n are drawn with the pdfs $P_a(a_n)$ and $P_b(b_n)$. Although the model (8.28) appears to be an apparently innocuous AR(1) (autoregressive of order 1) process, the *random and positive* multiplicative coefficient a_n lead to non-trivial intermittent behavior for a large class of distributions for a_n. This model has been applied to a large variety of situations, such as population dynamics with external sources, auto-catalytic processes, epidemics where it provides a rationalization for observed power law data, finance and insurance, immigration and investment portfolios and the internet (see [879] and references therein). One of its interesting features is that it provides a simple and general mechanism for generating power law distributions [501, 561, 858, 891] (see Chap. 14). Figure 8.2 presents a typical time evolution of X_n given by expression (8.28). For the parameters used in this figure, $\langle X_n \rangle = \langle b \rangle/(1 - \langle a \rangle) = 25$. Most of the time, X_n is significantly less than its average, while rare intermittent bursts propel it to very large values. Qualitatively, this behavior arises because of the occurrence of sequences of successive individual values of a_n that are larger than 1 which, although rare, are always bound to occur in this random process. The persistence of the temporal behavior has a decay which is $\exp[\langle \ln a \rangle \tau]$ on the average. The inverse of the decay rate defines a correlation time $1/|\langle \ln a \rangle| = 14.8$ governing the impact of past innovations $b(t - \tau)$ on the present value X_n. The distribution of X_n from a numerical realization with the properties above gives an histogram characterized by a power law tail

$$P(X) \sim X^{-(1+\mu)} , \quad \text{with } \mu \approx 1.5 . \tag{8.29}$$

It also contains a rolloff at smaller values of X_t, a result that is mandated by the fact that as $X(t) \to 0$ the process is dominated by the injection of new stock $0 < b < 1$, so that the population is repelled from a zero value.

Looking at Fig. 8.2, it is not obvious how to decide what is the arrow of time in this rather complex stochastic time series. To answer this question, we calculate the correlation function C_3 proposed by Pomeau and write it as $C_3 = C_1 - C_2$ where

$$C_1 \stackrel{\text{def}}{=} \langle X_n \, X_{n+1} \, X_{n+3} \rangle , \tag{8.30}$$

$$C_2 \stackrel{\text{def}}{=} \langle X_n \, X_{n+2} \, X_{n+3} \rangle , \tag{8.31}$$

where the brackets define as usual ensemble averages. The calculations presented in this section have been performed in collaboration with P. Jögi. These two expressions expands and simplifies to (assuming $\langle u_n \rangle = \langle u \rangle$),

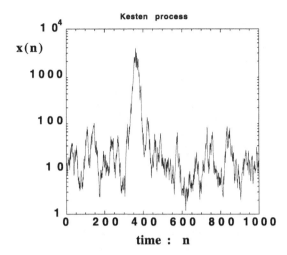

Fig. 8.2. Time evolution X_n of the Kesten variable defined by (8.28) with a_n uniformly taken in the interval $[0.48; 1.48]$ and b_t uniformly taken in the interval $[0; 1]$. Notice the intermittent large excursions

$$C_1 = \left(\langle b \rangle^2 + \langle a \rangle \langle b \rangle^2 + \langle a \rangle^2 \langle b^2 \rangle \right) \langle X \rangle$$

$$+ \left(\langle a \rangle \langle b \rangle + \langle a \rangle^2 \langle b \rangle + 2 \langle a \rangle^3 \langle b \rangle \right) \langle X^2 \rangle + \langle a \rangle^2 \langle a^2 \rangle \langle X^3 \rangle \,, \tag{8.32}$$

$$C_2 = \left(\langle b \rangle^2 + \langle a \rangle \langle b \rangle^2 + 2 \langle a \rangle^2 \langle b \rangle^2 + \langle a \rangle \langle b^2 \rangle + \langle a \rangle \langle a^2 \rangle \langle b^2 \rangle \right) \langle X \rangle$$

$$+ \left(\langle a \rangle^2 \langle b \rangle + 2 \langle a \rangle^3 \langle b \rangle + 2 \langle a \rangle^2 \langle a^2 \rangle \langle b \rangle \right) \langle X^2 \rangle + \langle a \rangle \langle a^2 \rangle^2 \langle X^3 \rangle \,. \tag{8.33}$$

We have

$$\langle X \rangle = \frac{\langle b \rangle}{1 - \langle a \rangle} \,, \tag{8.34}$$

$$\langle X^2 \rangle = \frac{2 \langle a \rangle \langle b \rangle^2 / (1 - \langle a \rangle) + \langle b^2 \rangle}{1 - \langle a^2 \rangle} \,, \tag{8.35}$$

$$\langle X^3 \rangle = \frac{3 \langle a \rangle \langle b \rangle \langle b^2 \rangle / (1 - \langle a \rangle)}{1 - \langle a^3 \rangle} \tag{8.36}$$

$$+ \frac{3 \langle a^2 \rangle \langle b \rangle \left(2 \langle a \rangle \langle b \rangle^2 / (1 - \langle a \rangle) + \langle b^2 \rangle \right) \Big/ (1 - \langle a^2 \rangle) + \langle a^3 \rangle}{1 - \langle a^3 \rangle} \,.$$

Note that, if $b = 0$, $C_1 = C_2$, qualifying the statistical time symmetry: this is expected since the map (8.28) then becomes a simple multiplicative process. We see that the additive term in the affine map (8.28) provides a powerful source of time reversal asymmetry. To explore this statement further, consider the case $\langle a^n \rangle = A^n$ (i.e. no randomness on the multiplicative factor), for which

$$C_3 = C_1 - C_2 = \frac{A^4 \left(2\langle b \rangle^3 - 3\langle b \rangle \langle b^2 \rangle + \langle b^3 \rangle\right)}{1 + A + A^2}. \tag{8.37}$$

This vanishes if b is a constant or for b_n uniformly distributied, i.e. $b_n \in \{\Theta(b_n) - \Theta(b_n - 1)\}$, since in this case we have $\langle b \rangle = 1/2, \langle b^2 \rangle = 1/3$ and $\langle b^3 \rangle = 1/4$. However, for the simple two-step "staircase" distribution

$$P_{b_n}(b_n) = p\left(\Theta(b_n) - \Theta(b_n - 1)\right)$$
$$+ \frac{1 - p}{B - 1} \left(\Theta(b_n - 1) - \Theta(b_n - B)\right) \tag{8.38}$$

(with $B > 1$), $C_3 = C_1 - C_2$ does not vanish and the time series is statistically non-time-reversal invariant:

$$C_1 - C_2 = \frac{A^4 \, p \, (1 - p) \, (pB - 1)}{4(1 + A + A^2)}. \tag{8.39}$$

The interesting condition that lets this expression vanish, $B = 1/p$, is exactly the condition that makes the pdf, (8.38), uniform.

The fact that $C_3 = C_1 - C_2$ vanishes for uniform distribution of the additive variable b_n does not necessarily imply that the property of statistical time-reversal invariance holds. Indeed, other qualifiers will detect statistical time-asymmetry. For instance, the qualifier

$$Q_m = \langle (X_{t+m} - X_t)^3 \rangle / \langle (X_{t+m} - X_t)^2 \rangle \tag{8.40}$$

has been proposed as a test for nonlinearities in time series [953]. It turns out that it is first a test for statistical time-reversal symmetry. Applied to the Kesten process with non-random $a_n = A$, we get

$$Q_1 = \frac{-3A\,(1 + A)\left(2\langle b \rangle^3 - 3\langle b \rangle \langle b^2 \rangle + \langle b^3 \rangle\right)}{2\,(1 + A + A^2)\left(\langle b \rangle^2 - \langle b^2 \rangle\right)}, \tag{8.41}$$

$$Q_2 = \frac{-3A^2\,(1 + A)\left(2\langle b \rangle^3 - 3\langle b \rangle \langle b^2 \rangle + \langle b^3 \rangle\right)}{2\,(1 + A + A^2)\left(\langle b \rangle^2 - \langle b^2 \rangle\right)}, \tag{8.42}$$

which have the same b moment polynomial already encountered in Pomeau's C_3. Q_1 and Q_2 as well as the other higher orders are vanishing for a constant b or for random b's taken from a uniform distribution, while they are non-zero for other more complex distributions of b. Notice that the so-called nonlinear probes Q_n can be non-zero and thus qualify the affine map (8.28) as apparently nonlinear. It has indeed been noted that the affine map behaves as an apparent nonlinear system [879].

Another possible probe of statistical time reversal symmetry is the "cubic" correlation measure

$$K_m = \langle X_{t+m}{}^3 X_t \rangle - \langle X_t{}^3 X_{t+m} \rangle \, . \tag{8.43}$$

Applied to Kesten's process with constant $a_n = A$, we get

$$
\begin{aligned}
K_1 = {}& -6 \frac{A^3 (1+A)}{(1+A^2)(1-A^3)} \langle b \rangle^4 - 3 \frac{A (1-3A^2-4A^3)}{(1+A^2)(1-A^3)} \langle b \rangle^2 \langle b^2 \rangle \\
& + 3 \frac{A}{1+A^2} \langle b^2 \rangle^2 + \frac{A (1-3A^2-4A^3)}{(1+A^2)(1-A^3)} \langle b \rangle \langle b^3 \rangle \\
& - \frac{A}{1+A^2} \langle b^4 \rangle \, .
\end{aligned}
\tag{8.44}
$$

With this probe, even with $b_n \in \{\Theta(b_n) - \Theta(b_n - 1)\}$ uniform,

$$K_1 = \frac{A}{120 \, (1+A^2)} > 0 \, , \tag{8.45}$$

i.e. this probe detects a violation of the statistical time-reversal symmetry even when C_3 does not.

In general, if $P_{b_n}(b_n)$ is sufficiently symmetric such that $\langle b \rangle = \langle b^3 \rangle = 0$, we find

$$K_1 = \frac{A}{1+A^2} (3\langle b^2 \rangle^2 - \langle b^4 \rangle) \, , \tag{8.46}$$

which for a Gaussian pdf will be zero.

The introduction of a random component in the multiplicative term a_n makes all these probes go even further from zero, showing that Kesten's process is not the same in general when interchanging the arrow of time. The generic situation thus appears to be statistical time-reversal asymmetry.

While the conceptual impact of a breakdown of statistical time-reversal symmetry is important, it is less clear how these measures provide a quantitative and practical insight into the properties of stochastic time series. Only a few works have addressed this question. The most studied application has been to probe whether time asymmetry can be detected in fluid flows and in particular in hydrodynamic turbulence. The outstanding question, still largely unresolved, is whether the turbulent "cascade", assumed to describe the emergence of turbulence [316], is a genuine dynamical process which implies a progression in time and thus a time arrow, for instance with increasing delays when going from large scales to small scales. Such a time-correlation structure accross scales has been documented in financial time series [34]. Alternatively, is the cascade more an instantaneous spatial structure? Testing for statistical time asymmetry may cast light on this fundamental question. Only a few investigations have been performed [754, 974, 1035] and seem to indicate that statistical time asymmetry is present and there is a time arrow felt in the statistical properties. In turbulent flows, third-order two-point triple correlations exhibit near antisymmetry in the time delay and their shapes determine the direction of energy transfer across the spectrum, which is a sort of asymmetry or preferred direction. In isotropic turbulence, the acceleration of the fluid has a positive skewness (more large positive values than large negative ones), associated (via Taylor's hypothesis) with the skewness of the gradient of the velocity field whose origin is vortex stretching in three dimensional flows. One sided ramp-like structures in the signal can lead to this behavior.

Another more recent domain of application is economy and finance. Ramsey and Rothman [760] have shown that economic time series exhibit statistical time-reversal asymmetry. As already mentionned, Arnéodo et al. [34] have shown, by using wavelets to decompose the volatility (standard deviation) of intraday S&P500 (the main US stock market index) return data across scales, that two-point correlation functions of the volatility logarithms across different time scales reveals the existence of a causal information cascade from large scales to fine scales.

8.4.3 Fractional Derivation and Long-Time Correlations

Derivatives of fractional order provide a compact description of long-range time correlations and memory in time series. Fractional-difference captures parsimoniously long-run structures that decay very slowly [395]. Indeed, it is always possible to account for long-range memories by introducing a high order derivative or high-order lags in an auto-regressive model. But this involves many parameters. The fractional derivative approach is solely defined in terms of a unique parameter, the fractional order of the derivation. This makes this approach an interesting candidate to fit time series with long-range correlations. Fractional derivatives have also more recently emerged in physics as generators of time evolutions of dynamical systems and as a tool for classifying phase transitions in thermodynamics by generalizing the classification scheme of Ehrenfest. We borrow from and refer to [420, 421, 619] for introductions and lists of useful references and for applications. Fractional integrals and derivatives are convolution operators with a power law kernel. This makes them natural tools in scaling theory.

Definitions. Consider a real function f that maps the interval $[a, b]$ onto the real axis. Recall that its n-th order integral is given by

$$(I_{a+}^n f)(x) = \int_a^x \int_a^{y1} \ldots \int_a^{y_{n-1}} f(y_n) \, dy_n \ldots dy_1 \qquad (x > a)$$

$$= \frac{1}{(n-1)!} \int_a^x (x-y)^{n-1} f(y) \, dy \qquad (8.47)$$

as is readily proven by induction. Generalizing (8.47) to noninteger n defines the *Riemann–Liouville fractional integral of order* $\alpha > 0$ as

$$(I_{a+}^\alpha f)(x) = \frac{1}{\Gamma(\alpha)} \int_a^x (x-y)^{\alpha-1} f(y) \, dy \,, \qquad (8.48)$$

for $x > a$ and f being L_1 integrable in the interval $[a, b]$. For $0 < a < 1$, these definitions are extended to the whole real axis for f being L_p integrable with $1 \leq p < 1/\alpha$.

A natural approach to define a fractional derivative tries to replace α with $-\alpha$ directly in the definition (8.48). However the resulting integral is divergent and needs to be regularized as follows:

$$
\begin{aligned}
(\partial_+^\alpha f)(x) &= \frac{d}{dx}(I_+^{1-\alpha} f)(x) \\
&= \frac{1}{\Gamma(1-\alpha)} \frac{d}{dx} \int_{-\infty}^{x} (x-y)^{-\alpha} f(y)\, dy \\
&= \frac{1}{\Gamma(1-\alpha)} \frac{d}{dx} \int_{0}^{\infty} t^{-\alpha} f(x-t)\, dt \\
&= \frac{\alpha}{\Gamma(1-\alpha)} \int_{0}^{\infty} f'(x-t) \int_{t}^{\infty} \frac{1}{z^{1+\alpha}}\, dz\, dt \\
&= \frac{\alpha}{\Gamma(1-\alpha)} \int_{0}^{\infty} \frac{f(x) - f(x-t)}{t^{1+\alpha}}\, dt
\end{aligned}
\tag{8.49}
$$

where the last equality is obtained by an integration by parts and serves to define $(\partial_+^\alpha f)$ for $0 < \alpha < 1$. f' denotes the first derivative of f with respect to its argument. These definitions (8.48) and (8.49) introduce fractional differentiation as an inverse operation to fractional integration, i.e. as integration of order $-\alpha$.

A more direct approach arises from the fact that derivatives are limits of difference quotients. Let T^h denote the translation by h:

$$
(T^h f)(x) = f(x-h)\,.
\tag{8.50}
$$

The finite difference of order α is defined as

$$
(\Delta_h^\alpha f)(x) = (1 - T^h)^\alpha f(x) = \sum_{k=0}^{\infty} (-1)^k \binom{\alpha}{k} f(x-kh)\,,
\tag{8.51}
$$

with the identity $\mathbf{1} = T^0$, and

$$
\binom{\alpha}{k} = \frac{(-1)^{k-1}\alpha\Gamma(k-\alpha)}{\Gamma(1-\alpha)\Gamma(k+1)}\,.
\tag{8.52}
$$

It reduces to the familiar finite difference of integer order when α is an integer. The *Grünwald fractional derivative of order* α is then defined as the limit of a fractional finite difference quotient:

$$
(\partial_\pm^\alpha f)(x) = \lim_{h \to +} h^{-\alpha}(\Delta_{\pm h}^\alpha f)(x)\,.
\tag{8.53}
$$

Let us also mention the frequency approach to fractional derivative. In the Fourier domain, a nth order differentiation of $f(x)$ amounts to multiplying its Fourier transform by $(i\omega)^n$. Making n non-integer provides a Fourier equivalent of Liouville's definition (8.48) of fractional derivative. Thus, a fractional derivative of order α is obtained by first Fourier transforming $f(x)$, then multiplying its Fourier transform by $(i\omega)^\alpha$. The last step is to take the inverse Fourier transform.

A very readable short guide to fractional derivation is [619], which demonstrates in details its application to anomalous diffusion and random walks with long memory.

Phase Transitions. Fractional derivatives have been employed in thermodynamics to generalize the Ehrenfest classification scheme for phase transitions discussed in Chap. 9. This generalization gives rise to a generalized form of static scaling at the transition, with amplitudes and scaling exponents which depend on the choice of the path C taken to reach the critical point.

This analysis is interesting for two reasons. First, it provides a systematic classification scheme that allows one to predict what cases can or cannot occur and what to expect in the scaling behavior in new problems. This approach complements the renormalization group theory of phase transitions which does not provide an exhaustive classification of the "fixed points", the corresponding universality classes and exponents. Second, this analysis leads to new predictions of scaling functions and amplitude ratios of static macroscopic phenomena [420].

Time Evolution with Long-Term Memory. Let us also mention the role of fractional derivation in the description of the time evolution of dynamical systems. Standard first-order time derivatives play a fundamental role in the time evolution for dynamical systems. This raises the question of the meaning and status of fractional time derivatives for the time evolution of dynamical systems. The answer is that it is indeed possible to use fractional derivatives as completely consistent generators of the time evolution of dynamical systems. Fractional derivatives possess the semigroup structure that is necessary for this purpose. Physically, they are associated with algebraic decay in time. To see this, consider the equation

$$\partial_+^\alpha \mu(t) = 0 \ , \tag{8.54}$$

describing the conservation of the phase space volume of a dynamical system under a fractional time-dependent process. The solution of (8.54) is

$$\mu(t) = C_0 \, t^{\alpha-1} \ . \tag{8.55}$$

This results from the fact that the fractional derivative of order α of x^{b-1} is $[\Gamma(b)/\Gamma(b-\alpha)] \, x^{b-\alpha-1}$, where $\Gamma(x)$ is the Gamma function. Setting $b = \alpha$ gives zero since $\Gamma(b-\alpha) \to \infty$ as $b \to \alpha$. Thus algebraic decay in time can be a sign of stationarity for induced dynamics on subsets of measure zero. The algebraic time decay shows that α plays the role of a dynamic scaling exponent.

Fractional Derivative of Fractals. The intuitive meaning of fractional derivatives and the previous applications suggest that taking the fractional integral of order D of a fractal function of dimension D may lead to non-fractal objects. Rocco and West have given form to this intuition [785] by showing that the fractional derivative (integral) of the generalized Weierstrass function $W(t)$ is another fractal function with a greater (smaller) fractal dimension:

$$\mathrm{Dim}[\partial_{\pm}^{\alpha} W] = \mathrm{Dim}[W] \pm \alpha . \tag{8.56}$$

The Weierstrass–Mandelbrot fractal function is defined by

$$W(t) = \sum_{n=-\infty}^{+\infty} \frac{1 - e^{i\gamma^n t}}{\gamma^{(2-D)n}} \, e^{i\phi_n} , \tag{8.57}$$

with $1 < D < 2$ and ϕ_n arbitrary. D is the Hausdorf–Besicovich fractal dimension of the graph of $W(t)$ (more precisely of the graphs of the real part or imaginary part of W). The result of Rocco and West [785] is that $\partial_{\pm}^{\alpha} W(t)$ takes the same form as (8.57) with D replaced by $D \pm \alpha$.

Fractional Diffusion Equation and Stable Lévy Laws. Diffusion models in physics describe the spreading of a cloud of tracer particles. In classical diffusion, particles spread out at a rate proportional to $t^{1/2}$ where $t > 0$ represents the time scale. Anomalous diffusion is characterized by a different rate of spreading, t^H, where $H < 1/2$ is subdiffusion and $H > 1/2$ is superdiffusion. Although the partial differential equations for diffusion are deterministic, they also govern the random particle jumps which are the physical cause of the diffusion. Let $X(t)$ be the position of a randomly selected particle along a line at time $t \geq 0$ and let $P(x, t)$ be the density of $X(t)$. If a sufficiently large ensemble of independent particles evolves according to this model, then $P(x, t)$ also represents the relative concentration of particles at location x at time $t > 0$. Saichev and Zaslavsky [806], Chaves [164] and Benson et al. [72] introduced the following modeling equation to describe anomalous diffusion

$$\frac{\partial P(x,t)}{\partial t} = -v \frac{\partial P(x,t)}{\partial x} + Bq \frac{\partial^{\alpha} P(x,t)}{\partial (-x)^{\alpha}} + Bp \frac{\partial^{\alpha} P(x,t)}{\partial x^{\alpha}} , \tag{8.58}$$

where $v \in R$, $B > 0$, $0 \leq p, q \leq 1$ and $p + q = 1$. $\partial^{\alpha}/\partial(\pm x)^{\alpha}$ are fractional derivatives of order $1 \leq \alpha \leq 2$, which are most easily defined in terms of Fourier transforms. Using the Fourier transform convention $\hat{f}(k) = \int e^{-ikx} f(x) \, dx$, we specify as above these fractional derivative operators by requiring that $\partial^{\alpha} f(x)/\partial(\pm x)^{\alpha}$ has the Fourier transform $(\pm ik)^{\alpha} \hat{f}(k)$, generalizing the familiar formula for integer order derivatives. Taking the Fourier transform of (8.58), we obtain

$$\frac{d\hat{P}(k,t)}{dt} = -v(ik)\hat{P}(k,t) + Bq(-ik)^{\alpha}\hat{P}(k,t) + Bp(ik)^{\alpha}\hat{P}(k,t) \tag{8.59}$$

so that, for the initial condition $\hat{P}(k,0) = 1$, we get

$$\hat{P}(k,t) = \exp\left[-vt(\mathrm{i}k) + Btq(-\mathrm{i}k)^\alpha + Btp(\mathrm{i}k)^\alpha\right] . \tag{8.60}$$

This initial condition corresponds to the assumption $X(0) = 0$ with probability 1. Using $(\mathrm{i}k)^\alpha = (\mathrm{e}^{\mathrm{i}\pi/2}\,k)^\alpha = |k|^\alpha \cos[\pi\alpha/2]\{1 + \mathrm{i}\,\mathrm{sign}(k)\tan[\pi\alpha/2]\}$, with $-\mathrm{e}^{\mathrm{i}\pi/2} = \mathrm{e}^{\mathrm{i}\pi/2-\mathrm{i}\pi} = e^{-\mathrm{i}\pi/2}$, we obtain

$$\hat{P}(k,t)$$
$$= \exp\left[-vt(\mathrm{i}k) + Bt|k|^\alpha \cos[\pi\alpha/2]\{1 + \mathrm{i}\beta\,\mathrm{sign}(k)\tan[\pi\alpha/2]\}\right] \tag{8.61}$$

and recognize the Fourier transform of a stable density with index α, skewness $\beta = p - q$, center vt and scale $\sigma^\alpha = -Bt\cos[\pi\alpha/2]$, as discussed in Chap. 4. Hence $P(x,t)$ is the density of an α-stable Lévy motion $\{X(t)\}$ with drift, where $1 \leq \alpha \leq 2$. If $\alpha = 2$, then $P(x,t)$ is a Gaussian density and (8.58) reduces to the classical diffusion equation for Brownian motion with drift, which was discussed in Chap. 2. In the symmetric case $\beta = 0$, the fractional diffusion equation (8.58) was considered by Metzler et al. [621] and Compte [178], and in the totally skewed case $\beta = 1$, (8.58) was treated by Zaslavsky [1054] and Compte [179]. Benson et al. [71, 73] present several applications of (8.58) in hydrology which illustrate the practical utility of the model. Schumer et al. [828] give a physical derivation of (8.58) which justifies the assumption $\alpha > 1$.

9. Phase Transitions: Critical Phenomena and First-Order Transitions

9.1 Definition

One of the most conspicuous properties of nature is the great diversity of size or length scales in the structure of the world. An ocean, for example, has currents that persist for thousands of kilometers and has tides of global extend; it also has waves that range in size from less than a centimeter to several meters; at much finer resolution, seawater must be regarded as an aggregate of molecules whose characteristic scale of length is roughly 10^{-8} centimeter. From the smallest structure to the largest is a span of some 17 orders of magnitude.

In general, events distinguished by a great disparity in size have little influence on one another: they do not communicate, and so the phenomena associated to each scale can be treated independently. The interaction of two adjacent water molecules is much the same whether the molecules are in the Pacific Ocean or in a teapot. What is equally important, an ocean wave can be described quite accurately as a disturbance of a continuous fluid, ignoring completely the molecular structure of the liquid. The success of almost all practical theories in physics depends on isolating some limited range of length scales. If it were necessary in the equations of hydrodynamics to specify the motion of every water molecule, a theory of ocean waves would be far beyond the means of 20th-century science.

A class of phenomena does exist, however, where events at many scales of length make contributions of equal importance.... Precisely at the critical point, the scale of the largest fluctuations becomes infinite, but the smaller fluctuations are in no way diminished. Any theory that describes (a system) near its critical point must take into account the entire spectrum of length scales.

K.G. Wilson [1021]

The word "critical" is used in science with different meanings. Here, we use it in the context of the critical phenomena studied in statistical physics in connection with phase transitions. In this framework, it describes a system at the border between order and disorder, which is characterized by an extremely large susceptibility to external factors and strong correlation between different parts of the system. These properties result from the cascade of correlations occurring at all existing scales in the system, in other words criticality is characterized by a self-similarity of the correlations. Examples are liquids at their critical point characterized by a specific value T_c of the temperature and of the pressure p_c and magnets at their Curie point at T_c under zero magnetic field, among many other systems. The interest in these critical phenomena lies in the (initially surprising but now well-understood but still fascinating) observation that simple laws acting at the microscopic level can produce a complex macroscopic behavior characterized by long-range correlations and self-similarity. Such critical behavior is fundamentally a cooperative phenomenon, resulting from the repeated interactions between "microscopic" elements which progressively "phase up" and construct a "macroscopic" self-similar state. At the mathematical level, the challenge is to understand and deal with the non-analytic behavior of the functions describing the response of the system.

One must distinguish between two types of collective phenomena. The existence of collective organizations is one of the first evidence when looking at the macroscopic behavior of systems emerging from the intrinsic chaotic motion of the atoms and molecules. The law of large numbers allows us indeed to derive the law of a perfect gas, the hydrodynamic equations, etc., which are all expression of a large scale coherence emerging from chaos at the atomic scale. However, the concept of criticality refers to a more subtle case of collective phenomena in which it is not possible to identify a specific correlation length above which averaging of the physical properties is possible, as is done for instance in deriving the Navier–Stokes hydrodynamic or the heat diffusion equations. This impossibility stems from the existence of a self-similar structure adopted by the critical system in which many length scales become relevant [1021].

We are interested in discussing the collective phenomena occurring in critical systems because there has been stimulating suggestions about the possible relevance of these concepts for natural phenomena, such as large earthquakes, which could be a signature of a kind of critical point [14], faulting with its hierarchical crustal fracture patterns [710, 711, 862], meteorology and turbulence [102, 316, 992], up to the large scale structure of the universe [177, 726].

We first present an illustration of the long-range correlations present in the organization of critical systems by reviewing some properties of spin systems using the statistical language and tools introduced in the previous chapters.

9.2 Spin Models at Their Critical Points

9.2.1 Definition of the Spin Model

Spin models are among the best known and simplest models of interacting elements which can exhibit critical phenomena. While apparently formulated in terms unrelated to geological problems, spin models in fact are much more general than their initial domain of definition as they incarnate the basic competition between an ordering tendancy from the action of interactions and the disordering influence of external stochastic noise. A spin represents the generic entity which can take only a finite discrete number of different states, the simplest case being two states 0 and 1 (or $-1/2$ and $+1/2$ often represented by an up and down arrows). You can thus think that a spin represents the local magnetization of anisotropic magnets, or the state "intact" or "damaged" of a material that will eventually rupture [324, 572]. Spin models have even been used to model the collective behavior of animal and human communities, where the two states represent two different opinions ("adoption" and "rejection", "positive" and "negative" votes, etc.) [132, 706].

A spin model is defined by 1) the lattice on which the spins are positionned (for instance at the nodes of a square or cubic lattice, or more generally of some graph) and 2) by the interactions between the spins. The spin representation can be used to model interacting elements or agents also subjected to random influences. Consider a network of elements: each one is indexed by an integer $i = 1, \ldots, I$, and $N(i)$ denotes the set of the elements which are directly connected to element i according to some graph. For simplicity, we assume that element i can be in only one of several possible states. In the simplest version called the Ising model, we can consider only two possible states: $s_i \in \{-1, +1\}$. We could interpret these states as 'positive' and 'negative', 'occupied' and 'empty', 'intact' and 'broken', etc. The general features of the model, such as the existence of critical points, remain robust when the number of states is modified. For q states where q is an arbitrary integer and when the interaction between spins are only of two types (positive for identical spins and negative for dissimilar spins), the model is known as the Potts model [1028]. This model has critical points for low values of q and exhibits abrupt transitions for large values of q. This shows that there is a large set of models and a broad range in parameter spaces for which critical behavior occurs.

To be specific, let us consider the Ising model with only two states such that the state of an element i is determined by:

$$s_i = \text{sign} \left(K \sum_{j \in N(i)} s_j + \sigma \varepsilon_i \right) \tag{9.1}$$

where the sign(x) function is equal to $+1$ (to -1) for positive (negative) argument x, K is a positive constant, ε_i is independently distributed according to the standard normal distribution and $N(i)$ is the number of relatives with whom the elements interacts significantly. This equation belongs to the class of stochastic dynamical models of interacting particles [565, 566] (Liggett, 1985, 1997), which have been much studied mathematically in the context of physics and biology.

In this model (9.1), the tendency towards imitation is governed by K, which is called the coupling strength; the tendency towards idiosyncratic behavior is governed by σ. Thus the value of K relative to σ determines the outcome of the battle between order and disorder. More generally, the coupling strength K could be heterogeneous across pairs of neighbors, and it would not substantially affect the properties of the model. Some of the K_{ij}'s could even be negative, as long as the average of all K_{ij}'s was strictly positive.

Note that (9.1) only describes the state of an element at a given point in time. In the next instant, new ε_i's are drawn, new influences propagate themselves to neighbors, and elements can change states. Thus, the best we can do is to give a statistical description of the states.

Many quantities can be of interest. The one that best describes the chance that a large group of elements finds itself suddenly of the same sign is called the *susceptibility* of the system. To define it formally, assume that a global influence term G is added to (9.1):

$$s_i = \text{sign} \left(K \sum_{j \in N(i)} s_j + \sigma \varepsilon_i + G \right). \tag{9.2}$$

This global influence term will tend to favour state $+1$ (state -1) if $G > 0$ ($G < 0$). Equation (9.1) simply corresponds to the special case $G = 0$: no global influence. Define the average state as $M = (1/I) \sum_{i=1}^{I} s_i$. In the absence of global influence, it is easy to show by symmetry that $\langle M \rangle = 0$: elements are evenly divided between the two states. In the presence of a positive (negative) global influence, elements in the positive (negative) state will outnumber the others: $\langle M \rangle \times G \geq 0$. With this notation, the susceptibility of the system is defined as:

$$\chi = \left. \frac{d(E[M])}{dG} \right|_{G=0} \tag{9.3}$$

In words, the susceptibility measures the sensitivity of the average state to a small global influence. The susceptibility has a second interpretation as a constant times the variance of the average state M around its expectation of zero caused by the random idiosyncratic shocks ε_i. Another related interpretation is that, if you consider two elements and you force the first one to be in a certain state, the impact that your intervention will have on the second element will be proportional to χ.

An alternative definition of the Ising model is obtained by writing that the energy of the system of spins is

$$E = -K \sum_{<ij>} \delta(\sigma_i \sigma_j), \tag{9.4}$$

where K is the coupling, the sum is taken over nearest neighbors and the delta function equals one if $\sigma_i = \sigma_j$, zero otherwise (this corresponds to the so-called Potts model [1028] which is equivalent to the Ising model when the number of states taken by the spins is two). As we said above, for $K > 0$ (attractive interaction), the interaction energy tends to make neighboring spins identical (failure favors failure, healing promotes healing) since this configuration minimizes energy. If the system is to obey an extremum principle so

as to minimize its energy, it will take the configuration where all the spins
are in the same direction. But, this tendency is opposed by the coupling of
the system to the exterior and also by the presence of stochasticity. Usually,
this stochastic "force" is quantified by the concept of a temperature. We have
shown in Chap. 7 how general this concept is and have presented a general
derivation that suggests its potential application outside the realm of thermal
equilibrium systems, for instance to quantify more generally the stochasticity of dynamical systems. We will use this framework in describing the spin
system.

Equation (9.4) gives the energy of a given spin configuration. The usual
assumption of thermal equilibrium means that the system is coupled to a thermal bath and the spin configurations evolve randomly in time and space in
response to thermal fluctuations with a probability to be in a configuration
with energy E proportional to the Boltzmann factor $e^{-\beta E}$, where β is the
inverse of the temperature (suitably dimensionalized by the Boltzmann constant).

9.2.2 Critical Behavior

As the simplest possible network, let us assume that elements are placed
on a two-dimensional grid in a Euclidean plane. Each agent has four nearest
neighbors: one each to the North, South, East and West. The relevant parameter is K/σ which measures the tendency towards imitation relative to the
tendency towards idiosyncratic behavior. In the context of the alignment of
atomic spins to create magnetization, this model (9.1) is identical to the two-
dimensional Ising model which has been solved explicitly by Onsager [703].
The two formulations (9.1) and (9.4) underline different aspects of the model:
(9.1) stresses the dynamical view point while (9.4) emphasizes the variational
view point usually taken in textbooks [351].

In the Ising model, there exists a critical point K_c that determines the
properties of the system. When $K < K_c$, disorder reigns: the sensitivity to
a small global influence is small, the clusters of elements that are in alignment
remain of small size, and imitation only propagates between close neighbors.
Formally, in this case, the susceptibility χ of the system is finite. When K
increases and gets close to K_c, order starts to appear: the system becomes
extremely sensitive to a small global perturbation, aligned elements form
large clusters, and imitation propagates over long distances. These are the
characteristics of *critical* phenomena. In this case, the susceptibility χ of the
system goes to infinity. The hallmark of criticality is the *power law*, and
indeed the susceptibility goes to infinity according to a power law:

$$\chi \approx A(K_c - K)^{-\gamma} \ . \tag{9.5}$$

where A is a positive constant and $\gamma > 0$ is called the *critical exponent* of the
susceptibility (equal to 7/4 for the 2-d Ising model).

9.2.3 Long-Range Correlations of Spin Models at their Critical Points

In this section, we follow [111] to present the main characteristics of critical phenomema within a probabilistic interpretation. The state of a spin system can be characterized by various quantities. As we said, the simplest one (called the magnetization) is

$$M = \sum_{i=1}^{L^d} S_i \,, \tag{9.6}$$

which is the sum of all spins in the system consisting of a cube of linear size L. This magnetization is the result of a competition between the interactions tending to decrease the energy and correlate the spins at long distance (making them all the same) and disorder (quantified by the temperature). At high temperature, the latter dominates and correlations between spins are short-ranged. M is thus zero on average with fluctuations of order

$$M = \sum_{i=1}^{L^d} S_i \sim \sqrt{L^d} \,. \tag{9.7}$$

More precisely, the central limit theorem applies in the usual Gaussian form:

$$P(M) \sim \frac{1}{\sqrt{2\pi\chi L^d}} e^{-M^2/2\chi L^d} \,, \tag{9.8}$$

with a variance proportional to the susceptibility χ, defined by $\chi = (\partial M/\partial H)|_T$, where H is the external field. When present, the magnetic field contributes to the energy (9.4) of the system by an additional term $-H\sum_i S_i$.

In contrast, in the low-temperature phase, perfect correlations are favoured (a finite fraction of all spins align parallel) and lead to a non-zero correlation function at large separations, equal to the magnetization value m:

$$M = \sum_{i=1}^{L^d} S_i \propto L^d m \pm \sqrt{L^d} \,. \tag{9.9}$$

The distribution $P(M)$ for large L is again essentially a Gaussian law centered at $L^d m$ (by symmetry, the same results hold for $-L^d m$).

These two regimes (high and low temperatures) cross over from one to the other at a critical point occurring at a special temperature T_c called the critical (or Curie) temperature. At this point, the spin correlations decay algebraically:

$$\langle S_0 S_r \rangle \sim \frac{1}{r^{d-2+\eta}} \,, \tag{9.10}$$

where the expression $d - 2 + \eta$ is due to historical reasons. According to Chap. 8 on long-range correlations, the system can be depicted as possessing N_{fam} effective families of N_{id} almost perfectly correlated spins, with

$$N_{\text{id}} \simeq \int^L \frac{r^{d-1} \, dr}{r^{d-2+\eta}} \sim L^{2-\eta} \ . \tag{9.11}$$

This expression (9.11) is the straightforward generalization of (8.9) to multi-dimensional sums. The fluctuations of the magnetization are thus of order

$$M \sim \sqrt{L^d N_{\text{id}}} \sim (L^d)^\nu, \ \text{with} \ \nu = (d + 2 - \eta)/2d \neq 1/2 \ , \tag{9.12}$$

where the exponent ν refers to the common usage in critical phenomena [19, 351, 922]. Thus, at the critical point, the total magnetization follows an anomalous power law as a function of the number of spins L^d with an anomalous diffusion exponent ν. It is the variable $M/(L^d)^\nu$ which has a limit distribution

$$P(M) = \frac{1}{L^{d\nu}} f\left(\frac{M}{(L^d)^\nu}\right) \ , \tag{9.13}$$

where $f(u) \sim \exp(-cu^a)$ for large u with $a = 1/(1 - \nu) \approx 6$ for $d = 3$ and $a = 16$ for $d = 2$ (exact result).

This probabilistic framework provides some insights on the neighborhood of the critical point T_c. The approach to the critical point is characterized by power laws describing the behavior of the energy F, the magnetization m, the correlation length ξ and the susceptibility χ, among other quantities:

$$F \sim L^d |T - T_c|^{2-\alpha} \ , \tag{9.14}$$

$$m \sim |T - T_c|^\beta \ , \tag{9.15}$$

(the exponent β should not be confused with the inverse temperature)

$$\xi \sim |T - T_c|^{-\nu_{\text{th}}} \ , \tag{9.16}$$

and

$$\chi \sim |T - T_c|^\gamma \ . \tag{9.17}$$

Notice that the correlation length diverges as $T \to T_c$. Close to T_c, the system can be viewed as made up of regions of size ξ in the critical state. There are $(L/\xi)^d$ such regions, which each contribute to the total free energy by an amount of order $k_B T$, hence

$$F/k_B T \simeq (L/\xi)^d \ , \tag{9.18}$$

which, using (9.14) and (9.16) gives the following relationship between the critical exponents

$$2 - \alpha = \nu_{\text{th}} d \ . \tag{9.19}$$

The total magnetization can be estimated for $T < T_c$ from (9.12) (replacing L by ξ) as

$$M = (\xi^d)^\nu (L/\xi)^d \sim \xi^{d(\nu-1)} . \tag{9.20}$$

This leads to the relation

$$\beta = d\nu_{\text{th}}(1 - \nu) . \tag{9.21}$$

For $T > T_c$, the fluctuations of the total magnetization (whose average is zero) are

$$\delta M \sim \xi^{d\nu} \sqrt{(L/\xi)^d} , \tag{9.22}$$

where the square root term is the expression of the central limit theorem and the term $\xi^{d\nu}$ is again obtained from (9.12) by replacing L by ξ. This expresses the fact that the susceptibility to an external magnetic field is enhanced by correlations measured by ξ. Since the susceptibility χ is proportional to $(\delta M)^2$, this yields another relationship between the critical exponents

$$\gamma = \nu_{\text{th}}(2 - \eta) . \tag{9.23}$$

These relations between the critical exponents are well-known [19, 922] and have been derived here using purely statistical considerations.

9.3 First-Order Versus Critical Transitions

9.3.1 Definition and Basic Properties

Many transitions are not critical (continuous) but are abrupt. According to Ehrenfest's classification, they are called "first-order" phase transitions because the first derivative of the free energy of the system with respect to the temperature at fixed volume, i.e. the entropy, is discontinuous. This discontinuity corresponds to the release or absorption of heat, called the latent heat. In contrast, critical phase transitions have continuous entropies as a function of the control parameter. Only the second or higher order derivative exhibit singularities, usually in the form of a power law divergence at the critical temperature.

The simplest parametric example of a first order transition is given by the van der Waals equation of state for a fluid:

$$\left(p + \frac{a}{V^2}\right)(V - b) = RT , \tag{9.24}$$

where p and V are the fluid pressure and volume, T is the temperature and $R = 8.314$ J mol^{-1} K^{-1} is the molar gas contant. The term a/V^2 is an additional effective pressure contribution due to interactions between fluid molecules. The interaction constant a can be negative (attractive case: leading to a reduction of the effective pressure $p + a/V^2$) or positive (repulsion). The excluded volume term $b > 0$ accounts for the fact that one cannot compress the fluid below a minimum residual volume. Figure 9.1 shows the pressure p as a function of the volume V at fixed temperature $RT = 1$ for different repulsive interaction strengths $a = 0.02$–0.04. One observes that for $a <$ 0.033, the isotherm $p(V)$ is no longer monotonic: the part of the curve with a positive slope $dp/dV|_T > 0$, corresponding to a negative compressibility i.e. the pressure increases as the fluid expands, is unstable and the fluid separates into two coexisting phases, one with low density (large volume) and the other with large density (low volume).

Figure 9.2 shows the pressure p as a function of the volume V at fixed repulsive interaction strength $a = 0.04$ for different temperatures $RT = 1 - 1.5$. For temperatures larger than $RT_c \approx 1.2$, the isotherms $p(V)$ are continuous and no phase transition occurs. For smaller temperatures, the isotherms $p(V)$ exhibit a range of volumes over which the compressibility $-(1/V)(dp/dv)|_T$ is negative, signaling an instability and a separation of the fluid into two phases. Maxwell's construction rule of equal areas fixes the volumes of the two coexisting phases: as shown for $RT = 1$, the area in the domain $A - B - C$ is equal to the area in the domain $C - D - E$. This geometrical construction reflects the equality of the chemical potential of the two phases. The locii of points for which $p(V)$ is either a minimum or a maximum, i.e. for which the compressibility vanishes, define the spinodal lines. The maximum of the spinodal line coincides with the maximum of the coexistence curve and is nothing but a critical point, as described above. In the present case, this critical point is positioned approximately at $p_c = 1.5$, $V_c = 0.032$ and $RT_c = 1.2$.

The hallmark of first-order transitions are the so-called van der Waals loops in pressure p versus volume V plots, corresponding to the domain of

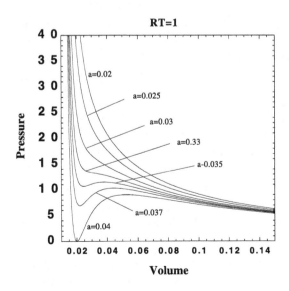

Fig. 9.1. Pressure p as a function of the volume V at fixed temperature $RT = 1$ for different repulsive interaction strengths $a = 0.02$–0.04. The isotherms $p(V)$ are solutions of (9.24)

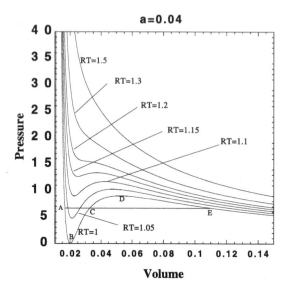

Fig. 9.2. Pressure as a function of the volume, at fixed repulsive interaction strength $a = 0.04$ for different temperatures $RT = 1$–1.5. The isotherms $p(V)$ are solutions of (9.24)

volumes where three solutions exist for the volume given a fixed pressure. The loop terminology comes from the hysteresis which is observed under a closed circuit in pressure. The size of a van der Waals loops is exactly equal to the free energy barrier for nucleation of the other phase. In large systems, van der Waals loops are to be taken as signs of first-order transitions only if their size vanishes in the thermodynamic limit as the inverse of the linear system size L.

The situation is complicated by the fact that the amount of phase that appears depends on the time-scale with which the experiment is performed. Consider the case $RT = 1$ in Fig. 9.2. As long as the pressure is above the pressure p_A of point A, the system is in a dense phase. When the pressure reaches p_A, the light phase with specific volume V_E starts to nucleate. The nucleation is a slow activation process requiring the spontaneous formation of sufficiently large droplets of the new phase. If the pressure is dropped sufficiently rapidly, the nucleation will not have time to appear and the system reaches point B which is a real instability: at this point, the dense phase undergoes a so-called spinodal decomposition with rapid critical-like dynamics without nucleation barriers. Similar behaviors occur when approaching the coexistence curve or the spinodal line from the large volume limit or from more tortuous paths in this phase diagram.

9.3.2 Dynamical Landau–Ginzburg Formulation

Upon entering the coexistence regime, a system witnesses the formation and evolution of correlated regions separated by "walls". Inside these regions, an ordered phase exists which eventually grows to become macroscopic in size. We follow [118] for a brief presentation of the theoretical background to describe these growth processes. The importance of these phenomena goes beyond the application to thermodynamic phase transitions as interesting analogies have been proposed between nucleation and abrupt rupture [416] on one hand and spinodal decomposition and breakdown of disordered media [867, 870, 1049] and earthquake rupture [511, 800] on the other hand.

Two important time scales determine if the transition occurs in or out of equilibrium:

- the relaxation time of long wavelength fluctuations (since these are the ones that order the system) $\tau_{\text{rel}}(k)$ and
- the inverse of the cooling rate $t_{\text{cool}} = T(t)/\dot{T}(t)$.

If $\tau_{\text{rel}}(k) \ll t_{\text{cool}}$, then these wavelengths are in local thermodynamical equilibrium, but if $\tau_{\text{rel}}(k) \gg t_{\text{cool}}$ these wavelengths fall out of thermodynamical equilibrium, freeze out, and the phase transition occurs in a quenched manner. These modes do not have time to adjust locally to the temperature change and for them the transition from a high temperature phase to a low temperature one occurs instantaneously.

Whereas the short wavelength modes are rapidly thermalized (typically by collisions), the long-wavelength modes with wavenumber $k \ll 1/\xi(T)$, with correlation length $\xi(T)$ (in the disordered phase), become critically slowed down. As T tends to the critical temperature T_c^+, the long wavelength modes relax very slowly, fall out of local thermodynamic equilibrium and any finite cooling rate causes them to undergo a quenched non-equilibrium phase transition. As the system is quenched from $T > T_c$ (ordered phase) to $T \ll T_c$ (disordered phase), ordering occurs instantaneously. The length scale of the ordered regions grows in time (after some initial transients) as the different broken symmetry phases compete to select the final equilibrium state. A dynamical length scale $\xi(t)$ typically emerges which is interpreted as the size of the correlated regions. This dynamical correlation length grows in time to become macroscopically large [550].

The phenomenological description of phase ordering kinetics begins with a coarse grained local free energy functional of a (coarse grained) local order parameter $M(\mathbf{r})$ [550], which determines the equilibrium states. In Ising-like systems, this $M(\mathbf{r})$ is the local magnetization (averaged over many lattice sites), in binary fluids or alloys it is the local concentration difference, in fluid-gas transitions it is the difference in density or of its inverse, the molar volume between the two phases, in rupture it is the local strain field, etc. The typical free energy is (phenomenologically) of the Landau–Ginzburg form:

$$F[M] = \int d^d\mathbf{x} \left\{ \frac{1}{2}[\nabla M(\mathbf{x})]^2 + V[M(\mathbf{x})] \right\} , \qquad (9.25)$$

$$V[M] = \frac{1}{2} r(T) M^2 + \frac{\lambda}{4} M^4 - H M ; \quad r(T) = r_0(T - T_c) , \qquad (9.26)$$

where H is the field conjugate to the order parameter M. Notice that equating to zero the derivative of $V[M]$ with respect to the order parameter M retrieves the third order equation (9.24), as it should since this corresponds to taking the mean field approximation.

The equilibrium states for $T < T_c$ correspond to the broken symmetry states with $M = \pm M_0(T)$ with

$$M_0(T) = \begin{cases} 0 & \text{for } T > T_c , \\ \sqrt{r_0/\lambda}(T_c - T)^{1/2} & \text{for } T < T_c . \end{cases} \qquad (9.27)$$

Below the critical temperature, the potential $V[M]$ presents a non-convex region with $\partial^2 V[M]/\partial M^2 < 0$ for

$$-M_s(T) < M < M_s(T) ; \quad M_s(T) = \sqrt{\frac{r_0}{3\lambda}}(T - T_c)^{1/2} \quad (T < T_c) . \quad (9.28)$$

This region is called the spinodal region which was already discussed in association with (9.24). It corresponds to thermodynamically unstable states.

1. The line $M_0(T)$ vs. T, given by (9.27), is known as the coexistence line. When crossing this line from the single stable phase, the system becomes metastable and starts to slowly nucleate to the other phase as metastable droplets that eventually run away in size as they grow above a critical value. The activated dynamics of the growth corresponds to that of an effective random walk that has to cross the nucleation barrier [617].
2. The line $M_s(T)$ vs. T, given by (9.28) is known as the classical spinodal line. When crossing this line from the previous metastable phase, the system becomes unstable and undergoes spinodal decomposition, which has similarities with dynamical critical phase transitions [428].

The dynamics of the phase transition and the process of phase separation can be described by a phenomenological but experimentally successful description, involving the Time Dependent Ginzburg–Landau theory (TDGL) where the basic ingredient is the Langevin dynamics [550]

$$\frac{\partial M(\mathbf{r}, t)}{\partial t} = -\Gamma[\mathbf{r}, M] \frac{\delta F[M]}{\delta M(\mathbf{r}, t)} + \eta(\mathbf{r}, t) , \qquad (9.29)$$

with $\eta(\mathbf{r}, t)$ a stochastic noise term which is typically assumed to be white (uncorrelated) and Gaussian. It obeys the fluctuation-dissipation theorem:

$$\langle \eta(\mathbf{r}, t)\eta(\mathbf{r}', t') \rangle = 2T\Gamma(\mathbf{r}) \, \delta^3(\mathbf{r} - \mathbf{r}')\delta(t - t') ; \quad \langle \eta(\mathbf{r}, t) \rangle = 0 , \qquad (9.30)$$

where the averages $\langle \cdots \rangle$ are over the Gaussian distribution function of the noise.

There are two important cases to distinguish.

- Non-conserved order parameter: then, $\Gamma = \Gamma_0$ is a constant independent of space, time and order parameter, and which can be absorbed in a rescaling of time.

- Conserved order parameter: this leads to

$$\Gamma[\mathbf{r}] = -\Gamma_0 \, \nabla_{\mathbf{r}}^2 \; ,$$

where Γ_0 could depend on the order parameter. In this case, the average over the noise of the Langevin equation can be written as a conservation law

$$\frac{\partial M}{\partial t} = -\nabla \cdot J + \eta \;\;\Rightarrow\;\; \frac{\partial}{\partial t} \left\langle \int d^3 \, \mathbf{r} \, M(\mathbf{r}, t) \right\rangle = 0 \; , \tag{9.31}$$

$$J \;\; = \nabla_{\mathbf{r}} \left[-\Gamma_0 \frac{\delta F[M]}{\delta M} \right] \equiv \nabla_{\mathbf{r}} \mu \; , \tag{9.32}$$

where μ is recognized as the chemical potential.

An example of a non-conserved order parameter is the magnetization in ferromagnets. A conserved order parameter could be the concentration difference in binary fluids or alloys.

For a quench from $T > T_c$ deep into the low temperature phase $T \rightarrow 0$, the thermal fluctuations are suppressed after the quench and the noise term is irrelevant. In this situation, the dynamics is now described by a deterministic equation of motion,

1. non-conserved order parameter:

$$\frac{\partial M}{\partial t} = -\Gamma_0 \, \frac{\delta F[M]}{\delta M} \; , \tag{9.33}$$

2. for conserved order parameter:

$$\frac{\partial M}{\partial t} = \nabla^2 \left[\Gamma_0 \frac{\delta F[M]}{\delta M} \right] \; , \tag{9.34}$$

which is known as the Cahn–Hilliard equation [550].

9.3.3 The Scaling Hypothesis: Dynamical Length Scales for Ordering

The process of ordering is described by the system developing ordered regions or domains that are separated by walls or other type of defects. The experimental probe to study the domain structure and the emergence of long range correlations is the equal time pair correlation function

$$C(\mathbf{r}, t) = \langle M(\mathbf{r}, t) M(\mathbf{0}, t) \rangle \; , \tag{9.35}$$

where $\langle \cdots \rangle$ stands for the statistical ensemble average in the initial state (or average over the noise in the initial state before the quench). It is convenient to expand the order parameter in Fourier components

$$M(\mathbf{r}, t) = \frac{1}{\sqrt{\Omega}} \sum_{\mathbf{k}} m_k(t) \, e^{i \mathbf{k} \cdot \mathbf{r}}$$

and to consider the spatial Fourier transform of the pair correlation function

$$S(\mathbf{k}, t) = \langle m_{\mathbf{k}}(t) m_{-\mathbf{k}}(t) \rangle , \qquad (9.36)$$

known as the *structure factor* or power spectrum which is experimentally measured by neutron scattering (in ferromagnets) or light scattering (in binary fluids) or sound scattering (in material breakdown). The scaling hypothesis introduces a dynamical length scale $L(t)$ that describes the typical scale of a correlated region and proposes that

$$C(\mathbf{r}, t) = f\left(\frac{|\mathbf{r}|}{L(t)}\right) \Rightarrow S(\mathbf{k}, t) = L^d(t) g(kL(t)) , \qquad (9.37)$$

where d is the spatial dimensionality and f and g are scaling functions. Ultimately, scaling is confirmed by experiments and numerical simulations. Theoretically, scaling emerges from a renormalization group approach to dynamical critical phenomena which provides a calculational framework to extract the scaling functions and the corrections to scaling behavior derived from finite size effects, deviations from criticality and the existence of other relevant perturbations [428]. This scaling hypothesis describes the process of phase ordering as the formation of ordered "domains" or correlated regions of typical spatial size $L(t)$. For non-conserved order parameters, the typical growth laws are $L(t) \approx t^{1/2}$ (with some systems showing weak logarithmic corrections) and $L(t) \approx t^{1/3}$ for scalar and $\approx t^{1/4}$ for vector order parameters in the conserved order parameter case [123].

10. Transitions, Bifurcations and Precursors

What do a gas pressure tank carried on a rocket, a seismic fault and a busy market have in common? Recent research suggests that they can all be described as self-organising systems which develop similar patterns over many scales, from the very small to the very large. And all three have the potential for extreme behaviour: rupture, quake or crash.

Are these events predictable?

The outstanding scientific question that needs to be addressed to guide prediction is how large-scale patterns of catastrophic nature might evolve from a series of interactions on the smallest and increasingly larger scales, where the rules for the interactions are presumed identifiable and known. For instance, a typical report on an industrial catastrophe describes the improbable interplay between a succession of events. Each event has a small probability and limited impact in itself. However, their juxtaposition and chaining lead inexorably to the observed losses. A commonality between the various examples of crises is that they emerge from a collective process: the repetitive actions of interactive nonlinear influences on many scales lead to a progressive build-up of large-scale correlations and ultimately to the crisis. In such systems, it has been found that the organization of spatial and temporal correlations does not stem, in general, from a nucleation phase diffusing accross the system. It results rather from a progressive and more global cooperative process occurring over the whole system via repetitive interactions.

For hundreds of years, science has proceeded on the notion that things can always be understood – and can only be understood – by breaking them down into smaller pieces, and by coming to know those pieces completely. Systems in critical states flout this principle. Important aspects of their behavior cannot be captured knowing only the detailed properties of their component parts. The large scale behavior is controlled by their cooperativity and the scaling up of their interactions. In [883, 886], this idea has been developed in four examples: rupture of engineering structures, earthquakes, stock market crashes [887, 888] and human parturition.

The classification of crises as bifurcations between a stable regime and a novel regime provides a first step towards identifying signatures that could be used for prediction [29, 477, 479, 481, 482, 889, 900, 913, 916].

10.1 "Supercritical" Bifurcation

The basic idea can be understood from the following simple analogy with the Rayleigh–Bénard experiment of the buyancy driven convective instability relevant to understanding mantle convection and atmospheric motion. Consider a liquid placed between two horizontal plates, the top one being at a temperature T_0 smaller than that $T_0 + \Delta T$ of the bottom plate. In a thermally expansive fluid, any temperature difference creates a density difference. The cold liquid, which is dense and located in the upper part of the system, tends to fall, whereas the lower part, warmer and less dense, tends to rise. Still, as long as the temperature difference ΔT remains small, no convective motion appears, due to the stabilizing effects of fluid viscous friction and thermal conductance. There is a critical value ΔT_c of the temperature difference at which an instability occurs:

- below ΔT_c, the fluid is at rest on the average and only small local thermal, density and velocity fluctuations occur;
- above ΔT_c, convection appears and the fluid motion becomes organized at the macroscopic scale, with the formation of a regular structure of rolls [77].

Many different systems can be characterized in a similar way when close enough to a transition between two regimes. This transition is called a bifurcation. For instance, parturition (the act of being born) can be seen in this way [890, 895], in which the control parameter analogous to the temperature difference ΔT is a maturity parameter (MP), roughly proportional to time, and the so-called order parameter (the fluid velocity in the convection problem) is the amplitude of the coherent global uterine activity in the parturition regime.

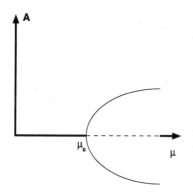

Fig. 10.1. Bifurcation diagram, near the threshold μ_c, of a "supercritical" bifurcation. The order parameter A bifurcates from a reference value 0 to a non-zero value $\pm A_s(\mu)$ represented by the two branches as the control parameter crosses the critical value μ_c. The value $A = 0$ represented by the *dashed line* becomes unstable for $\mu > \mu_c$

This idea is summarized in Fig. 10.1, representing the "bifurcation" diagram near the threshold of a so-called "supercritical" bifurcation, which is simply the mathematical counterpart of a critical phase transition discussed in Chap. 9. The abcissa μ represents the control parameter. The ordinate A

gives the average over time and space of the order parameter. Note the change of behavior from the value of the order parameter A which is small or even vanishing for $\mu < \mu_c$ to a large value for $\mu > \mu_c$.

This simple view is in apparent contradiction with the extreme complexity of typical natural systems. The basis of the bifurcation theory relies on many works [384, 385, 428, 935] in a variety of domains (mathematics, hydrodynamics, optics, chemistry, biology, etc.) which have shown that a lot of complex systems consisting of many nonlinear coupled subsystems or components may self-organize and exhibit coherent behavior of a macroscopic scale in time and/or space, under suitable conditions. The Rayleigh–Bénard experiment is one of the simplest paradigms for this type of behavior [77]. The coherent behavior appears generically when the coupling between the different components becomes strong enough to trigger or synchronize the initially incoherent sub-systems.

Mathematicians have proved [35, 954] that, under fairly general conditions, the local study of such bifurcations from a fixed point with a single control parameter can be reduced to a few archetypes. More precisely, it is proved that there exists reduction processes, series expansions and changes of variables of the many complex microscopic equations such that, near the fixed point (i.e. for A small), the behavior is described by a small number of ordinary differential equations depending only on one parameter μ. The result is non-trivial since a single effective number μ represents the values of the various physical variables and a single order parameter is sufficient to analyze the bifurcation instability.

This result applies both to bifurcation theory and catastrophe theory [35, 954], which are two areas within the field of dynamical systems. Both are studies of smooth systems, focusing on properties that are manifestly non-smooth. As we have seen, bifurcation theory is concerned with the sudden changes that occur in a system when one or more parameters are varied. Catastrophe theory became quite famous during the 1970's, mostly because of the sensation caused by applications of its principal ideas to other fields of science. Catastrophe theory is accurately described as singularity theory and its applications.

After having applied transformations that move the fixed point to the origin, the equations of the dynamical system become one of the classical "normal forms". The diagram represented in Fig. 10.1 corresponds to the pitchfork bifurcation of normal form given by [77]

$$\frac{\mathrm{d}A}{\mathrm{d}t} = (\mu - \mu_c)A - \frac{A^3}{A_s^2} \ . \tag{10.1}$$

For $\mu < \mu_c$, A is attracted to zero and A goes to

$$A = A_s \, (\mu - \mu_c)^{1/2} \ , \qquad \text{for } \mu \geq \mu_c \ , \tag{10.2}$$

where A_s gives the characteristic scale of the amplitude of A. This square root law predicts the rapid increase of the order parameter as the control parameter crosses its critical value. This law is very general in nature and has been observed in many different systems undergoing a similar critical transition from a quiescent state to a dynamical state. The exponent $1/2$ is called the "mean-field" value in the theory of critical phenomena, because it corresponds to neglecting the spatial dependence of the order paramater whose fluctuations may "renormalize" the exponent away from its mean-field value (see for instance the expressions (9.27) and (9.28) in Chap. 9). For larger μ, A is no longer described by (10.2) and instead saturates due to the existence of other limiting factors.

10.2 Critical Precursory Fluctuations

Below the instability threshold $\mu < \mu_c$, the order parameter is zero on average. However, this does not mean that the system is completely quiescent. It is in fact characterized by fluctuations in time and also in position within its structure. The bifurcation model allows one to predict the manner with which the amplitude of these fluctuations grows as μ approaches its critical value from below. Furthermore, a direct result of the model is that the average spatial extent within the system that is excited by a fluctuation grows according to a specific mathematical law that we now describe. The general mathematical formalism is taken from [428, 687]. We start from (10.1) and add a noise term $f(t)$, embodying the different sources of incoherent activity in the regime prior to the critical bifurcation:

$$\frac{dA}{dt} = (\mu - \mu_c)A - \frac{A^3}{A_s^2} + f(t) , \tag{10.3}$$

where $f(t)$ is Gaussian white noise with variance D. The term $f(t)$ is a random function of time with zero average and short time correlation. Expression (10.3) is called a Langevin equation, already encountered, for instance in (2.47).

Let us first neglect the nonlinear term. The solution of (10.3) is

$$A(t) = \int_0^t e^{-\delta(t-\tau)} f(\tau) \, d\tau , \tag{10.4}$$

where $\delta = \mu_c - \mu$ is taken positive (below the transition). From (10.4), we see that $A(t)$ is a linear sum of the random contributions $f(\tau)$ from zero to time t. Then, as $f(t)$ is a Gaussian noise, so is $A(t)$ with the variance

$$\langle [A(t)]^2 \rangle = \int_0^t d\tau \int_0^t d\tau' \, e^{-\delta(t-\tau)} e^{-\delta(t-\tau')} \langle f(\tau)f(\tau') \rangle \tag{10.5}$$

$$= D \int_0^t e^{-2\delta(t-\tau)} \, d\tau \quad \rightarrow \quad \frac{D}{2(\mu_c - \mu)} , \tag{10.6}$$

for $t \rightarrow +\infty$. We could have guessed this result by realizing from expression (10.4) that $A(t)$ is the sum of order $1/\delta$ uncorrelated random variables $f(t)$

with variance D. Its variance is thus proportional to this number multipled by the factor D. We recover the result (10.6) (apart from the numerical factor $1/2$). The distribution $p(A)$, measured over a large time interval, is

$$p(A) = \sqrt{\frac{\delta}{\pi D}}\ e^{-\delta A^2/D}\ . \tag{10.7}$$

These calculations show that the average standard deviation of the fluctuations of the order parameter diverges as the critical bifurcation point is approached from below $\delta \to 0^-$ ($\mu \to \mu_c^-$). $\langle [A(t)]^2 \rangle$ plays the role of the susceptibility, already encountered in the theory of critical phenomena in Chap. 9. This divergence of the susceptibility suggests a general predictability of critical bifurcations: by monitoring the growth of the noise on the order parameter, we get information on the proximity to the critical point. This method has been used in particular for material failure [29, 333], human parturition [890, 895], financial crashes [479, 481, 482, 900] and earthquakes [116, 477, 910].

These results (10.6) and (10.7) hold for $\mu_c - \mu \geq \sqrt{D/2A_s^2}$, i.e., not too close to the bifurcation point, for which the nonlinear correction is negligible. For $\mu_c - \mu < \sqrt{D/2A_s^2}$, we need to take into account the nonlinear saturation term. It can be shown [933] that, for $\mu < \mu_c$, the order parameter fluctuates randomly around a zero average with a probability distribution $p(A)$ given by

$$p(A) \sim \exp\left[-\frac{1}{D}\left(\delta A^2 + \frac{1}{2}\frac{A^4}{A_s^2}\right)\right]\ . \tag{10.8}$$

To get an intuitive feeling of the origin of this result, notice that (10.3) can be written

$$\frac{dA}{dt} = -\frac{dV}{dA} + f(t)\ , \tag{10.9}$$

where the potential $V(A)$ is defined by

$$V(A) = \frac{1}{2}(\mu_c - \mu)A^2 + \frac{A^4}{4A_s^2}\ . \tag{10.10}$$

As D is the variance of the noise, the standard result of statistical physics recalled in Chap. 7 leads to a distribution $p(A)$ proportional to the Boltzmann factor $\sim \exp[-V(A)/D]$, which indeed retrieves (10.8) up to a numerical factor in the exponential. Expression (10.8) can also be obtained as the stationary solution of the Fokker–Planck equation associated with the Langevin equation (10.3) (see Chap. 2). Close to or at the critical value μ_c, the variance $\langle [A(t)]^2 \rangle$ saturates to a value of order $A_s\sqrt{D}$. These are the general results within the present "mean-field" approach.

Depending upon the nature of the coupling between these different elements of the system, the mean field exponent $\gamma = 1$ relating $\langle [A(t)]^2 \rangle \sim \delta^{-\gamma}$ to δ in (10.6) may be modified [428]. The global picture nevertheless remains qualitatively the same: on the approach to the critical instability, one expects a characteristic increase of the fluctuations of the order parameter. Other quantities that could be measured and which are somehow related to the order parameter are expected to present a similar behavior. For instance, if we are able to produce a small spatially and temporally localized perturbation in the system, we expect that this perturbation may diffuse spatially with a characteristic diffusion coefficient $\mathcal{D}$ increasing as a power law

$$\mathcal{D} \sim (\mu_c - \mu)^{-\epsilon} \tag{10.11}$$

on the approach to the critical instability [556]. In the Rayleigh–Bénard case, the exponent ϵ is equal to $3/2$ and we can derive this result (10.11) simply as follows [875, 876]. The order parameter is identified as the average convection velocity and the fluctuations are associated with streaks or patches of nonzero velocity v occurring below the critical Rayleigh number (the critical Rayleigh value is such that a global convection starts off for larger control parameter values). The divergence of the spatial diffusion coefficient $\mathcal{D}(\xi)$ of the fluctuations can be understood as a consequence of the existence of the power law relation (10.6) describing the amplitude of velocity fluctuations prior to the instability. The diffusion coefficient is the product of an average square velocity times a characteristic time scale $\tau(\xi)$:

$$\mathcal{D}(\xi) \sim \alpha(\xi)\langle v^2 \rangle \, \tau(\xi) \, , \tag{10.12}$$

where $\xi \sim \delta^{-\nu}$ is the correlation length, $\tau(\xi) \sim \xi^z$ is the typical duration of a fluctuation of spatial extension ξ (here z is the dynamical critical exponent) and $\alpha(\xi)$ takes into account the viscous drag at the scale ξ. Using the previous result $\langle v^2 \rangle \sim \delta^{-\gamma}$, we thus get

$$\mathcal{D}(\xi) \sim \delta^{-(\gamma + z\nu)} \, . \tag{10.13}$$

Within mean field approximation which applies for the Rayleigh–Bénard supercritical instability, $\gamma = 1, \nu = 1/2$ and $z = 2$ which gives $\gamma + z\nu = 2$. For a hydrodynamic instability, one must also take into account the hydrodynamic drag which, from Stockes law, gives the correction $\alpha(\xi) \sim \xi^{-1}$. We thus recover the exponent $3/2$ of (10.11). The enhancement of the spontaneous fluctuations near the instability produces a singular diffusion coefficient that may be measurable [429, 940]. This singular behavior (10.11) reflects the existence of very large fluctuations in the velocity field on the approach to the critical point.

These results rely on the fact that the spatial extension ξ of the coherent domains of the order parameter flickering activity has the power law dependence

$$\xi \sim (\mu_c - \mu)^{-\nu} \, , \tag{10.14}$$

where we have already used the mean field value $\nu = 1/2$. This behavior is obtained by generalizing the above mathematical framework [428] to take into account the interactions between the many degrees of freedom participating in the dynamics of the order parameter. This law is crucial in the sense that it describes the progressive build-up of the cooperativity of the elements of the system to finally give the global synchronized activity at and above the critical threshold, at which the system acts coherently. Linked to this increase of the spatial correlation of the system activity is that of the susceptibility of the system with respect to external perturbations.

These results suggest that the noise level and its structure can be predictors of impending failure, seen as a super-critical bifurcation. Indeed, the variance of the fluctuations $\langle [A(t)]^2 \rangle$ exhibits a power law critical behavior (10.6) on the approach to the bifurcation point μ_c. Similarly, the effective diffusion coefficient $\mathcal{D}$ diverges according to (10.11). If one measures them and fits them to power laws, an approximate determination of μ_c can be obtained. This idea is at the basis of the so-called time-to-failure approach [181, 657, 996–998] which describes and tries to predict rate-dependent material failure, based on the use of an empirical power law relation obeyed by a measurable quantity, with many applications to damage, failure and even volcanic eruptions. Recently, these ideas have been applied to rupture of composite pressure vessels and to earthquakes [29, 477, 910, 979], with a generalization to include the possibility that the exponents possess an imaginary part, reflecting a log-periodic behavior (see Chap. 5 and [878] for a review).

We note that this idea on the possibility of using noise measurements in the analysis and prediction of systems is often considered for other systems, such as electronic devices and integrated circuits. The goals are to estimate the device reliability, to select reliable devices, to predict the device failure and to control and screen in order to provide the expected device quality and reliability during manufacture and to diagnose defects and failures [473]. Let us mention for instance the monitoring of electrical machines for the detection of faults [512, 825]. When collective effects become important, similar critical-like signatures may be expected and would be useful to look for as they provide one of the clearest signatures of the impending rupture (see also Chap. 13 on rupture of heterogeneous media).

A particularly interesting application of the above ideas has been suggested to explain the way our internal hearing organ, the cochlea, works [266]. The cochlea exhibits remarkable and quite special properties: a nonlinear response for arbitrary small amplitude of sollicitations, a compression of the dynamic range when the amplitude decreases, infinitely sharp tuning at infinitesimal input, and generation of combination tones. These properties derive naturally if the cochlea poises itself at a Hopf bifurcation, which then automatically maximizes tuning and amplification. The normal form for a Hopf bifurcation is similar to (10.3) but require a complex amplitude to account for oscillations:

$$\frac{\mathrm{d}A}{\mathrm{d}t} = (\mu - \mu_c + \mathrm{i}\omega_0)A - |A|^2 A + F\mathrm{e}^{\mathrm{i}\omega t} \ . \tag{10.15}$$

ω_0 is the natural angular frequency of the oscillations and the last term in the forcing with amplitude F and angular frequency ω. Exactly at the bifurcation point $\mu = \mu_c$, one can easily show that $A(t)$ is of the form $Re^{i\omega t}$ with

$$F^2 = R^6 + (\omega - \omega_0)^2 R^2 . \tag{10.16}$$

Thus, at resonance $\omega = \omega_0$, $R \sim F^{1/3}$, showing the intrinsic nonlinearity of the response, no matter how small F is. In the case of the cochlea, there is no audible sound soft enough not to evoke the nonlinear effects mentioned above. All the well-documented nonlinear aspects of hearing appear to be consequences of this same underlying Hopf bifurcation mechanism [266]. A possible mechanism for maintaining the hair bundle in the cochlea at the threshold of an oscillatory instability (self-tuned critical oscillations) involves the combination of two mechanisms that are known to modify the ionic current flowing through the transduction channels of the hair bundle [991]: (i) a rapid process involves calcium ions binding to the channels and (ii) a slower adaptation is associated with the movement of myosin motors.

10.3 "Subcritical" Bifurcation

"Subcritical" bifurcations are the mathematical analog of first-order phase transitions described in Chap. 9.

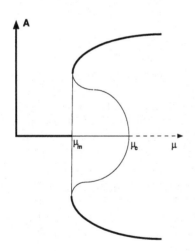

Fig. 10.2. Bifurcation diagram of a "subcritical" (first-order) bifurcation. For $\mu < \mu_m$, the only solution is $A = 0$. When μ reaches μ_m, the solution $A = 0$ becomes metastable and gives place to the thick branches under sufficiently large perturbations. From the condition of continuity of the behavior of the order parameter, the transition can be delayed up to $\mu = \mu_c$, corresponding to a genuine instability, at which the order parameter will abruptly switch to the upper branch as in a spinodal decomposition

A large class of observations seem to be left out from the description presented above, corresponding to the possible existence of abrupt transitions (as opposed to the more continuous bifurcation of supercritical type) and of hysteresis and delay. Indeed, a standard critical bifurcation of the type shown in Fig. 10.1 cannot explain delayed transitions, which is more characteristic of another class of bifurcations called sub-critical [77], shown in

Fig. 10.2. In order to address this problem, we must enrich the theory and introduce a second control parameter e, which takes into account the fact that the system may present a delayed (subcritical) bifurcation as well as a standard critical one, depending upon the value of the new control parameter e. This parameter e quantifies the tendency of the system to present delays and abrupt behavior and embodies the variables which are at the origin of this phenomenon. Starting from the normal form (10.3), we generalize it by allowing the coefficient of the A^3 term to vary:

$$\frac{dA}{dt} = (\mu - \mu_c)A - e\frac{A^3}{A_s^2} . \tag{10.17}$$

As long as e remains positive, the previous scenerio and results are not modified: the cubic term A^3 represents a non-linear feedback which tends to limit and saturates the amplitude of the order parameter beyond the critical instability. However, suppose that the variation in strength of an interaction leads to a change of sign of e which now becomes negative. Then, the cubic term A^3 is no longer a limiting factor but amplifies the instability described by the first term $(\mu - \mu_c)A$, for $\mu > \mu_c$. In order to get a physically meaningful (i.e. bounded) result, one then needs to add the next term in the expansion of the normal form giving dA/dt as a sum of powers of A:

$$\frac{dA}{dt} = (\mu - \mu_c)A - e\frac{A^3}{A_s^2} + A^5 . \tag{10.18}$$

This equation is represented by the sub-critical bifurcation diagram shown schematically in Fig. 10.2. For $\mu < \mu_m$, the only solution is $A = 0$. When μ reaches μ_m, the solution $A = 0$ becomes metastable, i.e. corresponds to a local minimum and not to the absolute minimum of the energy of the system. If the system is shocked sufficiently hard, A can jump to a more stable configuration represented by one of the two thick branches in Fig. 10.2. For weak perturbutions, the system will however be quenched in its metastable phase $A = 0$. This value μ_m represents the threshold for the "nucleation" of the coherent phase represented by the two thick branches. It is the analog of the coexistence point in first-order phase transitions discussed in Chap. 9. In the absence of large external fluctuations, it follows from the condition of continuity in the behavior of the order parameter that the transition will be delayed up to $\mu = \mu_c$, corresponding to a genuine instability, at which point the order parameter will abruptly switch to the upper branch. The threshold is the analog of the spinodal point in first-order phase transitions discussed in Chap. 9.

For such so-called "first-order" transitions, there are no evident precursors of order parameter activity similar to the ones described above for super-critical bifurcations. The system exhibits an abrupt jump from a no-activity state to the active coherent one. In the range $\mu_m < \mu < \mu_c$, the system may be activated to its coherent active regime by "large" external perturbations which produce the jump of A from 0 to the upper or lower branch represented in Fig. 10.2. We note that the case ($\mu = \mu_c$; $e = 0$) corresponds to a so-called bifurcation of "co-dimension two" (two control parameters are relevant simultaneously), in contrast to the critical and sub-critical bifurcation of "co-dimension one" (only one control parameter is relevant for the transition). Simply stated, these special values of the control parameters correspond to a system which "hesitates" between the two different regimes (critical and sub-critical). There are specific signatures of the spatial and temporal fluctuations of the order parameter, similar to, but different from those described above for the critical bifurcation, which can help establish a diagnostic of the approach of this co-dimension-two bifurcation.

10.4 Scaling and Precursors Near Spinodals

The transformation from one phase to another at a first-order phase transition usually occurs by a nucleation process. Nucleation near a spinodal appears to be very different from classical nucleation. Droplets appear to be fractal objects and the process of nucleation is due to the coalescence of these droplets, rather than the growth of a single one [642].

The theoretical description of homogeneous spinodal nucleation is based on the Landau–Ginzburg free energy of a spin system in the presence of an external magnetic field [643]. When the temperature is below the critical value, the free energy has the typical two-well structure. In the presence of an external magnetic field, one of the wells is depressed with respect to the other, which therefore represents the metastable state. The system must cross a free energy barrier to relax into the stable phase. When the external field is increased, this nucleation barrier decreases, eventually vanishing at the spinodal, as seen in Fig. 9.2. Using this formalism, it has been shown that the approach to the spinodal is characterized by scaling laws, analogous to critical phenomena. The magnetization or the order parameter M scales with the external field H as

$$M - M_s \propto (H - H_s)^{1/2} \ , \tag{10.19}$$

where M_s and H_s are, respectively, the order parameter and the field at the spinodal. This law implies a divergence of the quasistatic susceptibility χ

$$\chi = \frac{dM}{dH} \propto (H - H_s)^{-1/2} \ . \tag{10.20}$$

The fluctuations in the order parameter can be related to suitably defined clusters, whose sizes turn out to be power law distributed with an exponent $\mu = 1/2$, in mean-field theory. For finite-dimensional short-range models, this mean-field picture is expected to fail, since the system will nucleate before reaching the limit of metastability. On the other hand, mean-field behavior is expected to be valid in the presence of long-range interactions, and it has been numerically verified in Monte Carlo simulations of the long-range Ising model [763].

The limit of stability in a thermally activated homogeneous fracture has been proposed to correspond to a spinodal point [1049, 1050]. One should then be able to observe scaling laws consistent with those found in spinodal nucleation.

10.5 Selection of an Attractor in the Absence of a Potential

Let us generalize the problem, having in mind applications to meteorology, oceanography or population dynamics. We consider the general dynamical system

$$\frac{dx}{dt} = \mathbf{F}(\mathbf{x}) + \eta(t) ,\qquad(10.21)$$

which replaces (10.3) or (10.9). The quantity $\mathbf{x}$ is a d- dimensional vector of the d degrees of freedom characterizing the state of the system at a given time t. $\mathbf{F}$ is a generalized force which does not in general derive from a potential, in contrast to the previous situation (10.9) with (10.10). This situation is usually termed "non-variational". The noise source $\eta(t)$ is a Gaussian vector satisfying:

$$\langle \eta_i(t)\eta_j(t') \rangle = DQ_{ij}\delta(t - t') ,\qquad(10.22)$$

where Q is a symmetric matrix. The Langevin equation (10.21) is equivalent to the following Fokker–Planck equation for the probability $P(\mathbf{x}, t)$ that the system is in $\mathbf{x}$ at time t:

$$\frac{\partial P(\mathbf{x}, t)}{\partial t} = -\frac{\partial [F_i(\mathbf{x})P(\mathbf{x}, t)]}{\partial x_i} - \frac{D}{2}\frac{\partial^2 [Q_{ij}P(\mathbf{x}, t)]}{\partial x_i\, \partial x_j} ,\qquad(10.23)$$

where summation is performed over repeated indices.

In this general non-variational situation, the role of the potential is replaced by a so-called "pseudo-potential" $S(\mathbf{x})$ defined by expressing $P(\mathbf{x}, t)$ using a WKB approach

$$P(\mathbf{x}, t) \approx N(D)\, Z(\mathbf{x})\, e^{-S(\mathbf{x})/D} ,\qquad(10.24)$$

and using this expression in the Fokker–Planck equation (10.23). It is possible to get an asymptotic result in the limit of small D in the general case: the probability for the system to go from state $\mathbf{x}_i$ at time 0 to state $\mathbf{x}_f$ at time t is

$$P(\mathbf{x}_f|\mathbf{x}_i, t) \sim \exp\left(-\frac{1}{D}\min_T \int_{-t}^{0} d\tau \left[p_i \frac{dx_i}{dt} - H(\mathbf{p}, \mathbf{x})\right]\right) ,\qquad(10.25)$$

where

$$H(\mathbf{p}, \mathbf{x}) = F_i(\mathbf{x})p_i + \frac{1}{2}Q_{ij}p_ip_j \ , \tag{10.26}$$

and

$$p_i \equiv \frac{\partial S(\mathbf{x})}{\partial x_i} \ . \tag{10.27}$$

In (10.25), the minimum is taken over all continuous trajectories $\mathbf{x}(t)$ with $\mathbf{x}(-t) = \mathbf{x}_i$ and $\mathbf{x}(0) = \mathbf{x}_f$. The transition from one attractor or one state to another is thus controlled by the point on the domain separating the two attractors, analogous to a saddle point in the potential case, where $S(\mathbf{x})$ is minimum. In the potential case, the theory describing the transition from one state to another via the climbing of a potential barrier is known as reaction-rate theory or as the Kramers' problem [397, 617].

In the case where Q is the identity, the function $S(\mathbf{x})$ of the path $T\{\mathbf{x}(t)\}$ takes a simple form

$$S(\mathbf{x}) = \int_T \mathrm{d}t \ \left| \frac{\mathrm{d}\mathbf{x}}{\mathrm{d}t} - \mathbf{F}(\mathbf{x}) \right|^2 = \int_T \mathrm{d}t \ |\eta(t)|^2 \ . \tag{10.28}$$

The last equality shows that the optimal trajectory is the one that needs the smallest noise in order to go from the nearest deterministic trajectory to it. This recovers the principle of least dissipation introduced by Onsager [701, 702, 704] and generalized to strong noise [15, 282–284] (see Chap. 7). We refer to [155, 312, 360–362, 365, 499, 950] for the derivations and examples of applications.

Smelyanskiy et al. [850] have analyzed the probabilities of large infrequent fluctuations in nonadiabatically driven systems, i.e. when an external field is varied with a characteristic time scale comparable to the natural frequency to pass over the barrier. The change of the activation energy can be described in terms of an observable characteristic, the logarithmic susceptibility defined as the logarithm of the fluctuation probability. The change of activation energy is linear in the field and can be estimated from an instanton-like calculation using a field-perturbed optimal path approach as above [850]. The highlight of this calculation is that the effect of an oscillatory field with radial frequency ω is to lower the activation barrier, provided that ω exceeds the nucleation rate. Reversing the argument, in the seismological context, this result sheds light on why tides with a period of 12 h do not seem to exhibit noticeable correlations with earthquakes [990]: the earthquake nucleation rate may occur over a time scale smaller than 12 h!

11. The Renormalization Group

11.1 General Framework

Before presenting a specific model, we would like to revisit in more depth and from a more general perspective the renormalization group (RG) formalism previously introduced in Chap. 2. The RG analysis, introduced in field theory and in critical phase transitions, is a very general mathematical (and conceptual) tool, which allows one to decompose the problem of finding the "macroscopic" behavior of a large number of interacting parts into a succession of simpler problems with a decreasing number of interacting parts, whose effective properties vary with the scale of observation. The renormalisation group thus follows the proverb "divide to conquer" by organizing the description of a system scale-by-scale. It is particularly adapted to critical phenomena and to systems close to being scale invariant. The renormalisation group translates in mathematical language the concept that the overall behavior of a system is the aggregation of an ensemble of arbitrarily defined sub-systems, with each sub-system defined by the aggregation of sub-sub-systems, and so on.

Technically, this is done as we will see by defining a mapping between the observational scale and the distance $|T - T_c|$ from the critical point. The term "observational scale" usually refers to the physical scale of an observation. In the spin context, the observational scale rather refers to the size of the block of spins that one analyzes within the system. The usefulness of the RG approach is based on the existence of scale invariance and self-similarity of the observables at the critical point. The purpose of the RG is to translate in mathematical language the concept that a critical point results from the aggregate response of an ensemble of elements. In addition, the RG formalism can be used as a tool of model construction [1021]. In this presentation, we follow [808, 878].

Let us consider, for illustration, the behavior of the free energy F of the spin system. This is a suitable quantity to characterize the organization and cooperativity of the system of spins, which can also be measured (or its derivatives can be measured) experimentally. Using the RG formalism on the free energy amounts to assuming that F at a given temperature T is related to that at another temperature T' by the following transformations

$$x' = \phi(x) , \tag{11.1}$$

$$F(x) = g(x) + \frac{1}{\mu} F[\phi(x)] , \tag{11.2}$$

where $x = |T_c - T|$ is the absolute value of the distance to the critical point. The function ϕ is called the RG flow map. Here,

$$F(x) = F(T_c) - F(T) \tag{11.3}$$

such that $F = 0$ at the critical point and μ is a constant describing the rescaling of the free energy upon the rescaling of the temperature distance to its critical value. The function $g(x)$ represents the non-singular part of the function $F(x)$. We assume as usual [351] that the function $F(x)$ is continuous and that $\phi(x)$ is differentiable.

The critical point(s) is (are) described mathematically as the value(s) at which $F(x)$ becomes singular, i.e. when there exists a finite k-th derivative $d^k F(x)/dx^k$ which becomes infinite at the singular point(s). To remain simple, we consider $k = 1$. The formal solution of (11.2) is obtained by considering the following definitions:

$$f_0(x) \equiv g(x) , \tag{11.4}$$

and

$$f_{n+1}(x) = g(x) + \frac{1}{\mu} f_n [\phi(x)], \quad n = 0, 1, 2, ... \tag{11.5}$$

It is easy to show (by induction) that

$$f_n(x) = \sum_{i=0}^{n} \frac{1}{\mu^i} g \left[\phi^{(i)}(x) \right] , \quad n > 0 . \tag{11.6}$$

Here, we have used superscripts in the form "(n)" to designate composition, i.e.

$$\phi^{(2)}(x) = \phi[\phi(x)] ; \tag{11.7}$$

$$\phi^{(3)}(x) = \phi \left[\phi^{(2)}(x) \right] ; \tag{11.8}$$

etc. It naturally follows that

$$\lim_{n \to \infty} f_n(x) = F(x) , \tag{11.9}$$

assuming that it exists (for a more mathematical treatment, see for instance [261]). Note that the power of the RG analysis is to reconstruct the nature of the critical singularities from the embedding of scales, i.e. from the knowledge of the non-singular part $g(x)$ of the observable and the flow map $\phi(x)$ describing the change of scale. The connection between this formalism and critical points comes from the fact that the critical points correspond to the unstable fixed points of the RG flow $\phi(x)$. Indeed, the singular behavior emerges from the infinite sum of analytic terms if the absolute value of the eigenvalue λ defined by

$$\lambda = \mathrm{d}\phi/\mathrm{d}x \mid_{x=\phi(x)} \tag{11.10}$$

becomes larger than 1, in other words, if the mapping ϕ becomes unstable by iteration at the corresponding (critical) fixed point. The fixed point condition ensures that the same number appears in the argument of $g(.)$ in the series (11.6). In this case, the i-th term in the series for the k-th derivative of $F(x)$ will be proportional to $(\lambda^k/\mu)^i$ which may become larger than the unit radius of convergence for sufficiently large k since $\lambda > 1$, hence the singular behavior.

Thus, the qualitative behavior of the critical points and the corresponding critical exponents can be simply deduced from the structure of the RG flow $\phi(x)$. If $x = 0$ denotes a fixed point ($\phi(0) = 0$) and $\phi(x) = \lambda x + ...$ is the corresponding linearized transformation, then a solution of (11.2) close to $x = 0$ obeys

$$F(x) \sim x^m , \tag{11.11}$$

with m a solution of

$$\frac{\lambda^m}{\mu} = 1 . \tag{11.12}$$

This yields

$$m = \frac{\ln \mu}{\ln \lambda} .$$

The exponent f is thus solely controlled by the two scaling exponents μ and λ.

11.2 An Explicit Example: Spins on a Hierarchical Network

11.2.1 Renormalization Group Calculation

We now present a simple exact illustration of this general method, applied to a spin system in which the spins are put at the nodes of a hierarchical diamond lattice. The exposition is taken from [808]. The iterative rule to construct to network is shown in Fig. 11.1. Starting with a bond at magnification 1, we replace this bond by four bonds arranged in the shape of a diamond at magnification 2. Then, each of the four bonds are replaced by four bonds in the shape of a diamond and so on. At a given magnification 2^p, one sees 4^p bonds, and thus $(2/3)(2 + 4^p)$ sites (and spins).

In the same way that the lattice appears different at different scales from a geometrical point of view, one sees a different number of spins at different

p=0   p=1   p=2

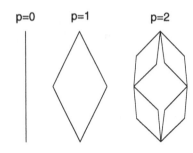

Fig. 11.1. First three steps of the iterative construction of the hierarchical diamond lattice

scales, and they interact in a scale dependent way as we will show. Suppose that for a given magnification $x = 2^p$, the spins we can see appear coupled with an interaction energy given by

$$E = -J \sum_{<ij>} \delta(\sigma_i \sigma_j) , \qquad (11.13)$$

where J is a coupling term, the sum is taken over nearest neighbors and the delta function equals one if arguments are equal, zero otherwise.

The full determination of the physical state of the system is found in principle once the partition function Z

$$Z(\beta) = \sum_{E} e^{-\beta E} , \qquad (11.14)$$

as defined by (7.6), is calculated. We have shown that its logarithm gives the free energy and, by suitable differentiation, all possible thermodynamic quantities of interest. A priori, the calculation of Z constitutes a formidable problem. However, since the spins are located at the vertices of the fractal network shown in Fig. 11.1, it turns out to be possible to write down an exact renormalization group equation and thus solve the problem exactly. We will not compute Z_p completely, but first perform a partial summation over the spins seen at one scale and which are coupled only to two other spins. This is how, in this particular example, one can carry out the program of the renormalization group by solving a succession of problems at different scales.

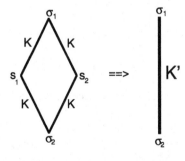

Fig. 11.2. Representation of the "decimation" process on an elementary diamond. Starting from a partition function defined from a sum over all possible configurations of all spins s_1, s_2, σ_1 and σ_2, the partial sum over all possible configurations of the spins s_1 and s_2 is performed, leading to a novel partition function defined solely as a sum over all possible configurations of σ_1 and σ_2 with a "renormalized" interaction

Let us isolate a particular diamond, call σ_1, σ_2 the spins at the extremities and s_1, s_2 the spins in between as in Fig. 11.2. The contribution of this diamond to $e^{-\beta E}$ is

$$K^{\delta(\sigma_1,s_1)+\delta(\sigma_2,s_1)+\delta(\sigma_1,s_2)+\delta(\sigma_2,s_2)} , \tag{11.15}$$

where we have defined $K = e^{\beta J}$. Since s_1, s_2 enter only in this particular product, we can perform the summation over them first when we compute Z_p. The final result depends on whether σ_1 and σ_2 are equal or different:

$$\sum_{s_1,s_2} K^{\delta(\sigma_1,s_1)+\delta(\sigma_2,s_1)+\delta(\sigma_1,s_2)+\delta(\sigma_2,s_2)}$$

$$= (2K + Q - 2)^2, \ \sigma_1 \neq \sigma_2 \tag{11.16}$$

$$= (K^2 + Q - 1)^2, \ \sigma_1 = \sigma_2 , \tag{11.17}$$

so we can write

$$\sum_{s_1,s_2} K^{\delta(\sigma_1,s_1)+\delta(\sigma_2,s_1)+\delta(\sigma_1,s_2)+\delta(\sigma_2,s_2)}$$

$$= (2K + Q - 2)^2 \left[1 + \left(\frac{(K^2 + Q - 1)^2}{(2K + Q - 2)^2} - 1 \right) \delta(\sigma_1, \sigma_2) \right]$$

$$= (2K + Q - 2)^2 K'^{\delta(\sigma_1,\sigma_2)} , \tag{11.18}$$

where we used the identity

$$K'^{\delta(\sigma_1,\sigma_2)} = 1 + (K' - 1)\delta(\sigma_1, \sigma_2) , \tag{11.19}$$

and we set

$$K' \equiv \left(\frac{K^2 + Q - 1}{2K + Q - 2} \right)^2 . \tag{11.20}$$

If we perform this partial resummation in each of the diamonds, we obtain exactly the system at a lower magnification $x = 2^{p-1}$. We see therefore that the interaction of spins transforms very simply when the lattice is magnified: at any scale, only nearest neighbor spins are coupled, with a scale dependent coupling determined recursively through the *renormalization group map*

$$K_{p-1} = \left(\frac{K_p^2 + Q - 1}{2K_p + Q - 2} \right)^2 \equiv \phi(K_p) . \tag{11.21}$$

This equation (11.21) provides an explicit realization of the postulated map $\phi(x)$ given by (11.1) written in the exposition of the RG method, where the coupling parameter $K - K_c$ plays the role of the control parameter x.

The spins which are "integrated out" in going from one magnification to the next simply contribute an overall numerical factor to the partition function, which is equal to the factor $(2K + Q - 2)^2$ per edge of (11.18). Indeed, integrating out the spins s_1 and s_2 leaves only σ_1 and σ_2 whose interaction weight is by definition $K'^{\delta(\sigma_1,\sigma_2)}$, denoting K' the effective interaction weight

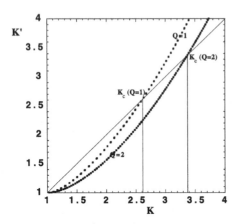

Fig. 11.3. Flow map $K'(K)$ given by (11.21) for two cases, $Q = 2$ (Ising spin model) and $Q = 1$ (Percolation model)

at this lower magnification 2^{p-1}. The additional numerical factor shows that the partition function is not exactly invariant with the rescaling but transforms according to

$$Z_p(K) = (2K + Q - 2)^{2 \cdot 4^p} Z_{p-1}[\phi(K)] , \qquad (11.22)$$

since there are 4^p bonds at magnification 2^p. Now the free energy, which is defined as the logarithm of the partition function per bond, reads

$$f_p(K) = \frac{1}{4^{p+1}} \ln Z_p(K) . \qquad (11.23)$$

From (11.22), we deduce the following

$$f_p(K) = g(K) + \frac{1}{4} f_{p-1}(K') , \qquad (11.24)$$

where

$$g(K) = \frac{1}{2} \ln(2K + Q - 2) . \qquad (11.25)$$

For an infinite fractal, the free energy for some microscopic coupling K satisfies the following renormalization group equation

$$f(K) = g(K) + \frac{1}{\mu} f(K') , \qquad \text{where } \mu = 4 . \qquad (11.26)$$

We thus recover exactly the functional form (11.2) postulated in Sect. 11.1.

This explicit calculation makes clear the origin of the scaling for the free energy: the interaction weights remain of the same functional form at each (discrete) level of magnification, up to a multiplicative factor which accounts for the degrees of freedom "left-over" when integrating from one magnification to the next. This is the physical origin of the function g in (11.2).

In addition to the geometrical aspect according to which different numbers of spins are seen at different magnifications, we now have a physical aspect: the coupling appears different at different magnifications as seen in (11.21).

There is still a fundamental difference between geometrical and physical aspects however. While the number of spins varies exponentially when one changes the scale, the coupling varies in general in a much more complicated way, which is represented in Fig. 11.3 for the particular case of spins with two states $(Q = 2)$ and for the percolation model (shown to be equivalent to the Potts model in the mathematical limit $Q \to 1$; see Chap. 12).

11.2.2 Fixed Points, Stable Phases and Critical Points

Stable Phases: Stable Fixed Points of the Renormalization Group.
Consider the map $K' = \phi(K)$ given by (11.21). It exhibits 3 fixed points [defined by $K' = K = \phi(K)$] located respectively at $K = 1, K = \infty$ and $K = K_c$ where K_c is easily determined numerically, for instance $K_c \approx 3.38$ for $Q = 2$ (see Fig. 11.3), and $K_c \approx 2.62$ for $Q = 1$. That $K = 1$ and $K = \infty$ are fixed points is obvious. The former corresponds to totally uncoupled spins, the latter to spins which are forced to have the same value. In both cases, the dynamics disappear completely, and one gets back to a purely geometrical problem. Observe that these two fixed points are attractive. This means that if we start with some coupling with say $K > K_c$ deep down in the system, that is for very large magnifications, when one diminishes the magnification to look at the system at macroscopic scales, spins appear almost always parallel and therefore are more and more correlated as one reduces magnification. Similarly if we start with $K < K_c$, spins are less and less correlated as one reduces magnification. The condition $K > K_c$ together with the definition $K = e^{\beta J}$ implies $\beta > \beta_c$, i.e. this is to the low-temperature regime dominated by the attractive energy. The physical meaning of the attraction of the renormalization group flow to the fixed point $K = \infty$, i.e. zero temperature, means that the macroscopic state of the spins is ferromagnetic with a macroscopic organization where a majority of spins have the same value. Similarly, the condition $K < K_c$ implies $\beta < \beta_c$, i.e. it corresponds to the high-temperature regime dominated by the entropy or thermal agitation. The physical meaning of the attraction of the renormalization group flow to the fixed point $K = 0$, i.e. infinite temperature, means that the macroscopic state is completely random with zero macroscopic magnetization.

In summary, within the renormalization group formalism, attractive fixed points describe stable thermodynamic phases. This is reminiscent of the behavior of the probability density function of the sum of N random variables converging to the Gaussian law in the limit of large N, where the Gaussian law is nothing but the fixed point of the corresponding renormalization group described in Chap. 2. Actually, generalizations of the probabilistic formulation of the renormalization group clarify the deep statistical significance of critical universality (see [487] and references therein). Technically, the problem is to generalize the central limit theorem to the case of dependent variables. The random fields appearing in these new limit theorems that apply to critical phenomena have scaling properties and some examples had already appeared in the probabilistic literature. The new challenging problem posed by the theory of phase transitions is the case of short range interactions producing at the critical point long range correlations whose scaling behaviour cannot be easily guessed from the microscopic parameters. A general theory of such limit theorems is still missing and so far rigorous progress has been obtained in situations which are not hierarchical but share with these the fact that some form of scaling is introduced from the beginning [487].

Phase Transitions: Unstable Fixed Points of the Renormalization Group. The intermediate fixed point K_c, which in contrast is repulsive, plays a completely different and very special role. It does not describe a stable thermodynamic phase but rather the transition from one phase to another. The repulsive nature of the renormalization group map flow means that this transition occurs for a very special value of the control parameter (the temperature or the coupling $K = K_c$). Indeed, if we have spins interacting with a coupling strength right at K_c at microscopic scales, then even by reducing the magnification we still see spins interacting with a coupling strength right at K_c! This is where spins must have an infinite correlation length (otherwise it would decrease to zero as magnification is reduced, corresponding to a different effective interaction): by definition, it is a *critical* point. Close to K_c, we can linearize the renormalization group transformation

$$K' = \phi(K) \approx K_c + \lambda(K - K_c) , \tag{11.27}$$

or

$$K' - K_c \approx \lambda(K - K_c) , \qquad \text{where } \lambda = \left|\frac{\mathrm{d}\phi}{\mathrm{d}K}\right|_{K_c} > 1 . \tag{11.28}$$

For couplings close enough to the critical point, as we increase magnification, the change in coupling becomes very simple; it is not the coupling that gets renormalized by a multiplicative factor, but the distance to K_c.

It is worth discussing the meaning of this renormalization a little more. We will restrict ourselves to the domain $K < K_c$ here. The fact that K_c is an unstable fixed point is expressed mathematically by the condition $\lambda > 1$. Therefore what (11.28) tells us is that if the system at microscopic scales looks "almost critical", nevertheless at larger scales it looks less and less critical since K moves away from K_c. This is natural since, being on one side of the transition, the system is in a well-defined physical state characterized macroscopically by the corresponding attractive fixed point towards which the renormalization group flows. The meaning of the critical point is that all these scales which are generally further from the singularity than smaller ones can actually all become critical at the same coupling when K equals K_c exactly. This is the reason and condition for long-range correlations to develop and allows for the appearance of a macroscopic behavior. Beyond this K_c fixed point, the system will be attracted to the other fixed point characterizing the other disordered state.

To summarize, the *renormalization group* theory is based on a sort of double rescaling structure: when one changes scale in space, one also changes the distance to criticality in coupling space (or in time).

11.2.3 Singularities and Critical Exponents

A critical point has other important properties. We go back to the renormalization group equations (11.21) and (11.24). They can be solved for the free energy and give, as we have already shown,

$$f(K) = \sum_{n=0}^{\infty} \frac{1}{\mu^n} g[\phi^{(n)}(K)] \ , \tag{11.29}$$

where $\phi^{(n)}$ is the n-th iterate of the transformation ϕ (e.g. $\phi^{(2)}(x) = \phi[\phi(x)]$). It is easy to show [220] that the sum (11.29) is *singular* at $K = K_c$. This stems from the fact that K_c is an unstable fixed point. Thus, the derivative of ϕ at K_c is λ which is larger than one. Therefore, if we consider the k-th derivative of f in (11.29), it is determined by a series whose generic term behaves as $\left(\lambda^k/\mu\right)^n$ which is greater than 1 for k large enough, so that this series diverges. In other words, high enough derivatives of f are infinite at K_c. Very generally, this implies that close to K_c, one has

$$f(K) \propto (K - K_c)^m \ , \tag{11.30}$$

where m is called a *critical exponent*. For instance if $0 < m < 1$, the derivative of f diverges at the critical point. Plugging this back in (11.24), we see that, since g is regular at K_c as can be checked easily from (11.25), we can substitute it in (11.24) and recover the leading critical behavior and derive m solely from the equation $(K - K_c)^m = (1/\mu)[\lambda(K - K_c)]^m$ involving only the singular part, with the flow map which has been linearized in the vicinity of the critical point. Therefore, the exponent satisfies

$$\lambda^m = \mu \ , \tag{11.31}$$

whose real solution is

$$m = \frac{\ln \mu}{\ln \lambda} \ . \tag{11.32}$$

The remarkable result obtained from the RG is that the "local" analysis of the RG flow map (captured in λ) and the rescaling of the observable measured by μ is enough to determine the critical exponent. In all these examples, λ is given by the slope of the renormalization group map at the unstable fixed point, and μ is the ratio of the number of degrees of freedom at two different scales connected by the renormalization group map. Intuitively, the critical exponent quantifies the self-similar rescaling structure of the observable close to the critical point which is completely determined by a local (in scale) analysis of the self-similarity.

We emphasize that g accounts for the degrees of freedom "left-over" when integrating from one magnification up to the next. It is a non-singular function because it is fully determined by local properties and does not "see" the building of long-range correlations. Remarkably however, g is all that is needed – together with the mapping ϕ – to determine the final free energy. The renormalization group analysis reconstructs the singularities from the embedding of scales, i.e. from the knowledge of the non-singular part of the observable and the flow map describing the effect of change of scale on the control parameter. The singular behaviour emerges from the infinite sum of analytic terms and corresponds to an infinite iteration of the local scaling which allows for long-range fluctuation correlations.

11.2.4 Complex Exponents and Log-Periodic Corrections to Scaling

In fact, there is a more general mathematical solution to (11.31):

$$m_n = \frac{\ln \mu}{\ln \lambda} + in\frac{2\pi}{\ln \lambda} \ . \tag{11.33}$$

To get expression (11.33), we have used the identity $e^{i2\pi n} = 1$. The existence of complex exponents results from the property of discrete scale invariance, captured mathematically by (11.24), which relates the free energy at two different scales in the ratio of a power of 2, as seen from Fig. 11.1. This is the same property discussed in Sect. 5.4 on complex dimensions of discretely scale invariant fractals.

We thus find that a critical phenomenon on a fractal exhibits *complex critical exponents*. Of course f is real, so the most general form of f close to the critical point should be

$$f(K) \approx (K - K_c)^m \left\{ a_0 + \sum_{n>0} a_n \cos[n\Omega \ln(K - K_c) + \Psi_n] \right\} , \tag{11.34}$$

where

$$m = \frac{\ln \mu}{\ln \lambda}, \qquad \Omega = \frac{2\pi}{\ln \lambda} \ , \tag{11.35}$$

hence exhibiting the log-periodic oscillations (corresponding to the cosine of the logarithm. This is represented in Figs. 11.4 and 11.5.

This result should be somewhat obvious. The geometry exhibits discrete scale invariance. Since, close to K_c, a rescaling in space corresponds to a rescaling of the distance to the critical point $K - K_c$, physical properties of the dynamics coupled to the geometry exhibit discrete scale invariance as a function of $K - K_c$. A discrete geometrical scale invariance implies a discrete scale invariance of the physical properties as a function of $K - K_c$ giving rise to log-periodic corrections. Although derived for the particular case of the diamond lattice, the previous results apply in principle to any fractal exhibiting discrete scale invariance.

The mathematical existence of complex critical exponents and therefore of log-periodic corrections were identified quite soon after the discovery of the renormalization group [486, 667, 683]. Soon after, they were rejected for translationally invariant systems, since a period (even in a logarithmic scale) implies the existence of one or several characteristic scales which is forbidden in these ergodic systems in the critical regime. In addition to their possible existence in hierarchical geometrical systems, they can appear in Euclidean systems possessing a quenched heterogeneity, such as in random dipolar spin systems [6], in spin glasses [165], in critical phenomena with random couplings [503] and long range correlations [117, 1009].

Recent works have documented evidence of log-periodic structures decorating the main power law behavior in acoustic emissions prior to rupture [29] and in the seismic precursory activity before large earthquakes [477, 910, 979]. Log-periodic oscillations have also been documented in the fractal structure of arrays of cracks in geological media [712]. These log-periodic structures are the expression of *discrete scale invariance* (DSI), the property of a system

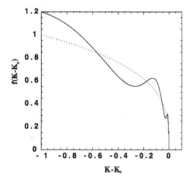

Fig. 11.4. Function f given by expression (11.34) truncated at the first oscillatory correction as a function of $K - K_c$ on the left of the critical point, with $m = 0.3$, $a_0 = 1$, $a_1 = 0.2$, $a_{n>1} = 0$, $\Psi_1 = 0$ and $\Omega = 2\pi$ corresponding to a prefered scaling ratio equal to $\exp(2\pi/\Omega) \approx 2.7$. The oscillations, which have their local frequency accelerating without bound when $K - K_c \to 0$, are not visible very close to the critical point due to lack of resolution in the graphic. The *dashed line* corresponds to the pure real power law $(K - K_c)^m$.

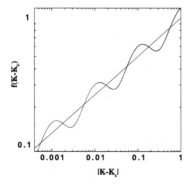

Fig. 11.5. Same as Fig. 11.4 in double logarithmic scale. The *straight line* corresponds to the pure real power law $(K - K_c)^m$. The *horizontal scale* is inverted compared to Fig. 11.4 since one approaches the critical point from right to left

which is invariant under a *discrete* set of dilatations only.[1] Unlike continuous scale invariance which is very common in all the critical phenomena of nature, until recently DSI was considered merely a man-made artifact or restricted to discrete fractals or hierarchical constructions. It is not necessarily so. Formally, discrete scale invariance corresponds to complex critical exponents, a situation which is actually possible in non-unitary systems like geometrical systems with non-local properties (percolation, polymers and their generalizations), or in models with disorder (spin-glasses) in *non-fractal* lattices [807]. DSI has also recently been seen quite clearly in the mass-radius scaling relationship of diffusion-limited-aggregation (DLA) clusters [899]. Log-periodic structures have also be proposed for turbulence [30, 316, 696, 697, 881], in biology [839, 1013] as well as in models of earthquake aftershocks [436, 555].

In growth and rupture phenomena, damage and precursory phenomena occur at particular discrete times and not in a continuous fashion, and these discontinuities reflect the localized and threshold nature of the mechanics of rupture and faulting. It is this "punctuated" physics which gives rise to the existence of scaling precursors modeled mathematically by the log-periodic correction to scaling.

The renormalization group equations of the form (11.26) or (11.36) have not been derived from first principles for growth, rupture and other out-of-equilibrium processes alluded to above, even if there are various attempts to develop approximate RG descriptions on specific models of these processes. It may thus seem a little premature to use this discrete renormalization group description for these systems. Actually, expressions (11.26) or more generally (11.36) below can be obtained without any reference to a renormalization group approach: as soon as the system exhibits a discrete scale invariance, the natural tool is provided by q-derivatives [279] from which it is seen that expression (11.36) is nothing but a Jackson q-integral [451–453, 955] of the function $g(x)$, which constitutes the natural generalization of regular integrals for discretely self-similar systems [279]. The way the Jackson q-integral is related to the free energy of a spin system on a hierarchical lattive was explained in [278].

It is always a marvel to realize that mathematical structures initially thought of as useless esoteric constructions of the intellect end up providing new applications and new insights. The history of the development of human science is full of such examples.[2] This gives us a sense of wonder as Einstein said: "How can it be that mathematics, being after all a product of human thought independent of experience, is so admirably adapted to the objects of reality?" E. Wigner emphasized [1015]: "The enormous usefulness of mathematics in the natural sciences is something bordering on the mysterious...

---

[1] It is crucial not to confuse DSI with the existence of a discrete scale. For instance, a square lattice is a discrete system, but does not have discrete scale invariance.

[2] Negative masses have been interpreted correctly as anti-particles by Dirac, complex (amplitude) probabilities lead to quantum interference, and so on.

The miracle of the appropriateness of the language of mathematics for the formulation of the laws of physics is a wonderful gift, which we neither understand nor deserve." The generalization of exponents to the complex plane provides another example.

11.2.5 "Weierstrass-Type Functions" from Discrete Renormalization Group Equations

Let us consider again expression (11.29) giving the formal solution of the renormalization group equation (11.26). Around fixed points K_c solution of $\phi(K_c) = K_c$, the renormalization group map can be expanded up to first order in $K - K_c$ as $\phi(K) = \lambda(K - K_c)$. Posing $x = K - K_c$, we have $\phi^{(n)}(x) = \lambda^n x$ and the solution (11.29) becomes

$$f(x) = \sum_{n=0}^{\infty} \frac{1}{\mu^n} g[\lambda^n x] \ . \tag{11.36}$$

In principle, (11.36) is only applicable sufficiently "close" to the critical point $x = 0$, such that the higher-order terms in the expansion $\phi(K) = \lambda(K - K_c)$ can be neglected. The effect of nonlinear corrections terms for $\phi(K)$ have been considered in [220, 223].

In the mathematical literature, the function (11.36) is called a *Weierstrass-type function*, to refer to the introduction by K. Weierstrass of the function [1007]

$$f_W(x) = \sum_{n=0}^{\infty} b^n \cos[a^n \pi x] \ , \tag{11.37}$$

corresponding to the special case $\mu = 1/b$, $\lambda = a$ and $g(x) = \cos[\pi x]$. To the surprise of mathematicians of the 19th century, Weierstrass showed that the function (11.37) is continuous but differentiable nowhere, provided $0 < b < 1, a > 1$ and $ab > 1 + 2/3\pi$. Note that, in the context of the renormalization group of critical phenomena, the condition $a = \lambda > 1$ implies that the fixed point K_c is unstable. Hardy was able to improve later on the last bound and obtain that the Weierstrass function (11.37) is non-differentiable everywhere as soon as $ab > 1$ [402]. In addition, Hardy showed that it satisfies the following Lipschitz condition (corresponding to self-affine scaling) for $ab > 1$, which is much more than just the statement of non-differentiability:

$$f_W(x + h) - f_W(x) \sim |h|^m \ , \quad \text{for all } x \text{ where } m = \ln[1/b]/\ln a \ . \tag{11.38}$$

Note that for $ab > 1$, $m < 1$, expression (11.38) shows that $f_W(x + h) - f_W(x) \gg |h|$ for $h \to 0$. As a consequence, the ratio $[f_W(x + h) - f_W(x)]/h$

has no limit which recovers the property of non-differentiability. Continuity is obvious from the fact that $f_W(x + h) - f_W(x) \to 0$ as $h \to 0$ since $m > 0$. For the border case $a = b$ discovered by Cellerier before 1850, f_W is not non-differentiable in a strict sense since it possesses infinite differential coefficients at an everywhere dense set of points [847]. Richardson is credited with the first mention of the potential usefulness for the description of nature of the continuous everwhere non-differentiable Weierstrass function [777]. Shlesinger and co-workers [440, 509, 648, 840] have previously noticed and studied the correspondence between (11.36) and the Weierstrass function.

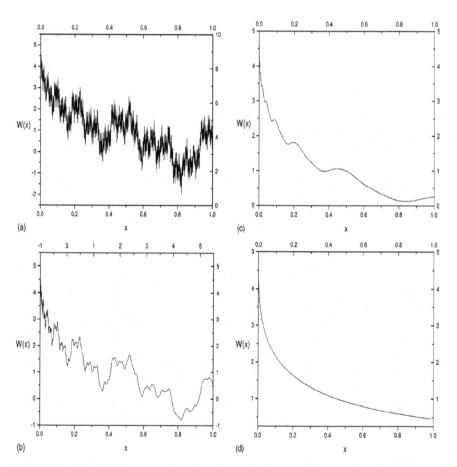

Fig. 11.6. Quasi-Weierstrass function for $m = 0.25, \omega = 7.7$, using $N = 32$ terms to estimate the sum (11.45). *Top-left panel* (**a**): $\alpha = \pi/2$ (genuine Weierstrass function); *bottom-left panel* (**b**): $\alpha = 0.993\pi/2 = 1.56$; *top-right panel* (**c**): $\alpha = 0.9\pi/2 = 1.414$; *bottom-right panel* (**d**): $\alpha = 0$. Reproduced from [343]

If one is interested in the non-regular (or non-analytic) behavior only close to the critical point $x = 0$, the regular part can be dropped and the analysis of

$$f(x) = \frac{1}{\mu} f(\lambda x) \tag{11.39}$$

is sufficient. It is then easy to show by simple verification that the most general solution of (11.39) is (see [878] and references therein)

$$f(x) = x^m P\left(\frac{\ln x}{\ln \lambda}\right) , \tag{11.40}$$

where m is given by (11.35) and $P(y)$ is an arbitrary periodic function of its argument y of period 1. Its specification is actually determined by the regular part $g(x)$ of the renormalization group equation, as shown for instance in the explicit solution (11.36). The scaling law $f(x) \sim x^m$ implied by (11.40) is a special case of (11.38) obtained by putting $x = 0$ and replacing h by x in (11.38).

The Laplace transform $f_L(\beta)$ of $f(x)$ defined by (11.36) also obeys a renormalization equation of the type (11.26). Denoting $g_L(\beta)$ the Laplace transform of the regular part $g(x)$, we have

$$f_L(\beta) = \sum_{n=0}^{\infty} \frac{1}{(\mu\lambda)^n} g_L[\beta/\lambda^n] , \tag{11.41}$$

and

$$f_L(\beta) = g_L(\beta) + \frac{1}{\mu\lambda} f_L\left(\frac{\beta}{\lambda}\right) . \tag{11.42}$$

The general solution of (11.42) takes the same form as (11.40):

$$f_L(\beta) = \frac{1}{\beta^{1+m}} P_L\left(\frac{\ln \beta}{\ln \lambda}\right) , \tag{11.43}$$

where $P_L(y)$ is a periodic function of its argument y of period 1.

As an example, let us consider the regular part $g(x)$ of the renormalization group equation defined as

$$g(x) = e^{-\cos(\alpha) x} \cos(x \sin(\alpha)) , \quad \text{with} \quad \alpha \in \left[0, \frac{\pi}{2}\right] . \tag{11.44}$$

The parameter α quantifies the relative strength of the oscillatory structure of $g(x)$ versus its "damping": for $\alpha = \pi/2$, (11.36) with (11.44) recovers the initial function (11.37) introduced by Weierstrass with $b = 1/\mu$, $a = \lambda$ and $\cos(\pi x)$ replaced by $\cos(x)$; for $\alpha = 0$, $g(x) = \exp[-x]$ has no oscillation anymore and corresponds to a pure exponential relaxation considered in [620].

Plugging (11.44) in (11.36) gives

$$f(x) = \sum_{n=0}^{\infty} \frac{1}{\lambda^{(2-D)n}} e^{-\cos(\alpha)\lambda^n x} \cos(\lambda^n x \sin(\alpha)) , \tag{11.45}$$

where

$$D = 2 - m = 2 - \frac{\ln \mu}{\ln \lambda} \; . \tag{11.46}$$

The exponent D turns out to be equal to the fractal dimension of the the graph of the Weierstrass function obtained for $\alpha = \pi/2$. Recall that the fractal dimension quantifies the self-similarity properties of scale invariant geometrical objects. Note that $1 < D < 2$ as $1 < \mu < \lambda$ which is the condition of non-differentiability found by Hardy [402] for the Weierstrass function. The graph of the Weierstrass function is thus more than a line but less than a plane. For $\alpha < \pi/2$, $f(x)$ is smooth and non-fractal ($D = 1$) and its graph has the complexity of the line. Actually, there are several fractal dimensions. It is known that the box counting (capacity, entropic, fractal, Minkowski) dimensions and the packing dimensions of the Weierstrass function are all equal to D [494] given by (11.46) for $\alpha = \pi/2$. It is conjectured but not proved that the Hausdorff fractal dimension of the graph of the Weierstrass function obtained for $\alpha = \pi/2$ is also equal to D given by (11.46). It is known that the Hausdorff dimension of the graph of $f(x)$ does not exceed D but there is no satisfactory condition to estimate its lower bound [433].

Figure 11.6 shows the function (11.45) for $\alpha = \pi/2 = 1.5708$ (pure Weierstrass function), $\alpha = 0.993\pi/2 = 1.56$, $\alpha = 0.9\pi/2 = 1.414$ and and $\alpha = 0$. Reference [343] uses this example and many others to address the question of the origin of the strength of log-periodic oscillations in certain systems. For instance, for Ising or Potts spins with ferromagnetic interactions on hierarchical systems, the relative magnitude of the log-periodic corrections are usually very small, of order 10^{-5} [223]. In growth processes (DLA), rupture, earthquake and financial crashes, log-periodic oscillations with amplitudes of the order of 10% have been reported. The "technical" explanation for this 4-order-of-magnitude difference is found in the properties of the "regular function" $g(x)$ embodying the effect of the microscopic degrees of freedom summed over in a renormalization group approach leading to equations such as (11.26). It is found that the "Weierstrass-type" solutions of the renormalization group can be put into two classes characterized by the amplitudes A_n of the power law series expansion obtained from a Mellin transform of (11.36). These two classes are found to be separated by a novel "critical" point.

A known example of a system of the first class is the q-state Potts model with *antiferromagnetic* interactions [220, 613]. Another example is the statistics of closed-loop self-avoiding walks per site on a family of regular fractals with a discrete scale-invariant geometry such as the Sieirpinsky gasket [725]. A known example of the second class is the q-state Potts model with ferromagnetic interactions [223].

Growth processes (DLA), rupture, earthquake and financial crashes thus seem to belong to the first class and to be characterized by oscillatory or bounded regular microscopic functions $g(x)$ that lead to a slow power law decay of A_n, giving strong log-periodic amplitudes. If in addition, the phases of A_n are ergodic and mixing, the observable presents self-affine non-differentiable properties. In contrast, the regular function $g(x)$ of statistical physics models with "ferromagnetic"-type interactions at equilibrium involves unbound logarithms of polynomials of the control variable that lead to a fast exponential decay of A_n giving weak log-periodic amplitudes and smoothed observables.

11.3 Criticality and the Renormalization Group on Euclidean Systems

The reader should not have the false impression that the renormalization group works only on hierarchical or discrete scale invariant fractals. Hierarchical systems are the exceptions as physics is usually defined in Euclidean space. It is true that hierarchical systems provide most of the examples which can be solved exactly within the renormalization group formalism. However, the renormalization group has initially been introduced as an efficient approximation scheme for calculations of critical properties of Euclidean systems.

In these systems, one defines a change of scale by an arbitrary factor $b > 1$ and spins (or more generally degrees of freedom) are "integrated out" going from magnification L to L/b, where L is the system size. In general, some approximations are called for to carry on the renormalization group program but the procedure and the end results are similar to those obtained above in equations (11.21) and (11.24), with $\mu = b^d$, where d is the space dimension. One can also show that the scaling factors λ of the relevant control parameters are powers of the scale factor b. As a consequence, the real part of the critical exponents, which involve the ratio $\ln \mu / \ln \lambda$ are independent of the arbitrary choice of b. However, from (11.24) with a finite $\mu = b^d$ associated with a fixed scale change $b > 1$, the critical exponents take an imaginary part, equal to $2\pi/\ln \lambda$, which depends explicitly on b (through λ). Thus, the associated log-periodic corrections are created and controlled by the specific choice of the scaling factor b. In a homogeneous Euclidean system, there is a priori no preferable choice for b, therefore the period of the log-periodic corrections can take any value, controlled by the corresponding choice of b. In other words, the results of the renormalization group calculation for the critical exponents must be independent of the choice of b since the physics is independent of it. As a consequence, the imaginary part which depend explicitly on b is not allowed. The coefficients a_n for $n > 0$ in (11.34) must be zero in this case.

This result can be retrieved from the previous analysis leading to (11.24) by noting that, in a Euclidean system, the most natural choice for b is $b = 1 + \mathrm{d}l$, with $\mathrm{d}l$ infinitesimal, i.e. at each step of the renormalization group, one integrates out an infinitesimal fraction of the degrees of freedom (instead of half the spins as in the case of the diamond lattice). This choice ensures that all b's play the same role since any scaling factor B greater than

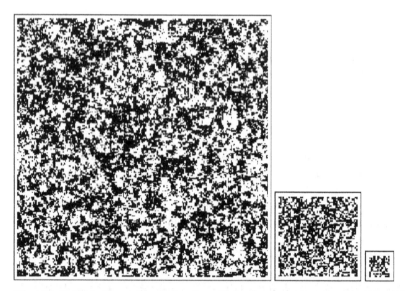

Fig. 11.7. This figure illustrates the effect of renormalisation for $K < K_c$ of the Ising model, which corresponds to the disordered regime. The two discording states are encoded in white and black. Starting from a square lattice with some given configuration of spins on the *left*, two successive applications of the renormalisation group are shown on the *right panels*. Repeated applications of renormalisation group change the structure of the lattice with more and more disorganization. All the shorter range correlations, quantified by the typical sizes of the black and white domains, are progressively removed by the renormalisation process and the system becomes less and less ordered, corresponding to an effective decrease in the interaction strength K. Eventually, upon many iteration of the renormalisation group, the distribution of black and white squares becomes completely random. The system is driven away from criticality by the renormalisation. The renormalisation group thus qualifies this regime as disordered under change of scales

1 can be obtained by $(\ln B)/\mathrm{d}l$ iterations of the infinitesimal scaling $1 + \mathrm{d}l$. Then, the renormalization group discrete flow equation (11.21) becomes an ordinary differential equation $\mathrm{d}K/\mathrm{d}l = \phi(K)$ and similarly for (11.26). The modification from discrete equations to ordinary differential equations rule out the existence of complex critical exponents if a technical condition is met, namely the renormalization group relations may be represented by gradient flows [1002]. This is usually born out for translational invariant systems, such as homogeneous Euclidean systems.

The renormalization group has been applied to a variety of problems, such as percolation (see Chap. 12), polymers, transition to chaotic behavior, turbulence, and many out-of-equilibrium systems. It is also a fundamental tool for theories of fundamental interactions and particle physics where it was firsd developed with a different formulation [1008].

To illustrate how the renormalization group works for an Euclidean system, let us consider a collection of spins, which each has one out of two possible states (up or down, 0 or 1, yes or no). The renormalisation group then works as follows.

1. The first step is to group neighboring spins into small groups. For instance, in a two-dimensional square lattice, we can group spins in clusters of size equal to 9 spins corresponding to squares of side 3 by 3.

2. The second step is to replace the cacophony of spin states within each group of 9 spins by a single representative spin, resulting from a chosen majority rule. Doing this "decimation" procedure obviously lowers the complexity of the problem since there are 9 times fewer spin states to keep track of.

3. The last step is to scale down or shrink the super-lattice of squares of size 3 by 3 to make them of the same size as the initial lattice. Doing this, each cluster is now equivalent to an effective spin endowed with a spin representing an average of the spins of the 9 constitutive spins.

One loop involving the three steps applied to a given system transforms it to a new system which looks quite similar, but is different in one impor-

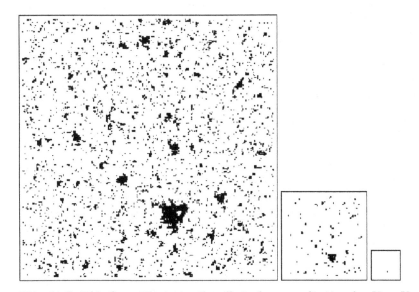

Fig. 11.8. This figure illustrates the effect of renormalisation for $K > K_c$ of the Ising model, which corresponds to the ordered regime in which one spin state (*white*) dominates (the two spin states are encoded in *white* and *black*). Starting from a square lattice with some given configuration of spins, two successive applications of the renormalisation group are shown on the right panels. We observe a progressive change of the structure of the lattice with more and more organization (one color, i.e., spin state, dominates more and more). All the shorter range correlations are removed by the renormalisation process and the system becomes more and more ordered, corresponding to an effective increase in the interaction strength K. The system is driven away from criticality by the renormalisation. The renormalisation group thus qualifies this regime as ordered under change of scales

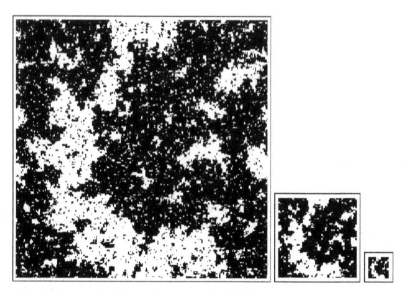

Fig. 11.9. This figure illustrates the effect of renormalisation for $K = K_c$ of the Ising model, which corresponds to the critical point. The two different spin states are encoded again in *white* and *black*. Repeated applications of renormalisation group leave the structure of the lattice invariant statistically. All the shorter range correlations are removed by the renormalisation process, nevertheless the system keeps the same balance between order and disorder and the effective interaction strength remains unchanged and fixed at the critical value K_c. The system is kept fixed at criticality by the renormalisation. The renormalisation group thus qualifies this regime as critical, which is characterized by the symmetry of scale invariance. In other words, the system of clusters of spins is fractal

tant aspect: the distribution and spatial organization of the spins have been modified as shown in Figs. 11.7, 11.8 and 11.9.

Three situations can occur which are illustrated in the Figs. 11.7, 11.8 and 11.9. We use the context of a model of spins with positive interactions tending to align the spins according to a coupling strength K. A large K leads to strong organization where most of the spins have the same state. A small K corresponds to a population which is split in half between the two spin states such that the spatial organization of spins is disorganized. In between, there exists a critical value K_c separating these two extreme regimes at which the system is critical, i.e., scale invariant. The renormalisation group makes these statements precise as shown in the Figs. 11.7, 11.8 and 11.9.

Except for the special critical value K_c, the application of the renormalisation group drives the system away from the critical value. It is possible to use this "flow" in the space of systems to calculate precisely the critical exponents characterizing the divergence of observables when approaching the critical points. Critical exponents play the role of control functions of this flow, i.e., describe the speed of separation from the critical point. The specific calculations can be performed exactly in some special instances, as for hierarchical lattices discussed above in this chapter. In most other cases, one must resort to approximations, whose precision can however in general be systematically controlled and improved [302, 1022].

11.4 A Novel Application to the Construction of Functional Approximants

11.4.1 General Concepts

In the theory of condensed matter, in statistical physics, and particle physics and in many other fields, one usually obtains physical quantities in the form of expansions in powers of some parameters. The expansions assume that these parameters are small. However, as a rule, the physically relevant regimes correspond to values of these parameters which are not small and can even be very large. Many popular expansions are asymptotic and lead to reasonable results only in the low orders and then diverge at higher orders. Moreover, they become wrong in the close vicinity of phase transition or, generally speaking, when critical phenomena of different nature are involved, where fractional indices appear instead of the integer ones, typical of those expansions. When a number of terms in a divergent series is known, one may invoke resumation techniques, such as Pade and Pade–Borel [49] ones. However, the knowledge of only a few first terms does not permit to use these techniques.

Another method for extracting information from the perturbative series is a renormalization-group approach [97, 1022]. Since renormalization group is nothing but a kind of a dynamical system in which time is replaced by scale, the approach can be formulated in the language of dynamical theory [1038], where the (semi-)group property is expressed as a kind of a self-similarity relation for the sought function conserving its form under the change of an appropriately chosen group variable often expressing a coarse-graining operation. In such a dynamical system description, the role of time can be played by momentum in the case of quantum field theory, or by space-scale as in fractals.

In the case of the self-similar approximation theory [344, 345, 1038], the role of time defined in discrete steps in the dynamical system description is played by nothing else but the approximation number. Motion with respect to this discrete time corresponds to transfer from one approximation to another. This makes it possible to define a dynamical system whose trajectory is bijective to the sequence of approximations. Such a dynamical system with discrete time has been called the approximation cascade. Convergence of a sequence of approximations is equivalent to stability of a dynamical trajectory.

The stability and, respectively, convergence are governed by control functions. The fixed point of a trajectory defines the sought function. These ideas result in a practical way of extrapolation also allowing to calculate critical indices for various problems appearing in the context of the critical phenomena, which is presented pedagogically in the next section. Recent works show that this technique is in general far superior to competing methods such as the Padé resummation [346, 347]. See also [21, 342, 344, 345, 348, 654, 1037–1041] for a full exposition and for applications.

11.4.2 Self-Similar Approximants

The self-similar renormalization method discussed here defines effective sums of asymptotic series or effective limits of iterative sequences. It is based on the ideas of optimal control theory, renormalization group theory, and general dynamical theory. The principal concepts of this approach are as follows.

- The first pivotal step is the introduction of control functions whose role is to govern an optimal convergence of approximation sequences. This results in the optimized perturbation theory which has been widely employed for a variety of applications.
- The second idea is to consider the passage from one successive approximation to another as the motion on the manifold of approximants, where the approximant order plays the role of discrete time. The recurrent relations, representing this motion, are formalized by means of group self-similarity.
- A dynamical system in discrete time, whose trajectory is bijective to an approximation sequence, is called the approximation cascade. Embedding the latter into an approximation flow makes it possible to derive differential and integral equations of motion, whose fixed points represent self-similar approximants to the sought function. The stability of the calculational procedure is characterized by local multipliers.
- Another key point of the approach is the introduction of control functions with the help of fractal transforms. This allows one to analyse asymptotic series by transforming them into their fractal counterparts and then invoking all the machinery of the self-similar approximation theory for the transformed series. In this way, the approximants, possessing a nice self-similar structure, were obtained, such as the self-similar exponential approximants and self-similar root approximants.

Assume that we are interested in finding a function $f(x)$ of a real variable $x \in (-\infty, +\infty)$. Without loss of generality, the function $f(x)$ may be considered to be real, since the case of a complex function can be always reduced to that of two real functions. Let perturbation theory give for the function $f(x)$ approximations $p_k(x)$ with $k = 0, 1, 2, \ldots$ enumerating the approximation number. The standard form of $p_k(x)$ is a series in powers, not necessarily integer, of x. The algebraic transform is defined as

$$P_k(x, c) = x^c p_k(x) \, , \tag{11.47}$$

with c real. This transform changes the powers of the series $p_k(x)$, changing by this the convergence properties of the latter. Effectively, the approximation order increases from k to $k + c$ as a result of (11.47). The transform inverse to (11.47) is $p_k(x) = x^{-c} P_k(x, c)$.

To construct an approximation cascade, we define the expansion function $x = x(f, c)$ by the equation $P_0(x, c) = f$, where P_0 is the first available expression from (11.47). Substituting $x(f, c)$ back into (11.47), we get

$$y_k(f, c) \equiv P_k(x(f, c), c) \, . \tag{11.48}$$

The left-hand side of (11.48) represents a point of the approximation-cascade trajectory corresponding to approximation (11.47). The transformation inverse to (11.48) reads $P_k(x, c) = y_k(P_0(x, c), c)$.

Consider the family $\{y_k : k \in \mathbf{Z}_+\}$ as a dynamical system in discrete time. Since the trajectory of this dynamical system is bijective to the approximation sequence $\{P_k\}$, this system is called the approximation cascade. In order to simplify the consideration, let us pass from discrete time to continuous one. To this end, embed the approximation cascade into an approximation flow, which means that the trajectory $\{y(\tau, f, c)\}$ of the flow has to pass, when $\tau = k = 0, 1, 2, ...$, through all points $y(k, f, c) = y_k(f, c)$ $(k = 0, 1, 2, ...)$ of the cascade trajectory. The evolution equation

$$\frac{\partial}{\partial \tau} y(\tau, f, c) = v(y(\tau, f, c), c) \tag{11.49}$$

for the approximation flow, where $v(f, c)$ is the velocity field, can be integrated for an arbitrary time interval, say, from $\tau = k - 1$ to $\tau = k^*$, which gives

$$\int_{y_{k-1}}^{y_k^*} \frac{df}{v(f, c)} = k^* - k + 1 \, ; \tag{11.50}$$

here $y_k = y(k, f, c)$, $y_k^* = y(k^*, f, c)$. The upper limit in (11.50) corresponds to an approximation $P_k^*(x, c) = y(k^*, P_0(x, c), c)$. The moment $\tau = k^*$ is chosen so that to reach the approximation P_k^* by the minimal number of steps. That is, we require that the right-hand side of (11.50) be minimal, $\tau_k \equiv \min(k^* - k + 1)$. Let us note that the differential form (11.49) of the evolution equation, or its integral form (11.50), are equivalent to the functional relation

$$y(\tau + \tau', f, s) = y(\tau, y(\tau', f, s), s) \, . \tag{11.51}$$

In physical applications, the latter is labeled as the functional self-similarity relation, which explains the term we use. The self-similarity, in general, can occur with respect to motion over different parameters. In our case, this is the motion over the steps of a calculational procedure, the number of steps playing the role of effective time.

To find P_k^* explicitly, we need to concretize in (11.50) the velocity field $v(f, c)$. This can be done by the Euler discretization of (11.49) yielding the finite difference

$$v_k(f, c) = y_k(f, c) - y_{k-1}(f, c) \;. \tag{11.52}$$

Thus, using (11.52), the evolution integral (11.50) can be written as

$$\int_{P_{k-1}}^{P_k^*} \frac{\mathrm{d}f}{v_k(f, c)} = \tau_k \;, \tag{11.53}$$

where $P_k = P_k(x, c), P_k^* = P_k^*(x, c)$. When no additional restrictions are imposed, the minimal number of steps for reaching a quasifixed point is, evidently, one, $\min \tau_k^* = 1$. Unless τ_k is introduced explicitly, its value will be set to one automatically. It is worth noting that the evolution equation (11.53) is generally nonlinear and can have several different solutions leading to different self-similar approximations. In such a case, to select a physically meaningful solution, we need to involve additional conditions as constraints. The role of the latter can be played, e.g., by properties of symmetry, by asymptotic properties at $x \to 0$ or $x \to \infty$, or by some other physically important point such as sum rules or other relations containing some known information on the character of the sought solution. Such additional constraints narrow down the set of possible solutions to a class with desired properties.

In this way, the sole quantity that is not yet defined is the parameter c of the transformation (11.47). Recall that the aim of the method is to find an approximate fixed point of the cascade trajectory, a quasi-fixed point, which, by construction, represents the sought function. Therefore, the power c of the transform in (11.47) is to be chosen so as to force the trajectory of the approximation dynamical system to approach an attracting fixed point. Therefore, c is nothing but a kind of control function and can thus be defined by a fixed-point condition. The definition of the fixed point would be to require the velocity to be zero. After the stabilizer c is found from the condition on the fixed point, we substitute it into expression for P_k^* and, using the inverse transformation (11.48), we obtain the self-similar approximation

$$f_k^*(x) = x^{-c_k(x)} P_k^*(x, c_k(x)) \tag{11.54}$$

for the sought function. The procedure of calculating the self-similar approximations (11.54), starting from a perturbative series $p_k(x)$ is now completely defined. The power c one chooses, that is, the effective order we need to take into account, is dictated by the condition on the fixed point, which selects the most stable trajectory of the approximations cascade. In particular, it may happen that $c = 0$, and one does not need to proceed further, or, vice versa, one may have to go to the limit of $c \to \infty$, thus having to take into account all approximation orders. In each concrete case, the effective order one needs to reach depends on how good is the perturbative sequence $\{p_k(x)\}$ one start with and also on how much information can be extracted from its first terms by means of the self-similar renormalization.

As an illustration of the the self-similar renormalization method, consider explicitly a perturbative series

$$p_k(x) = \sum_{n=0}^{k} a_n x^n, \quad a_0 \neq 0 , \tag{11.55}$$

containing integer powers of x, although, as is mentioned above, the procedure works for arbitrary noninteger powers. This form is the most frequently met in physical applications. Let us write the algebraic transform $P_k(x,c) = \sum_{n=0}^{k} a_n x^{n+c}$ of (11.55). As is seen, this transform corresponds to an effectively higher perturbation order, $k + c$, as compared with the initial series (11.55) of order k. The equation for the expansion function $x(f,c)$ now reads $P_0(x,c) = a_0 x^c = f$, from where $x(f,c) = (f/a_0)^{1/c}$. Repeating now all steps described above, after some straightforward calculations, the initial perturbative expansion (11.55) is transformed into the self-similar approximation

$$f_k^*(x) = p_{k-1}(x) \left[1 - \frac{k a_k}{c a_0^{1+k/c}} x^k p_{k-1}^{k/c}(x) \right]^{-c/k} . \tag{11.56}$$

The stabilizer $c(x)$ is determined by the minimal difference condition

$$\Delta_k(x,c) = f_k^*(x,c) - f_{k-1}^*(x,c) , \tag{11.57}$$

or the minimal sensitivity condition: $(\partial/\partial c)f_k^*(x,c) = 0$. Asymptotically, as $x \to 0$, the behavior of $p_k(x)$ in (11.55) and $f_k^*(x)$ coincides up to the linear terms, while the higher order terms are renormalized.

There are several variants and extensions of the self-similar renormalization method. In particular the method of "self-similar factor approximants" is based on the self-similar approximation theory, with an additional trick consisting in transforming, first, a series expansion into a product expansion and in applying the self-similar renormalization to the latter rather to the former [347, 1040]. This results in self-similar factor approximants extrapolating the sought functions from the region of asymptotically small variables to their whole domains with improved efficiency. This provides a general methodology for constructing crossover formulas and for interpolating between small and large values of variables.

11.5 Towards a Hierarchical View of the World

Stretching the concept, the renormalization group can be thought of as a construction scheme or "bottom-up" approach to the understanding and even to the design of large-scale hierarchical structures. At a qualitative level, it can be thought of as an embodiement of the famous message of P.W. Anderson: "More is different" [26], of which we reproduce the main message.

The reductionist hypothesis does not by any means imply a "constructionist" one: the ability to reduce everything to simple fundamental laws does not imply the ability to start from those laws and reconstruct the universe. In fact, the more the elementary particle physicists tell us about the nature of the fundamental laws, the less relevance they seem to have to the very real problems of the rest of science, much less to those of society. The constructionist hypothesis breaks down when confronted with the two difficulties of scale and complexity. The behavior of large and complex aggregates of elementary particles, it turns out, is not to be understood in terms of a simple extrapolation of the properties of a few particles. Instead, at each level of complexity entirely new properties appear and the understanding of the new behaviors requires research which I think is as fundamental in its nature as any other. That is, it seems to me that one may array the sciences roughly linearly in a hierarchy, according to the idea: the elementary entities of science X obey the laws of science Y.

X	Y
solid state	elementary particle
many-body physics	physics
chemistry	many-body physics
molecular biology	chemistry
cell biology	molecular biology
.	.
.	.
.	.
psychology	physiology
social sciences	psychology

But this hierarchy does not imply that science X is "just applied Y." At each stage, entirely new laws, concepts, and generalizations are necessary, requiring inspiration and creativity to just as great a degree as in the previous one. Psychology is not applied biology, nor is biology applied chemistry. In my own field of many-body physics, we are, perhaps, closer to our fundamental, intensive underpinnings than in any other science in which non-trivial complexities occur, and as a result we have begun to formulate a general theory of just how this shift from quantitative to qualitative differentiation takes place. This formulation, called the theory of "broken symmetry", may be of help in making more generally clear the breakdown of the constructionist converse of reductionism.

12. The Percolation Model

Percolation [928] is a very simple but powerful model of heterogeneous media. It is more than half a century old, and thousands of papers as well as several books have been written about it. Most of the recent progress has occurred in applications, particularly in geophysics and transport of fluids in porous media [803]. The basic foundations are well established but some key results are still emerging with some surprises [927], notably in the number of percolating clusters and their description by simple renormalization group techniques.

12.1 Percolation as a Model of Cracking

To motivate the percolation model, let us consider a piece of rock which is progressively altered as a consequence of the application of stress in the presence of water (corrosion) and possibly other processes. A qualitative physical picture for the progressive damage of a system leading to global failure is as follows: at first, single isolated microcracks appear, and then with the increase of load or time of loading they both grow and multiply leading to an increase of the density of cracks per unit volume. As a consequence, microcracks begin to merge until a "critical density" of cracks is reached at which the main fracture is formed.

The basic idea is that the formation of microfractures prior to a major failure plays a crucial role in the fracture mechanism. The simplest statistical model containing these facts is the percolation model. The percolation model is a purely static statistical model which assumes that microcracks are uniformly and independly distributed in the system, for all crack concentrations. Each concentration of microcracks can then be characterized by the distribution of microcrack cluster sizes and shapes for which many exact results are known [225, 375, 928].

The connection between rupture and the percolation model is made by assuming that as the load or time increases, the density of cracks also increases monotonically. However, in contrast to more elaborate theories of rupture some of which are described in Chap. 13, it is assumed that new cracks appear in a random and uncorrelated manner in the system, so as to always obey the rules of the percolation model.

The percolation model exhibits a critical density at which point an infinite cluster of microcracks appears which disconnects the system into at least two pieces. This critical percolation point is identified with the global rupture point.

The term "infinite" cluster refers here to the ideal situation where the physical system is taken of infinite size, a situation often refered to as the "thermodynamic limit". In practice, systems are of finite size and the "infinite" cluster acquires a size limited by that of the system.

This simple percolation model for rupture can be shown to be a correct description of material breakdown in the limit of "infinite disorder", for which the effect of stress enhancement at the largest crack tips remains small compared to the heterogeneity of the medium. In general however, the percolation model is bound to become incorrect close enough to the global rupture when stresses at the tip of the cracks ($\sim \sqrt{L}$ where L is the length of a given crack) become larger than the typical scale of the strength of heterogeneities of the system. Nothwithstanding this fact, the percolation model of rupture has been at the basis of a significant theoretical effort to model material rupture in general and earthquakes in particular.

The properties of percolation clusters are the following.

• Infinite percolation clusters have in general a complex form with bends and hook-like configurations. As a consequence, global rupture requires a higher density of cracks such that the infinite clusters are free of "locks" preventing uncoupling. For instance in tension, the relevant percolation threshold is that of so-called "directed percolation" (see [865] and [339] for a discussion of this aspect as well as the application of percolation theory to cracks put at random in a continuum as opposed to the discrete lattice models generally used).

• As a consequence of the critical nature of the percolation point, one can represent experimental data on acoustic emission, dilatancy and other possible precursors of rupture and of earthquakes using the power laws of percolation theory. These power laws are special cases of power laws more generally associated with critical phenomena. They can be used for prediction purposes; see for instance [14] for a general discussion and [325, 861, 904] where this idea has been explicitly used for the case of telluric voltage precursors [980, 981], using a percolation model of heterogeneous piezo-electricity.

These power laws, which describe the behavior of various physical quantities when the critical percolation point is approached, are based on the fact that the correlation length ξ exhibits a power law behavior

$$\xi \sim |p - p_{\mathrm{c}}|^{-\nu} , \qquad (12.1)$$

(where ν is a critical exponent) and diverges at $p = p_{\mathrm{c}}$. The correlation length roughly represents the typical size of the largest clusters in the system [928] as shown in Fig. 12.1. As it becomes larger and larger as $p \to p_{\mathrm{c}}$, cooperative

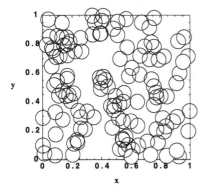

Fig. 12.1. Continuous percolation: disks are dropped at random on the square such that the coordinates x and y of their centers are uniformely distributed between 0 and 1. This picture corresponds to a concentration of disks just below the percolation threshold. Notice the existence of large clusters of size comparable to but smaller than the system size. For continuous percolation of disks, the product of the concentration by the square of their radius defines a percolation threshold which is independent of the disk radius and corresponds to a well-defined disk coverage fraction

effects appear of pure geometrical origin. Notwithstanding the fact that these correlations are purely geometrical and of statistical origin, the electrical or mechanical behavior of a system close to the percolation point exhibits large responses and a high susceptibility to perturbations. For example, the elastic energy stored in the system at constant stress diverges as $|p - p_c|^{-\tau}$, with $\tau \simeq 3.5$ in three dimensions, as $p \to p_c$. This kind of behavior characterizes essentially any physical quantity whose response depends on the connectivity and the geometry of the large clusters.

A growing body of work has shown that the percolation model is suitable as a first approximation as soon as the property of connectivity is the dominating mechanism. Many domains have been explored which include flows in porous media with many geophysical applications, microemulsion structures and mixed reverse micelles, phases transitions in high-energy nuclear collisions, structure and stability of foam, spin-glasses, biological population dynamics, conductivity of YBaCuO ceramics (used in high-T_c supraconductors), transport properties of metal-dielectric composites, transport in doped polymers, photoluminescence of porous silicon, catalytic solid-phase reactions, theory of liquid water, of city growth and of stock market price fluctuations, oil recovery, etching of random solids, coarsening in epitaxial thin film growth, modeling of geosphere–biosphere interactions and of habitat fragmentation, solid state ionics, the human vasculature system, earthquake nucleation, rupture in random media, permeability of partially molten upper mantle rock, diffusion and viscosity in silicate liquids, iceberg calving, model of magnetopause current layer in asymmetric magnetic fields and many others.

12.2 Effective Medium Theory and Percolation

When facing a situation where some heterogeneity is present, the natural tendency is to try to average out the disorder and replace the disordered medium with an effective homogeneous medium with similar "averaged" properties. This is the strategy followed by methods known as "effective medium" theories, "homogeneous" approaches and "mean field" approximations [262, 868]. At the core of these approaches is the idea that a heterogeneous medium can be seen as an equivalent homogeneous medium, i.e. its properties are the same as those of a homogeneous medium. This strategy is valid for small disorder but is bound to fail at large disorder, for which the percolation model becomes a good paradigm.

Let us illustrate this by considering the electric conductance of a random network of conductances. Consider the simple case of a cubic lattice of electric bonds. Each bond of the lattice carries a conductance g, taken at random from a pdf $P(g)$. For instance, we will use

$$P(g) = p\delta(g - 1) + (1 - p)\delta(g) , \tag{12.2}$$

corresponding to a mixture of conducting bonds with unit conductance with concentration p and of isolating (or ruptured) links with concentration $1 - p$. The effective medium theory (also called more generally CPT for "coherent potential theory" in the context of wave scattering [262, 868]), amounts to replacing this system by a uniform network of conductances all equal to g_m, where g_m is chosen so that the global properties of the random system is best mimicked. The strategy of the effective medium approach is to introduce a single defect g on a bond of the effective medium and determine g_m such that the average of the perturbations brought by the single defect vanishes.

To solve the problem, we will use Thevenin's theorem of electrokinetics: the effect of the network as seen by the added conductance on a bond AB is the same as an effective conductance G'_{AB}. This effective conductance G'_{AB} can be calculated by inserting a current i_0 at A (which flows away to infinity

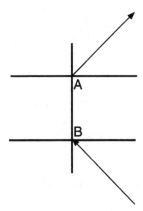

Fig. 12.2. Use of Thevenin's theorem to calculate the equivalent conductance seen by the bond AB of an infinite lattice of identical conductances surrounding the bond AB

in the plane) and by extracting a current out at B (coming from infinity) along the bond on which we put a conductance g_m, as shown in Fig. 12.2. The superposition of these two currents cancel out at infinity and lead to a closed electric circuit. Due to the square symmetry, the current inserted at A gives a contribution equal to $i_0/4$ in the bond AB. Another contribution $i_0/4$ comes from the current extracted at B, leading to a total current $i_0/2$ in the bond AB. The potential drop V_{AB} between A and B is thus

$$V_{AB} = \frac{i_0}{2} g_m^{-1} \equiv \frac{i_0}{G_{AB}} \,, \tag{12.3}$$

where G_{AB} is the total conductance seen between A and B. This gives $G_{AB} = 2g_m$ and the conductance G'_{AB} of the total lattice, excluding the bond AB is thus

$$G'_{AB} = G_{AB} - g_m = g_m \,. \tag{12.4}$$

We can now put a conductance g between A and B. Using the equivalent circuit of two conductance G'_{AB} and g associated in parallel and fed by a current i_0 entering in A and going out from B, we obtain

$$V_{AB} = \frac{i_1}{g} = \frac{i_0 - i_1}{G'_{AB}} \,, \tag{12.5}$$

where i_1 is the current flowing in the conductance g. This yields

$$i_1(g) = \frac{g}{g + G'_{AB}} i_0 \,. \tag{12.6}$$

The current perturbation brought by replacing g_m by g between A and B is thus

$$i_1(g) - i_1(g_m) = \frac{1}{2} \frac{g_m - g}{g_m + g} \,. \tag{12.7}$$

The effective medium conductance g_m is then determined by the condition that this perturbation is on average zero, i.e.

$$\int dg \, P(g) \frac{g_m - g}{g_m + g} = 0 \,. \tag{12.8}$$

For $P(g)$ given by (12.2), this gives

$$g_m = 2 \left(p - \frac{1}{2} \right) \,. \tag{12.9}$$

This calculation correctly predicts the existence of a "percolation" threshold $p_c = 1/2$ at which the conductance of the total network vanishes due to an absence of connectivity. This non-trivial prediction turns out to be correct quantitatively for the square bond lattice. In general, this is not the case for other topologies: only the existence of the percolation threshold is correctly predicted but its quantitative value is not often accurate. The linear dependence of the effective medium conductance as a function of $p - p_c$ is incorrect sufficiently close to p_c: it has been established that the global conductance of the system goes to zero like $(p - p_c)^t$, with $t \approx 1.300$ in two dimensions and $t \approx 2.5$ in three dimensions [688]. The effective medium theory is unable to capture this non-analytic power law behavior, which reflects a critical behavior in which holes of all sizes are present as shown in Fig. 12.1 and which modify the exponent t from its mean field value 1 to its true value larger than one. The effective medium approach, being a "one-body" method, is unable to capture these phenomena associated with a collective connectivity effect. The problem is more severe for other lattices such as the triangular lattice, for which the effective medium approach is even wrong in its prediction of p_c. We refer to [508] for a discussion and extension to other lattices in three dimensions.

There is a huge literature on effective medium theories and homogeneization approaches. For weak heterogeneities, corresponding to dilute inclusions with a weak constrast of properties between the matrix and the inclusions, these theories are in general quite precise and validated by many experiments. The problems appear when the concentration increases and the contrast of properties is strong. In this case, clusters of inclusions appear and the "one-particle" approximation explicit in these methods become more and more problematic. In this regime where percolation becomes relevant, it is better to think of the properties of the heterogeneous system as resulting from the contribution of many clusters of particles of various sizes and shapes. As we will see, the theory of percolation makes precise statements on the statistics of clusters as a function of the concentration of inclusions. This allows one to develop quantitative methods to calculate the properties of such systems [782–784]. We note that effective medium theories are, by construction, unable to take into account the possible existence of multiple modes of deformation or propagation. Again, addressing the physical mechanisms at the scale of the clusters allows one to make significant progress. These cluster methods can be applied to the description of electrical, mechanical, piezoelectrical and rupture properties of complex heterogeneous systems.

12.3 Renormalization Group Approach to Percolation and Generalizations

Recall that the renormalization group (RG) method discussed in Chaps. 2 and 11, amounts basically to decomposing the general problem of finding the behavior of a large number of interacting elements into a succession of simpler problems with a smaller number of elements, possessing effective properties varying with the scale of observation. It is based on the existence of a scale invariance or self-similarity of the underlying physics at the critical point.

In the real-space version of RG which is the best suited for percolation, one translates literally, in real space, the concept that rupture at some scale results from the aggregate response of an ensemble of ruptures at a smaller scale. This idea is particularly clear in faulting for instance, where it is well-known that large faults in the crust actually consist of anastomosed faults (i.e. smaller interacting structures) with varying local arrangements.

Real-space RG is a particular implementation of the technique which usually requires the discretization of space into cells. The real-space RG consists of replacing a cell of sites or bonds by a single super-site, provided that the linear dimension l of the cell is much smaller than the correlation length ξ. Of course one looses information, but since scaling relies on the fact that all cells of size l are similar to each other, one should obtain the relevant information for describing the critical point.

Quantitatively, one must specify the rules governing how this renormalization of cells to sites is to be performed. Similar to the renormalization of the coupling factor K in the Potts model studied in Chap. 11, the concentration p' of occupied supersites will in general be different from the concentration p of the original sites. Only right at the critical point do we have $p' = p = p_c$, since the correlation length is infinite and the system is perfectly self-similar implying that the geometry looks statistically the same after any scale reduction.

12.3.1 Cell-to-Site Transformation

One should stress that, although intuitively clear and appealing, there is no unique prescription for renormalizing in real space and a considerable number of approximate renormalization techniques have been developed, with varying degrees of success.

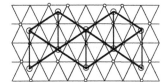

Fig. 12.3. A triangular lattice of nodes represented by *small open circles* is renormalized into another triangular lattice of sites represented by *large open circles*, by grouping the three nodes of the same triangle into a single super-site. This construction is such that each site of the initial triangle lattice belongs to one and only one super-site. The lengths of the bonds of the renormalized triangular lattice are equal to $\sqrt{3}$ times those of the initial lattice

Maybe the simplest model is to consider a triangular lattice with a fraction p of its nodes occupied, defining the "site-percolation" model. Then, a successful real space RG recipe is to replace each triangle with its three nodes by a unique supersite at its center as shown in Fig. 12.3. We now ask for the probability p' of such a supersite to be occupied, knowing the probability p for each of the triangle nodes to be occupied. The criterion that we use is that the supersite is occupied if a spanning cluster exists. For a triangle, this is the case if either all three sites are occupied (probability p^3), or if two neighbouring sites are occupied (probability $3p^2(1-p)$, the factor 3 stems from the three possible combinaisons). One obtains [772–774]:

$$p' = M(p) \equiv p^3 + 3p^2(1-p) \ . \tag{12.10}$$

The mapping $p' = M(p)$ is shown in Fig. 12.4. Its fixed points (corresponding to intersections with the diagonal) are $0, 1/2$ and 1. Close to a fixed point, one can linearize the mapping and get

$$p' - p_c \simeq \lambda(p - p_c) \ , \tag{12.11}$$

where $\lambda = \mathrm{d}p'/\mathrm{d}p = 3/2$ at $p = p_c = 1/2$. The fixed point $1/2$ is thus unstable, while the two other fixed points 0 and 1 are stable as can be seen in Fig. 12.4. They correspond respectively to the empty and occupied phases. Starting from some $p < 1/2$, the renormalized p flows towards 0 meaning that, at larger and larger scales, the system is more and more similar to an empty lattice. Inversely, starting from some $p > 1/2$, the renormalized p flows towards 1 meaning that, at larger and larger scales, the system is more and more similar to a completely occupied lattice. These two stable fixed points thus represent stable phases while the unstable fixed point describes the transition between these two stable phases. Note that $p_c = 1/2$ is the exact value for the two-dimensional triangular lattice.

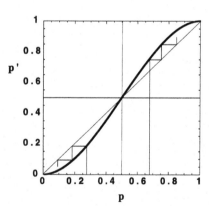

Fig. 12.4. Renormalization group mapping represented by the *thick line* giving the renormalized occupancy parameter p' of the renormalized triangular lattice as a function of the occupancy parameter p of the initial lattice. The *thin lines*, bouncing between the *thick line*, representing the mapping, and the diagonal, corresponding to trajectories obtained by iterating the decimation procedure, exhibit the attractive property of the two fixed points 0 and 1, and the repulsive nature of the fixed point $1/2$

The basic equation of the RG is

$$l|p' - p_c|^{-\nu} = |p - p_c|^{-\nu} \ , \tag{12.12}$$

which simply states that the correlation length ξ should remain physically unaltered after one operation of the RG. If the bonds of the initial triangular lattice are of length one, we call $l = \sqrt{3}$ the bond length of the renormalized triangular lattice shown in thick lines in Fig. 12.3. Taking the logarithm of this RG equation yields

$$\frac{1}{\nu} = \frac{\ln(|p' - p_c|/|p - p_c|)}{\ln l} = \frac{\ln \lambda}{\ln l} . \tag{12.13}$$

Using the values of λ and l, we find

$$\nu = \frac{3^{1/2}}{\ln(3/2)} = 1.355 , \tag{12.14}$$

which is very close to the exact value $\nu = 4/3$ in 2D. One can proceed similarly and obtain the critical behavior of other quantities.

Such "real space" RGs have been reinvented in the geophysical community and have subsequently become popular. The percolation model and the real space RG has been applied to a three dimensional cubic lattice [14]. In this reinterpretation of the percolation model as a rupture model, p becomes the probability for an elementary domain to be fragile as a consequence of the increasing number of small cracks in response to an increasing applied stress. A non-trivial (i.e. different from 0 or 1) fixed point is found with essentially the same properties as shown in Fig. 12.4. The unstability of the mapping $p' = M(p)$ at this fixed point illustrates the notion of the rupture instability: if $p > p_c$, $p' > p$ and by successive iteration of the RG (i.e. by going to larger and larger scales), p converges to 1, describing the global rupture at large scales. On the other hand, if $p < p_c$, $p' < p$ and by successive iteration of the RG, p converges to 0, corresponding to the undeteriorated phase: the RG sees the partially damaged system, which however will not break, as fundamentally undamaged at the macroscopic scale when the details are washed out.

12.3.2 A Word of Caution
on Real Space Renormalization Group Techniques

The probability $R(p)$, that a lattice with L^d sites in d dimensions percolates in the sense of having at least one cluster spanning from top to bottom, is such that for large enough lattices $R(p < p_c) = 0$, $R(p > p_c) = 1$. Finite-size scaling theory predicts and computer simulation confirms over more than two decades [928] that, in a small transition region

$$\Delta p \propto L^{-1/\nu} , \tag{12.15}$$

the spanning probability moves from close to zero to close to one. One may define a size-dependent $p_c(L)$ by the condition that the spanning probability be 50%: $R[p_c(L)] = 1/2$. As used above, small-cell renormalization theory [772–774, 928] traditionally determines the critical point as a fixed point: $R(p_c) = p_c$. If we insert the bulk threshold value p_c, does this equation become valid for large enough cells?

It does not. Ziff [1061] showed numerically that in a special square lattice case $R(p_c)$ is $1/2$ and different from $p_c = 0.592746$. More generally, it differs from p_c (and also from $1/2$) in three to six dimensions and becomes unity in seven (see [927] and references therein). Does this mean that all the numerical works based on the now wrong idea $R(p_c) = p_c$ were wrong? It turns out that there is not a single estimate which needs to be revised. The reason is that $R(p)$ for $L \to \infty$ jumps from zero to one. Thus, any constant C between zero and one can be used to define a size dependent $p_c(L)$ through $R[p = p_c(L)] = C$; for large enough lattices this $p_c(L)$ converges to the proper bulk threshold. The validity of the "wrong" standard renormalization picture [772–774, 928] can be restored by just replacing cell-to-site renormalization (from L to 1) by cell-to-cell renormalization (from L to $L/2$): then, $R(p_c)$ for lattice size L agrees with $R(p_c)$ for lattice size $L/2$, provided L is large enough [432].

Another recent result is also of interest in light of the central limit theorem discussed in Chap. 2 and the effects of long-range correlations discussed in Chap. 8. Until quite recently, it was still assumed that the derivative $dR(p)/dp$ approaches a Gaussian for large lattices. This derivative gives the probability that a lattice starts to percolate at a concentration p. For a fixed sequence of random numbers, there is a well-defined onset of percolation when we increase p, and it is plausible that this onset follows a normal distribution for large systems. The width σ, defined as the standard deviation of this distribution of the L-dependent thresholds,

$$\sigma^2 = \langle p_c^2 \rangle_L - \langle p_c \rangle_L^2 , \tag{12.16}$$

varies as $L^{-1/\nu}$ and is proportional to the shift $\langle p_c \rangle_L - \langle p_c \rangle_\infty$. This is a convenient numerical tool to determine the bulk p_c by extrapolation without assuming any critical exponent ν [977]. Again, the (unprecisely defined) convergence to a Gaussian turned out to be poor when tested numerically: the distribution has a finite skewness (third-order cumulant) $\langle (p_c - \langle p_c \rangle)^3 \rangle / \langle (p_c - \langle p_c \rangle)^2 \rangle^{3/2}$ which shows no intention in vanishing for large lattices (see [927] and references therein). Mathematical theorems now establish that Gaussians are not correct. However, the determination of thresholds and exponents ν by the Gaussian assumption did not give incorrect results since $\sigma \propto L^{-1/\nu}$ remains valid and only the proportionality factor is influenced by the deviations from a Gaussian.

The reason why the distribution of thresholds in a finite system does not follow the usual central limit theorem and its Gaussian distribution is that the property of spanning is a collaborative effort of the whole lattice and not the sum of more or less independent contributions from smaller parts of the whole lattice. Therefore, the number of sites in the just-spanning cluster is not a self-averaging quantity with relative fluctuations vanishing for $L \to \infty$.

12.3.3 The Percolation Model
on the Hierarchical Diamond Lattice

We come back to the spin model on the diamond hierarchical lattice studied in Chap. 11. Observe that, although the spin model we originally defined on the diamond lattice makes sense for Q integer only (where Q is the number of states that any spin can take), all equations involve Q as a parameter, and therefore make sense for Q real as well. The value $Q = 1$, or more precisely, the limit $Q \to 1$, is of particular interest: it has been shown long ago [1028] that this limit corresponds to the *bond percolation problem*, where the probability of a bond to be occupied is

$$p = 1 - \frac{1}{K} \, . \tag{12.17}$$

Rather than recall the proof of this statement, which is rather long, let us focus on the renormalization equation (11.21). In the case $Q = 1$, it reads, after taking the inverse of both sides and expanding the powers,

$$1 - \frac{1}{K'} = 1 - \frac{4}{K^2} + \frac{4}{K^3} - \frac{1}{K^4} \, . \tag{12.18}$$

Fig. 12.5. The set of bond occupancies in an elementary diamond cell for which the global connectivity from *top* to *bottom* is ensured

Now observe that a configuration where a bond is occupied at some magnification corresponds at the next magnification to three possible situations depicted in Fig. 12.5. Their respective probabilities are

$$\text{lower left:} \quad p_1 = p^4 = \left(1 - \frac{1}{K}\right)^4 \tag{12.19}$$

$$\text{four upper cases:} \quad p_2 = (1 - p)p^3 = \frac{1}{K}\left(1 - \frac{1}{K}\right)^3 \tag{12.20}$$

$$\text{two cases at the lower right:} \quad p_3 = (1 - p)^2 p^2$$

$$= \frac{1}{K^2}\left(1 - \frac{1}{K}\right)^2 \tag{12.21}$$

and they have multiplicities $1, 4, 2$. By direct calculation, we obtain

$$p' = 1 - \frac{1}{K'} = p_1 + 4p_2 + 2p_3 , \tag{12.22}$$

thus demonstrating that the general renormalization equation makes sense in terms of percolation once the probability has been identified through (12.17).

It is easy to bridge our quite general discussion of the renormalization group method presented in Chap. 11 with the problem of percolation [14]. The present problem maps exactly onto percolation where an occupied bond occurs with probability p. The critical point corresponds to the appearance of an infinite cluster of occupied bonds. It occurs at

$$p_c = 1 - \frac{1}{K_c(Q = 1)} \approx 0.618 . \tag{12.23}$$

A non-trivial (i.e. different from 0 or 1) fixed point is found with essentially the same properties as discussed above and also in Chap. 11. The instability of the mapping $p' = \phi(p)$ at this fixed point illustrates the notion of the percolation instability. If $p > p_c$, $p' > p$ and by successive iteration of the renormalization group (i.e. at larger and larger scales), p converges to 1, describing the percolating (dense) phase. If $p < p_c$, $p' < p$ and by successive iteration of the renormalization group, p converges to 0, corresponding to the non-percolating phase. The renormalization group sees the partially non-percolating system, which is not connected at large scale, as fundamentally empty at the macroscopic scale. Since $Q = 1$, the free energy f defined in the context of the spin model of Chap. 11 is always zero. However, the derivative of f with respect to Q at $q = 1$ is non trivial: it corresponds to the average size of clusters [1028]. Since $f = 0$, one sees, by taking the derivative of (11.24) with respect to Q, that the derivative satisfies the same type of renormalization group equation but with a different function g. Hence all the properties given in Chap. 11 apply, and in particular the power law properties. Notice that, due to the discrete scale invariant structure of the hierarchical diamond lattice, the average size of clusters as well as the distribution of clusters possess log-periodic correction to scaling.

12.4 Directed Percolation

12.4.1 Definitions

Geometrical Definition. Among all critical phenomena, directed percolation has been associated with an extraordinarilly large variety of phenomena. Directed percolation is defined exactly as percolation discussed above except that an additional condition is added, namely that the connecting paths must be directed and must follow the direction of a prefered chosen direction as shown in Fig. 12.6. In other words, in directed bond percolation, each bond

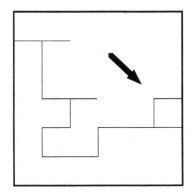

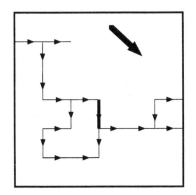

Fig. 12.6. *Top*: a square lattice on which a percolating cluster connects the *left side* to the *right side*. Note that the cluster is not directed, i.e. a walker following the backbone (obtained by removing the dangling branches) of the cluster will have to walk one step in the direction opposite to the arrow indicating the prefered direction. This case correponds to a system above the (usual) percolation threshold but below the "directed" percolation threshold. *Bottom*: by adding one bond (shown in *thick line*) to the figure at the top, the percolating cluster now possesses a directed backbone. This case thus correponds to a system above the "directed" percolation threshold

carries an arrow and directed percolation corresponds to the connectivity between bonds mediated by oriented bonds such that one can walk accross a connected directed percolation cluster by always following the oriented arrows.

As shown in Fig. 12.7, below a certain threshold p_c, all clusters are finite. Above p_c, some of the clusters may become infinite. Just at p_c, there are only finite clusters but the probability to generate a cluster of s sites decreases slowly as a power law $s^{-\tau}$ for large s.

Contact Processes. Another alternative definition is through the so-called "contact processes" [565] defined as evolutionary rules on a population of particles: each site of the d-dimensional cubic lattice, $\mathbf{Z}^d$, is either vacant or is occupied by a particle. The transition rules are easily stated: a vacant site with n occupied nearest neighbors becomes occupied at the rate $\lambda n/2d$, while particles disappear at unit rate, independent of their surroundings. Evidently, the vacuum is absorbing, in the sense that the state with zero particles remains with zero particles. The active phase, characterized by a nonzero stationary particle density, exists only for sufficiently large creation rate λ (and, strictly speaking, only in the infinite-volume limit). There is a continuous transition from the vacuum to the active phase at

position i

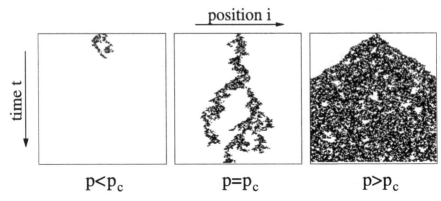

$$p<p_c \qquad p=p_c \qquad p>p_c$$

Fig. 12.7. Typical directed percolation cluster starting from a single seed below, at, and above the critical point p_c. The preferred direction can be interpreted as time. Reproduced from [426] with permission

a critical value λ_c, corresponding to the "directed" percolation threshold. The intuition for this correspondence is that the spatial dimension along the prefered direction in directed percolation can be interpreted as the time axis. Then, the space representation of the clusters as in Fig. 12.6 in dimension D corresponds to the $(d+1)$ space–time representation of the evolution of the particle population. Thus, the d-dimensional contact process corresponds to the geometrical definition of directed percolation in $d+1$ dimensions. In one space dimension (two "space-time" dimensions), $\lambda_c \simeq 3.2978$ for a square lattice. This transition is said to belong to the universality class of directed percolation, because the same exponents describe the two processes.

12.4.2 Universality Class

The universality class covered by directed percolation is very broad and contains many seemingly unrelated problems. This stems from the fact that there are many interpretations to the preferred direction of the directed percolation model:

- The prefered direction is a spatial direction. This was proposed to apply to material and charge transport in disordered media under the influence of external forces. It can also model the propagation of epidemics and forest fires under some directional bias, e.g. strong wind.
- The preferred direction is time. Here, the primary interpretation could be an epidemic without immunization, the so-called "contact process" [565] or the "simple epidemic" [640].
- A very early application (even if it took rather long until it was understood as such [136, 371]) was to "reggeon field theory", a theory for ultrarelativistic particle collisions popular in the seventies [652]. Here, the preferred

direction is that of "rapidity", while the directions transverse to it are provided by the impact parameter plane. This connection is interesting since it was through it that the first precise estimates of critical exponents and amplitudes were obtained for directed percolation [652].

- Another realization of the directed percolation transition occurs in simple models of heterogeneous catalysis [11, 70, 464, 470, 1060, 1062].
- Surface reaction models provide another application [377, 472, 1062].
- Branching and annihilating random walks with odd parity have also been shown to belong to the directed percolation class [465, 466, 943].
- Assorted multiparticle processes is another example [39, 40, 231, 233, 723].
- The transition from laminar to turbulence in boundary layer flows by interactions between local subcritical bifurcations has also been argued to be a critical phenomenon in the universality class of directed percolation [159, 160, 744]. T. Bohr et al. [98] have shown that the transition from laminar to active (spatio-temporal intermittent) behavior in chaotic systems can in fact vary from a continuous transition in the universality class of Directed Percolation with infinitely many absorbing states to what appears as a first order transition due to finite lifetime "solitons."
- The parallel (or annealed) version of the Bak–Sneppen model for extremal evolution dynamics can be mapped exactly onto a site directed percolation critical point [42, 372, 894], as will be shown in Chap. 15.

There is ample evidence for the suggestion [364, 459] that the universality class of directed percolation contains all continuous transitions from a "dead" or "absorbing" state to an "active" one with a single scalar order parameter, provided the dead state is not degenerate and that some technical points are fulfilled: short range interactions both in space and time, nonvanishing probability for any active state to die locally, translational invariance [absence of "frozen" randomness], and absence of multicritical points.

If the dead state is degenerate, for example with twofold degeneracy with conservation laws which prevent some active states from dying, it is clear that any transition – if it occurs at all – has to be in a different universality class [366, 370, 943]. More generally, it seems that models can be generically in the directed percolation class even if they have an absorbing state with positive entropy. The physical argument is that such a state is essentially unique on a coarse scale, provided its evolution is ergodic and mixing and provided it does not involve long range correlations (long correlations should be entirely due to patches of "active" states). Since only coarse-grained properties should influence critical behavior, this would suggest that such transitions are in the directed percolation class. Recent simulations support this conjecture [467–469, 471, 618]. Despite its theoretical success, no fully demonstrative experiment or natural system is known to clearly reproduce the critical exponents of directed percolation [426], probably due to the fact that, in Nature, an absorbing state will often exhibit some residual fluctuation and will not be fully absorbing.

A typical contact process model is an interacting particle system characterized by rules for elementary processes such as creation, annihilation, and diffusion. Looking only at the rules, there is little to tell us what sort of critical behavior to expect, nor why it is universal. The understanding of universality emerges instead from the study of coarse-grained formulations which capture the large-scale features essential to critical behavior. In such field theories, the microscopic picture of particles on a lattice is replaced by a set of densities which evolve via stochastic partial differential equations (SPDEs). At this level, renormalization group methods may be applied. A basis for universality appears if one can show that the continuum descriptions for various models differ only by irrelevant terms. At present, there are many models known that have directed percolation critical behavior (as far as numerical simulations can tell) but only a few have been confirmed rigorously to belong to the universality class of directed percolation using field theory. Useful continuum descriptions of multiparticle processes, for example, have yet to be devised.

12.4.3 Field Theory: Stochastic Partial Differential Equation with Multiplicative Noise

Janssen [459] proposed a continuum description of the contact processes and corresponding directed percolation models which reads as follows:

$$\frac{\partial \rho(\mathbf{x}, t)}{\partial t} = a\rho(\mathbf{x}, t) - b\rho^2 - c\rho^3 + \cdots + D\nabla^2\rho + \eta(\mathbf{x}, t) , \qquad (12.24)$$

where $\rho(\mathbf{x}, t) \geq 0$ is the coarse-grained particle density. The ellipsis represents terms of higher order in ρ. $\eta(\mathbf{x}, t)$ is a Gaussian noise which respects the absorbing state ($\rho = 0$), i.e. vanishes with ρ,

$$\overline{\eta(\mathbf{x}, t)\eta(\mathbf{x}', t')} \propto \rho(\mathbf{x}, t)\delta(\mathbf{x} - \mathbf{x}')\delta(t - t') , \qquad (12.25)$$

and is delta-correlated in time and space.

This form can be justified by coarse graining the contact process, in the limit of large bin size. Let n_i be the number of particles in bin i, and Δn_i the change in this number during a brief interval. The latter has expectation $\overline{\Delta n_i} \propto a\, n_i + \mathcal{O}(n_i^2)$, (with $a \propto \lambda - 1$), and under the customary assumption of Poissonian statistics for reaction systems, its variance equals $\overline{\Delta n_i}$. For sufficiently large bins, we may approximate the distribution of Δn_i by a Gaussian law. Thus, since reactions in different bins are uncorrelated, coarse-graining the original model leads to a stochastic field theory with Gaussian noise whose autocorrelation is proportional to the local density. There is also noise due to the fluctuating diffusive current. But diffusive noise does not affect the critical behavior in the present case, so we shall ignore it in the interest of simplicity. Since (12.24) involves multiplicative noise, one must decide upon an interpretation (Ito versus Stratonovich, see Chap. 2). In the present case, the Ito interpretation of (12.24) is demanded by physical considerations.

Without noise, (12.24) is a reaction–diffusion equation, which exhibits a mean-field critical point. It is perhaps surprising that driving a reaction–diffusion equation with multiplicative noise leads to the proper exponents. Of course the condition expressed in (12.25) is crucial in this regard.

In the mean-field approximation (the spatially uniform, noise-free version of (12.24)), the vacuum becomes unstable when $a = 0$, and for $a, b > 0$ there is an active state. When fluctuations are taken into account, the critical point shifts to $a_c > 0$, and the critical behavior is nonclassical. For example, the stationary density scales as $\propto (a - a_c)^\beta$, with $\beta \simeq 0.277$ in one dimension, while in mean-field theory, $\beta = 1$. Field-theoretic analysis [459] reveals that the cubic and higher-order terms are irrelevant to critical behavior, so long as $b > 0$. The situation is analogous to that in equilibrium critical phenomena, where the Ising universality class is generic for models with a scalar order parameter and short-range interactions.

12.4.4 Self-Organized Formulation of Directed Percolation and Scaling Laws

There is a very different way to define the percolation model, which is reminiscent of the correspondence between the Fokker–Planck formulation and the Langevin equations of stochastic processes discussed in Chap. 2. Its philosophy is to transform a probabilistic *discrete* process into a deterministic *continuous* one, modulo the introduction of a random function. By discrete and continuous, we refer to the ensembles in which the variables $x_{i,t}$ that are attributed to each site i, t take their value: these variables live on a discrete space–time lattice. To be equivalent to the site directed percolation on the square lattice, we need an equation of evolution for the $x_{i,t}$ which can be easily connected to the directed percolation connectivity probability p belonging to $]0, 1[$. In this goal, we assume that a site i, t is wetted (resp. empty) if its variable $x_{i,t}$ is less (resp. larger) than p.

Consider the three sites $i + 1, t$, i, t and $i - 1, t$ which are connected to site $i, t + 1$ in the lattice as shown in Fig. 12.8. The evolution equation is a formula specifying $x_{i,t+1}$ as a function of $x_{i+1,t}$, $x_{i,t}$ and $x_{i-1,t}$. The rules are the following:

- if $x_{i+1,t}$, $x_{i,t}$ and $x_{i-1,t}$ are all larger than p (empty), then the site $i, t + 1$ must be empty according to the directed percolation (DP) rule, i.e. $x_{i,t+1}$ must be larger than p;

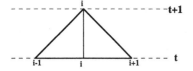

Fig. 12.8. Three sites $i + 1, t$, i, t and $i - 1, t$ are connected to site $i, t + 1$ and control its value

- if $x_{i+1,t}$, $x_{i,t}$ or $x_{i-1,t}$ or all are less than p (wetted), then the site $i, t+1$ is wetted with probability p, i.e. $x_{i,t+1}$ is smaller than p with probability p.

These rules are embodied in the following evolution equation:

$$x_{i,t+1} = \max(\eta_{i,t+1}, \min(x_{i+1,t}, x_{i,t}, x_{i-1,t})) \; . \tag{12.26}$$

where the η's are random numbers, independently drawn for each site i, t from a uniform density between 0 and 1. Equation (12.26) provides the desired stochastic equation of evolution, equivalent to the directed percolation probabilistic model.

We have gained something in the construction process: indeed, notice that p does not appear explicitly in the evolution equation (12.26)! It is thus equivalent to directed percolation for an *arbitrary* p. In other words, following the evolution (12.26) with arbitrary initial conditions allows one to simulate directed percolation for all p's at the same time! The same set of $x_{i,t}$ allows one to reconstruct the cluster statistics for all p's: for $p < p_c$, only isolated finite clusters of sites with $x_{i,t} < p$ exist (non-percolating phase), while for $p > p_c$, an infinite cluster of sites with $x_{i,t} < p$ appears (percolating phase). This situation is remarkable in the sense that there is no control parameter for the time evolution of the $x_{i,t}$. The parameter p serves only as a threshold to distinguish wetted from empty sites. This is completely analogous to the "sea-level" formulation of (non-directed) percolation: a random number between 0 and 1 is attributed to each lattice site. One then introduces an arbitrary p (the sea level) and selects all sites whose number is less than p. For $p < p_c$ (where p_c is the percolation threshold), only isolated finite lakes exist, whereas for $p > p_c$, an ocean bounded in extent only by the system size appears.

There is however an important difference between standard percolation and directed percolation: in standard percolation, the $x_{i,t}$ are uniformly distributed and do not provide any information on the critical properties; in directed percolation corresponding to (12.26), the $x_{i,t}$ have a distribution $P(x)$ which is singular at the directed percolation threshold $x = p_c$. The evolution (12.26) thus describes the subtle long-range correlation which is intrinsic to directed percolation.

A study of the distribution of $x_{i,t}$ gives important information on the critical properties of DP. To see this fact, simply notice that

$$\rho(p) = \int_0^p P(x) \, dx \; , \tag{12.27}$$

where $\rho(p)$ is the density of DP growth sites. From the known singular properties of $\rho(p)$ [367], we deduce that near p_c we should have the scaling behavior

$$P(x) \approx t^{1/\nu_\parallel - \delta} g\left[(x - p_c)t^{1/\nu_\parallel}\right] \; , \tag{12.28}$$

where $\nu_\parallel = 1.7336$ and $\delta = 0.1596$ in $1 + 1$ dimensions, leading to

$$P(x) \approx (x - p_c)^{-(1 - \delta\nu_\parallel)} \approx (x - p_c)^{-0.7233} \quad \text{for large } t \; . \tag{12.29}$$

This stems from the relation

$$\rho(p) \approx (p - p_c)^\beta F\left((p - p_c)t^{1/\nu_\|}\right) \approx (p - p_c)^\beta \qquad (12.30)$$

for t large, together with the use of the scaling relation $\beta = \delta\nu_\|$ [367]. Using (12.27), we find $p_c = 0.7056 \pm 0.0003$, in agreement with the value obtained by much more painful series expansions in [280]. This evolution (12.26) provides an extremely efficient method for a numerical estimation of p_c and of critical exponents. To conclude, all the results obtained above are not specific to the $1 + 1$ dimensional case discussed here and generalize straigthforwardly to higher space dimensions.

13. Rupture Models

The damage and fracture of materials are technologically of enormous interest due to their economic and human cost. They cover a wide range of phenomena like the cracking of glass, aging of concrete, the failure of fiber networks in the formation of paper and the breaking of a metal bar subject to an external load. Failure of composite systems are of utmost importance in naval, aeronautics and space industries [769]. By the term composite, we refer to materials with heterogeneous microscopic structures and also to assemblages of macroscopic elements forming a super-structure. Chemical manufacturing and nuclear power plants suffer from cracking due to corrosion either of chemical or radioactive origin, aided by thermal and/or mechanical stress.

Despite the large amount of experimental data and the considerable efforts undertaken by material scientists [564], many questions about fracture have not yet been answered. There is no comprehensive understanding of rupture phenomena but only a partial classification in restricted and relatively simple situations. This lack of fundamental understanding is indeed reflected in the absence of reliable prediction methods for rupture, based on a suitable monitoring of the stressed system. Not only is there a lack of non-empirical understanding of the reliability of a system, but the empirical laws themselves have often limited value. What we need are models that incorporate the underlying physics to identify and use relevant precursory patterns. Recent models developed to address this question are based on two key concepts: the role of heterogeneity and the possible existence of a hierarchy of characteristic scales.

Many material ruptures occur by a "one crack" mechanism and a lot of effort is being devoted to the understanding, detection and prevention of the nucleation of the crack [194, 301, 587, 602]. Exceptions to the "one crack" rupture mechanism are heterogeneous materials such as fiber composites, rocks, concrete under compression and materials with large distributed residual stresses. The common property shared by these systems is the existence of large inhomogeneities that often limit the use of homogeneization or effective medium theories for the elastic and more generally the mechanical properties. In these systems, failure may occur as the culmination of a progressive damage involving complex interactions between multiple defects and growing micro-cracks. In addition, other relaxation, creep, ductile, or plastic behaviors, possibly coupled with corrosion effects may come into play. Many important practical applications involve the coupling between mechanic and chemical effects with a competition between several characteristic time scales. Application of stress may act as a catalyst of chemical reactions [341] or, reciprocally, chemical reactions may lead to bond weakening [1014] and thus promote failure. A dramatic example is the aging of present aircrafts due to repeated loading in a corrosive environment [664]. The interaction between multiple defects and the existence of several characteristic scales present a considerable challenge to the modeling and prediction of rupture. Those are the systems and problems that guide the modeling efforts recapitulated in this chapter.

13.1 The Branching Model

13.1.1 Mean Field Version or Branching on the Bethe Lattice

The general class of branching models provides a simple and general tool for describing the notion of a cascade that may either end after a finite number of steps or diverge, depending upon the value of a control parameter known as the branching probability. For instance, it has been applied to the description of material failure and earthquakes, seen as resulting from a succession of events chained through a causal connection. In this class of models, it is supposed that a crack does not propagate in a single continuous movement but through a series of steps or branches invading the elastic medium [983]. The branching model is also a mean-field version of the percolation model discussed in Chap. 12. The meaning of "mean-field" refers here to the fact that there are no loops, i.e. paths closing on themselves. This defines a so-called Bethe lattice.

Let us consider the simplest version of the branching model. The general case is easily obtained [404]. At each discrete step, the earthquake is assumed to

- stop, with probability C_0,
- propagate over a unit length (i.e. remain a single branch), with probability C_1,
- or develop two branches, with probability C_2.

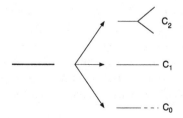

Fig. 13.1. Definition of the branching model: starting from an existing branch, with probability C_0 the branch stops at the next step; with probability C_1, the branch continues to grow at the next step; with probability C_2, it develops two branches

Each branch is taken to describe the dissipation of a unit amount of elastic energy. A cascade of such branches can develop into many generations to create a large earthquake. In this model, the emphasis is put on the stochastic character of the propagation of rupture, assumed to stem from the large geometrical and material heterogeneity of the crust. From one generation to the next, the number of branches on average increases by a factor $C_1 + 2C_2$. The value $C_1 + 2C_2 = 1$ corresponds to a critical point:

1. if $C_1 + 2C_2 < 1$, the average number of generations is finite and all earthquakes are finite. Their size distribution is an exponential.
2. if $C_1 + 2C_2 > 1$, the probability to generate a "run away" (i.e. an event of infinite size) becomes nonzero. This is similar to being above the threshold in the percolation model presented in Chap. 12.
3. if $C_1 + 2C_2 = 1$, the system is critical and the size distribution of events is a power law as we show below.

The critical condition $C_1 + 2C_2 = 1$ together with the normalization $C_0 + C_1 + C_2 = 1$ yields the condition $C_0 = C_2$. Using the theory of generating functions, one can show [404, 983] that the probability $P(E)$ that a branching process releases an energy E is, for large E:

$$P(E) \simeq A \mathrm{e}^{-aE} E^{-(1+\mu)} , \tag{13.1}$$

where

$$A = \left(\frac{2C_0 + C_1 (C_0/C_2)^{1/2}}{4\pi C_2} \right)^{1/2} , \tag{13.2}$$

$$a \simeq \frac{(C_0 - C_2)^2}{4C_0} , \tag{13.3}$$

and

$$\mu = \frac{1}{2} . \tag{13.4}$$

Equation (13.1) gives a power law distribution $P(E) \sim E^{-(1+\mu)}$ up to a maximum size $E_{\max} \sim 1/a \sim |C_0 - C_2|^{-2}$, above which the distribution crosses over to an exponential (this distribution (13.1) is sometimes known as the Gamma distribution (see Chap. 6) [491, 912]). Note that for $C_1 + 2C_2 - 1 \equiv C_2 - C_0 > 0$ for which a run away occurs, $P(E)$ describes only the finite events.

Such branching processes has been proposed as "mean field" descriptions of avalanches in sandpile models with self-organized criticality [16, 17]. The value $\mu = 1/2$ is the characteristic "mean field" exponent of the frequency–size distribution of earthquakes and is found generally in such classes of models where the spatial dimensions are neglected or rather homogenized over. Note that its value 1/2 is general and does not depend upon the details (for instance the number of branches per generation) of the branching model.

13.1.2 A Branching–Aggregation Model Automatically Functioning at Its Critical Point

Consider a population of clusters with a distribution $D(s,t)$ giving the normalized number of clusters of size s at time t. Following [100], we think of a cluster as a group of animals. Such groups are widespread throughout the animal kingdom and are known as groups, herds, schools or flocks. The question we address is what determines the distribution $D(s,t)$ of their sizes. In particular, one would like a simple explanation for the observation that $D(s,t)$ is often a power law distribution with small exponent μ close to $1/2$, with a truncation or fast fall-off for the largest group sizes. A simple mean-field model of the formation of animal groups is as follows [100].

Animal clusters can occupy N distinct discrete locations. When a cluster of size s_1 visits an occupied cluster of size s_2, the two cluster merge to form a larger cluster of size $s_1 + s_2$. Animal clusters of size s are injected randomly on the N sites with a probability $p_{\text{inj}}(s)$. The probability for a cluster to move to a specific site and fuse with its cluster is assumed to be $1/N$. From these ingredients, we get the following evolution equation for $D(s,t)$:

$$
D(s,t+1) = \sum_{r=0}^{N} \binom{N}{r} \left(\frac{1}{N}\right)^r \left(1 - \frac{1}{N}\right)^{N-r}
$$

$$
\times \sum_{s_1=0}^{\infty} \cdots \sum_{s_r=0}^{\infty} \sum_{s_{\text{inj}}=0}^{\infty} p_{\text{inj}}(s_{\text{inj}}) D(s_1,t)\ldots D(s_r,t)\delta(s - (s_1 + s_2 + \ldots + s_r + s_{\text{inj}})).
$$

The combinatorial factor accounts for the number of ways one can have r clusters moving and merging together. The Dirac function ensures that the merging of the r clusters together with the injection produce a cluster of size s. In the spirit of mean-field theory, we take the limit $N \to \infty$ and $r \ll N$ for which it is easy to verify that $\binom{N}{r}(1/N)^r(1-1/N)^{N-r} \to (1/e)/r!$ and becomes independent of N. The characteristic function $\hat{D}(k,t)$ of $D(s,t)$, defined as

$$
\hat{D}(k,t) \equiv \sum_{s=0}^{\infty} D(s,t)e^{iks} , \tag{13.5}
$$

then obeys the following equation

$$
\hat{D}(k,t+1) = \hat{p}_{\text{inj}}(k) \sum_{r=0}^{\infty} \frac{1}{er!} \left[\hat{D}(k,t)\right]^r
$$

$$
= \hat{p}_{\text{inj}}(k) \exp\left[\hat{D}(k,t) - 1\right] , \tag{13.6}
$$

where $\hat{p}_{\text{inj}}(k)$ is the characteristic function of the injection probability $p_{\text{inj}}(s)$. The characteristic function of the asymptotic distribution for large times is thus solution of

$$
\hat{D}(k) = \hat{p}_{\text{inj}}(k) \exp\left[\hat{D}(k) - 1\right] , \tag{13.7}
$$

Generically, $\hat{p}_{\text{inj}}(k)$ can be expanded in powers of k as

$$\hat{p}_{\text{inj}}(k) = 1 + ik\langle s_{\text{inj}}\rangle + O(k^2) \ . \tag{13.8}$$

Assuming a functional form for small k as

$$\hat{D}(k, t \to \infty) = 1 - A|k|^{\mu} \ , \tag{13.9}$$

we get from (13.7)

$$1 - A|k|^{\mu} = (1 + ik\langle s_{\text{inj}}\rangle)\exp\left(-A|k|^{\mu}\right) \ . \tag{13.10}$$

Expanding the exponential in powers of k, we find that we need to expand up to second order in $|k|^{\mu}$, as the l.h.s. of (13.10) cancels out with the two first term of the expansion of the exponential in the r.h.s. We finally get $\mu = 1/2$ which recovers the result given in [73]. The determination of the scale factor A requires to take into account higher-order terms in the expansion (13.9). From Chap. 4, we know that an expansion like (13.9) with $\mu = 1/2$ corresponds to a power law distribution $D(s) \sim s^{-3/2}$. It is noteworthy that, due to the injection of new individuals, the process is non-stationary and the total number of individuals increases without bounds, but this does not prevent the existence of a well-defined limit distribution. Extension of this model by inclusion of a splitting probability leads to the same power tail with the addition of a cross-over at large sizes to a faster decaying law. Related models of coagulation and fragmentation, based on Smoluchowsky rate equation including scale invariant break-up kernels give similar results [167, 765, 976].

13.1.3 Generalization of Critical Branching Models

These models can be extended in many different ways. A general description of critical branched structures is as follows [110]. We first assume that the branched crack possesses a well-defined backbone on which branches can grow and develop new branches. One can thus define the probability per unit length that the backbone of the crack gives birth to a new offspring as $1/l_1$. The probability that this new branch survives at least for a length $r - r'$ will be denoted $K(r - r')$. The number of branches this new structure contains will be denoted $N(r - r')$. If the whole structure is self-similar (i.e. the same kernels $K(r - r')$ and $N(r - r')$ describe the branching processes everywhere along the backbone), then the total number of branches after a length r is given by

$$N(r) = \int_0^r dr' \frac{1}{l_1} K(r - r')N(r - r') \ , \tag{13.11}$$

which leads to the following differential equation for $N(r)$:

$$\frac{dN(r)}{dr} = K(r)N(r) \ . \tag{13.12}$$

Now, if $K(r)$ decays faster than $1/r$, $N(r)$ tends to a constant at large r. If $K(r)$ decays as $r^{-\alpha}$ with $\alpha < 1$, then $N(r)$ grows very fast with r, as $N(r) \sim$

$\exp(r^{1-\alpha})$. Hence, $\alpha = 1$ appears as a critical value where the branched structure barely survives. Writing $K(r) = l_2/r$, one finds

$$N(r) \sim r^{l_2/l_1} . \tag{13.13}$$

It is interesting to note that the exponent l_2/l_1 measures the ratio of two important length scales: the typical length l_2 of one branch and the distance l_1 between branches.

Following a similar line of arguments, it is possible to derive the general form for the probability $P(z, r)$ that the point at coordinates (z, r) belongs to the branched structure, knowing that $(0,0)$ does (r lies in the average plane of the branched crack and z is the direction perpendicular to it). One finds

$$P(z, r) = r^{-\zeta} F(z/r^\zeta) , \tag{13.14}$$

with $F(u) \sim u^{-1}$ implying that $P(z, r) \sim z^{-1}$ independently of r as soon as $r^\zeta < z < z_{\max}$ where $z_{\max}$ is the total width of the structure. The branched crack has thus a self-affine backbone (with roughness exponent $\zeta \simeq 0.8$ in three dimensions [110]), decorated by a density of branches decaying as z^{-1}. This geometrical structure is characteristic of the critical branching condition $K(r) \sim r^{-1}$. It is also found analytically in various tree structures studied numerically (see [110] and references therein). It is also recovered in the directed percolation model, which can be considered as a minimal model for branching in two dimensions. Last but not least, this geometrical structure has been successfully tested in various experiments on rupture in the laboratory [108].

The mechanism selecting the special kernel $K(r) \sim r^{-1}$ is not discussed in these approaches and remains to be found. It probably involves some self-organizing or feedback process of the types discussed in Chap. 15.

13.2 Fiber Bundle Models and the Effects of Stress Redistribution

This class of models represents a significant improvement over branching models in that the assumed stochastic rules of branching are now derived from the mechanism of stress redistribution. In an elastic medium, when there is a rupture somewhere, the stress elsewhere is modified according to the law of elasticity. This can in turn trigger more ruptures later on, which can cascade to the final failure of the system. As a first step, the fiber bundle models are characterized by simple geometries and simple rules for stress redistribution, thought to approximate or bound the real laws of stress redistribution derived from elasticity. The interest in these models is the possibility to obtain exact results, providing bounds and references for more involved modeling.

13.2.1 One-Dimensional System of Fibers Associated in Series

In order to put the problem in suitable perspective, let us first consider the simplest possible case of a one-dimensional system composed of links associated in series with randomly distributed breakdown thresholds. Global failure is then associated with that of the weakest link and can be described mathematically using the theory of probability of extreme order statistics discussed in Chap. 1.

Let us call $X_1 \leq X_2 \leq \ldots \leq X_N$ the failure thresholds of the N links forming the 1D-system, ordered in increasing values. We assume that they are identically distributed according to the pdf $P(X)\,dX$ giving the probability for the rupture threshold of a link to be equal to X to within dX. We are interested in determining the weakest threshold X_1, which determines the global rupture. The order of magnitude of X_1 is easily obtained from the equation $N \int_0^{X_1} P(X)\,dX \simeq 1$, expressing that there is typically one value of X smaller than or equal to X_1 out of the N links. If $P(X)$ is a power law $P(X)\,dX = CX^{\alpha-1}\,dX$ (with $\alpha > 0$ to ensure normalization of the probability for small X's), this yields $X_1 \sim N^{-1/\alpha}$, illustrating that the system strength decreases as its size increases (a particular illustration of the well-known size effect in rupture).

One can be more precise and derive the exact probability $F(X_1)\,dX_1$ that the weakest rupture threshold be equal to X_1 to within dX_1:

$$F(X_1)\,dX_1 = \binom{N}{1}\left[1 - \int_0^{X_1} P(X)\,dX\right]^{N-1} P(X_1)\,dX_1 \, . \tag{13.15}$$

If $P(X)$ is a power law, the distribution of system strengths is a Weibull distribution

$$F(X_1)\,dX_1 = NCX_1^{\alpha-1}e^{-[C(N-1)/\alpha]X_1^{\alpha}}\,dX_1 \, , \tag{13.16}$$

with Weibull exponent α. In the mechanical literature, most Weibull exponents are found between 2 and 15. For $\alpha > 1$, $F(X_1)$ presents a peak at $X_1^* = ((\alpha - 1)/(C(N - 1)))^{1/\alpha}$, which recovers the announced scaling $X_1 \sim N^{-1/\alpha}$. Around this maximum, $F(X_1)$ can be written as

$$F(X_1) \simeq F(X_1^*) - \frac{1}{2}\left|\frac{d^2F(X_1)}{dX_1^2}\right|(X_1 - X_1^*)^2 \, , \tag{13.17}$$

allowing one to get an estimation of the dispersion of the rupture threshold X_1 through the calculation of $\Delta X_1 \equiv \langle(X_1 - \langle X_1\rangle)^2\rangle^{1/2}$. We obtain

$$\frac{\Delta X_1}{X_1^*} = \frac{1}{\sqrt{\alpha}} \, , \tag{13.18}$$

showing that the relative fluctuations of the system strength are of order 1 and do not decrease with the system size. This is characteristic of extreme order statistics of power law distributions.

This simple calculation provides a mechanism for the existence of large fluctuations based on the sensitivity of extremes to disorder realizations. This should be kept in mind for the modelling of rupture and of earthquakes.

13.2.2 Democratic Fiber Bundle Model (Daniels, 1945)

The "democratic fiber bundle model" (DFBM) represents the other extreme case, in which the fibers are associated in parallel and not in series. The fundamental assumption of the model is that, at all steps, the total load is "democratically" shared between all the remaining unbroken fibers. This can be considered as a mean-field model since all elements are coupled to each other through the global stress applied to the system. We shall see that, in contrast to the previous case, the rupture properties are not controlled by extreme order statistics but by a central limit theorem (see Chap. 2), indicating a cooperative behavior.

We introduce the cumulative probability

$$P_<(s) = \int_0^s P(X)\,\mathrm{d}X \tag{13.19}$$

of finding a rupture threshold smaller than or equal to s. Under a total load F applied to the system, by definition of the cumulative distribution, a fraction $P_<(s = F/N)$ of the threads will be subjected to more than their rated strength and will fail immediately. After this first step, the total load will be redistributed by the transfer of stress from the broken links to the other unbroken links. This transfer will in general induce secondary failures which in turn induce tertiary ruptures and so on.

The properties of this rupture problem are obtained by noting that the total bundle will not break under a load F if there are n links in the bundle each of which can withstand F/n. In other words, if the first $k - 1$ weakest links are removed, the bundle will resist under a force smaller than or equal to $F_k = (N - k + 1)X_k$, since there remains $N - k + 1$ links of breaking strength larger than or equal to X_k. The strength F_N of the bundle is then given by

$$F_N = \max_k X_k(N - k + 1) \quad \text{for } 1 \le k \le N , \tag{13.20}$$

which converges to

$$N\max_x \{x[1 - P_<(x)]\} , \tag{13.21}$$

in the limit of very large system size N, as can be seen by replacing the discrete variable X_k by x and $(k - 1)/N$ by its continuous approximation $P_<(x)$.

It can be shown [326] that F_N obeys a central limit theorem according to which the probability that the global failure threshold F_N be equal to F is

$$P_G(F_n = F) \simeq \frac{1}{(2\pi N)^{1/2}x_0} \, e^{-(F-N\theta)^2/2Nx_0^2} , \tag{13.22}$$

and thus F_N converges to $N\theta$ where

$$\theta = x_0\big[1 - P_<(x_0)\big] \tag{13.23}$$

is the unique maximum of $x\,[1 - P(x)]$ at $x = x_0$. The strength does not decreases as in the 1D-model but increases with the system size (the average strength per fiber goes to a constant). The variable x has the physical meaning of being the force per surviving fiber.

Many other properties of this model can be studied [733, 734, 832, 833, 867, 938]. For instance, as the applied stress s (i.e. force per fiber) is increased, as long as it is smaller than θ, the system holds while having some of its fibers break down simultaneously. Indeed, starting from a stable configuration corresponding to some value F_k, a simultaneous rupture of Δ fibers, which can be called an event or burst of size Δ, occurs if $F_n < F_k$ for $k+1 \leq n \leq k+\Delta$ and $F_{k+\Delta+1} \geq F_k$.

The function F_k can be shown to undergo a kind of random walk excursion as k increases. It turns out that this random walk model gives not only the correct qualitative behavior but also enables one to get the quantitatively correct exponents for the distribution of burst sizes [867, 870]. Using this random walk picture, it is easy to obtain the probability that a burst of size Δ occurs after k fibers have been broken [400, 415, 514, 867]. This corresponds to the probability $p_1(\Delta)$ of first return to the origin of a random walker. In the absence of any bias, the probability to be found at the origin after Δ steps decays as $\Delta^{-1/2}$ and the probability to return for the first time to the origin is $p_1(\Delta) \sim \Delta^{-3/2}$. Thus, the local differential distribution $d(\Delta)$ of bursts of size Δ is given by (see also Chap. 14)

$$d(\Delta) \sim p_1(\Delta) \sim \Delta^{-3/2} \ . \tag{13.24}$$

This recovers the previously derived mean field exponent $\mu = 3/2 - 1 = 1/2$. This power law distribution holds for $\Delta \leq \Delta_{\max}(x)$, where $\Delta_{\max}(x) \sim (x_0 - x)^{-2}$, due to bias of the random walk of F_k created by the slow increase in the average strength as the weaker fibers fail.

This regime holds up to the global failure threshold, occurring after a finite fraction $k_N = NP_<(x_0)$ of the fibers have failed. The remaining $N[1-P_<(x_0)]$ fibers break down suddenly in one sweeping run away event when s reaches θ at which the stress $x(s)$ per surviving fiber reaches x_0. The DFBM is thus not critical in the usual sense and ressembles a first-order phase transition, characterized by fluctuations preceeding the abrupt rupture, followed by the catastrophic growth of the instability beyond threshold. This is reminiscent of the phenomenon of spinodal decomposition discussed in Chap. 9. Due to the long range nature of the interaction between the fibers (resulting from the democratic load sharing rule), the fluctuations (simultaneous fibers ruptures) are distributed according to a power law. In the presence of a finite interaction range, the power law will be truncated beyond this interaction range.

Note that the existence of $\Delta_{\max}(x)$ implies that an event of size Δ may occur only when $\Delta_{\max}(x) \geq \Delta$, i.e. $x_0 - x \leq \Delta^{1/2}$. This is reminiscent of Omori's law for foreshocks often observed for real earthquakes, stating that on the approach to a large earthquake, the rate of energy release increases. In the DFBM, we find $\Delta_{\max}(x) \sim (\theta - s)^{-1}$.

Due to this cut-off $\Delta_{\max}(x)$ going to infinity as the global rupture point is approached, the total (over the whole stress history) *differential* number $D(\Delta)$ of bursts of size Δ can be shown to be $D(\Delta) \sim \Delta^{-5/2}$, with a larger exponent than for the local distribution $d(\Delta)$. This results from two ingredients:

- due to the cut-off $\Delta_{\max}(x)$, bursts of size larger or equal to Δ occur only when x is sufficiently close to x_0, i.e. close to the global rupture;
- as the global rupture is approached, there are fewer and fewer bursts since they are larger and larger.

It is remarkable to find the coexistence of a local $d(\theta)$ with an exponent $\mu_d = 1/2$ and a global $D(\theta)$ with an exponent $\mu_D = 3/2$. Applied to earthquakes, this result suggests that there is no contradiction in observing a small "b-value" $(= \beta/2)$ in a restricted time interval (which necessarily samples only relatively small earthquakes) and a larger "b-value" $(= 3\beta/2)$ when the time interval is extended up to the occurrence of the greatest earthquake. Here, β is an exponent appearing in the conversion from Δ to energy. This result may also suggest a clue for the observed drift of b-values often observed before an impending earthquake.

The DFBM can also be used to find the rate dE/dt of elastic energy release as the run away event is approached. If the system is driven at a constant strain rate, dE/dt goes to a constant at global rupture. If, on the other hand, the system is driven at a constant stress rate, the rate of elastic energy release diverges when approaching global failure: using the local distribution

$$d(\Delta) \simeq \Delta^{-3/2} e^{-\Delta/\Delta_{\max}} , \tag{13.25}$$

we find

$$\frac{dE}{dt} \sim \int_1^{\Delta_{\max}} \Delta d(\Delta)\, d\Delta \sim \Delta_{\max}^{1/2} \sim (\theta - s)^{-1/2} . \tag{13.26}$$

This corresponds to a marked average increase of rupture activity prior to the run away analogous to Omori's law for foreshocks and similar to the power law acceleration of the susceptibility preceeding supercritical instabilites, as discussed in Chap. 10. These results also underline the sensitivity of the behavior with respect to the loading path. In most models of earthquakes, a constant strain rate is chosen. It is not clear whether this is the case in nature since a given fault is surrounded by an elastic medium deteriorated by many other faults which interact, leading to ill-defined boundary conditions. Therefore, the loading path of a real fault is probably intermediate to the pure constant strain or stress rate, which might explain why increased foreshock activity is not always observed before a main large earthquake.

A key parameter is the degree and nature of disorder. This was considered early by Mogi [637], who showed experimentally on a variety of materials that, the larger the disorder, the stronger and more useful are the precursors to rupture. For a long time, the Japanese research effort for earthquake prediction and risk assessment was based on this very idea [638]. The DFBM exhibits an interesting transition as a function of the amplitude of disorder: there exists a tri-critical transition [5], from a Griffith-type abrupt rupture (first-order) regime to a progressive damage (critical) regime as the disorder increases. This transition is also observed in other models with limited stress amplification, such as in spring-block models with stress transfer over limited range [22, 905].

Let us finally mention a series of works dealing with various arrangements of fiber bundles [611, 851, 852], including local load sharing for which upper bounds for the system strength have been derived [403, 540, 733, 734] and where exact results were obtained in some specific versions [354].

13.3 Hierarchical Model

Hierarchical models can be thought of as self-similar mixtures of associations of links in series and in parallel. We have already discussed the branching problem on the Bethe hierarchical lattice. Many authors have studied hierarchical models of rupture and earthquakes, because

1. they contain by construction a hierarchical scale invariance, thought to be crucial to describe the failure of heterogeneous media as well as earthquakes in the brittle earth,
2. they are also by construction amenable to renormalization group methods.

Let us mention in particular the Russian school: in addition to developing models for which the failure probability distributions can be obtained from renormalization group recursion relationships, they also explored specific realizations on discrete hierarchical systems or "trees" [662, 663, 841, 968]. Moreover, they compared in detail the model results with the scalings found in earthquake catalogs.

13.3.1 The Simplest Hierarchical Model of Rupture

This model has been introduced by Shnirman and Blanter [842] to demonstrate the crucial role played by heterogeneity. Consider a hierarchical system with branching number $n = 3$ rather than $n = 2$ which has been shown in Fig. 13.1. The fact that the branching ratio is at least three is crucial. If it was two, the following results would not hold. If it is larger than three, the results remain valid. Each element at the level $l + 1$ corresponds to a group of three elements of the previous level l. Each element of the system can take one of two states, either broken or unbroken. The state of an element at the level $l + 1$ is determined by the number of broken elements in the corresponding group of three elements at the previous level l. One assumes that each element is characterized by a critical number k such that when the number of broken elements among the three elements at level l linked to level $l + 1$ is larger than or equal to k, the element at level $l + 1$ is also broken. If this number is less than k, the element at level $l + 1$ is unbroken. The critical number k can be $1, 2$ or 3. The new ingredient in this kind of branching model is to introduce heterogeneity by distributing randomly and independently the threshold values $1, 2$ or 3 on all elements according to the probabilities a_1, a_2 and $a_3 = 1 - a_1 - a_2$.

The problem simplifies by looking solely for self-similar behavior of rupture, which implies that the probabilities a_k are the same at all levels l. The density of broken elements at the level $l + 1$ is

$$p(l + 1) = F[p(l)] , \tag{13.27}$$

where $F(p)$ denotes the probability to obtain a broken element at level $l+1$ if the density of broken elements at level l is p. The assumption of self-similarity is again invoked to assume that F is the same at all levels of the system. $F(p)$ is given by the explicit expression

$$F(p) = \sum_{k=1}^{3} a_k W_k(p) , \quad \text{where} \quad W_k = \sum_{m=k}^{3} \binom{3}{m} p^m (1 - p)^{3-m}. \tag{13.28}$$

The term $\binom{3}{m}p^m(1 - p)^{3-m}$ is the probability to have exactly m broken elements in a group of three elements at the same level. Therefore, the sum W_k over m from k to 3 expresses the condition that the number of broken elements at level l in a group of three is larger than the critical value k. Then, the sum over k weighted by the corresponding probabilities a_k is carried over all possible critical values k. This simple calculation uses the assumption of independence of the rupture thresholds. We get explicitly

$$F(p) = 3a_1 p(1 - p)^2 + 3(a_1 + a_2)p^2(1 - p) + p^3 . \tag{13.29}$$

$F(p)$ is the renormalization group flow for the density of broken bonds. As explained in Chaps. 11 and 12, we can characterize the system by studying its fixed points $F(p) = p$ and the stability of these fixed points. As usual, it is easy to check that $p = 0$ (no broken elements) and $p = 1$ (all elements broken) are fixed points. Their stability is checked by calculating

$$\left.\frac{\mathrm{d}F}{\mathrm{d}p}\right|_{p=0} = 3a_1 , \tag{13.30}$$

$$\left.\frac{\mathrm{d}F}{\mathrm{d}p}\right|_{p=1} = 3a_3 = 3(1 - a_1 - a_2) . \tag{13.31}$$

This shows the existence of four cases summarized in the Fig. 13.2:

1. Stable: $a_1 < 1/3$ and $a_3 > 1/3$, i.e. $a_1 + a_2 < 2/3$; by iteration, $p(l)$ converges to $p = 0$. The system is stable and unbroken at large scale, independent of the initial distribution of broken elements at the first level [different from $p(l) = 1$].

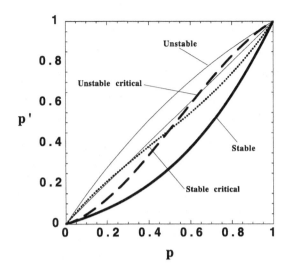

Fig. 13.2. The renormalization group map $p' = F(p)$ defined by (13.29) for different parametric domains organized according to the values $(\mathrm{d}F/\mathrm{d}p)|_{p=0}$ and $(\mathrm{d}F/\mathrm{d}p)|_{p=1}$ with respect to 1: Stable: $a_1 < 1/3$ and $a_3 > 1/3$; Unstable: $a_1 > 1/3$ and $a_3 < 1/3$; Unstable critical: $a_1 < 1/3$ and $a_3 < 1/3$; Stable critical: $a_1 > 1/3$ and $a_3 > 1/3$

2. Unstable: $a_1 > 1/3$ and $a_3 < 1/3$, i.e. $a_1 + a_2 > 2/3$; by iteration, $p(l)$ converges to $p = 1$. The system is unstable and broken at large scale whatever the initial distribution of broken elements at the first level.

3. Unstable critical: $a_1 < 1/3$ and $a_3 < 1/3$; there is a third fixed point $0 < p^* < 1$ which is unstable $(\mathrm{d}F/\mathrm{d}p)|_{p=p^*} > 1$. This corresponds to an unstable critical rupture. If $p(1) < p^*$, by iteration the system converges to the unbroken state $p = 0$ at large scale, while if $p(1) > p^*$, the system converges to the broken state $p = 1$. The transition at $p = p^*$ is a critical phase transition as discussed in Chaps. 11 and 12. We retrieve qualitatively the same map diagram as for the Potts model and the percolation transition.

4. Stable critical: $a_1 > 1/3$ and $a_3 > 1/3$; there is also a third fixed point $0 < p^* < 1$ but this time it is stable $(\mathrm{d}F/\mathrm{d}p)|_{p=p^*} < 1$. This corresponds to a situation where, independently of the initial small scale distribution of broken elements, the large scale structure is characterized by a non-zero fraction of broken bonds. We can call this regime intrinsically heterogeneous and robust: it does not break fully but does not homogenize either as in previous cases. This is a disorder fixed point. Note that this situation occurs when there is enough disorder at the small scale, as seen from the fact that the smallest (probability $a_1 > 1/3$) and largest (probability $a_3 > 1/3$) threshold concentrations must be sufficiently large. The fact that the critical point can be an attractor of the renormalization group puts this case in the class of "self-organized critical" models [44, 46] (see Chap. 15).

This very simple model recovers the finding [22, 889, 905], based on analysis of the democratic fiber bundle model and of sliding and failing spring-block models, that rupture becomes critical for sufficiently large heterogeneity. The importance of heterogeneity to control the nature of rupture processes seems an ubiquitous property.

13.3.2 Quasi-Static Hierarchical Fiber Rupture Model

We now discuss extensions of the fiber bundle models to hierarchical geometries. Exact results on the failure properties can also be obtained. Physically, the hierarchical structure of the model is also intended to capture the proposed fractal structure of fault patterns within the crust. One thus hopes to study its mechanical consequences. Some caution must be exercised in the interpretation of the results since a regular hierarchical model is bound to give too strong an imprint of its regular geometry on the mechanical properties. A better model would involve a random hierarchy.

First, we recall the definition of the model given in [679]. Consider an assembly of blocks with no bonding or friction between them. We refer to the individual blocks as blocks of order 0. Now suppose that these blocks are grouped sequentially in groups of m blocks and consider each such group as though it were itself a block, which we refer to as a block or group of order 1. Suppose in turn that these groups of order 1 are also grouped sequentially in groups of m to form groups of order 2, and so on. In this way, one obtains a hierarchical structure where a group of order n is made of m^n individual blocks.

One can reformulate this model so as to be more appealing from a geological point of view, by starting from the largest scale such as for a fractal with a small magnification and then increasing this magnification. In this approach, one would rather describe the group as made up of one block at magnification 1 which turns out to be made of m blocks at the next magnification, each of them in turn made of m blocks at the next magnification and so on. Restricting to $m = 4$, we can consider a given source region as crossed by 4 faults shown in Fig. 13.3. Each subregion can in turn be considered to be crossed by 4 faults, and so on. The geometry of the faults is not of crucial importance here, just the nested structure. The failure properties of such systems can be solved recursively since at each iteration, a link at the nth generation is replaced by a system involving an association in series and in parallel of a finite number of bonds of the $(n + 1)$-th generation.

We consider the hierarchical diamond lattice represented in Fig. 11.1 in Chap. 11. Suppose that we have stopped its iterative construction at the N-th generation. The lattice contains therefore $L = 2^N$ links in parallel between the upper and lower nodes. Suppose that a force F is applied at the upper and lower nodes and let us denote $s = F/2^N$ the stress on each link. Due to their hierarchical association, we can solve for the failure probability distribution for one element supporting the stress $s'(= 2s)$ which is made up of 4 bonds for the diamond hierarchical lattice. Then, the process can be repeated since at the next level of the hierarchical lattice, the link at the n-th generation becomes one of the 4 bonds of a link at the $(n-1)$-th generation. After $n = N$

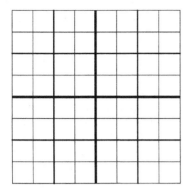

Fig. 13.3. Idealized hierarchical network of faults. The *thickest lines* represent the original four faults

successive iterations, we obtain a very simple system consisting of a single link whose strength is obtained from N iterations of a recursion relation (13.32). Then, the strength probability distribution of this single "renormalized" link is exactly that of the hierarchical lattice made up of 2^N links.

Let us denote $p_n(s)$ the probability that the link does not break under s, i.e. that a rupture in a link at the n-th generation occurs at a stress value larger than s. We then have the following *exact* recursion relation:

$$p_{n-1}(s) = \left[p_n \left(\frac{s}{2} \right) \right]^4 + 2 \left[p_n(s) \right]^2 \left\{ 1 - \left[p_n \left(\frac{s}{2} \right) \right]^2 \right\} . \qquad (13.32)$$

The first term of the r.h.s of (13.32) is the probability that all four bonds do not break under $s/2$ and the second term is the probability that two bonds along a line hold under s while one of the bonds of the other line has failed under $s/2$. This recursion relation (13.32), which is nothing but a renormalization group equation, is a mapping in the space of probability distribution functions $p(s)$. From a given function $p(s)$, (13.32) gives a new function $p'(s)$. The space in which the renormalization flow occurs is a space of functions, instead of being a space of coupling parameters of finite dimensionality as was the case in the usual critical phenomena that we have discussed in the percolation model in Chap. 12. Such a situation is in fact the general case: under application of the renormalization group, one has in general an infinite number of new "coupling" coefficients appearing. What is remarkable is the fact that, in general, the critical behavior is completely controlled by only a few of them. These coefficients are said to be "relevant", while the other parameters are "irrelevant" in the large scale limit.

It is not clear a priori whether such an exact renormalization group equation (13.32) will present non trivial fixed points, associated with genuine critical behavior. A certain number of authors [848, 849, 866, 971] have used different approximations to solve the renormalization group equation (13.32). These approximations led to the prediction of the existence of a rupture critical point: in the limit of large systems, there would be a well defined stress threshold above which the system breaks down, and below which the system is stable. It turns out that this picture is slightly incorrect: there is no transition because subtle logarithmic corrections appear which produce a universal scaling relationship [679] giving the stress at rupture

$$s \sim \frac{1}{\ln N} \sim \frac{1}{\ln \ln \text{Mass}} \, , \tag{13.33}$$

where N is the total number of iterations of the construction of the hierarchical lattice and Mass $\sim e^{aN}$ is the total number of fibers. This exact result shows that the average strength per fiber always decreases and goes to zero in the large size limit. But it does so extremely slowly, as $1/\ln \ln \text{Mass}$. This may explain why an exact treatment was called for since any approximation may destroy this subtle law. Furthermore, it turns out that this scaling law is preserved for any hierarchical organization of fiber bundles.

13.3.3 Hierarchical Fiber Rupture Model with Time-Dependence

All the previous models are quasistatic with no time dependence. Recently, Newman et al. [680, 681] have generalized the previous hierarchical model to encompass dynamical effects. They assume that, under an arbitrary stress history (as long as the stress remains non vanishing), all fibers must break eventually given sufficient time due to a kind of stress corrosion or "fatigue". The fundamental determinant of failure is the integral over time of some measure of the stress-dependent rate of corrosion. Due to the sequence of ruptures, each fiber possesses its own load history $s(t'), t' \geq 0$. Then, they assume that this fiber possesses a random failure time t distributed according to

$$P_0(t) \equiv \int_0^t p_0(t') \, dt' = 1 - \exp\left\{ -\kappa \int_0^t [\sigma(t')]^\rho \, dt' \right\} \, . \tag{13.34}$$

The stress intensity factor $\kappa(s)$ is taken to be a power law with positive exponent ρ. An exact Monte Carlo calculation of the probability distribution of failure times of hierarchical systems indicates that the distribution of failure times becomes a staircase (or jumps from 0 to 1) at a well-defined non-zero critical time t^*. In contrast to the static case, the exact renormalization group gives a non-trivial critical point. If it had not been the case, this would have meant that the time-to-failure of a larger system would have converged to zero asymptotically for very large systems.

It turns out that it is possible to explicitly write a renormalization group equation for this problem [808]. We consider the probability of a given bundle to fail betweem time t and $t + dt$. For simplicity, consider a hierarchical tree with $m = 2$ branches at each successive iteration, as in Fig. 13.4.

We can then use a key identity given in the appendix of Newman et al. [522] that can be derived easily as follows. Let us consider quantitatively the effect of the rupture of one fiber at time t_1 on the other fiber which would have broken at time t_2 without this additional load transfer. For a population

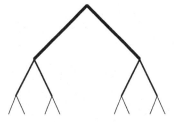

Fig. 13.4. Idealized hierarchical tree structure

of such pairs of fibers, the distribution of the time-to-failure for the remaining fiber is obtained from (13.34) by taking the failure function equal to σ up to t_1 and equal to 2σ from t_1 up to the second rupture:

$$P_0(t) = 1 - \exp\left\{-\kappa\sigma^\rho[t_1 + 2^\rho(t - t_1)]\right\} \ . \tag{13.35}$$

Doing this calculation for the ensemble, the population of fibers must be the same since the population is homogeneous at this level and $P_0(t)$ should therefore also be equal to $1 - \exp\left(-\kappa\sigma^\rho t_2\right)$. Identifying this expression with (13.35), we get the fundamental result that the time-to-failure of a fiber is modified from its initial value t_2 to an earlier time t_{12} by the influence of the other fiber which has failed at the earlier time t_1, according to:

$$t_{12} = t_1 + 2^{-\rho}(t_2 - t_1) \ . \tag{13.36}$$

The inequality $2^{-\rho} \leq 1$ (for $\rho > 0$) ensures that $t_1 \leq t_{12} \leq t_2$. This corresponds to a genuine cooperative process as the time-of-failure of the second fiber is decreased by the load transfer from the first fiber. Expression (13.36) was first proven by Newman et al. [522]. This remarkable result holds for any realization of the stochastic process. Let us stress that t_1 and t_2 are the lifetimes of the two uncoupled bundles and (13.36) describes the effect of the rupture of the first bundle on the second one which gets an additional loading at time t_1.

Now, we consider the model starting from the last order, say N (so there are initially 2^N fibers). The hierarchical bundle is made of two bundles of order $N - 1$. The crucial ingredient of the demonstration is that these two bundles are completely decoupled, no matter what occurs within each of them, as long as both withstand the applied stress and do not fail completely. It is only at the time when one of the bundles breaks down that the other one gets a load increase of a factor 2. Call $p_{N-1}(t)$ the probability density for such a bundle to fail between t and $t + dt$ if it were not bundled with the other one. Then from (13.36), one has

$$p_N(t) = 2\int_0^t dt_1 \int_{t_1}^\infty dt_2 p_{N-1}(t_1)p_{N-1}(t_2)\delta[t - t_1 - \alpha(t_2 - t_1)]$$

$$= \frac{2}{\alpha}\int_0^t dt_1 \ p_{N-1}(t_1)p_{N-1}\left(\frac{t - (1 - \alpha)t_1}{\alpha}\right) \ , \tag{13.37}$$

where the factor 2 occurs because either of the two bundles of order $N-1$ can fail first. Note that this equation (13.37) fully takes into account the modification of the distribution function of the surviving bundle by the increase in load that has occurred as a result of the failure of the first bundle.

Of course, each bundle of order $N-1$ is made of two bundles of order $N-2$ and we can now write a similar recursion relation with N replaced by $N-1$, and so on. So we conclude that N does not play any special role in (13.37), which is therefore nothing else but the renormalization group equation for the probability of failure of the whole hierarchical fiber bundle. Starting with P_0 given in (13.34), we obtain the corresponding rupture probability for the whole hierarchical system by N iterations of (13.37).

A critical point – that is a rupture of the bundle – corresponds to $p_N(t)$ converging to a delta function as N becomes large since this means that a well-defined lifetime exists for the system in the large size limit:

$$p_N(t) \to p_\infty(t) \equiv \delta(t - t_c) \text{ as } N \to \infty . \tag{13.38}$$

The existence of this limit has in fact been proven rigorously (W.I. Newman, private communication; [808]), for instance by showing that the Delta function is a solution of the renormalization group equation, in other words, it is a "fixed point". The critical time t_c can also be obtained [808].

The case $\alpha = 1/2$ is particularly simple to discuss since (13.36) becomes $t_{12} = (t_1 + t_2)/2$, which when iterated over N levels of the hierarchy gives the time t_{2^N} to rupture the 2^N fibers as the average of the time-to-failure of the individual fibers. This implies that t_{2^N} obeys the central limit theorem and converges to $\langle t_i \rangle$ with a standard deviation going to zero as $1/\sqrt{2N}$. Similar results hold for the more general case $\alpha < 1/2$.

The existence of such a critical point at a finite time has possibly important implications for failure and earthquake predictions because criticality is always associated with some precursory phenomena such as an increasing susceptibility or specific foreshock patterns, as we observed in Chap. 10. We believe that the conclusions reached for these hierarchical models may have a broader domain of applications: the crucial condition is that the stress released by failed units be redistributed over regions of comparable size (hierarchical systems automatically obey this condition). In nature, faults are more complicated but it is possible that different mechanisms may conspire to make this property hold approximately.

13.4 Quasi-Static Models in Euclidean Spaces

A vast class of quasi-static models have been invented and studied since 1985 in the statistical physics community, hoping to unravel some (if they exist) universal feature of rupture in random media [158, 416]. These works have allowed a partial classification of some possible different regimes of rupture and in particular have illuminated the links between the physics of fracture and fractal growth phenomena [923], thus providing new insight into this field.

Rupture properties of inhomogeneous media are difficult to determine due to the complex interplay between the role of the quenched (i.e. fixed) disorder and the growth aspects of the rupture controlled by the long range elastic interactions [204, 896]. This last aspect depends upon the existence of screening and enhancement effects occurring on large defects and which can have a long range [340]. A complete unifying picture of the failure properties of random systems does not yet exist but an important step is to recognize that the study of breakdown problems can be roughly divided into two main areas:

- the statistics and statistical mechanics of breakdown in random media,
- the geometrical patterns emerging in rupture related to crack growth and fractal branching.

In the first area, which is the one on which most of the works have been focused on, the question of the behavior of the strength S of the system as a function of its size and disorder is very important theoretically and also for obvious practical applications. In this respect, only very partial results exist, either based on numerical simulations [203, 416], or on bounds obtained from local configurations analysis with extreme order statistics [256, 257, 493, 577] or also from studies of special systems which can be exactly solved but which are far from being realistic. Some of these models have been described above.

The models we describe below can be thought of as an improvement over the percolation and hierarchical models already discussed in that they incorporate the correlated growth of cracks induced by the elastic interactions. We shall not review the vast literature but only point out the most important points.

The basic motivation, in this class of studies, is to explore the nature of the organization of the damage leading eventually to global failure, when putting together the long-range elastic forces (be it scalar or tensorial) with initial *quenched* disorder, and to ask whether there could exist universal aspects of rupture. Dynamical effects are neglected for simplicity. Putting together disorder and elasticity in rupture produces already a very rich behavior, whose exploration turns out to be quite involved. These studies can thus be seen as first attempts to explore the wealth of phenomena associated with rupture, including the realistic euclidean connectivity and the long range elasticity.

A typical numerical simulation goes as follows. In a two-dimensional system, each link represents an element possessing an ideal brittle rheology, either a clamped bar or beam, a spring or an electrical fuse. Each element is characterized by its spring coefficient and rupture threshold, one or both being distributed according to a distribution $p(x)$. After the set of elastic coefficients and rupture thresholds have been chosen once for all at the beginning of the computation, the algorithm consists of solving for the force field on each element of the network and determining the link closest to its rupture threshold. Then, an increase of the load so that only this most critical element fails is followed by a recalculation of the force field on all remaining elements, excluding the one that has failed which can no longer carry any stress. Then, one iterates until complete rupture (qualified by the fact that the network cannot support any load). The results, obtained by averaging over many different initial disorder configurations, refer to the global characteristics (force, displacement) of the network, the local quantities such as the force on each link as well as the geometrical structures (number of cracks, their size, their geometry...). One of the major interests is in the characterization of size effects of rupture.

The major result is the observation of a progressive statistical growth of damage preceeding the impending rupture. In general, this damage is first diffuse and uncorrelated, reflecting the geometrical structure of the initial quenched disorder. Then, it organizes itself into larger structures presenting specific long-range correlations. These observations can thus be seen as reflecting a type of organization akin to that observed in percolation models, except that here the growth of cracks becomes progressively not random but rather reflects in itself the progressive geometrical organization of the damage. It is thus a kind of bootstrap mechanism in which the growth is controlled by the geometry which itself evolves due to the damage evolution. In the language of statistical physics, in a homogeneous medium, rupture is an abrupt, first order transition, whereas, loosely speaking, it is a "critical point" in a heterogeneous system. By "critical", we mean that the correlation length (corresponding to the largest crack sizes) increases and important "critical" fluctuations of various quantities (such a jumps in the elastic coefficient or acoustic emissions...) are observed on the approach to the "critical" rupture. In this vain, it has been proposed [804] that there could exist only a limited number of different classes for this critical behavior, each characterized by a well-defined universal ratio of its elastic modulii, independent of microscopic features of the system. One could use the convergence towards the universal ratio on the approach to the critical rupture point to predict it. These facts have important consequences on the nature of the localization of the damage eventually leading to the global rupture ("localization criterion") on the one hand, and to the possible identification of precursory phenomena on the other.

Most importantly, effort has been made in attempting to classify the different scaling laws observed in the statistics of rupture [401]. The concept of a singularity (of strength $1/2$ which is the exponent characterizing the divergence of the stress at the crack tip) of the stress field at a crack tip in a homogeneous medium has been generalized using the statistical multifractal analysis. In a disordered system, the stress field becomes scale dependent and develops a continuous set of singularities when the critical rupture point is approached from below, each of them being characterized by the fractal dimension of its geometrical set. The experimental or numerical analysis of the multifractal spectrum allows a clear classification of scale effects in fracture in terms of simple characteristics of the disorder medium. We refer to Chap. 5 which has presented the general multifractal formalism.

The existing classification of the different classes of rupture based on the multifractal method goes as follows. Suppose first that we deal with a *scale-independent* distribution $p(x)$. Such a distribution can nevertheless be characterized by the multifractal method, which, as we shall show, exhibits the statistics of the field extremes. This will be useful in order to compare with the multifractal analysis of the stress field just before complete rupture. Suppose that

$$p(x) \sim x^{\phi_0 - 1} \quad \text{for} \quad x \to 0 \ , \tag{13.39}$$

and

$$p(x) \sim x^{-(1+\phi_\infty)} \quad \text{for} \quad x \to +\infty \ . \tag{13.40}$$

Then, the probability that, out of N trials (independent of any scale), x is of the order of

$$x \sim N^\alpha \ , \tag{13.41}$$

is of the order of

$$p(x \sim N^\alpha) N^\alpha \ , \tag{13.42}$$

where the second term N^α stems from the fact that, in an x-interval proportional to x, the same scaling $x \sim N^\alpha$ holds approximately. From the definition of

$$f(\alpha) = \frac{\ln \left[N p(x \sim N^\alpha) N^\alpha \right]}{\ln N} \tag{13.43}$$

as the ratio of the logarithm of the number of trials which gives $x \sim N^\alpha$ to the logarithm of the total number of trials, we get the so-called multifractal spectrum

$$f(\alpha)|_{\alpha > 0} = 1 - \phi_\infty \alpha \ . \tag{13.44}$$

A similar calculation yields

$$f(\alpha)|_{\alpha < 0} = 1 + \phi_0 \alpha \ . \tag{13.45}$$

The singularity spectrum $f(\alpha)$ is thus piecewise linear. This result shows clearly the meaning and use of the multifractal method in classifying the extreme events of a distribution since $f(\alpha)$ reflects the structure of the tails of the $p(x)$ distribution.

Applied to the stress field associated with the singularity at a crack tip in a two-dimensional system, an extension of the previous discussion yields

$$f(\alpha)|_{\alpha > 0} = 2 - 4\alpha \ , \tag{13.46}$$

showing that the capacity dimension of the set of points with non-singular stress is 2, the space dimension, and that there is a single point (the crack tip) of maximum strength singularity $1/2$ [whose fractal dimension is thus $f(\alpha = 2) = 0$].

Similarly, a diffuse damage in a weakly heterogeneous system would yield

$$f(\alpha) = d\delta(\alpha) \ , \tag{13.47}$$

where $\delta(\alpha)$ is the Dirac function, d is the space dimension and $N = L^d$ is replaced by the linear scale L of observation, showing that all points are non-singular ($\alpha = 0$). Of course, we do not learn anything by applying the multifractal technique to these previous cases, but this is useful as a way of comparison and of classification of the various regimes.

In contrast to these examples, the stress field prior to complete rupture in a heterogeneous system is no longer scale independent but varies with the size of the system. The multifractal method allows the characterization of the scale dependence by defining a singularity strength α by

$$\alpha = \frac{\ln s}{\ln L} \ , \tag{13.48}$$

where s is the stress at some point and L is the system size. The rupture threshold involves the ratio s/s_c, where s_c is the stress threshold, which can be represented by the difference $\alpha - \alpha_c$, where $\alpha_c = \ln s_c / \ln L$ corresponds to the singularity of the stress threshold. This suggests to examine the multifractal singularity spectrum $f(\alpha_c)$ of the distribution of rupture thresholds and expect that the nature of the rupture could be classified according to the structure of the initial quenched disorder described solely in terms of $f(\alpha_c)$.

This turns out to be the case [401]. From numerical simulations and perturbation expansions, three main regimes have been unravelled. There exists two critical values ϕ_0^c and ϕ_∞^c for the exponents ϕ_0 and ϕ_∞ describing the tails $s_c \to 0$ (resp. $s_c \to +\infty$) of the distribution of rupture thresholds.

1. Weak disorder: for $\phi_0 > \phi_0^c$ and $\phi_\infty > \phi_\infty^c$, all the systems have simple scaling laws: the number of broken elements at the stress maximum does not increase with the system size ($N^* \sim L^0$), the number of broken bonds just before complete rupture is $N_f \sim L$ and the maximum stress does not increase or decrease with the system size ($s^* \sim L^0$).

2. Rupture controlled by the weak elements: for $\phi_0 \leq \phi_0^c$ and $\phi_\infty > \phi_\infty^c$, $N^* \sim L^\gamma$ with $\gamma \simeq 1.75$ in two dimensions, $N_f \sim L^\gamma$ and $s^* \sim L^{-\beta}$, with $\beta \simeq 0.25$, implying an important size effect. The damage is diffuse but presents a structure at large scales. The stress-strain characteristics becomes system size independent when written in terms of the reduced variables $NL^{-\gamma}$ and sL^β.

3. Rupture controlled by the strong elements: for $\phi_0 > \phi_0^c$ and $\phi_\infty \leq \phi_\infty^c$, $N^* \sim L^2$ in two dimensions, $N_f \sim L^2$ and $s^* \sim L^0$. Cracks are nucleated at defects as in the weak disorder case, but are stopped by strong barriers. The final damage is diffuse and the density of broken elements goes to

a non-vanishing constant. This third case is very similar to the percolation models of rupture discussed above, as it should, since we have seen that percolation is retrieved by taking the limit of very large disorder.

These results control the level of precursory damage prior to the large event. It is very low in the weak disorder case, and all the larger when the intrinsic disorder increases. For earthquakes and for rupture in heterogeneous media, this implies that the larger the heterogeneity, the stronger and more numerous the precursory phenomena will be: we are led to the rather paradoxical conclusion that the more complex the system is, the more predictable it should be.

The stability of these results with respect to the incorporation of a genuine dynamics is an open question and constitutes possibly the most important problem in this area of research.

In the next section, we present what may be the simplest genuine dynamical extension to this class of statistical models of rupture and show that the situation may be much more complicated. In particular, delay and relaxation effects become important and give birth to a wealth of behaviors with fractal crack patterns whose structure depends continuously on the damage law. For instance, we find $N^* \simeq L^{d_f}$, with d_f going from 1.9 to 1, even for bound rupture threshold distributions ($\phi_0 \to +\infty$ and $\phi_\infty \to +\infty$) as the damage exponent b defined in (13.49) goes from 0 to $+\infty$. The above classification is thus not robust against certain dynamical perturbations. Incorporating the full elasto-dynamic solution, in addition to quenched disorder in an extended system, remains an open and computationally demanding problem.

13.5 A Dynamical Model of Rupture Without Elasto-Dynamics: the "Thermal Fuse Model"

The model is essentially the same as the models described in the previous section, apart from one essential ingredient [889, 913, 914, 916, 978]: it is not the element which has the largest s/s_c which is brought quasi-statically to failure. Instead, in addition to a force and displacement variable, each element n is also characterized by a damage variable D_n, reacting dynamically to the force applied on this element according to the following equation:

$$\frac{\mathrm{d}D_n}{\mathrm{d}t} = \frac{1}{g_n} s_n^b - a D_n \ . \tag{13.49}$$

b is a damage exponent describing the sensitivity of the damage to the applied stress, the larger is b, the more catastrophically the damage reacts to an increasing stress. The second term $-a D_n$ describes a form of healing or a work hardening term. Thus, if the stress on an element is reduced, the damage rate $\mathrm{d}D/\mathrm{d}t$ can become negative. In this case, the element becomes less likely to rupture as time goes on unless the stress level is increased once again. In the context of earthquakes, this law (13.49) stems from water-assisted corrosion and recrystallization for instance.

As long as all elements have their damage variable below 1, the network geometry is not changed and the stress field is not modified. Thus, the same force remains constant in the damage equation (13.49) written for each element, as long as no link reaches the failure threshold. When, for some element, D_n reaches 1, this element fails. As a consequence, it is not able to support any load and the stress must redistribute to all other surviving elements according to equilibrium elasticity. This model does not make use of elastodynamics. Its simplicity stems from the fact that elasticity and failure are coupled only through the modification of the network geometry occurring after each bond failure. This model, while presenting a genuine dynamics, still uses the static formulation for the elasticity. This relies on the fundamental assumption that the relaxation of the stress field occurs on a much shorter time scale than the time scales involved in the time evolution of the damage variable. There are two important time scales, in addition to the possible time scale of the driving load at the system borders:

- the characteristic damage times, g_n/s_n^b, which are functions of the local loads on each element,
- the healing time $1/a$.

The interesting consequence of the competition between these two time scales is that failure often occurs, not instantaneously on the most stressed element (such that s_n is maximum) but, on the element whose stress history has maximized its damage. This model mimics progressive damage in a material subjected to a contant stress, leading to failure at sub-critical stress levels if stressed for long enough, as shown in Fig. 13.5. The surrounding material responds to the rupture of an element in an elastic fashion on short time scales. The redistribution of elastic stresses can bring other elements to rupture although only after some time delay required for damage to accumulate.

There are similarities between this model and the time-dependent hierarchical models studied by Newman et al. [679] (see our above discussion), in which it is assumed that the fundamental determinant of failure is the integral of time of some measure of the stress-dependent rate of failure. In both models, rupture is also found to be a "critical" point, in the sense that the rate of elastic energy stored in the system at constant applied stress increases as a power law of the time to failure ($\sim |t_f - t|^{-\alpha}$) as shown in Fig. 13.6. Furthermore, the crack patterns upon rupture are found to be self-similar. Both the exponent α and the fractal dimension of the crack patterns are continuously dependent upon the damage exponent b. These power laws result naturally from the many-body interactions between the small cracks forming before the impending great rupture.

Low values of the exponent b tend to smooth out the effect of the heterogeneity of the stress field and thus of the stress enhancement effects on the

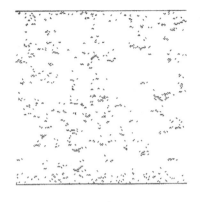

f=50% t/t_r=0.9912

a)

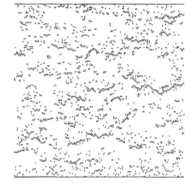

f=80% t/t_r=0.9982

b)

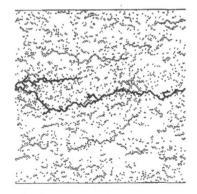

f=100% t/t_r=1.

c)

Fig. 13.5. Typical crack patterns at three characteristic times of the rupture dynamics (13.49) of the same system for $b = 2$ in a square lattice of size 180×180 tilted at $45°$. For clarity of the figure, only the broken bonds are displayed. A anti-plane (mode III) shear stress is applied at the top and bottom bus bars at time $t = 0$ and the system is then let to evolve towards global rupture under this constant stress. The coefficients g_n are uniformly sampled in the interval $[0.9, 1.1]$. (**a**) Fraction of ruptured elements equal to 50% of the total number (3423) needed to complete the global rupture ($t/t_r = 0.9912$ where t_r is the time of global rupture). One observes essentially isolated independent breakdown events leading to a continuously increasing damage of the system. One may however notice the existence of a few relatively large clusters of broken elements which tend to dominate the rupture in the continuing rupture process. Note that the damage occurs rather late in the dynamics. For instance, the first element breaks down at $t/t_r = 0.886$, the second one at $t/t_r = 0.895$ and so on. (**b**) Fraction of ruptured elements equal to 80% of the total number needed to complete the global rupture ($t/t_r = 0.9982$). Many large cracks are competing and, from their observation, it is very difficult to predict where will be the chosen path of the macroscopic rupture. A small change in the initial disorder realization may drastically change the final rupture pattern. (**c**) System at complete rupture (100% broken elements needed to deconnect the system in two pieces and $t/t_r = 1$). The main crack which cuts the network in two pieces is represented by thick bonds. Reproduced from [978]

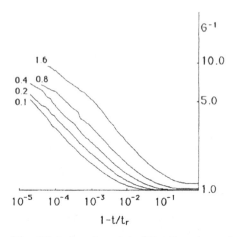

Fig. 13.6. Log–log plot of the time dependence of the integral over time of the total elastic energy stored in the system as a function of $(t_r - t)/t_r$, where t_r is the time of macroscopic rupture of the network for $b = 2$ and $a = 0$. Each curve corresponds to a different value of the range Δg of disorder of the coefficients g_n which are drawn uniformly at random in the interval $[1 - \Delta g/2, 1 + \Delta g/2]$. The lowest (respectively highest) curve corresponds to $\Delta g = 0.1$ (respectively $\Delta g = 1.6$). Each curve is the ensemble average over 25 realizations with the same parameters of the model. The average slope of the curves in their linear portion in the log–log plot gives the exponent $\alpha - 2 \approx 0.3$, where α is defined in the text as the exponent characterizing the power law dependence of the rate of energy released as a function of time. In a real experiment, the rate of energy release corresponds to the rate of the acoustic emission energy. Reproduced from [916]

damage evolution. In the presence of quenched disorder, the dynamics is then controlled by the heterogeneity of the coefficients g_n. In the limit $b = 0$, one can show [889, 913, 914, 916, 978] that the rupture dynamics is spontaneously attracted to the critical state of the bond percolation model. In this limit, both the geometrical fractal properties of the set of broken bonds and the dynamical exponent describing the time evolution of the elastic modulus of the network can be predicted exactly, using percolation theory. Their values are in good agreement with direct numerical estimations. On the contrary, large values of b tend to favor the rupture of those bonds which present the largest stress. In the limit $b \to +\infty$, the model recovers the quasi-static rupture model described in the preceeding section. The fractal dimension of the connected macroscopic crack which disconnects the network into at least two pieces varies continuously and monotonically from the 2D percolation value 1.9 $(b = 0)$ to 1 $(b \to +\infty)$ as the damage exponent b is varied. Similarly, the exponent α describing the behavior of the global elastic energy stored in the network varies continuously and monotonically from $\alpha = t + 1 \simeq 2.3$ $(b = 0)$ in two dimensions, where t is the conductivity percolation exponent, to 1 $(b \to +\infty)$. Note that the exponent α is independent of the amount of initial disorder within a broad interval. In the language of the classification of the previous section, these results have been obtained for a weak disorder regime defined by $\phi_0 \geq 1$ (constant or vanishing probability to get a zero threshold) and $\phi_\infty \to +\infty$ (existence of a finite upper bound for the thresholds). The dependence of the crack patterns on the exponent b is illustrated in Fig. 13.7.

Let us mention the particular case $b = 1$ and $a = 0$, which turns out to present exceptional properties. Integration of (13.49) yields that rupture occurs on a given element after a time t_r such that the integrated stress over this time t_r reaches a constant, characteristic of this element. Since the integrated stress over this time t_r is simply proportional to the elastic displacement accumulated over this time t_r, the rupture criterion simplifies to a threshold criterion on the cumulative elastic displacement u. Then, it is easy to show [121, 122, 1029] that the average time-to-failure $\langle T_f \rangle$ is extremely well approximated by $\langle T_f \rangle \simeq n(\Gamma_s) t_0$, where $\langle \ \rangle$ denotes an average over initial disorder configurations, $n(\Gamma_s)$ is the number of elements composing the shortest (in some metric) crack or path Γ_s cutting the system into two pieces (for instance, if the initial disorder corresponds to dilution, $n(\Gamma_s)$ is just the number of bonds along the shortest path Γ_s). The characteristic time t_0 is defined by $t_0 = u_0/S$, where S is the total force applied on the system and u_0 is the displacement threshold for rupture. In this case, there is a clear geometrical interpretation to the fractal and scaling behaviors close to rupture: it is related to the geometrical structure of minimal paths within a certain metric. Minimal paths in random systems are related to the rich statistical physics of optimal manifolds in random media, random directed or non-directed polymers in random media and to the physics of spin glasses [388]. Remarkably, this same concept of optimal paths have been found to apply to fault structures in a simple 2D dynamical model of a tectonic plate with long range elastic forces and quenched disorder. In this model, the interplay between long-range elasticity, threshold dynamics, and the quenched featureless small-scale heterogeneity allows one to capture both the spontaneous formation of fractal fault structures by repeated earthquakes, the short-time spatio-temporal chaotic dynamics of earthquakes, and the Gutenberg–Richter power law of earthquake size distribution. The faults are mapped onto a minimal interface problem, which in 2D corresponds to the random directed polymer problem and are thus self-affine with a roughness exponent $2/3$ [185, 627, 907]. The geometrical configurations of these optimal directed polymers are given in Fig. 5.15 in Chap. 5.

13.6 Time-to-Failure and Rupture Criticality

13.6.1 Critical Time-to-Failure Analysis

Most of the recently developed mechanical models [681, 805, 913, 1050, 1055] and experiments [29, 333] on rupture in strongly heterogeneous media (which is also the relevant regime for the application to the earth) view rupture

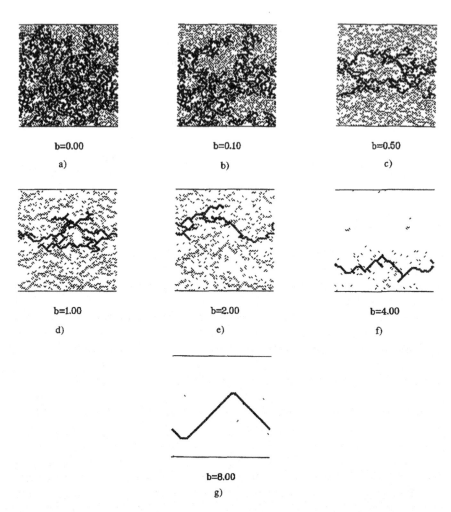

Fig. 13.7. Final rupture patterns in the limiting case $a = 0$ (i.e., for very large applied stresses or no healing), for **(a)** $b = 0$, **(b)** $b = 0.1$, **(c)** $b = 0.5$, **(d)** $b = 1$, **(e)** $b = 2$, **(f)** $b = 4$ and **(g)** $b = 8$. The coefficients g_n are uniformly sampled in the interval $[0.9, 1.1]$. As in Fig. 13.5, only the broken bonds are displayed. The main crack which deconnects the network is outlined in *thick lines*. As discussed in the text, the pattern of broken bonds for $b \to 0$ retrieves the percolation model. In the other limit $b \gg 1$, the system breaks by a single crack running away, with no diffuse damage. The different morphologies of the crack patterns can be quantified by fractal dimensions. The capacity fractal dimension of the main crack deconnecting the system in two pieces changes from the percolation value $D \approx 1.9$ of the infinite cluster in two dimension for $b \to 0$ to the value $D = 1$ of a straight line for b large. Reproduced from [978]

as a singular "critical" point [22]: the cumulative damage D, which can be measured by acoustic emissions, by the total number of broken bonds or by the total surface of new rupture cracks, exhibits a diverging rate as the critical stress σ_c is approached, such that D can be written as an "integrated susceptibility"

$$D \approx A + B(\sigma_c - \sigma)^z , \tag{13.50}$$

The critical exponent $0 < z < 1$ is equal to $1/2$ in mean field theory [870, 1050] and can vary, depending on the amplitudes of corrosion and healing processes. In addition, it has been shown [29, 478, 899] that log-periodic corrections decorate the leading power law behavior (13.50), as a result of intermittent amplification processes during the rupture (see Chaps. 5 and 11). They have also been suggested for seismic precursors [910]. This log-periodicity introduces a discrete hierarchy of characteristic time and length scales with a prefered scaling ratio λ [878]. As a result, expression (13.50) is modified into

$$D \approx A + B\left(\sigma_c - \sigma\right)^z + C\left(\sigma_c - \sigma\right)^z \cos\left(2\pi f \ln\left(\sigma_c - \sigma\right) + \phi\right) , \tag{13.51}$$

where $f = 1/\ln(\lambda)$. Empirical [29], numerical [478, 899] as well as theoretical analyses [878] point to a prefered value $\lambda \approx 2.4 \pm 0.4$, corresponding to a frequency $f \approx 1.2 \pm 0.25$ or radial frequency $\omega = 2\pi f \approx 7.5 \pm 1.5$. The value λ close to 2 is suggested on general grounds from a mean field calculation for an Ising or Potts model on a hierarchical lattice in the limit of an infinite number of neighbors [878]. It also derives from the mechanisms of a cascade of instabilities in competing sub-critical crack growth [434, 899].

This hypothesis that rupture of heteregeneous systems is a critical phenomenon has been tested on real composite structures in engineering [29]. This critical behavior may correspond to an acceleration of the rate of energy release or to a deceleration, depending on the nature and range of the stress transfer mechanism and on the loading procedure. Based on the above general considerations on the nature of the experimental signatures of critical rupture, the power law behavior of the time-to-failure analysis should be corrected for the presence of log-periodic modulations [29]. This method has been tested extensively by the French Aerospace company Aérospatiale on pressure tanks made of kevlar-matrix and carbon-matrix composites carried aboard the European Ariane 4 and 5 rockets. The method consists of recording acoustic emissions under constant stress rate. The acoustic emission energy as a function of stress is fitted by the log-periodic critical theory, as shown in Fig. 13.8. One of the parameters is the time-to-failure and the fit thus provides a "prediction" without the sample being brought to failure in the first test [29]. Unpublished improvements of the theory and of the fitting formula were applied to about 50 pressure-tanks. The results indicate that a precision of a few percent in the determination of the stress at rupture is obtained using acoustic emission recorded up to 20% below the stress at rupture. This success has warranted the selection of this non-destructive eveluation technique as the de-

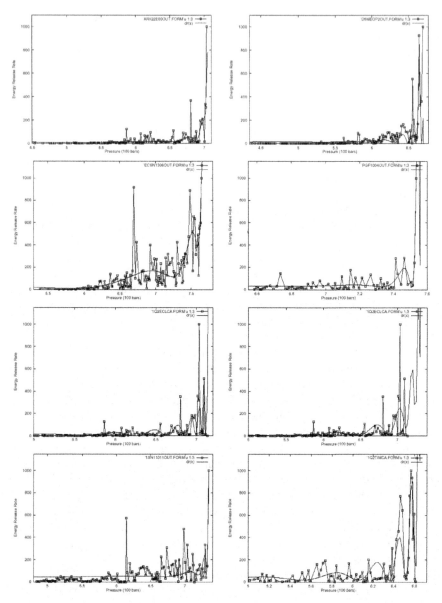

Fig. 13.8. Analysis of the acoustic emissions recorded during the pressurisation of spherical tanks of kevlar or carbon fibers pre-impregnated in a resin matrix wrapped up around a thin metallic liner (steel or titanium) fabricated and instrumented by Aérospatiale-Matra Inc (now EADS). These experiments are performed as part of a routine industrial procedure which tests the quality of the tanks prior to shipment. Eight acoustic emission recordings of eight pressure tanks are shown here, together with their fit with (13.51). All pressure tanks were brought to rupture. The acoustic emission rates all exhibit clear acceleration in agreement with a power law "divergence" expected from the critical point theory. In addition, there is strong evidence of log-periodic corrections that quantify the intermittent succession of accelerating bursts and quiescent phases of the acoustic emissions on the approach to rupture. The log-periodic oscillations allow to account for an accelerating rate of bursts on the approach of the rupture. Reproduced from [480]

facto standard in the industrial fabrication process. This is a nice example of the direct relevance of an abstract theoretical concept in a concrete real-life engineering context. Numerical simulations [478, 805] have recently confirmed that, near the global failure point, the cumulative elastic energy released during fracturing of heterogeneous solids follows a power law with log-periodic corrections to the leading term. A recent experimental study in a well-controlled situation has also confirmed the existence of critical precursors [333].

13.6.2 Time-to-Failure Behavior in the Dieterich Friction Law

A critical time-to-failure behavior similar to (13.50) also describes the early stage of sliding motion between two solid surfaces preceding the elastodynamic rupture instability, as for instance applied to earthquakes [235]. In a sense, this regime is similar to the critical nucleation discussed in [1049]. This comes from the dependence of the solid friction coefficient μ on the slip δ and the slip velocity $\dot{\delta}$. There are several forms of rate/state-variable constitutive laws that have been used to model laboratory observations of solid friction. The version currently in best agreement with experimental data, known as the Dieterich–Ruina or slowness law, is expressed as

$$\mu = \mu_0 + A \ln \frac{\dot{\delta}}{\dot{\delta}_0} + B \ln \frac{\theta}{\theta_0} \, , \tag{13.52}$$

where the state variable θ is usually interpreted as the surface of contact between asperities of the two surfaces. μ_0 is the friction coefficient for a sliding velocity $\dot{\delta}_0$ and a state variable θ_0. The state variable θ also evolves according to

$$\frac{d\theta}{dt} = 1 - \frac{\theta \dot{\delta}}{D_c} \, , \tag{13.53}$$

where D_c is a characteristic slip distance, usually interpreted as the typical size of asperities. We note that (13.53) can be rewritten

$$\frac{d\theta}{d\delta} = \frac{1}{\dot{\delta}} - \frac{\theta}{D_c} \, . \tag{13.54}$$

After a sufficiently long time in the accelerating phase such that the frictional velocity $\dot{\delta}$ has become significantly larger than D_c/θ, one can neglect the first term $1/\dot{\delta}$ in the right-hand-side of (13.54). This yields

$$\theta = \theta_0 \exp\left(-\delta/D_c\right) \, , \tag{13.55}$$

which means that θ evolves towards an ever diminishing state, corresponding to a renewal of contacts between asperities, as the total slip distance δ increases.

We start from the usual Coulomb solid friction law $\tau(t) = \mu\sigma$, expressing that the shear stress τ is proportional to the normal stress with a coefficient of proportionality defining the friction coefficient:

$$\frac{\tau(t) - k\delta}{\sigma} = \mu_0 + A \ln \frac{\dot\delta}{\dot\delta_0} - \frac{B\delta}{D_c} , \qquad (13.56)$$

where we have inserted (13.55) into (13.52). We have also substracted the stress $k\delta$ due to fault slip in the presence of a shear elastic coefficient k.

Consider the simple situation where the load $\tau(t)$ increases linearly with time $\tau(t) = \tau_0 + \dot\tau t$. Equation (13.56) then implies that $\dot\delta$ is proportional to the exponential of δ. As a consequence, δ explodes to infinity in finite time as

$$\delta(t) \sim \ln \frac{t_0}{t_c - t} , \qquad (13.57)$$

and the sliding velocity diverges as

$$\dot\delta(t) \sim \frac{t_0}{t_c - t} , \qquad (13.58)$$

where t_0 and t_c are constant of integration determined from the initial conditions and the physical constants of the friction process. Of course, this divergence does not unravel completely as the elasto-dynamic sliding instability will take over when $\dot\delta(t)$ reaches a finite fraction of the shear wave velocity. For the application to earthquakes, since $\dot\delta(t)$ starts from a typical value of about 10^{-9} m/s (corresponding to the tectonic loading velocity of a few centimeter per year) and the shear wave velocity in the crust is about 3 km/s, we see that (13.57) and (13.58) apply over a large range of velocities and time scales.

14. Mechanisms for Power Laws

Power law distributions are ubiquitous statistical features of natural systems and are found in many different scientific disciplines. Indeed, many natural phenomena have power law size distributions reading, in the notation of Chap. 4,

$$P(x) \propto \frac{1}{x^{1+\mu}} \qquad (14.1)$$

up to some large limiting cut-off [7, 592, 781]. In expression (14.1), $P(x)\,dx$ is the probability to observe the variable in the range between x and $x + dx$. Power laws seem to also describe a large ensemble of social and economic statistics [33, 288, 590, 599, 600, 718, 1004, 1065].

The specific statistical properties of power law distributions have been studied in Chap. 4. The question we address here is a theme recurring again and again in the scientific literature: what is(are) the mechanism(s) responsible for power laws in nature and the social sciences?

A somewhat related question concerns the ubiquitous observation of $1/f$ noise ("one-over-f noise", occasionally called "flicker noise" or "pink noise") in nature, which is a type of noise whose power spectra $P(f)$ as a function of the frequency f behaves like $P(f) = C/f^a$, where the exponent a is close to 1. It has been found in electronic devices, biology, network traffic and traffic flows, music and speech, astronomy, economics and financial markets, ecological systems, magnetic systems, granular flow, texture images, geophysical records, dynamical systems, etc [562]. Many different mechanisms can lead to $1/f$ noise [1012].

In a similar vain, the ubiquitous observation of power law distributions in nature suggests that an underlying universal mechanism could be found. The goal of this chapter is to present several reasonable mechanisms that can lead to similar power law distributions. By getting an understanding of the main possible mechanisms, it will then be possible for the reader to identify which is the one most relevant to a given problem. This chapter is thus intended as a dictionary of mechanisms to refer to when in contact with a given observation. This chapter must be read as both an introduction to and a complement of Chap. 15, which describes the concept of self-organized criticality introduced as a universal mechanism for the observed power law distributions in nature [44, 46]. However, the more we learn about complex out-of-equilibrium systems, the more we realize that the concept of universality developed from critical phenomena at equilibrium has to be enlarged to embody a more qualitative meaning: the critical exponents defining the universality classes are often very sensitive to many (but not all) details of the models [324]. In addition, as this chapter shows, there are many physical and/or mathematical mechanisms that generate power law distributions and self-similar behavior. Understanding how a mechanism is selected by the microscopic laws constitute an active field of research.

The set of mechanisms presented below must be complemented by mechanisms relying on collective effects, such as in percolation, criticality and self-organized criticality. Already, the random field Ising model discussed below illustrates an interplay between frozen randomness at the microscopic scales and interactions between flipping spins that may result in a large distribution of avalanches. Apart from this special case that finds its place in the present chapter because it relies on "plain old criticality" rather than on self-organizing principles, the other mechanisms involving collective effects are treated in other chapters: critical phenomena in Chap. 9 and some of its applications to rupture models in Chap. 13, percolation in Chap. 12 and self-organized criticality in Chap. 15. We refer to Sect. 13.1 for a generalization of the branching model which is automatically tuned at its critical point and thus exhibits a power law distribution of cluster sizes with mean field exponent $\mu = 1/2$. Here, we focus on mechanisms that are simpler but are often underestimated as possible sources of power law distributions.

14.1 Temporal Copernican Principle and $\mu = 1$ Universal Distribution of Residual Lifetimes

Suppose that there is a phenomenon that has a beginning, or birth, at time t_0 and an end, or death, at time $t_0 + T$, where T is the duration of the phenomenon which is distributed according to some prior density distribution $w(T)$. Knowing that the phenomenon is in progress but being ignorant of the duration t_p already spent since the beginning, we show below that the probability that the remaining lifetime t_f is larger than $Y t_p$, is given by the universal Gott's law

$$P(t_f > Y t_p) = \frac{1}{1 + Y} \, . \tag{14.2}$$

This problem has been proposed by Caves [149] to correct the previous incorrect proposal by Gott [357] that a temporal version of the Copernican principle would allow one to predict future longevity based on present age of essentially any possible cases, such as the longevity of individuals, of journals or of the human species.

The probability $P(t_f > Y t_p)$ that the remaining lifetime t_f is larger than $Y t_p$, conditioned on knowing that the process is in progress but not knowing t_p, is given by

$$P(t_f > Y t_p) = \int_0^{+\infty} dt_p \int_{Y t_p}^{+\infty} dt_f \, p(t_p, t_f | I) \ , \tag{14.3}$$

where $p(t_p, t_f | I)$ is the distribution of t_p and t_f conditioned on the fact that the phenomenon is in progress. From the definition of $w(T)$, we can guess that

$$p(t_p, t_f | I) = C w(t_p + t_f) \ , \tag{14.4}$$

where C is a normalizing constant such that

$$\int_0^{+\infty} dt_p \int_0^{+\infty} dt_f \, p(t_p, t_f | I) = 1 \ . \tag{14.5}$$

Solving for this normalization equation gives $C = 1/\langle T \rangle = 1/\int_0^\infty T w(T) \, dT$, leading to

$$P(t_f > Y t_p) = \int_0^{+\infty} dt_p \int_{Y t_p}^{+\infty} dt_f \, \frac{w(t_p + t_f)}{\int_0^\infty T w(T) \, dT} \ . \tag{14.6}$$

This result can be obtained directly from a Bayesian analysis using Bayes' theorem derived in Chap. 1, as shown in [149]. As explained by Caves [149], the expression for $p(t_p, t_f | I)$ used in (14.6) is the mathematical embodiment of the temporal Copernican principle for the phenomenon known to be in progress: if you know that the phenomenon is in progress but do not know its present age, the temporal Copernican principle imposes to treat equivalently the past duration t_p and future lifetime t_f. This amounts to split uniformly the total duration T into past and future. This means that the same probability $w(T)$ is assigned to all possible ways of splitting T into past and future.

Performing the change of variable $t_f \rightarrow t_p + t_f$ in the second integral of (14.6) and integrating by part leads to the universal result (14.2) independently of the prior lifetime distribution $w(T)$. Note that, to get this result (14.2), it is essential that one does not know the present age t_p: technically, this implies that $P(t_f > Y t_p)$ is calculated by integrating also over all possible values of t_p as seen in (14.3). It is this summation over all possible ages (i.e. time scales) that gives the scale-free power law (14.2).

If we are able to measure the present age t_p, the distribution $p(t_p, t_f | I)$ is changed from

$$\frac{w(t_p + t_f)}{\int_0^\infty T w(T) \, dT} \tag{14.7}$$

to

$$\frac{w(t_p + t_f)}{\int_0^\infty w(T) \, dT} \ . \tag{14.8}$$

This can be seen from the fact that, again, we expect the distribution of remaining lifetimes, that we now denote $p_{t_p}(t_f)$ to stress that t_p is known, to be proportional to $w(t_p + t_f)$ but with the different normalization condition $\int_0^{+\infty} dt_f\, p_{t_p}(t_f) = 1$. This last situation has been analyzed in details in [902] where a general classification is given according to the question whether it is true or not that "The longer it has been since the last event, the longer the expected time till the next?." The Poisson exponential distribution is the unique fixed point of the transformation $w(T) \to p_{t_p}(t_f)$, expressing its absence of memory. Distributions with fatter (resp. thinner) tails give a positive (resp. negative) answer to the question.

Another deceptively simple example is given by the distribution of car platoons driving on a single lane: the probability that a cluster contains N or more than N cars is exactly $1/N$, giving a power law pdf with exponent $\mu = 1$ [292, 672]. In a nutshell, a car platoon forms when a slow car is in the front. Among N cars, the probability that the slowest one is in the front is $1/N$.

14.2 Change of Variable

A simple and powerful mechanism to generate power law distributions relies on the fundamental identity

$$P(x)\,dx = P(y)\,dy \ , \tag{14.9}$$

expressing the conservation of probability under a change of variable $x \to y$: in other words, the objective estimation of the probability of an event is invariant under a change of mathematical description of this event.

14.2.1 Power Law Change of Variable Close to the Origin

Therefore, if

$$y = x^{-1/\alpha} \ , \tag{14.10}$$

then

$$P(y) = \alpha \frac{P(x(y))}{y^{1+\alpha}} \ . \tag{14.11}$$

Suppose for instance that $P(x)$ goes to a constant for $x \to 0$, then the distribution of y for large y is a power law (14.1) with a tail exponent

$$\mu = \alpha \ . \tag{14.12}$$

The uniform fluctuation of x close to zero lead to scale-free and arbitrarily large fluctuations of its inverse power y. The power law form is kept obviously (with a modification of the value of μ) if $P(x)$ itself goes to zero or diverges close to the origin as a power law.

Continuous Percolation, Holtsmark's Distribution and Vortices.
There are many physically relevant situations where this mechanism occurs.
For instance, the distribution of transport coefficients such as conductance,
permeability and rupture thresholds and of necks between random holes or
random cracks, are power laws generated by this mechanism [294, 387, 865].
Another example encountered in Chap. 3 is the pdf $P(u)$ of velocities u
due to vortices given by the expression (3.104): $P(u) \sim 1/u^3$. The Holtsmark's distribution of gravitional forces created by a random distribution of
stars in an infinite universe is a stable Lévy law with exponent $3/2$ and
its power law tail results directly from the inversion mechanism (14.10),
where y is the force and x is the distance from the measurement point
to the closest start (see Chap. 17 for a detailled discussion). This result
generalizes and applies to any other field with a suitable modification of
the exponent, such as electric, elastic or hydrodynamics, with a singular
power law dependence of the force as a function of the distance to the
source.

**Distribution of the Relative Changes of Magnetization in the Ising
Model.** A nice illustration of the inversion mechanism (14.10) has been proposed by Jan et al. [455]. It also exemplifies the widespread and misled belief
that power laws are equivalent to critical behavior. For instance, the Ising
model right at the critical temperature $T = T_c$ has a distribution $P(M)$ of
magnetization, not with a power law tail but, with very "thin" exponential
tails of the form (9.13)

$$P(M) \propto M^{(\delta-1)/2} \exp(-\text{const } M^{\delta+1}) \tag{14.13}$$

with $\delta + 1 \approx 5.8$ for the 3D Ising universality class [127]. The central part
of the distribution $P(M)$ is extremely well represented by the following
ansatz [967]

$$P(M) \propto \exp\left[-\left(\frac{M^2}{M_0^2} - 1\right)^2 \left(a\frac{M^2}{M_0^2} + c\right)\right] . \tag{14.14}$$

This ansatz (14.14) is motivated by the observation that the effective potential in the 3D Ising universality class can be in many cases well approximated
by a polynomial consisting of M^2, M^4 and M^6 terms. This is exactly what
appears in the exponent in (14.14). In addition, the distribution of changes
ΔM of the magnetization M under a fixed number of Monte Carlo steps per
site is a Gaussian at $T = T_c$ [90, 929, 966]. However, the distribution of the
relative changes $X \equiv \Delta M/M$ is a power law with exponent -2 [455].
    The reason for this last result is the following. The distribution of
$\Delta M = M_i - M_f$ is Gaussian, where M_i and M_f are the initial and final
magnetisations over a time interval from 2 to 500 Monte Carlo steps per site.
We are interested in $P(X \to \infty)$, and large values of X come from the limit
$M \to 0$ rather than $\Delta M \to \infty$. The probability $P(M)$ is a constant δ for

$M \rightarrow 0$, while ΔM can be approximated by the width Δ of the Gaussian; thus:

$$P(X)\,dX = P(\Delta M/M)\,d(\Delta M/M) = P(M)\,dM \approx \text{const}\,dM \quad (14.15)$$

and

$$P(X) = P(M)/(dX/dM) = \text{const}/[d(\Delta/M)/dM] \propto 1/X^2 \quad (14.16)$$

in agreement with the simulations (Metropolis or Heat Bath) of Jan et al. [455]. The $1/X^2$ power law is in fact not restricted to the critical point and is very general since it results simply from the inversion mechanism (14.10). However, in practice, the power law distribution of $\Delta M/M$ is clearly seen only near the critical point.

Student's Distribution. A related mechanism involves the fluctuations close to zero of the estimated standard deviation S in a denominator. Small values of S then lead to large fluctuations that turn out to be distributed according to the Student's distribution which possesses a power law tail. The Student's distribution with μ degrees of freedom has the following density function [485]

$$P_\mu(w) = \frac{\Gamma((\mu+1)/2)}{\sqrt{\mu\pi}\,\Gamma(\mu/2)}\,\frac{1/s}{\left[1 + (w/s\sqrt{\mu})^2\right]^{(1+\mu)/2}}\,, \quad (14.17)$$

and is defined for $-\infty < w < +\infty$. The Student's distribution $P_\mu(w)$ has a bell-like shape like the Gaussian (and actually tends to the Gaussian in the limit $\mu \rightarrow \infty$) but is a power law $C/w^{1+\mu}$ (see Chap. 4) for large $|w|$ with a tail exponent equal to the number μ of degrees of freedom defining the Student's distribution and with a scale factor given by

$$C_\mu = \frac{\Gamma((\mu+1)/2)}{\sqrt{\mu\pi}\,\Gamma(\mu/2)}\,\mu^{(1+\mu)/2}s^\mu\,. \quad (14.18)$$

The parameter s represents the typical width of the Student's distribution.

If $x_1, x_2, \ldots, x_n$ are independent random variables with the same normal distribution with mean $\langle x \rangle$ and standard deviation σ, then $\sqrt{n}(\bar{x} - \langle x \rangle)/\sigma$ with $\bar{x} = (1/n)\sum_{j=1}^{n} x_j$ has a centered normal distribution with unit variance. This statistics is often used in the construction of tests and confidence intervals relating to the value $\langle x \rangle$ if σ is known. If σ is not known, it is reasonable to replace it by the estimator $S = [(n-1)^{-1}\sum_{j=1}^{n}(x_j - \bar{x})^2]^{1/2}$ and study the statistics of

$$T = \frac{\sqrt{n}(\bar{x} - \langle x \rangle)}{S} = \frac{\sqrt{n}(\bar{x} - \langle x \rangle)}{[(n-1)^{-1}\sum_{j=1}^{n}(x_j - \bar{x})^2]^{1/2}}\,. \quad (14.19)$$

In 1908, Student has derived the distribution of T as given by (14.17) with $w = T$, $\mu = n - 1$ is the number of degrees of freedom and $s = 1$.

Here is a simple argument that retrieves the power law shape $C/w^{1+\mu}$ with $\mu = n - 1$ of the tail of Student's distribution (14.17). A value of T larger or equal to X typically arises when all terms $|x_j - \bar{x}|$ in the denominator of (14.19) are smaller than a value proportional to $1/X$. The probability that this occurs is proportional to the integral of their probability density from 0 to $1/X$ for each of the variable. As the Gaussian part of the distribution goes to a constant for $1/T \to 0$, the probability that $|x_j - \bar{x}|$ be smaller than $1/X$ is only controlled by the width of the interval and thus proportional to $1/X$. Since there are only $n - 1$ independent variables in the sum defining the denominator S, the probability that T is larger than X is proportional to $1/X^{n-1}$, hence the power law tail with exponent $\mu = n - 1$.

$-7/2$ Power Law pdf of Density in the Burgers/Adhesion Model and Singularities. The mechanism to generate power law distributions by a change of variable close to the origin also operates in the determination of the tail behavior of the pdf of mass density within the one and d-dimensional Burgers/adhesion model used, e.g., to model the formation of large-scale structures in the Universe after baryon–photon decoupling [317, 502]. Here, we attempt to give the flavor of the derivation for the 1d case by following [317] who have shown that large densities are localized near singularities (preshocks or nascent shocks in 1d and extremities of shock lines in 2d).

The Lagrangian coordinate is denoted by a, the velocity by u, the random initial velocity is $u_0(a) = -d\psi_0(a)/da$ (where ψ is the velocity potential) and the deterministic and uniform initial background density is ρ_0. At regular points (outside of shocks), the Eulerian velocity and density are given implicitly by

$$u(x,t) = u_0(a) \;, \qquad \rho^{(\mathrm{E})}(x,t) = \frac{\rho_0}{\partial_a x} \;, \tag{14.20}$$

$$x = a + t u_0(a) \;. \tag{14.21}$$

From (14.20) and (14.21) we have, at regular points,

$$\rho^{(\mathrm{E})}(L_t a, t) = \frac{\rho_0}{1 - t\, d^2\psi_0(a)/da^2} \;, \tag{14.22}$$

where $L_t a$ is the Lagrangian map. Large values of $\rho^{(\mathrm{E})}$ are thus obtained in the neighborhood of Lagrangian points with vanishing Jacobian, where $d^2\psi_0(a)/da^2 = 1/t$. The only points with vanishing Jacobian at the boundary of regular regions are obtained at preshocks, that is when a new shock is just born at some time t_*. Such points, denoted by a_*, are local negative minima of the initial velocity gradient, characterized by the following relations

$$\frac{d^2\psi_0}{da^2}(a_*) = \frac{1}{t_*} \;, \qquad \frac{d^3\psi_0}{da^3}(a_*) = 0 \;, \qquad \frac{d^4\psi_0}{da^4}(a_*) < 0 \;, \tag{14.23}$$

as well as by an additional global regularity condition that the preshock point a_* has not been captured before t_* by a mature shock. By Taylor expanding the Lagrangian potential and the Lagrangian map near the space-time location (a_*, t_*), one obtains the following "preshock normal forms"

$$\varphi(a, t) \simeq \frac{\tau a^2}{2} + \zeta a^4 , \tag{14.24}$$

$$x(a, t) \simeq -\tau a - 4\zeta a^3 , \tag{14.25}$$

$$J(a, t) \simeq -\tau - 12\zeta a^2 , \tag{14.26}$$

where

$$\tau \equiv \frac{t - t_*}{t_*} , \qquad \zeta \equiv \frac{t_*}{24} \frac{\mathrm{d}^4 \psi_0}{\mathrm{d}a^4}(0) < 0 . \tag{14.27}$$

From (14.26) we see that the density ρ_0/J has a a^{-2} singularity in Lagrangian coordinates at $t = t_*$ ($\tau = 0$). Since, by (14.25), the relation between a and x is cubic at $\tau = 0$, the density $\rho^{(\mathrm{E})}(x, t_*) \propto |x|^{-2/3}$ which is unbounded. For any $t \neq t_*$ the density remains bounded, except at the shock location. For $\tau < 0$, this follows immediately from (14.26), which implies $\rho^{(\mathrm{E})} \leq \rho_0/|\tau|$. For $\tau > 0$, the exclusion of the shock interval requires $|a| > a_+$. Hence, $\rho^{(\mathrm{E})} \leq \rho_0/(2\tau)$. It is clear that large densities are obtained only in the immediate neighborhood of the preshock. More precisely, it follows from (14.25) and (14.26) that $\rho^{(\mathrm{E})} > \rho$ requires simultaneously

$$|\tau| < \frac{\rho_0}{\rho} \quad \text{and} \quad |x| < (-12\zeta)^{-1/2} \left(\frac{\rho_0}{\rho}\right)^{3/2} , \tag{14.28}$$

which become very small intervals around the spatio-temporal location of preshocks when ρ is large.

The cumulative probability to have a large density at a *given* Eulerian point x and a given time t is defined by

$$P^>(\rho; x, t) \equiv \mathrm{Prob}\left[\rho^{(\mathrm{E})}(x, t) > \rho\right] . \tag{14.29}$$

In the random case, each preshock has a random Eulerian location x_*, occurs at a random time t_* and has a random $\zeta < 0$ coefficient. Only those realizations such that x_* and t_* are sufficiently close to x and t will contribute large densities. Denoting by $p_3(x_*, t_*, \zeta)$ the joint pdf of the three arguments, which is understood to vanish unless $\zeta < 0$, we have

$$P^>(\rho; x, t) = \int_{\rho^{(E)}(x, t) > \rho} p_3(x_*, t_*, \zeta) \, \mathrm{d}x_* \, \mathrm{d}t_* \, \mathrm{d}\zeta . \tag{14.30}$$

Because of the very sharp localization near preshocks implied by (14.28), for large ρ's, we may replace $p_3(x_*, t_*, \zeta)$ by $p_3(x, t, \zeta)$. Using then, in a suitable frame, the normal forms (14.24)–(14.26), we can rewrite (14.30) as an integral over local Lagrangian variables a and τ and obtain

$$P^>(\rho; x, t) \simeq \int_D t \left(-\tau - 12\zeta a^2\right) p_3(x, t, \zeta) \, \mathrm{d}a \, \mathrm{d}\tau \, \mathrm{d}\zeta . \tag{14.31}$$

Here, the domain D is the set of (a, τ, ζ) such that

$$\frac{\tau}{-4\zeta} < a^2 < \frac{1}{-12\zeta} \left(\frac{\rho_0}{\rho} + \tau \right) . \tag{14.32}$$

The right inequality of (14.32) expresses that the density exceeds the value ρ, while the left one excludes the shock interval $]a_-, a_+[$. In (14.31), the factor $-\tau - 12\zeta a^2$ is a Jacobian stemming from the change to Lagrangian space variables and the factor t stems from the change of temporal variables. The integration over a and τ yields

$$P^>(\rho; x, t) \simeq C(x, t) \left(\frac{\rho_0}{\rho} \right)^{5/2} , \tag{14.33}$$

$$C(x, t) \equiv At \int_{-\infty}^0 |\zeta|^{-1/2} p_3(x, t, \zeta) \, d\zeta , \tag{14.34}$$

where A is a positive numerical constant. Thus, for any x and t, the cumulative probability of the density follows a $\rho^{-5/2}$ law. Hence, the pdf $p(\rho; x, t) \propto \rho^{-7/2}$, as $\rho \to \infty$, which establishes the $-7/2$ law. The 2d case follows a similar treatment.

Berry's "Battles of Catastrophes". Berry's "battles of catastrophes" [85] involves similar contributions but from other singularities. Berry considers integrals of the type

$$I(t) = \int \int dx \, dy \, \delta \left[t - H(x, y, \{C_i\}) \right] , \tag{14.35}$$

where $H(x, y, \{C_i\})$ is a function which depends on a certain set of parameters $\{C_i\}$. Expression (14.35) determines the dependence of I

$$I = f(\{C_i\}) \tag{14.36}$$

as a function of the set of parameters $\{C_i\}$. As $\{C_i\}$ vary, I develops strong variations, especially when the function H is such that singularities appear. These strong variations have power law dependence of I as a function of the distance in the $\{C_i\}$ space to the catastrophe, similar to (14.10). As a consequence, following a similar reasoning as above, Berry has shown that, for unrestricted functions H, the density distribution $P(I)$ has a generic heavy tail decaying as a power law with exponent $\mu = 8$. This result is obtained by using a scaling argument involving Arnold's classification of catastrophes. If H is only quadratic in y, $\mu = 9$. If the integral defining I in (14.35) is one-dimensional, $\mu = 2$. If it is three-dimensional, the limited available classification of all known catastrophes and singularities provides only a bound $\mu < 47$. The probability distribution of dI/dt decays generically as a power law with $\mu = 1$. This last result comes from the fact that the generic saddle contributes a logarithmic divergence to I so that dI/dt diverges as the inverse to the distance to the saddle point, corresponding to the case $\alpha = 1$ in (14.10), where I plays the role of y and x is the distance to the saddle node.

The motivation for investigating the distribution of a quantity like I defined in (14.35) comes from the fact that it can represent circulation times of fluid particles in the plane, orbital periods or semi-classical densities of states for one-dimensional Hamiltonian systems, spectral densities for two-dimensional crystals, or the strength of a wave pulse produced by the propagation of a deformed step discontinuity [85].

14.2.2 Combination of Exponentials

Consider the exponential distribution $P(x) = \mathrm{e}^{-x/x_0}/x_0$, for $0 \le x < +\infty$, with average and standard deviation equals to x_0. Let us assume that y is exponentially large in x:

$$y = \mathrm{e}^{x/X} \; , \tag{14.37}$$

where X is a constant. Then by (14.9), we get

$$P(y) = \frac{\mu}{y^{1+\mu}} \; , \quad \text{with} \quad \mu = \frac{X}{x_0} \; \text{for } 1 \le y < +\infty \; . \tag{14.38}$$

The exponential amplification (14.37) of the fluctuations of x compensates the exponentially small probability for large excursions of x.

A simple statistical implementation of this mechanism is the following. Let ξ be a binomial random value:

$$\xi = d < 1 \; , \quad \text{with probability } 1 - p \tag{14.39}$$

$$\xi = D > 1 \; , \quad \text{with probability } p \; . \tag{14.40}$$

Let us construct the process $X = \xi^\kappa$ where κ is a discrete random variable with geometrical pdf:

$$P(\kappa) = a^{\kappa-1}(1-a) \; , \quad \kappa = 1, 2, 3 \dots \; , \quad 0 < a < 1 \; . \tag{14.41}$$

The tail of the distribution of X is then given, for large z, by

$$\mathcal{P}_>(X > z) = (1-p)\mathcal{P}_>(d^\kappa > z) + p\mathcal{P}_>(D^\kappa > z)$$
$$\sim p\mathcal{P}_>(D^\kappa > z) = p\mathcal{P}_>(\kappa > \ln z/\ln D)$$
$$\simeq p \sum_{k=M}^{+\infty} (1-a)a^{k-1} \; , \tag{14.42}$$

where M is the integer part of $\ln z/\ln D$. This yields $\mathcal{P}_>(X > z) \sim 1/z^\mu$ with $\mu = \ln(1/a)/\ln D$. Miller [625] used this type of processes to demonstrate that the power law behavior of word frequency arises even without the underlying optimization problem proposed by Mandelbrot [589] (see Sect. 14.11.1). In the thought experiment imagined by Miller [625], a monkey types randomly on a keyboard with n characters and a space bar. A space is hit with probability q; all other characters are hit with equal probability $(1-q)/n$. A space is used to separate words. Then, by an argument very similar to that just given, the frequency distribution of words is a power law. This provides another warning to the recurrent theme of this chapter: just because one finds a compelling mechanism to explain a power law does not mean that there are not other, perhaps, simpler explanations.

Consider a population of agents who, from birth to death, accumulate wealth with an average growth of their fortune described by an exponential law $F = F_0 \exp(t/T_1)$, where F_0 is the initial endowment taken uniform over all agents (to simplify the exposition). Let us assume that the death rate of these agents is a constant $1/T_2$, implying that the fraction of fortunes older than t is $\exp(-t/T_2)$. Then, the distribution of wealth is a power law with exponent $\mu = T_1/T_2$. Indeed, the probability to find a fortune larger than F is

$$
\begin{aligned}
\text{Prob}_>(\text{fortune} > F) &= \text{Prob}_>(F_0 \exp(t/T_1) > F) \\
&= \text{Prob}_>(t > T_1 \ln(F/F_0)) \\
&= \exp\left(-[T_1 \ln(F/F_0)]/T_2\right) \\
&= 1/(F/F_0)^{T_1/T_2} \ .
\end{aligned}
\tag{14.43}
$$

A fit to the wealth distribution of households in the United Kingdom gives a power law distribution with exponent $T_1/T_2 \approx 2.2$. If fortunes grow at the rate of 3% per year ($T_1 = 34$ years), we obtain the estimate $T_2 = 15$ years. Interestingly, this mechanism was originally proposed by Fermi [295] in his theory of cosmic radiation. Fermi introduced a model for cosmic ray acceleration in terms of the motion of a ball bouncing between a fixed and an oscillating wall. The ball could be heated to very high energies by the impact of the oscillating wall up to a random time of escape. Fermi's theory also gave way to the realization that while the motion of the ball is chaotic at low energies, the phase space has an intricate fractal structure and there is an adiabatic limit to the heating.

The mechanism of the combination of exponentials operates in the problem of the thermal crossing times for random barriers (activated escape of a system from a well in the case where the well may have many different heights) [975]. This belongs to the classical Kramers' problem [397, 617] (see Chap. 10) for which it is well-known that the typical residence time of a random walking particle scales as

$$
\tau \propto \tau_0 e^{\beta \Delta E} \ ,
\tag{14.44}
$$

where β is the inverse temperature defined in Chap. 7 and ΔE is the height of the barrier to cross. Expression (14.44) is known as the Arrhenius activation law.

Now, assume that the system can be prepared initially in many different wells, with an initial distribution of barrier heights given by the Poisson distribution $P(E) \propto e^{-\Delta E/E_0}$. The result (14.38) implies a power law distribution of residence times $P(\tau) \propto \tau^{-1-\mu}$, with $\mu = 1/\beta E_0$. The cooler the system, the larger the inverse temperature β, the smaller the exponent μ

and the "fatter" the tail of the power distribution of residence times. A good example of this situation occurs in magnetic systems with quenched random impurities exhibiting the phenomenon of "aging" [109].

In this class of systems, the inverse temperature β_g at which $\mu = 1$ can be interpreted as a glass transition, since lower temperatures correspond to smaller exponents such that the average residence time becomes infinite (see Chap. 4). This infinite average residence time implies non-stationarity and "aging". To see this, let us calculate the cumulative number $m(t)$ of transitions from one well to another during a time interval t. Consider m successive transitions separated in time by $\tau_i, i = 1, \ldots, m$, where

$$\tau_1 + \tau_2 + \ldots + \tau_m = t = m\langle\tau\rangle \,, \tag{14.45}$$

By definition,

$$\langle\tau\rangle \sim \int^{\tau_{max}} d\tau \, \frac{\tau}{\tau^{1+\mu}} \sim \tau_{max}^{1-\mu} \,. \tag{14.46}$$

Since the maximum residence time τ_{max} among m trials is typically given by

$$m \int_{\tau_{max}}^{\infty} \frac{dt'}{t'^{1+\mu}} \sim 1 \,, \tag{14.47}$$

we have $\tau_{max} \sim m^{1/\mu}$. Thus $t = m\langle\tau\rangle \sim m^{1/\mu}$, i.e. $m \sim t^{\mu}$, which is valid for $\mu < 1$. For $\mu > 1$, we recover $m \sim t$. Thus, for $\mu \leq 1$, the rate of transitions is non-stationary as m/t is not constant and decays with time. Because of the self-similarity embodied in the power-law distributions, we can state that the longer since the last transition, the longer the expected time till the next [902]. In other words, any expectation of a transition that is estimated at any given time depends on the past in a manner which does not decay. This is a hallmark of aging.

14.3 Maximization of the Generalized Tsallis Entropy

As we have seen in Chap. 2, the Gaussian distribution plays a special role because it is the attractor of a large class of distributions under the repetitive action of the convolution operator, expressing the central limit theorem for the sum of random variables. The Gaussian distribution possesses another remarkable property of "parsimonious" description of uncertainty: knowing only the mean and variance of a random variable, the Gaussian distribution is the solution of an optimization process, namely it maximizes the entropy defined by $-k \int_{-\infty}^{+\infty} dx \, p(x) \ln p(x)$. This expresses that the Gaussian distribution assumes the minimum amount of information in addition to that given by the mean and variance. Conditioned on the sole knowledge of the mean and the variance, it is therefore the best choice to represent the distribution of random variables.

In this spirit, it is interesting to find that power laws naturally emerge under a similar entropy maximization principle [13, 746, 963, 964], where the entropy is generalized in the following way. In this brief exposition, we follow [963]. Consider the generalized entropy S_q defined by [962]

$$S_q[p(x)] = k \frac{1 - \int_{-\infty}^{\infty} (\mathrm{d}x/\sigma) \, [\sigma p(x)]^q}{q - 1} \, , \tag{14.48}$$

where x is a random variable and $\sigma > 0$ is the characteristic length of the problem. Note that, for $q \to 1$, $S_q[p(x)] \to_{q \to 1} -k \int_{-\infty}^{+\infty} \mathrm{d}x \, p(x) \ln p(x)$, i.e. recovers the standard definition of the entropy.

We optimize (maximize if $q > 0$, and minimize if $q < 0$) S_q with the normalization condition $\int_{-\infty}^{\infty} \mathrm{d}x \, p(x) = 1$ as well as with the constraint

$$\langle\langle x^2 \rangle\rangle_q \equiv \frac{\int_{-\infty}^{\infty} \mathrm{d}x \, x^2 \, [p(x)]^q}{\int_{-\infty}^{\infty} \mathrm{d}x \, [p(x)]^q} = \sigma^2 \, . \tag{14.49}$$

Note that $\langle\langle x^2 \rangle\rangle_q$ recovers the definition of the variance for $q = 1$, as it should.

The following distribution is obtained:

if $q > 1$:

$$p_q(x)$$

$$= \frac{1}{\sigma} \left(\frac{q-1}{\pi(3-q)} \right)^{1/2} \tag{14.50}$$

$$\times \frac{\Gamma(1/(q-1))}{\Gamma((3-q)/(2(q-1)))} \frac{1}{(1 + (q-1)/(3-q)(x^2)/(\sigma^2))^{1/(q-1)}} \, .$$

If $q = 1$:

$$p_q(x) = \frac{1}{\sigma} \left(\frac{1}{2\pi} \right)^{1/2} e^{-(x/\sigma)^2/2} \, . \tag{14.51}$$

If $q < 1$:

$$p_q(x)$$

$$= \frac{1}{\sigma} \left(\frac{1-q}{\pi(3-q)} \right)^{1/2} \frac{\Gamma((5-3q)/(2(1-q)))}{\Gamma((2-q)/(1-q))} \tag{14.52}$$

$$\times \left(1 - \frac{1-q}{3-q} \frac{x^2}{\sigma^2} \right)^{1/(1-q)} \, ,$$

if $|x| < \sigma[(3-q)/(1-q)]^{1/2}$ and zero otherwise.

The case (14.51) recovers the standard result mentioned above that the Gaussian law maximizes the usual entropy, given the variance. The two other

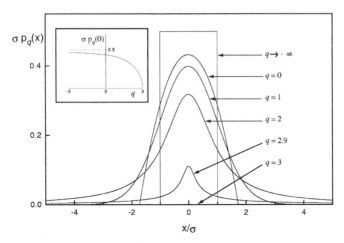

Fig. 14.1. The distributions $p_q(x)$ for typical values of q. The $q \to -\infty$ distribution is the uniform one in the interval $[-1, 1]$; $q = 1$ and $q = 2$ respectively correspond to Gaussian and Lorentzian distributions; the $q \to 3$ is completely flat. For $q < 1$ there is a cut-off at $|x|/\sigma = [(3 - q)/(1 - q)]^{1/2}$. Reproduced from [963]

cases for $q \neq 1$ extend this result. We see that the support of $p_q(x)$ is compact if $q \in (-\infty, 1)$ while $p_q(x)$ becomes a power-law tail for $q > 1$, with

$$p_q(x) \sim (\sigma/x)^{2/(q-1)} \tag{14.53}$$

in the limit $|x|/\sigma \to \infty$.

We can also check that $\langle\langle x^2 \rangle\rangle_1 = \langle x^2 \rangle_1 = \int_{-\infty}^{\infty} dx\, x^2 p_q(x)$ is finite if $q < 5/3$ and diverges if $5/3 \leq q \leq 3$ (the normalization condition cannot be satisfied if $q \geq 3$). It is interesting to observe that the Gaussian solution for $q = 1$ is recovered in both limits $q \to 1^+$ and $q \to 1^-$ by using the $q > 1$ and the $q < 1$ solutions given respectively by (14.50) and (14.52). This family of solutions is illustrated in Fig. 14.1.

The nonextensive Tsallis statistics has been applied to systems for which the standard Boltzmann–Gibbs statistical mechanics and thermodynamics present serious difficulties or anomalies. Examples are (see [963] and references therein) point systems involving long-range interactions (e.g., $d = 3$ gravitation), long-range microscopic memory (e.g., nonmarkovian stochastic processes, and conservative (e.g., Hamiltonian) or dissipative systems which in one way or another involve a relevant space–time and phase space which has a (multi)fractal-like structure (pure-electron plasma two-dimensional turbulence, Lévy anomalous diffusion, granular systems, phonon–electron anomalous thermalization in ion-bombarded solids, solar neutrinos, peculiar velocities of galaxies, inverse bremsstrahlung in plasma and black holes), etc.

A wealth of works has shown that the nonextensive statistical mechanics retains much of the formal structure of the standard theory and many important properties have been shown to be *q-invariant*. Among them, we mention (i) the Legendre transformations structure of thermodynamics, (ii) the H-theorem (macroscopic time irreversibility), (iii) the Ehrenfest theorem (correspondence principle between classical and quantum mechanics), (iv) the Onsager reciprocity theorem (microscopic time reversibility), (v) the Kramers and Wannier relations (causality), (vi) the factorization of the likelihood function (Einstein' 1910 reversal of Boltzmann's formula), (vii) the Bogolyubov inequality, (viii) thermodynamic stability (i.e., a definite sign for the specific heat) and (ix) the Pesin equality. In contrast, the following quantities depend on q: (i) the specific heat, (ii) the magnetic susceptibility, (iii) the fluctuation–dissipation theorem, (iv) the Chapman–Enskog expansion, the Navier–Stokes equations and related transport coefficients, (v) the Vlasov equation, (vi) the Langevin, Fokker–Planck and Lindblad equations, (vii) stochastic resonance, (viii) the mutual information or Kullback–Leibler entropy. Various theoretical tools have been developed in this nonextensive statistical framework, which include (i) Linear response theory, (ii) perturbation expansion (iii) Variational method (based on the Bogoliubov inequality), (iv) many-body Green functions, (v) path integral and Bloch equation, (vi) quantum statistics and those associated with the Gentile and the Haldane exclusion statistics, (vii) simulated annealing and related optimization, Monte Carlo and Molecular dynamics techniques.

14.4 Superposition of Distributions

14.4.1 Power Law Distribution of Widths

General Case. Consider the situation where the variable y, conditioned on a characteristic width σ, is distributed according to the "bare" distribution

$$P_\sigma(y) = C_1 e^{-f(y/\sigma)} , \tag{14.54}$$

where $f(y/\sigma) \to 0$ sufficiently fast for large y/σ to ensure the normalization of the distribution. Let us now assume that one measures a set of realizations of the variable y which is a mixture of different values of σ. This could result from a drift of the system or various other causes. Let us call

$$\Sigma(\sigma) \equiv C_2 e^{-g(\sigma)} \tag{14.55}$$

the distribution of widths σ. Then, the observed distribution of y is the following mixture of distributions

$$P_{\mathrm{ren}}(y) = \int_0^\infty \mathrm{d}\sigma\, \Sigma(\sigma) P_\sigma(y) = C_1 C_2 \int_0^\infty \mathrm{d}\sigma\, e^{-f(y/\sigma)-g(\sigma)} . \tag{14.56}$$

We ask what should be the form of $g(\sigma)$ in order for $P_{\text{ren}}(y)$ to be asymptotically of the form (14.1) for large y. For such large y, the integral can be evaluated by the saddle-node method [69], which yields

$$P_{\text{ren}}(y) \sim \frac{1}{\sqrt{F''[\sigma^*(y)]}} \, e^{-F[\sigma^*(y)]} \, , \tag{14.57}$$

where

$$F(\sigma) \equiv f(y/\sigma) + g(\sigma) \tag{14.58}$$

and $\sigma^*(y)$ is the solution of the saddle-node condition

$$\frac{y}{\sigma} \, f'(y/\sigma) = \sigma g'(\sigma) \, . \tag{14.59}$$

Here, the primes indicate the derivative of the functions with respect to their argument. The only solution of (14.59), such that $P_{\text{ren}}(y)$ is asymptotically proportional to $y^{-1-\mu}$ can be shown to be $\sigma^*(y) \propto y$, which implies from (14.57) and by solving (14.59) that $g(\sigma)$ must be of the form

$$g(\sigma) = (2 + \mu) \ln y + \text{const} \, . \tag{14.60}$$

Therefore, the distribution of the widths must be an asymptotic power law distribution with an exponent larger than the required one by one unit, due to the effect of the $1/\sqrt{F''\sigma^*(y)}$ factor in (14.57):

$$g(\sigma) \sim \frac{1}{\sigma^{2+\mu}} \, . \tag{14.61}$$

This mechanism holds as long as $P_\sigma(y)$ falls off faster than a power law at large y. This ensures that large realizations of y correspond to large width $\sigma \propto y$ occurrences.

Superposition of Exponential Distributions. Let us consider the exponential distribution $h(x) = \exp(-x/\lambda)$ and ask what should be the form of the fluctuations of the parameter $1/\lambda$ such that, upon averaging, the exponential $h(x)$ is transformed into

$$H(x) = \left(1 + \frac{x}{\lambda_0} \frac{1}{\alpha}\right)^{-(1+\mu)} \, , \tag{14.62}$$

which has a power law tail with characteristic exponent μ. We adopt the parameterization (14.62) because it has the form introduced by Tsallis (see the previous section) in his non-extensive statistics [962, 964] as a generalization of Boltzmann statistics discussed in Chap. 7 to account for long tail distributions occurring in many systems in nature. Tsallis' approach has been applied to many problems, from astrophysics to biological systems.

The central formula introduced by Tsallis is the following power-like distribution:

$$H_q(x) = C_q \left[1 - (1 - q)\frac{x}{\lambda}\right]^{1/(1-q)} \, , \tag{14.63}$$

which is just a one-parameter generalization of the Boltzmann–Gibbs exponential formula to which it converges for $q \to 1$: $H_{q=1} = h(x) = c\exp(-x/\lambda)$. Expression (14.63) gives (14.62) with the definition

$$1 + \mu = \frac{1}{q - 1} \, . \tag{14.64}$$

When $H_q(x)$ is used as a probability distribution of the variable $x \in (0, \infty)$, the parameter q is limited to $1 \leq q < 2$. For $q < 1$, the distribution $H_q(x)$ is defined only for $x \in [0, \lambda/(1 - q)]$. For $q > 1$, the upper limit comes from the normalization condition for $H_q(x)$ and from the requirement of positivity of the resulting normalisation constant C_q. However, if one demands in addition that the mean value of $H_q(x)$ is well defined, i.e., that $\langle x \rangle = \lambda/(3 - 2q) < \infty$ for $x \in (0, \infty)$, then q is further limited to $1 \leq q < 1.5$ (corresponding to $\mu > 1$).

Our aim is thus to deduce the form of the distribution $g(1/\lambda)$ of the fluctuations of the parameter $1/\lambda$ of the exponential $h(x)$ with mean value $1/\lambda_0$, which transforms it into the distribution:

$$\left(1 + \frac{x}{\lambda_0} \frac{1}{1 + \mu}\right)^{-(1+\mu)} = \int_0^\infty \exp\left(-\frac{x}{\lambda}\right) g\left(\frac{1}{\lambda}\right) \mathrm{d}\left(\frac{1}{\lambda}\right) \, . \tag{14.65}$$

From the representation of the Euler gamma function, we have

$$\left(1 + \frac{x}{\lambda_0} \frac{1}{1 + \mu}\right)^{-(1+\mu)}$$
$$= \frac{1}{\Gamma(1 + \mu)} \int_0^\infty \mathrm{d}\xi \, \xi^\mu \exp\left[-\xi \left(1 + \frac{x}{\lambda_0} \frac{1}{1 + \mu}\right)\right] \, .$$

Changing variables under the integral in such a way that $(\xi/\lambda_0)(1/(1+\mu)) = 1/\lambda$, one immediately obtains (14.65) with $g(1/\lambda)$ given by the following Gamma distribution in terms of the variable $1/\lambda$

$$g\left(\frac{1}{\lambda}\right) = \frac{(1 + \mu)\lambda_0}{\Gamma(1 + \mu)} \left(\frac{(1 + \mu)\lambda_0}{\lambda}\right)^\mu \exp\left(-\frac{(1 + \mu)\lambda_0}{\lambda}\right) \, , \tag{14.66}$$

with mean value

$$\left\langle \frac{1}{\lambda} \right\rangle = \frac{1}{\lambda_0} \tag{14.67}$$

and variation

$$\left\langle \left(\frac{1}{\lambda}\right)^2 \right\rangle - \left\langle \frac{1}{\lambda} \right\rangle^2 = \frac{1}{1 + \mu} \lambda_0^2 \, . \tag{14.68}$$

The distribution (14.66) is found to be exactly of the power law form (14.61) by noting that $g(1/\lambda) \sim 1/\lambda^\mu$ for large λ and that the distribution of λ is $(1/\lambda^2)g(1/\lambda) \sim 1/\lambda^{2+\mu}$.

14.4.2 Sum of Stretched Exponentials (Chap. 3)

A related but less trivial mechanism has been discussed in Chap. 3, in relation to the regime of "extreme deviations" and its application to fragmentation. In this mechanism, the distribution of fragments is a superposition similar to (14.56) but where the sum over the widths is replaced by a sum over the number N of fragmentation generations (3.83), $P_\sigma(y)$ is replaced by stretched exponentials with exponent inversely proportional to N and $\Sigma(\sigma)$ is replaced by the exponentially large number of fragments generated as a function of the generation order N. The interplay between the sub-exponential decay at a fixed generation number N with the exponential growth of the number of fragments as a function of N results in a power law distribution with an exponent determined from a transcendental equation (3.85).

14.4.3 Double Pareto Distribution by Superposition of Log-Normal pdf's

We now discuss a variation on the multiplicative generative model which also yields a power law behavior. Recall that in the simplest version of a multiplicative model, if we begin with some value X_0 and every step yields an independent and identically distributed multiplier from a lognormal distribution, then the resulting distribution X_t after t steps is lognormal. Actually, Kolmogorov was able to show that the log-normal distribution appears asymptotically as the consequence of a much more general class of multiplicative processes, as discussed in Chap. 16.

Suppose, however, that instead of examining X_t for a specific value of the number t of multiplications, we examine the random variable X_T where T itself is a random variable. As an example, when considering income distribution, in seeing the data, we may not know how long each person has lived. If different age groups are intermixed, the number of multiplicative steps each person may be thought to have undergone may be as a random variable. This effect was noticed by Montroll and Schlesinger [646, 647]. They showed that a mixture of lognormal distributions based on a geometric distribution would have essentially a lognormal body but a power law distribution in the tail. In the case where the time T is an exponential random variable, the resulting distribution of X_T has an asymptotic power law distribution [438, 439]. This is easily seen from the formula

$$P(X) = \int_0^{+\infty} dT \, \frac{X_0}{\sigma_0 X \sqrt{2\pi T}} e^{-(\ln(X/X_0))^2/2\sigma_0^2 T} e^{-T/T_0}$$

$$\sim \frac{1}{X^{1+\mu}} \,, \tag{14.69}$$

with

$$\mu = \sigma_0 \sqrt{\frac{2}{T_0}} \,, \tag{14.70}$$

as seen from a saddle-node argument. Huberman and Adamic suggest that this result can explain the power law distribution observed for the number of pages per site in the World Wide Web [438, 439]. As the Web is growing exponentially, the age of a site can roughly be thought of as distributed like an exponential random variable. If the growth of the number of pages on a Web site follows a multiplicative process, the above result suggests a power law distribution. Variations of the distribution of T away from the exponential pdf, incomplete sampling or finite-size effects may then provide mechanisms for empirical distributions that are found between log-normal and power laws.

The change of variable $T = u^2$ allows us to rewrite the integral in (14.69) as

$$P(X) = \frac{2X_0}{T_0 \sigma_0 X \sqrt{2\pi}} \int_0^{+\infty} du \, e^{-(\ln(X/X_0))^2/2\sigma_0^2 u^2} e^{-u^2/T_0} \ . \qquad (14.71)$$

Let us recall the useful identity

$$\int_0^{+\infty} dz \, e^{-az^2 - b/z^2} = \frac{1}{2} \sqrt{\frac{\pi}{a}} e^{-2\sqrt{ab}} \ . \qquad (14.72)$$

Comparing (14.71) with (14.72), in the exponent $2\sqrt{a\,b}$ of the identity, we have $b = (\ln(X/X_0))^2/2\sigma_0^2$. Because of this, there are two different behaviors, depending on whether $X/X_0 \geq \sigma_0$ or $X/X_0 < \sigma_0$. For $X/X_0 \geq \sigma_0$, $P(X)$ is a power law $\sim 1/X^{1+\mu}$ with exponent μ already given in (14.70). For $X/X_0 < \sigma_0$, $P(X)$ has also a power law branch $\sim 1/X^{1-\mu}$ with μ also given in (14.70). This double Pareto distribution seems to fit better empirical distribution of incomes [766–768]. Similar double Pareto distributions have been advocated to account for the distribution of earthquake energies, but with both exponents of the pdf larger than 1 [903].

14.5 Random Walks: Distribution of Return Times to the Origin

On motivation for studying the distribution of return times of random walks is that many self-organized critical (SOC) models discussed in Chap. 15, including the minimal model of Bak and Sneppen [48], can be mapped in the limit of infinite space dimension onto a critical branching process. As a consequence, this allows us to relate the exponent describing the fluctuation distribution (see Chap. 15 for definitions) to the first return time of a one-dimensional random walk [48]. The result is that we expect these models to possess an event size distribution characterized by an exponent of value $3/2$ for all dimensions above some upper critical dimension, and lower exponent values for dimensions below this.

14.5.1 Derivation

Taking into account all possible walk trajectories, the probability density for an unbiased random walker starting at the origin to be found at the position x along a line after a time t is given by (2.42) with $v = 0$. Thus, the probability density to return to the origin at time t (without excluding that previous returns to the origin might have occurred before) is

$$P_G(t) = \frac{1}{\sqrt{2\pi}} \frac{1}{\sqrt{2Dt}} \sim t^{-1/2} \,, \tag{14.73}$$

where D is the diffusion coefficient.

The probability $F(t)$ to return to the origin for the first time at time t after starting from the origin at time 0 is given by the relationship

$$P_G(t) = \delta(t) + \int_0^t dt' \, P_G(t') \, F(t - t') \,. \tag{14.74}$$

The first term $\delta(t)$ of (14.74) expresses the fact that the walker is at the origin at time zero with certainty and thus corresponds to the initial condition. The integral quantifies the fact that, in order to be found at the origin at time t, the walker can follow any trajectory that brings him back to the origin at a time t' earlier or at most equal to t and then come back at most once to the origin after an additional interval of time $t - t'$. All possible scenarios with all possible intermediate t' are to be considered. The integral in (14.74) is a convolution. Since there is an initial time $t = 0$, it can be dealt with by taking the Laplace transform $\hat{P}_G(\beta) \equiv \int_0^\infty dt \, P_G(t) e^{-\beta t}$, yielding

$$\hat{P}_G(\beta) = 1 + \hat{P}_G(\beta)\hat{F}(\beta) \,, \tag{14.75}$$

where we have used the fact that $\int_0^\infty dt \int_0^t dt' = \int_0^\infty dt' \int_{t'}^\infty dt$, as can be checked graphically in the plane (t', t). The solution of (14.75) is

$$\hat{F}(\beta) = \frac{\hat{P}_G(\beta) - 1}{\hat{P}_G(\beta)} \,, \tag{14.76}$$

whose inverse Laplace transform yields $F(t)$. To go further, we need to estimate $\hat{P}_G(\beta)$. We are interested in the form of $F(t)$ for large waiting times t, which corresponds to small conjugate variables β. The regime of small β's is thus controlled by the large t for $P_G(t)$ which is given by (14.73). Thus,

$$\hat{P}_G(\beta) \approx_{\beta \to 0} \int_{t_{\min}}^\infty dt \, \frac{e^{-\beta t}}{\sqrt{4\pi Dt}} \approx \frac{1}{C\beta^{1/2}} \,, \tag{14.77}$$

where $1/C = \int_0^\infty dx e^{-x}/\sqrt{4\pi Dx}$ is a constant. Inserting (14.77) in (14.76) gives

$$\hat{F}(\beta) \approx_{\beta \to 0} 1 - C\beta^{1/2} \approx e^{-C\beta^{1/2}} \,. \tag{14.78}$$

Using the results of Sect. 4.4, its inverse Laplace transform has a tail

$$F(t) \sim_{t \to \infty} t^{-3/2} . \tag{14.79}$$

An exact calculation gives

$$F(t) = \frac{C}{\sqrt{2\pi}t^{3/2}} e^{-C^2/2t} . \tag{14.80}$$

Expression (14.80) is nothing but the stable Lévy distribution with exponent $\mu = 1/2$ discussed in Chap. 4. The tail behavior (14.79) can be retrieved by a direct examination of (14.74) upon remarking that the integral for large t is controlled by the behavior of the integrand for t' small for which it reduces to $\sim \int_0 dt' (t-t')^{-\alpha}$, where we have assumed a power law dependence of $F(t) \sim t^{-\alpha}$. Then, the integration around the lower bound 0 retrieves (14.79), i.e. $\alpha = 3/2$, by the condition that the integral is proportional to $P_G(t) \sim t^{-1/2}$. The method of images also allows us to recover this result. It amounts to introducing an anti-source of walkers a small distance a from the origin which anihilates all walks crossing 0 before t. It is easy to see that this leads to

$$F(t) \sim \frac{\partial P_G(x,t)}{\partial x}\bigg|_{x=a} , \tag{14.81}$$

corresponding to a dipole source of walkers.

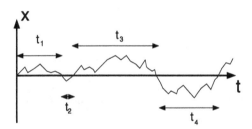

Fig. 14.2. Definition of the return times $t_1, t_2, t_3, \ldots$ to the origin of a random walk

The function $F(t)$ given by (14.79) and (14.80) is nothing but the distribution of waiting times to return to the origin. Consider the situation where a random walker is launched from the origin and we measure the time he takes to return for the first time to the origin. When this occurs, we continue to monitor his trajectory and measure the time needed to again cross the origin, and so on, as shown in Fig. 14.2. The distribution of the return times $t_1, t_2, t_3, \ldots$ is exactly given by expression (14.80).

14.5.2 Applications

The fact that the exponent $\mu = 1/2$ of this distribution is smaller than one has an important consequence, namely the average return time $\langle t \rangle = \int_0^\infty dt\, t F(t)$

is infinite. Physically, this means that it is controlled by the largest return time sampled in a given time span. Practically, this has a number of consequences. For instance, this implies a long-memory and thus anomolous aggregation effects as discussed in Chap. 8. Anecdotically, this also allows the rationalization of the overwhelming despair of frustrated drivers in dense highways that neighboring lanes always go faster than their lane because they often do not see a car return that was previously adjacent to them: assuming that we can model the differential motion of lanes in a global traffic flow by a random walk, this impression is a direct consequence of the divergence of the expected return time!

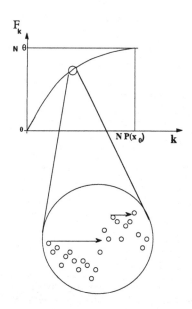

Fig. 14.3. Strength F_k of a bundle of N fibers as a function of the number k of broken fibers; the magnified view of the dependence of F_k illustrates the random walk in the space of forces F_k, the role of time being played by k

An interesting application of this distribution of return times to the origin is the distribution of rupture bursts in the democratic fiber bundle model discussed in Sect. 13.2.2. When looking at the fiber scale, simultaneous failure of many fibers can occur due to "small scale" fluctuations in the strength of bundle subsets. Indeed, the sequence $\{F_k\}$ of external loads at which the fibers would fail do not form a monotonically increasing sequence as shown in Fig. 14.3. The random variables defined as the fiber rupture thresholds X_k are indeed put in increasing order but they are multiplied by a monotically decreasing factor $(N + 1 - k)$ of unfailed fibers, as k increases. Starting from a stable configuration corresponding to some value F_k, a simultaneous rupture of Δ fibers, which can be called an event or burst of size Δ, occurs if $F_n < F_k$ for $k+1 \leq n \leq k+\Delta$ and $F_{k+\Delta+1} \geq F_k$. Since the set of failure thresholds X_k are independent random variables, the function F_k undergoes a random walk where the displacement is F_k and the time is k (see Fig. 14.3). This random walk is biased by a drift equal to $\langle F_{k+1} - F_k \rangle$ which vanishes at the instability threshold. The bias is large far away from the instability threshold and only small bursts occur there. As the global rupture is approached, $\langle F_{k+1} - F_k \rangle$ goes to zero and the fluctuations completely dominate the burst occurrence. In the random walk picture, it is easy to obtain the probability that a burst of size Δ occurs after a total of k fibers have been broken, i.e. at some value F_k. This corresponds to the probability $F(\Delta)$ of first return to the origin of a random walker (see Fig. 14.3). In the absence of any bias, we find that the local differential distribution $d(\Delta)$ of bursts of size E is given by [400, 415, 514, 867]

$$d(\Delta) \sim \Delta^{-3/2} \ . \tag{14.82}$$

This law holds sufficiently close to the global rupture point so that the bias in the random walk is not felt. Cumulating the measurements of the bursts over the entire rupture history, we have shown in Chap. 13 that the distribution $d(\Delta)$ is renormalized by the effect of the changing bias as one approaches rupture with an apparent exponent $5/2$ instead of $3/2$ [867, 870].

14.6 Sweeping of a Control Parameter Towards an Instability

This mechanism has been pointed out to show that several claims [133, 730] about evidence of self-organized criticality could be actually explained by the simpler mechanism of the slow sweeping of a control parameter towards a critical point [834, 835, 875, 876]. The idea is best illustrated in the context of material rupture.

In a typical experiment where one brings a material to rupture, a "control" parameter is varied which progressively stresses the system up to its global rupture. As the stress increases, the number and amplitude of acoustic emissions is usually found to increase. It has been known for decades in the engineering literature that the resulting distribution of energies of the acoustic emission bursts, measured over a complete experiment up to failure, is a power law with exponent μ close to one [741]. This phenomenon has been rediscovered in the physics literature [29, 133, 236, 333, 431, 994] and has become fashionable due its apparent connection with fractals and self-organized criticality.

In fact, the origin of the power law probably lies elsewhere. Notice that a typical system is subjected to an increasing stress or strain until it fails irreversively and is thus not in a stationary state. Rather, it undergoes a sweeping of its control parameter up to the critical rupture value. Systems are also often brought to rupture under constant stress due to some kind of corrosion mechanism, plastic deformation or aging. A plausible first-order description of these observations is as follows [553, 555]. Stressed elements undergo stress corrosion with exponent α that lead to a continuous decay of their threshold. When a threshold reaches the stress level, an abrupt rupture occurs with a stress drop ratio γ and with a partial (non-conservative) distribution of stress to neighboring elements with dissipation factor β and the threshold heals. The model is thus of the class of sandpile models discussed in Chap. 15 with absence of conservation and in addition dynamical thresholds coupled to the stress variable. Numerical simulations in 1d, 2d and 3d lattices and a mean-field approximation [555] determine that the average stress decays in a punctuated fashion, with a characteristic Omori's power law one-over-time dependence, with in addition events occurring at characteristic times increasing as a geometrical series with multiplicative factor $\lambda = [1 - \gamma + \gamma(1 - \beta)]^{-\alpha}$ which is a function of the stress corrosion exponent α, the stress drop ratio γ and the degree of dissipation β. This behavior is independent of the discrete nature of the lattice and stems from the interplay between the threshold dynamics and the power law stress relaxation. This discrete geometrical series can lead to observable log-periodic signatures.

Let us examine more generally this phenomenon of "sweeping" of a control parameter towards a "critical" point, or towards a global bifurcation, as it provides a robust mechanism for the appearance of power law distribution of events. Indeed, this can be traced back to the cumulative measurements of fluctuations, which diverge at the approach of the critical instability.

Consider the simplest possible examples of the Ising model or the percolation model. These models are among the simplest archetypes leading to a critical transition reached by finely tuning the temperature (Ising) or the concentration (percolation) to their critical values. As criticality is reached for some value of the control parameter, they are not self-organized. The thermal Ising problem and geometrical bond percolation are closely related since they can be put in one-to-one correspondence, using Coniglio and Klein's recipe [180]. For a given value of the control parameter p, spontaneous fluctuations occur. These fluctuations in both models correspond to the clusters defined by a connectivity condition [510]. These fluctuations can be visualized as spatial cooperative domains of all sizes between the microscopic scale up to the correlation length ξ in which the order parameter takes a non-zero value over a finite duration depending on the domain size. Their size distribution is given by [928]

$$P_p(s)\,\mathrm{d}s \sim s^{-a} f\left(\frac{s}{s_0(p)}\right)\,\mathrm{d}s\;, \tag{14.83}$$

with

$$a = 2 + \frac{1}{\delta} \quad (= 2.05 \quad \text{with} \quad \delta = 91/5 \quad \text{in 2D percolation})\;. \tag{14.84}$$

The size s is the number of spins or sites belonging to a given cluster. $s_0(p)$ is a typical cluster size given by

$$s_0 \sim |\,p_\mathrm{c} - p\,|^{-1/\sigma}\;, \tag{14.85}$$

with Fisher's notation

$$\frac{1}{\sigma} = \gamma + \beta \ . \tag{14.86}$$

$s_0(p)$ must be distinguished from the mean cluster size $\langle s \rangle(p)$, which scales as

$$\langle s \rangle(p) \sim |p_c - p|^{-\gamma} \ . \tag{14.87}$$

γ is the susceptibility exponent defined by the number of spins which are affected by the flip of a single spin, which corresponds to, in the language of percolation, the mean cluster size. β is the exponent characterizing the way the order parameter goes to zero as $|p_c - p|^{\beta}$ as $p \to p_c$. The scaling function $f(s/s_0(p))$ decays rapidly (exponentially or as an exponential of a power law) for $s > s_0$. Thus, the cluster or fluctuation size distribution is a power law $P_p(s) \, ds \sim s^{-a}$ for $s < s_0(p)$ and $P_p(s)$ is negligibly small for $s > s_0(p)$.

Now, suppose that one monitors the fluctuation amplitudes (i.e. cluster sizes) as the control parameter p is swept across its critical value p_c, say from the value $p = 0$ to $p = 1$. The results below remain true if the interval is reduced to $[p_c - c_1; p_c + c_2]$, with c_1 and c_2 both positive or zero. This sweeping can be done for instance by increasing the applied stress. The total number of clusters of size s which are measured is then proportional to

$$N(s) = \int_0^1 P_p(s) \, dp \ , \tag{14.88}$$

which can be written as

$$N(s) = \int_0^{p_c} P_p(s) \, dp + \int_{p_c}^1 P_p(s) \, dp \ . \tag{14.89}$$

In writting this expression, we have used the fact that the full distribution $P_p(s)$ of fluctuations can be sampled for each value p of the control parameter. In other words, the evolution of the control parameter is adiabatic. The change of variable $p \to s_0(p)$ in the above integral gives

$$N(s) = s^{-a} \int_1^{+\infty} s_0^{-\sigma(1+1/\sigma)} f\left(\frac{s}{s_0(p)}\right) ds_0$$

$$\simeq s^{-a} \int_s^{+\infty} s_0^{-(1+\sigma)} \, ds_0 \ , \tag{14.90}$$

using the fact that $f[s/s_0(p)]$ is negligible for $s_0(p) < s$. Here, the symbol $\sim$ is taken as meaning the leading behavior in decreasing powers of s. This finally yields the power law

$$N(s) \simeq s^{-(a+\sigma)} \ , \tag{14.91}$$

as its leading behavior. Note that we have not restricted p to stop at its critical value p_c but have allowed for a spanning of an arbitrary interval containing p_c.

This expression (14.91) demonstrates that a continuous monitoring of events or fluctuations up to and above a critical point yields a power law even if a similar measurement for a fixed value of p would only give a truncated power law (with a smaller exponent). Thus, by varying the temperature (Ising) or concentration (percolation), say, linearly in time from a value below the critical point to a value above the critical point and by integrating over all fluctuations observed during this entire time interval, one gets a power law in the distribution of fluctuations. Since only right at the critical point, large clusters exist, one gets a power law without cut-off for the time-integrated cluster number even if we do not stop at the critical point. Note the value of the renormalized exponent $a + \sigma$, stemming from the relatively smaller weight of large clusters which are found only in a narrow interval of the control parameter.

This mechanism works for the Democratic Fiber Bundle model just discussed above and in Chap. 13. As we have seen, this model exhibits a "local" differential power law distribution of bursts of size Δ with an exponent $a = 3/2$, with a cut-off exponent $\sigma = 1$, and a power law distribution of the total number of bursts of size Δ with an exponent $a + \sigma = 5/2$, in agreement with the above derivation. The exponent $5/2$ reflects the occurrence of larger and larger events when approaching the total breakdown instability. The mechanism of "sweeping a control parameter towards an instability" applies more generally to rupture phenomena and many other systems whose conditions vary with time. The relevance of this mechanism to a variety of models and experiments has been discussed in [875] such as in the Burridge–Knopoff model of earthquakes, foreshocks and acoustic emissions, impact ionization breakdown in semiconductors, the Barkhausen effect, charge density waves, pinned flux lattices, elastic strings in random potentials and real sandpiles.

14.7 Growth with Preferential Attachment

The mechanism for the generation of power laws often refered to as "preferential attachment" has a long and interesting history. It has been rediscovered many times in several disciplines. A flurry of activity and rediscovery occurred recently in computer science in which power law distributions are also found to be pervasive. For instance, computer file sizes are believed to be governed by a power law distribution. Consider the World Wide Web, which can naturally be thought of as a graph, with pages corresponding to vertices and hyperlinks corresponding to directed edges. Empirical work has shown that indegrees and outdegrees of vertices in this graph obey power law distributions. Recall that the indegree (resp. outdegree) is the number of inward (resp. outward) directed graph edges from a given graph vertex in a directed graph. Most models of random graphs developed to understand these observations are variations of the following theme, whose presentation borrows from a recent review by M. Mitzenmacher [635].

Let us start with a single page, with a link to itself. At each time step, a new page appears, with outdegree 1. With probability $p < 1$, the link for the new page points to a page chosen uniformly at random. With probability $1 - p$, the new page points to a page chosen proportionally to the indegree of the page. This model exemplifies preferential attachment: new objects tend to attach to popular objects. In the case of the Web graph, new links tend to go to pages that already have links. A simple argument deriving the power law distribution of page degrees follows. Let $X_j(t)$ be the number of pages with indegree j when there are t pages in the system. Then, for $j > 1$, the probability that X_j increases is

$$\frac{pX_{j-1}}{t} + \frac{(1-p)(j-1)X_{j-1}}{t} . \tag{14.92}$$

The first term is the probability that a new link is chosen at random and chooses a page with indegree $j - 1$. The second term is the probability that a new link is chosen proportionally to the indegrees and chooses a page with indegree $j - 1$. Similarly, the probability that X_j decreases is

$$\frac{pX_j}{t} + \frac{(1-p)jX_{j-1}}{t} . \tag{14.93}$$

Hence, for $j > 1$, the growth of X_j is approximated by

$$\frac{dX_j}{dt} = \frac{1}{t} \left[p(X_{j-1} - X_j) + (1-p)\left((j-1)X_{j-1} - jX_j\right) \right] . \tag{14.94}$$

This intuitively appealing use of a continuous differential equation to describe what is clearly a discrete process can be justified more formally using martingales [542, 1026]. The case of X_0 must be treated specially, since each new page introduces a vertex of indegree 0:

$$\frac{dX_0}{dt} = 1 - \frac{pX_0}{t} . \tag{14.95}$$

In the steady state limit, we can look for a solution of the form $X_j(t) = c_j t$, defining c_j as the steady-state fraction of pages with indegree j. Then, we can successively solve for the c_j. Equation (14.95) becomes $dX_0/dt = t(dc_0/dt) + c_0 = 1 - pc_0$, which yields $c_0 = 1/(1+p)$. More generally, using the equation for dX_j/dt, we obtain for $j \geq 1$,

$$c_j(1 + p + j(1 - p)) = c_{j-1}(p + (j-1)(1-p)) . \tag{14.96}$$

This recurrence can be used to determine the c_j exactly. Asymptotically for large j, we obtain

$$\frac{c_j}{c_{j-1}} = 1 - \frac{2-p}{1+p+j(1-p)} \sim 1 - \frac{2-p}{1-p}\frac{1}{j} . \tag{14.97}$$

Expression (14.97) implies the power law distribution

$$c_j \sim \frac{C}{j^{1+\mu}} , \quad \text{with } \mu = \frac{1}{1-p} , \tag{14.98}$$

for some constant C. Interestingly, the exponent μ is a function of the "preferential attachment" parameter $1 - p$: the larger is the preferential bias (the larger is $1 - p$), the smaller is μ, up to a point when, in the limit of very strong preferential attachment, μ tends to 1.

Although the above argument was described in terms of degree on the Web graph, this type of argument is very general and applies to any sort of preferential attachment. In fact, the first similar argument dates back to at least 1925. It was introduced by Yule [1042] to explain the distribution of species among genera of plants, which had been shown empirically by Willis to satisfy a power law distribution. Mutations cause new species to develop within genera, and more rarely mutations lead to entirely new genera. Mutations within a genus are more likely to occur in a genus with more species, leading to the preferential attachment. A clearer and more general development of how preferential attachment leads to a power law was given by Simon [844] in 1955. Simon listed several applications of this type of model in his introduction: distributions of word frequencies in documents, distributions of numbers of papers published by scientists, distribution of cities by population, distribution of incomes, and distribution of species among genera.

The mechanism of "preferential attachment" can also be invoked to rationalize very early observations of power law distributions, such as the Pareto distribution of income distribution in 1897 [718]. The first known attribution of the power law distribution of word frequencies appears to be due to Estoup in 1916 [281], although generally the idea (and its elucidation) are attributed to Zipf [1063–1065]. Similarly, Zipf is often credited with noting that city sizes appear to match a power law, although this idea can be traced back further to 1913 [38] (see [322] for a modern account). Lotka (circa 1926) found in examining the number of articles produced by chemists that the distribution followed a power law [571] (see [547] for an alternative viewpoint); indeed, power laws of various forms appear in many places in informetrics [101].

In graph theory, the preferential attachment argument has been developed as part of the study of random trees. Specifically, consider the following recursive tree structure. Begin with a root node. At each step, a new node is added; its parent is chosen from the current vertices with probability proportional to one plus the parent's number n of children. This is just another example of preferential attachment; indeed, it is essentially equivalent to the simple Web graph model described above with the probability p of choosing a random node equal to $1/2$, since there exists a factor f such that $f(1 + n) = p + n(1 - p)$ only for $p = 1/2$ [578, 854].

Modern works on random graph models have led to many new insights. Perhaps most important is the development of a connection between Simon's model, which appears amenable only to limiting analysis based on differential equations as shown above, and purely combinatorial models based on random graphs [99, 578, 854]. Such a connection is important for further rigorous analysis of these structures. Also, current versions of Simon's arguments based on martingales provide a much more rigorous foundation. More recent work has focused on greater understanding of the structure of graphs that arise from these kinds of preferential attachment model [241, 242]. It has been shown that in the Web graph model described above where new pages copy existing links, the graphs have community substructures [542], a property not found in random graphs but amply found in the actual Web. See also [537, 538] for exact results on the pdf of cluster sizes and other applications.

14.8 Multiplicative Noise with Constraints

14.8.1 Definition of the Process

Consider the simple multiplicative recurrence equation

$$x_{t+1} = a(t)x(t) \; , \tag{14.99}$$

where $a(t)$ is a stochastic variable with probability distribution $\Pi(a)$. With no other ingredient, expression (14.99) generates an ensemble of values $x(t)$ over all possible realizations of the multiplicative factors $a(0), a(1), a(2), \ldots, a(t)$, which is distributed according to the log-normal distribution [8, 350, 764] in the large t limit. Indeed, from the logarithm of (14.99), we see that $\ln x$ is the sum of t random variables. As seen in Chap. 2, we get, for large times t,

$$P(x) = \frac{1}{\sqrt{2\pi Dt}} \frac{1}{x} \exp\left[-\frac{1}{2Dt}(\ln x - vt)^2 \right] \; , \tag{14.100}$$

where $v = \langle \ln a \rangle \equiv \int_0^\infty da \ln a \Pi(a)$ and $D = \langle (\ln a)^2 \rangle - \langle \ln a \rangle^2$.

As we have seen in Chap. 4, expression (14.100) can be rewritten

$$P(x) = \frac{1}{\sqrt{2\pi Dt}} \frac{1}{x^{1+\mu(x)}} e^{\mu(x)vt} \; , \tag{14.101}$$

with

$$\mu(x) = \frac{1}{2Dt} \ln \frac{x}{e^{vt}} \; . \tag{14.102}$$

Since $\mu(x)$ is a slowly varying function of x, this form shows that the log-normal distribution can be mistaken for an apparent power law with an exponent μ slowly varying within the range x which is measured. However, notice that $\mu(x) \to \infty$ far in the tail $x \gg e^{(v+2D)t}$ and the log-normal distribution is *not* a power law.

One needs additional ingredients to transform (14.101) into a genuine power law distribution [205, 215, 501, 561, 858, 879, 880, 891, 944].

- The first ingredient is that the random map (14.99) must be contracting on average, i.e. $x(t) \to 0$ for large t if there is no other constraint, which is equivalent to the condition $v \equiv \langle \ln a \rangle < 0$.

- The second condition prevents this contraction to zero by ensuring that x remains larger than a minimum value $x_0 > 0$, or at least exhibits finite fluctuations even for $t \to +\infty$. One way to ensure this condition is to introduce an additive term in (14.99) leading to the Kesten process, already discussed in Sect. 8.4.2.

M. Mitzenmacher [635] discusses further the close relationship between the log-normal and power law distributions, which an emphasis on applications in computer science.

14.8.2 The Kesten Multiplicative Stochastic Process

Let us consider the process defined in (8.28), $X_{n+1} = a_n X_n + b_n$, where the stochastic multiplicative a_n and additive b_n variables are drawn with the pdf's $P_a(a_n)$ and $P_b(b_n)$. Such processes have been analyzed in depth by Kesten [501], Vervaat [986], and Goldie [352] and have recently found some interest in physics because of the intermittent character of the ensuing fluctuations [879, 880, 891, 944]. Very general and important results on the statistical behavior of solutions of stochastic difference equations of the form (8.28) with multiplicative stochastic coefficients are already obtained in [501], Proposition 5. Let us summarize the essential results from the papers by Kesten, Vervaat, and Goldie. The latter sections provide simple physical derivations and extensions.

Theorem (Kesten, Goldie):

- if a_n and b_n are i.i.d. real-valued random variables and if $\langle a_n \rangle < 0$, then X_n converges in distribution and has a unique limiting distribution.
- If, additionally, $b_n/(1 - a_n)$ is nondegenerate (that is, b_n is not a constant times $(1 - a_n)$), and if there exists some $\mu > 0$ such that

  1. $0 < \langle |b_n|^\mu \rangle < +\infty$,
  2. $\langle |a_n|^\mu \rangle = 1$, and
  3. $\langle |a_n|^\mu \ln^+ |a_n| \rangle < +\infty$,

then the tail of the limiting distribution of X_n is asymptotic to a power law $\mathcal{P}_>(X_n > x) \simeq c/x^\mu$.

The available theory on multiplicative random processes thus ensures power law behavior for a large class of stochastic processes under relatively mild and general conditions. Intuitively, the time-varying multiplicative coefficient a_n yields some sort of intermittent amplification which leads to the heavy power law tails of the distribution while the additive noise term b_n preserves the motion from dying out in the course of events. Note that the existence of a stationary distribution does not hinge on existence of the second or even first moment as μ may assume arbitrary positive values depending on the process under study and only moments of $\mathcal{P}_>(X_n > x)$ of order smaller than μ do exist.

As an extension of this theorem, let us consider the situation in which the tail of $\mathcal{P}_>(X_n > x)$ is now controlled by the additive variable b_n. This occurs when $P_b(b_n)$ is a power law $\sim C_b/b_n^{1+\mu_b}$ with $0 < \mu_b < \mu$, where μ is the real positive solution of $\langle |a_n|^\mu \rangle = 1$. In this case, the condition 1. $0 < \langle |b_n|^\mu \rangle < +\infty$ is violated. Then, $\mathcal{P}_>(X_n > x)$ can be shown to remain a power law, as in Kesten's theorem, but the exponent of its tail is now μ_b. To see why, let us first consider the very simple case where the multiplicative factors a_n are no more stochastic and are all equal to some constant $a < 1$. Iterating (8.28), we obtain

$$X_n = b_n + a\, b_{n-1} + a^2 b_{n-2} + a^3 b_{n-3} + \dots \tag{14.103}$$

From the calculation tools on power laws given in Sect. 4.4, we obtain immediately that

$$\lim_{n\to+\infty} \mathcal{P}_>(X_n > x) \simeq \frac{C}{x^{\mu_b}}, \quad \text{with } C = \frac{C_b}{1 - a^{\mu_b}}. \tag{14.104}$$

This result (14.104) shows that the origin of the power law pdf of X_n is no more to be found in the intermittent stochastic multiplicative amplification but results from the power law pdf of the additive terms b_n. The autoregressive nature of (8.28) has solely the effect of renormalizing the scale factor from C_b to C by the factor $1/(1 - a^{\mu_b})$.

Let us now consider the case where the positive a_n's are random with $\langle |a_n|^{\mu_b} \rangle < 1$ (this is true if $0 < \mu_b < \mu$). Then, expression (14.103) is replaced by

$$X_n = b_n + a_n b_{n-1} + a_n a_{n-1} b_{n-2} + a_n a_{n-1} a_{n-2} b_{n-3} + \dots \tag{14.105}$$

We can use Breiman's theorem [125] which states that, for two independent random variables ϕ and $\chi > 0$ with $\text{Prob}(|\phi| > x) \simeq c/x^\kappa$ and $\langle \chi^{\kappa+\epsilon} \rangle < +\infty$ for some $\epsilon > 0$, the random product $\phi\chi$ obeys $\text{Prob}(|\phi\chi| > x) \simeq \langle \chi^\kappa \rangle \times \text{Prob}(|\phi| > x)$ for $x \to +\infty$. This generalizes the results on the calculation tools on power laws given in Sect. 4.4.

The distribution of each term $A_i \equiv a_n a_{n-1} \dots a_{n-i+1} b_{n-i}$ in (14.105) is thus $\mathcal{P}_>(A_i > x) \simeq \langle a_j^{\mu_b} \rangle^i C_b/x^{\mu_b}$, where we have used the property of independence between the a_i's. For a finite number n of terms in (14.105), the distribution of X_n is thus

$$\mathcal{P}_>(X_n > x) \simeq \frac{C_b}{x^{\mu_b}} \sum_{i=0}^{n-1} \langle a_j^{\mu_b} \rangle^i. \tag{14.106}$$

Rigorous bounds show that we can extend this reasoning for $n \to +\infty$ (X. Gabaix and H. Kesten, private communications), which leads to expression (14.104) with a^{μ_b} replaced by $\langle a_j^{\mu_b} \rangle$.

14.8.3 Random Walk Analogy

To intuitively understand the effect of the interplay between the average contraction $\langle a(t) \rangle < 0$ of the multiplicative process and the barrier preventing complete collapse to 0 (for instance by the action of the additive term in the Kesten process), let us use the variables $y(t) = \ln x(t)$ and $l = \ln a$. Then, (14.99) defines a random walk in y-space with steps l (positive and negative) distributed according to the density distribution $\Pi(l) = e^l \Pi(e^l)$. The distribution of the position of the random walk is similarly defined: $\mathcal{P}[y(t), t] = e^{y(t)} P(e^{y(t)}, t)$.

For $v \equiv \langle l \rangle < 0$, for which the random walk drifts towards the barrier, the qualitative picture is represented schematically in Fig. 14.4: steady-state $(t \to \infty)$ establishes itself and is characterized by the property that the net drift to the left is balanced by the reflection at the barrier. The random walk becomes trapped in an effective cavity of size of order D/v with an exponential tail $e^{-\mu y}$ with $\mu \approx |v|/D$ (see below). Its incessant motion back and forth and repeated reflections off the barrier and diffusion away from it lead to the build-up of an exponential probability profile. The exponential profile must be distinguished from the usual Gaussian distribution found for a non-constrained random walk. y is the logarithm of the random variable x and one then obtains a power law distribution for x of the form $\sim x^{-(1+\mu)}$.

Let us derive this result following [891]. In this goal, we write the master equation, which is equivalent to the random walk process $y(t+1) = y(t) + l(t)$, as

$$\mathcal{P}(y, t+1) = \int_{-\infty}^{+\infty} \Pi(l) \mathcal{P}(y - l, t)\, dl\ , \qquad (14.107)$$

which gives the probability $\mathcal{P}(y, t)$ to find the walker at position y to within dy at time t. As seen in Chap. 2, the exact master equation can be approximated by its corresponding Fokker–Planck equation. Usually, the Fokker–Planck equation becomes exact in the limit where the variance of $\Pi(l)$ and the time interval between two steps go to zero while keeping a constant finite ratio defining the diffusion coefficient [780]. In our case, this corresponds to taking

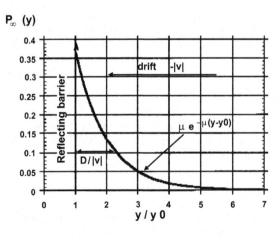

Fig. 14.4. Exponential steady-state profile of the probability density of the position of the random walk with a negative drift and a reflecting barrier. Taken from [891]

the limit of very narrow $\Pi(l)$ distributions. In this case, we can expand $\mathcal{P}(x - l, t)$ up to second order

$$\mathcal{P}(y - l, t) = \mathcal{P}(y, t) - l \frac{\partial \mathcal{P}}{\partial x}\bigg|_{(y,t)} + \frac{1}{2} l^2 \frac{\partial^2 \mathcal{P}}{\partial x^2}\bigg|_{(y,t)} \tag{14.108}$$

leading to the Fokker–Planck formulation

$$\frac{\partial \mathcal{P}(y, t)}{\partial t} = -\frac{\partial j(y, t)}{\partial y} = -v \frac{\partial \mathcal{P}(y, t)}{\partial y} + D \frac{\partial^2 \mathcal{P}(y, t)}{\partial y^2} , \tag{14.109}$$

where v and D have been defined above and are the leading cumulants of $\Pi(\ln a)$. $j(y, t)$ is the flux defined by

$$j(y, t) = v \mathcal{P}(y, t) - D \frac{\partial \mathcal{P}(y, t)}{\partial y} . \tag{14.110}$$

Expression (14.109) is nothing but the conservation of probability. It can be shown that this description (14.109) is generic in the limit of very narrow Π distributions: the details of Π are not important for the large t behavior; only its first two cumulants control the results [780]. v and D introduce a characteristic "length" $\hat{y} = D/|v|$. In the overdamped approximation, we can neglect the inertia of the random walker, and the general Langevin equation $m \, \mathrm{d}^2 y/\mathrm{d}t^2 = -\gamma \, \mathrm{d}y/\mathrm{d}t + F + F_{\text{fluct}}$ reduces to

$$\frac{\mathrm{d}y}{\mathrm{d}t} = v + \eta(t) , \tag{14.111}$$

which is equivalent to the Fokker–Planck equation (14.109). η is a noise of zero mean and delta correlation with variance D. This form illustrates the competition between drift $v = -|v|$ and diffusion $\eta(t)$.

The stationary solution of (14.109), $\partial \mathcal{P}(y, t)/\partial t = 0$, is immediately found to be

$$\mathcal{P}_\infty(y) = A - \frac{B}{\mu} e^{-\mu y} , \tag{14.112}$$

with

$$\mu \equiv \frac{|v|}{D}. \tag{14.113}$$

A and B are constants of integration. Notice that, as expected in this approximation scheme, μ is the inverse of the characteristic length $\hat{y}$. In absence of the barrier, the solution is obviously $A = B = 0$ leading to the trivial solution $P_\infty(x) = 0$, which is indeed the limit of the log-normal form (14.100) when $t \to \infty$. In the presence of the barrier, there are two equivalent ways to deal with it. The most obvious one is to impose normalization

$$\int_{x_0}^\infty \mathcal{P}_\infty(y) \, \mathrm{d}y = 1 , \tag{14.114}$$

where $y_0 \equiv \ln x_0$. This leads to

$$\mathcal{P}_\infty(y) = \mu e^{-\mu(y-y_0)} .$$ (14.115)

Alternatively, we can express the condition that the barrier at y_0 is *reflective*, namely that the flux $j(y_0) = 0$. Let us stress that the correct boundary condition is indeed of this type (and not absorbing for instance) as the rule of the multiplicative process is that we return $x(t)$ to x_0 when it becomes smaller than x_0, thus ensuring $x(t) \geq x_0$. An absorbing boundary condition would correspond to killing the process when $x(t) \leq x_0$. Substituting (14.112) in (14.110) with $j(y_0) = 0$, we retrieve (14.115) which is automatically normalized. Reciprocally, (14.115) obtained from (14.114) satisfies the condition $j(y_0) = 0$.

There is a faster way to get this result (14.115) using an analogy with a Brownian motion in equilibrium with a thermal bath. The bias $\langle l \rangle < 0$ corresponds to the existence of a constant force $-|v|$ in the $-x$ direction. This force derives from the linearly increasing potential $V = |v|y$. In thermodynamic equilibrium, a Brownian particle is found at the position x with probability given by the Boltzmann factor $e^{-\beta|v|y}$. This is exactly (14.115) with $D = 1/\beta$ as it should from the definition of the random noise modeling the thermal fluctuations.

Translating in the initial variable $x(t) = e^y$, we get the power law distribution

$$P_\infty(x) = \frac{\mu x_0^\mu}{x^{1+\mu}} ,$$ (14.116)

with μ given by (14.113):

$$\mu \equiv \frac{|\langle \ln a \rangle|}{\langle (\ln a)^2 \rangle - \langle \ln a \rangle^2} .$$ (14.117)

More generally, consider a linear or nonlinear multiplicative noise process

$$\frac{dx}{dt} = f(x) + g(x)\eta(t) ,$$ (14.118)

where η is a i.i.d. white noise. Then, the change of variable $x \to y = \int^x du/g(u)$ transforms (14.118) into the following additive noise process [396]

$$\frac{dy}{dt} = h(y) + \eta(t) , \quad \text{where} \quad h(y) = \frac{f(x)}{g(x)}\bigg|_{x(y)} .$$ (14.119)

14.8.4 Exact Derivation, Generalization and Applications

In the case where the barrier is absent, we have already pointed out that the random walk eventually escapes to $-\infty$ with probability one. However, it will wander around its initial starting point, exploring maybe to the right and left sides for a while before escaping to $-\infty$. For a given realization, we can thus measure the rightmost position y_{max} it ever reached. What is the distribution $\mathcal{P}_{\max}[\text{Max}(0, y_{\max})]$? The question has been answered in the mathematical literature using renewal theory ([293], p. 402) and the answer is

$$\mathcal{P}_{\max}(\text{Max}(0, y_{\max})) \sim e^{-\mu y_{\max}}, \tag{14.120}$$

with μ given by

$$\int_{-\infty}^{+\infty} \Pi(l)e^{\mu l}\, dl = \int_0^{+\infty} \Pi(\lambda)\lambda^\mu\, d\lambda = 1. \tag{14.121}$$

The proof can be sketched in a few lines [293] and we summarize it because it will be useful in the following. Consider the probability distribution function $M(y) \equiv \int_{-\infty}^y \mathcal{P}_{\max}(y_{\max})\, dy_{\max}$, that $y_{\max} \le y$. Starting at the origin, this event $y_{\max} \le y$ occurs if the first step of the random walk verifies $y_1 = Y \le y$ together with the condition that the rightmost position of the random walk starting from $-y_1$ is less or equal to $y - Y$. Summing over all possible Y, we get the Wiener–Hopf integral equation

$$M(y) = \int_{-\infty}^y M(y - Y)\Pi(Y)\, dY . \tag{14.122}$$

It is straightforward to check that $M(y) \to e^{-\mu y}$ for large y with μ given by (14.121). We refer to [293] for the questions of uniqueness and to [315, 651] for classical methods for handling Wiener–Hopf integral equations.

How is this result useful for our problem? Intuitively, the presence of the barrier, which prevents the escape of the random walk, amounts to reinjecting the random walker and enabling it to sample again and again the large positive deviations described by the distribution (14.120). Indeed, for such a large deviation, the presence of the barrier is not felt and the presence of the drift ensures the validity of (14.120) for large x. These intuitive arguments are exact for a broad class of processes.

Let us now generalize (14.99) with the barrier at x_0 into

$$x(t + 1) = e^{f(x(t), \{a(t), b(t)...\})}a(t)x(t) , \tag{14.123}$$

where $f(x(t), \{a(t), b(t)...\}) \to 0$ for $x(t) \to \infty$ and $f(x(t), \{a(t), b(t)...\}) \to \infty$ for $x(t) \to 0$.

The model (14.99) is the special case $f(x(t), \{a(t), b(t)...\}) = 0$ for $x(t) > x_0$ and $f(x(t), \{a(t), b(t)...\}) = \ln(x_0/a(t))$ for $x(t) \le x_0$. In general, we can consider a process in which at each time step t, after the variable $a(t)$ is generated, the new value $a(t)x(t)$ is readjusted by a factor $e^{f(x(t), \{a(t), b(t)...\})}$ reflecting the constraints imposed on the dynamical process. It is thus reasonable to consider the case where $e^{f(x(t), \{a(t), b(t)...\})}$ depends on t only through the dynamical variables $a(t)$ (in special cases, there is another variable b_t).

The existence of a limiting distribution for $x(t)$ obeying (14.123), for a large class of $f(x(t), \{a(t), b(t)...\})$ decaying to zero for large x and going to infinity for $x \to 0$, is ensured by the competition between the convergence of x to zero and the sharp repulsion from it. We shall also suppose in what follows that $\partial f\left(x(t),\{a(t),b(t)\ldots\}\right)/\partial x \to 0$ for $x \to \infty$, which is satisfied for a large class of smooth functions already satisfying the above conditions. It is an interesting mathematical problem to establish this result rigorously, for instance by the method used in [315, 561, 858]. Assuming the existence of the asymptotic distribution $P(x)$, we can determine its shape, which must obey

$$V \equiv axe^{-f(x,\{a(t),b(t)\ldots\})} \; , \tag{14.124}$$

where $\{a, b, \ldots\}$ represents the set of stochastic variables used to define the random process. The expression (14.124) means that the l.h.s. and r.h.s. have the same distribution. We can thus write

$$P_V(V) = \int_0^{+\infty} \mathrm{d}a\, \Pi(a) \int_0^{+\infty} \mathrm{d}x\, P_x(x)\delta(V - ax)$$

$$= \int_0^{+\infty} \frac{\mathrm{d}a}{a} \Pi(a) P_x\left(\frac{V}{a}\right) \; . \tag{14.125}$$

Introducing $W = \ln V$, $y \equiv \ln x$ and $l \equiv \ln a$, we obtain

$$P(W) = \int_{-\infty}^{+\infty} \mathrm{d}l\, \Pi(l) P_x(W - l) \; . \tag{14.126}$$

Taking the logarithm of (14.124), we have $W = y - f(y, \{a, b, \ldots\})$, showing that $W \to y$ for large $y > 0$, since we have assumed that $f(y, \{\lambda, b, \ldots\}) \to 0$ for large y. We can write $P(W)\,\mathrm{d}W = P_y(y)\,\mathrm{d}y$ leading to

$$P(W) = \frac{P_y[y(W)]}{1 - \partial f(y, \{a, b, \ldots\})/\partial y} \to P_y(W) \; , \quad \text{for } y \to \infty \; . \tag{14.127}$$

We thus recover the Wiener–Hopf integral equation (14.122) yielding the announced results (14.120) with (14.121) and therefore the power law distribution (14.116) for $x(t)$ with μ given by (14.121).

This derivation explains the origin of the generality of these results for a large class of convergent multiplicative processes repelled from the origin.

Let us now compare the two results (14.117) and (14.121) for μ. It is straightforward to check that (14.117) is the solution of (14.121) when $\Pi(l)$ is a Gaussian, i.e. $\Pi(a)$ is a log-normal distribution. The expression (14.117) can also be obtained perturbatively from (14.121): expanding $e^{\mu l}$ as $e^{\mu l} = 1 + \mu l + (1/2)\mu^2 l^2 + \ldots$ up to second order, we find that the solution of (14.121) is (14.117). This was expected from our previous discussion in Chap. 2 of the approximation involved in the use of the Fokker–Planck equation.

An interesting example of the class (14.123) is given by the map already discussed in Chap. 8, representing for instance the population of fishes in a lake: consider the number of fish X_t in a lake in the t-th year and let X_{t+1} be related to the population X_t through

$$X_{t+1} = a\, X_t + b, \tag{14.128}$$

where $a = a(t) > 0$ and $b = b(t) > 0$ are drawn from some probability density function. The additive contribution b ensures the repulsion from the origin and thus allows for intermittent bursts distributed according to the power law distribution (14.116). The growth rate a depends on the rate of reproduction and the depletion rate due to fishing or predation, as well as on environmental conditions, and is therefore a variable quantity. The quantity b describes the input due to restocking from an external source such as a fish hatchery in artificial cases, or from migration from adjoining reservoirs in natural cases; b can but need not be constant. In addition to the modeling of population dynamics with external sources, multiplicative maps of the type (14.123) and in particular (14.128) can be applied to epidemics in isolated populations [879], to finance and insurance applications with relation to ARCH(1) processes [215], immigration and investment portfolios [879], the Internet [2, 12, 437, 438, 879, 946, 947, 1019] and directed polymers in random media [684].

In Chap. 16, we present a fragmentation model which produces a power law distribution of fragment sizes. The mechanism also involves a multiplicative process; strictly speaking, there is not repulsion from the origin as discussed above; however, the fact that fragments can become unbreakable plays a similar role. We can thus consider the fragmentation model in the same class as the models discussed in this section. This is confirmed by the fact that (16.51) for the fragment distribution has exactly the same structure as (14.125).

To sum up, power law distributions are generated from multiplicative noise if

- the distribution of the amplitudes of the noise allows for intermittent amplification while being globally contracting and
- there is a "reinjection" mechanism of the variable to a non-vanishing value, so that it remains susceptible to the intermittent amplifications.

14.9 The "Coherent-Noise" Mechanism

Another class of models with a simple and robust mechanism for producing power law distributions has been introduced by Newman and Sneppen [677]. These so-called "coherent-noise" models consist of a large array of sites or elements which are forced to reorganize at each time step under an externally imposed random stress that acts on all agents at the same time. In their simplest form, there are no interaction between the elements, which makes the coherent-noise mechanism fundamentally different from self-organized criticality discussed in Chap. 15. While they are self-organized, coherent-noise systems are not critical. The self-organization stems from the competition between two processes described below which leads to a statistical steady-state. Notwithstanding the absence of criticality, these models exhibit a power-law distribution of event sizes, each event corresponding to the reorganization of the stress thresholds of the elements under the action of the externally imposed stress. There is a wide range of different exponents [856], depending on the special implementation of the basic mechanism. Moreover, coherent-noise models display power-law distributions in several other quantities, e.g., the life-time distribution of the agents. These models have been suggested to be relevant to earthquakes [677], rice piles [677], and biological extinction [673–675, 1017].

The model is defined as follows. Consider a system consisting of N units. Every element i has a threshold x_i again external stress. The thresholds are initially chosen at random from some probability distribution $p_{\text{thresh}}(x)$. The dynamics of the system is as follows:

(i) A stress η is drawn from some distribution $p_{\text{stress}}(\eta)$. All agents with $x_i \leq \eta$ are given new random thresholds, again from the distribution $p_{\text{thresh}}(x)$.

(ii) A small fraction f of the agents is selected at random and also given new thresholds.

(iii) The next time-step begins with (i).

The most common choices for the threshold and stress distributions are a uniform threshold distribution and some stress distribution that is falling off quickly, like the exponential or the Gaussian distribution. Under these conditions (with reasonably small f), it is guaranteed that the distribution of reorganization events that arises through the dynamics of the system will be a power law.

In this simplest version of the model, the elements are entirely non-interacting. This is why step (ii) is necessary to prevent the model from grinding to a halt. Without this random reorganization, the thresholds of the agents would after some time be well above the mean of the stress distribution and the average stress could not hit any agent anymore. This model is quite similar to the Bak–Sneppen model [48] discussed in Chap. 15, and especially to its parallel version presented in Sect. 15.4.4. The difference is essentially in the nature of the driving: infinitesimal increments in the Bak–Sneppen model versus random imposed stresses in the coherent-noise model. The later produces power law pdf's without need for interactions between elements in the coherent-noise model. If interactions are introduced, for instance between nearest neighbors, step (ii) can be omitted because the interactions ensure that elements with large thresholds will become destabilized.

There are indeed several possibilities for extending the model to make it more general, without loss of the basic features. Two extensions that have been studied by Sneppen and Newman [856] are the following.

• A lattice version where the agents are put on a lattice and, with every agent hit by stress, its nearest neighbours undergo reorganization even if their threshold is above the current stress level.

- A multi-trait version where, instead of a single stress, there are M different types of stress, i.e. the stress becomes a M-dimensional vector $\boldsymbol{\eta}$. Accordingly, every element carries a vector of thresholds $\mathbf{x_i}$. An element has to move in this model whenever at least one of the components of the threshold vector is exceeded by the corresponding component of the stress vector.
- Another extension is to biological evolution and to the dynamics of mass extinctions [1017], in which the number of elements (species in this case) is not kept constant because species go extinct or are created by mutation.

To show how power law distributions of event sizes emerge, we follow Sneppen and Newman [856]. First, one solves for the mean (time-averaged) distribution $\bar{\rho}(x)$ of the threshold variables x_i. In any small interval $[x, x+dx]$ of stress thresholds, the rate at which elements are reorganized is the sum of two contributions: (1) the random selection process with rate $f\bar{\rho}(x)$; (2) the stress event with rate given by $\bar{\rho}(x)$ times the probability that the stress level η will be greater than x, which is

$$p_{\text{move}}(x) = \int_x^\infty p_{\text{stress}}(\eta)\, d\eta \; . \tag{14.129}$$

The total rate $\bar{\rho}(x)(f + p_{\text{move}}(x))$ of reorganization in any small interval $[x, x + dx]$ is equal to the average rate of repopulation, equal to a constant A independent of x and determined by normalization, giving

$$\bar{\rho}(x) = \frac{A}{f + p_{\text{move}}(x)} \; . \tag{14.130}$$

To give a concrete example, consider the simple case where the stress η is restricted to non-negative values, $p_{\text{thresh}}(x)$ is uniformly distributed over the interval between 0 and 1 and $p_{\text{stress}}(\eta)$ is the normalized exponential distribution

$$p_{\text{stress}}(\eta) = \frac{1}{\sigma} \exp(-\eta/\sigma). \tag{14.131}$$

In this case, $p_{\text{move}}(x) = \exp(-x/\sigma)$ and

$$\bar{\rho}(x) = \frac{A}{f + \exp(-x/\sigma)} \; , \tag{14.132}$$

with

$$A = \frac{f}{\sigma}\left[\log \frac{f\exp(1/\sigma) + 1}{f + 1}\right]^{-1} \; . \tag{14.133}$$

Numerical simulations confirm this prediction (14.132) with good accuracy [856].

$\bar{\rho}(x)$ given in (14.130) has the shape of a smooth step, centered at a value x_c such that the two terms in the denominator of (14.130) are equivalent: $p_{\text{move}}(x_c) = f$. For $x > x_c$, $\bar{\rho}(x)$ is dominated by the constant term f

in the denominator, so that $\bar{\rho}(x)$ goes to a constant for large x. Below x_c, the second term $p_{move}(x)$ dominates, leading to

$$\bar{\rho}(x) \approx \frac{A}{p_{move}(x)} \; . \tag{14.134}$$

In the exponential example, this becomes $\bar{\rho}(x) \approx A \exp(x/\sigma)$. As we will see, it is this part of the distribution of stress thresholds which is responsible for the power-law distribution of event sizes in the model.

The size of the event which corresponds to a stress of magnitude η is simply given by (for positive stresses)

$$s(\eta) = \int_0^\eta \rho(x) \, dx \; . \tag{14.135}$$

In general, the threshold distribution at any particular time t differs from the time-averaged distribution $\bar{\rho}(x)$. In order to solve analytically for the event size distribution, we make the approximation that $\rho(x)$ is close to its time average $\bar{\rho}(x)$. Expression (14.135) thus gives

$$s(\eta) = \int_0^\eta \bar{\rho}(x) \, dx = \int_0^\eta dx \, \frac{A}{f + p_{move}(x)} \; . \tag{14.136}$$

Then, the pdf of event sizes s is given by the standard change of variable

$$p_{event}(s) = p_{stress}(\eta)\frac{d\eta}{ds} = \frac{p_{stress}(\eta(s))}{\bar{\rho}(\eta(s))} \; , \tag{14.137}$$

where we have calculated $d\eta/ds$ by differentiating expression (14.136). The function $\eta(s)$, which is the stress required to produce an event of size s, is given by the functional inverse of (14.136).

As we mentioned above, the power-law emerges due to the low-x part below x_c of the threshold distribution. In this regime, expressions (14.134) together with (14.129) and (14.137) lead to

$$p_{event}(s) = (1/A)p_{stress}(\eta(s)) \int_{\eta(s)}^\infty p_{stress}(x) \, dx \; . \tag{14.138}$$

Making the same approximations in (14.136), we get

$$s(\eta) = \int_0^\eta \frac{A \, dx}{\int_x^\infty p_{stress}(\eta') \, d\eta'} \; . \tag{14.139}$$

Between them, these two equations (14.138) and (14.139) define the event size distribution.

The crucial condition, which must be fulfilled to obtain power law distributions of event sizes, is that the value of $p_{stress}(\eta)$ in the tail should fall off fast enough that the integral of $p_{stress}(\eta)$ from η to ∞ should be dominated by the value of the function near η. A sufficient condition is

$$\int_\eta^\infty p_{stress}(x) \, dx \sim [p_{stress}(\eta)]^\alpha \; . \tag{14.140}$$

Substituting this condition into (14.138) and (14.139), we get

$$p_{\text{event}}(s) \sim [p_{\text{stress}}(\eta(s))]^{\alpha+1}, \tag{14.141}$$

and

$$s(\eta) \sim \int_0^\eta [p_{\text{stress}}(x)]^{-\alpha} \, \mathrm{d}x = \int_{p_{\text{stress}}(\eta)}^1 [p_{\text{stress}}]^{-\alpha} \frac{\mathrm{d}x}{\mathrm{d}p_{\text{stress}}} \, \mathrm{d}p_{\text{stress}}$$

$$\sim \frac{1}{p_{\text{stress}}(\eta)}, \tag{14.142}$$

where we have employed (14.140) to evaluate the derivative. Combining (14.141) and (14.142) gives

$$p_{\text{event}}(s) \sim 1/s^{1+\alpha}. \tag{14.143}$$

For most choices of $p_{\text{stress}}(x)$, $\int_\eta^\infty p_{\text{stress}}(x) \, \mathrm{d}x \sim p_{\text{stress}}(\eta)$ up to sub-dominant corrections, which implies $\alpha \approx 1$. This is true in general for $p_{\text{stress}}(x) \sim \exp[-(x/\sigma)^c]$ with arbitrary $c > 0$, including exponential, Gaussian and stretched exponential pdf's. If $p_{\text{stress}}(x)$ is a power law $\sim 1/x^\gamma$, we obtain $\alpha = 1 - 1/\gamma$. Of course, some stress distributions may not satisfy (14.140), such as any distribution which does not have a tail (we say that the pdf has a compact support), in which case the pdf of event sizes is not a power law. As long as the pdf of stresses has no finite upper limit, we can expect a power law event size distributions. The prediction (14.143) is verified accurately by numerical simulations [856].

The coherent-noise model also predict the occurrence of "aftershocks," somewhat similarly to earthquakes [856], characterized by power law distributions of event rates as a function of the time since a large event [1016, 1018]. This power law $P_{\text{relax}}(t) \sim 1/t^p$ of the relaxation of the event rate looks quite similar to the Omori law for earthquake aftershocks [973]. Together with the power law pdf of event sizes which is similar to the Gutenberg–Richter law, this has led these authors to suggest that the coherent-noise model might have value in modeling earthquakes. However, in the coherent-noise models, the exponent p of $P_{\text{relax}}(t)$ is found to decrease with the size of the mainshock and to be a function of the lower magnitude cut-off of the aftershocks. These two properties are not observed in real seismic catalogs, in which a remarkable self-similarity is observed: the Omori law is characterized by approximately the same exponent p for all earthquake magnitudes between 3 and 7 [410, 413] (there is not enough data for larger magnitudes and the catalogues are incomplete for lower magnitudes). The coherent-noise model is able to reproduce both the Gutenberg–Richter distribution of earthquake magnitudes and the Omori law of aftershock rate decay, which are both one-point statistics. However, it predicts two-point statistics which are in strong disagreement with empirical data. There is an interesting lesson here on the danger to qualify a model solely on the basis of its one-point statistical properties; higher-order statistics are very important and the dependence between multiple events in general reveal more.

14.10 Avalanches in Hysteretic Loops and First-Order Transitions with Randomness

In many first-order phase transitions, one observes hysteresis loops. An hysteresis loop is the graph of the response (say, the magnetization M of the material) which lags behind the force (say, an external magnetic field H). In many materials, the hysteresis loop is composed of small bursts, or avalanches, which cause acoustic emission (crackling noises); in magnets, they are called Barkhausen noise. These bursts result from some kind of inhomogeneity or disorder in the material that pins the deformation within it. In many systems where the heterogeneity is found at the microscopic scale size of a grain, the observed bursts are nevertheless observed over a wide range of sizes, from over three to six decades in size in a typical experiment (see [195, 196, 229, 730–732, 834] and references therein). Since the grains in the material do not come in such a variety of sizes, one can conclude that many grains must be activated at once, coupled together in a kind of avalanche.

Having events of all sizes has a profound meaning: if the coupling between grains is weak compared to the disorder, the grains will tend to flip independently, leading to small avalanches; if the coupling is strong, a grain which flips will give a large kick to its neighbors, likely flipping several of them, which will flip several more, thus leading to one large avalanche. In real experiments and in the following model, this is precisely what happens. The large range of avalanches is associated with a critical value of the disorder R_c relative to the coupling: when the avalanches can not decide whether to be huge or small, they come in all sizes!

The classic model for a first order transition [303, 751] is the Ising model in an external field H at $T < T_c$: as H passes through zero, the equilibrium magnetization reverses abruptly. This model is in sharp contrast with real first-order transitions as studied by materials scientists and metallurgists: solid material which transforms from one crystalline or magnetic form to another under the influence of temperature, external stress, or applied field often has no sharp transition at all and hysteresis becomes the dominating phenomenon. The ingredient added by Sethna et al. [195, 196, 229, 730–732, 834] to this idealized Ising model is disorder. This is performed by adding a random field f_i at each site of the Ising model

$$\mathcal{H} = -\sum_{ij} J_{ij} s_i s_j - \sum_i (f_i s_i + H s_i) \ . \tag{14.144}$$

The rule of the model is that, as the external field H is changed, each spin will flip when the direction of its total local field

$$F_i \equiv \sum_j (J_{ij} s_j + f_i + H) \tag{14.145}$$

changes. The deterministic dynamics corresponds to taking the limit of zero temperature. The random local fields f_i are taken from a Gaussian distribution $P(f)$ with zero mean and variance R^2. R is thus the control parameter quantifying the amount of disorder. Let us vary the disorder and the field to demonstrate the existence of a critical point induced by the disorder.

For small R, an infinite avalanche occurs. For sufficiently large R, only finite avalanches are seen due to the dominating effect of the local fields. On raising the field from the down state, there is a field $H_c^u(R)$ at which an infinite avalanche occurs and this transition at $H_c^u(R)$ is abrupt for $R < R_c$. On the other hand, as one approaches the critical field $H_c^u(R_c)$ at the critical disorder R_c, the transition appears to be continuous: the magnetization $M(H)$ has a power-law singularity, and there are avalanches of all sizes. As one approaches this "endpoint" at $(R_c, H_c^u(R_c))$ in the (R, H) plane, we find diverging correlation lengths and universal critical behavior.

We now recall how these critical properties can be exactly solved within mean-field theory [834]. Suppose every spin in (14.144) is coupled to all $N-1$ other spins with coupling J/N. The effective field acting on a site is $JM + f_i + H$, where $M = \sum_i s_i/N$ is the average magnetization. Spins with $f_i < -JM - H$ will point down, the rest will point up. Thus, the magnetization $M(H)$ is given implicitly by the expression

$$M(H) = 1 - 2 \int_{-\infty}^{-JM(H)-H} P(f) \, df \ . \tag{14.146}$$

This equation has a single-valued solution unless $R \leq R_c$ (which in the case of a Gaussian distribution corresponds to $P(0) \geq 1/2J$), at which point hysteresis and an infinite avalanche begin. Near the endpoint, the jump in the magnetization ΔM due to the avalanche scales as r^β, where $r \equiv (R_c - R)/R_c$. As one varies both r and the reduced field $h \equiv [H - H_c^u(R_c)]$, the magnetization scales as

$$M(h,r) \sim |r|^\beta \mathcal{M}_\pm(h/|r|^{\beta\delta}) \ , \tag{14.147}$$

where $\pm$ refers to the sign of r. In mean-field theory, $\beta = 1/2$, $\delta = 3$ and $\mathcal{M}_\pm$ is given by the smallest real root $g_\pm(y)$ of the cubic equation $g^3 \mp (12/\pi)g - (12\sqrt{2}/\pi^{3/2}R_c)y = 0$.

The avalanche size distribution near the critical point shown in the inset of Fig. 14.5 can be obtained within this mean-field approach. The probability $D(s,t)$ of having an avalanche of size s, where $t \equiv 2J\rho(-JM - H) - 1$ measures the distance to the infinite avalanche, is found as follows [834]. To have an avalanche of size s triggered by a spin with random field f, one must have precisely $s - 1$ spins with random fields in the range $\{f, f + 2Js/N\}$. The probability for this to happen is given by the Poisson distribution. In addition, they must be arranged so that the first spin triggers the rest. This occurs with probability precisely $1/s$, which one can see by putting periodic boundary conditions on the interval $\{f, f + 2Js/N\}$ and noting that there would be exactly one spin of the s which will trigger the rest as a single avalanche. This leads to the avalanche size distribution

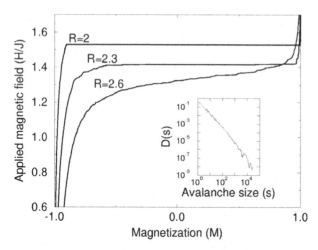

Fig. 14.5. Varying the Disorder: three $H(M)$ curves for different levels of disorder for a 60^3 system. The estimate of the critical disorder is $R_c = 2.16J$ [835] (J is set equal to 1 in the figure). At $R = 2 < R_c$, there is an infinite avalanche which seems to appear abruptly. For $R = 2.6 > R_c$, the dynamics is macroscopically smooth, although of course microscopically it is a sequence of sizable avalanches. At $R = 2.3$, near the critical level of disorder, extremely large events become common. *Inset:* Log–Log Plot of the avalanche-size distribution $D(s)$ vs. avalanche size s for the 60^3 system at $R = 2.3$ for $1.3 < H < 1.4$, averaged over 20 systems. Here $D(s) \sim s^{-1.7}$, compared to the mean-field exponent τ of $3/2$. Reproduced from [834]

$$D(s,t) = \frac{s^{s-2}}{(s-1)!}(t+1)^{s-1}e^{-s(t+1)} . \tag{14.148}$$

To put this in a scaling form, we must first express t as a function of r and h: $t \sim r\left[1 \mp (\pi/4)g_\pm(h/|r|^{3/2})^2\right]$. Using some simple expansions and Stirling's formula, one obtains D in the scaling form

$$D(s,r,h) \sim s^{-\tau}\mathcal{D}_\pm(s/|r|^{-1/\sigma}, h/|r|^{\beta\delta}) , \tag{14.149}$$

where the mean-field calculation gives $\tau = 3/2$, $\sigma = 1/2$, and the universal scaling function

$$\mathcal{D}_\pm(x,y) = \frac{1}{\sqrt{2\pi}}e^{-x\left[1\mp(\pi/4)g_\pm(y)^2\right]^2/2} . \tag{14.150}$$

Extraction of the critical exponents both numerically and via ϵ-expansion has been performed in [195, 196, 229, 730–732]. The results on the distribution of avalanches seem to be confirmed in real experimental magnetic systems.

In summary, the random field Ising model under a varying external magnetic field at zero temperature offers an attractive understanding of why the noise pulses in magnets and in other systems can span a large range of scales: the reason is that they are probably near a critical point where the hysteresis loop develops a jump. This model rationalizes the experimental observation of power laws as resulting from an anomalously large "critical domain": unlike more traditional phase transitions, this model has a large *critical range*: 4% away from the critical point, one observes six decades of scaling, and a factor of two away one still has two decades. In this approach, one does not need mechanisms to tune a system precisely to the critical point as in self-organized criticality (see Chap. 15). Here, due to the broadness of the critical region of the disorder critical point, it seems likely that the experimentalists could pick their sample inside this large range without tuning. The physical mechanism for the existence of such a large critical range is not obvious. Three contributing factors have been proposed [835]:

1. The critical exponent $\nu = 1.42$ in the random field Ising model, while it is 0.63 in the three-dimensional Ising model. Thus, getting twice as close to R_c makes the length spanned by an avalanche grow by a factor of 2.7, whereas getting twice as close to T_c for the Ising model makes the correlation length grow only by a factor of 1.55.
2. The size S of an avalanche is more like a volume than a length. Six decades of scaling in S should be thought of as roughly two decades in length scale. Actually, since the avalanches are not space filling in three dimensions, the volume scales as $S \sim \xi^{1/\sigma\nu} \sim \xi^{2.6}$, so that six decades in size S gives 2.3 decades in the length scale ξ.
3. Three dimensions is "close" to two dimensions. The behavior in three dimensions is far removed from the mean-field behavior of the model in six and higher dimensions. The fluctuations are extremely important; in two dimensions, one believes that they almost completely dominate, perhaps even preventing an infinite avalanche from ever occurring.

The mechanism has been applied to the breakdown of disordered media in [1049, 1050].

14.11 "Highly Optimized Tolerant" (HOT) Systems

The acronym HOT stands for highly optimized tolerance and refers to a mechanism for generating power law distributions which involves a global optimization principle [140]. It is motivated by optimization processes developed in engineerings and may also apply to some biological problems. In a nutshell, the idea is that an input with characteristic scales may lead to power law statistics of outputs after a global optimization of the system yield has been performed. The importance of the global optimization is reminiscent of the role played by a global conservation law in some version of self-organized critical models (see Chap. 15). Possible domains of applications are biology and epidemiology, aeronautical and automotive design, forestry and environmental studies, the Internet, traffic, and power systems.

14.11.1 Mechanism for the Power Law Distribution of Fire Sizes

To illustrate the HOT global optimization mechanism, let us first formulate the simplest possible collective optimization problem. Consider a variable X taking different values $x_1, \ldots, x_N$ with arbitrary but fixed probabilities $p_1, \ldots, p_N$. Let us assume that X is a function of another parameter r through the relation

$$X = Cr^{-\beta} , \tag{14.151}$$

where C and β are two constants. The optimization problem consists in minimizing

$$L = \sum_{i=1}^{N} p_i x_i(r_i) \tag{14.152}$$

with respect to the r_i variables, with the constraint

$$\sum_{i=1}^{N} r_i = \kappa . \tag{14.153}$$

Using the method of Lagrange multipliers to solve this optimization problem with constraint leads to

$$p_i = \frac{D}{x_i^{1+1/\beta}} , \tag{14.154}$$

where D is a constant. The global optimization together with the power law dependence (14.151) leads automatically to a power law distribution with exponent $\mu \equiv 1/\beta$. In other words, for a fixed set of probabilities p_i's, the optimization "compresses" the variables r_i's so that the variables x_i's are stretched just enough to automatically adjust to a power law with exponent $1/\beta$. In contrast, replacing (14.151) by $X = C \ln r$ leads to an exponential distribution $p_i \propto \exp(-cx_i)$, where c is a constant. We thus see that the global optimization *together with* the power law dependence (14.151) is responsible for the generation of the power law distribution.

This mechanism is reminiscent of Mandelbrot's classical result on the distribution of the length of words in an optimum language [589]. Suppose that you want to design the optimum language, that is, the optimum set of frequencies $f_1, f_2, \ldots, f_n$ assigned to n words. In an alphabet of d letters, for instance if we think of English text, the cost of a word might be thought of as the number of letters plus the additional cost of a space. Hence, a natural cost for using the i-th word is $\ln_d i \sim \ln i$ (the logarithm appears by the mechanism that the number of digits to represent a given number is close to its logarithm in base 10). The optimum language is expressed by optimizing the information transmitted, which is the entropy $-\sum_i f_i \ln f_i$ of the set of frequencies, divided by the expected transmission cost $\sum_i f_i \ln i$ taken as the average of the lengths of the words. Mandelbrot's result [589] is that frequencies that achieve the optimum correspond to a power law pdf of the lengths i. Notice that this mechanism is based on two key assumptions: (i) the cost of the i-th word is assumed to be propotional to $\ln i$ and (ii) the use of the entropy function involving logarithms of the frequencies f_i. Thus, the underlying mechanism involves a combination of logarithms, in the spirit of the combination of exponentials discussed in Sect. 14.2.2.

Let us now consider a forest fire problem in which spontaneous ignition (sparks) occurs preferentially in some part of the forest, in other words the spatial distribution of sparks is not homogeneous. This heterogeneity is an essential ingredient to obtain power laws as we shall see. To a given geometrical structure of firewalls corresponds a specific size and spatial distribution of tree clusters. When a spark falls on a tree, the whole connected cluster of trees delimited by the firewalls bounding it burns entirely. The optimal management of the forest consists in building firewalls in such a way that the yield after fires be maximum, in other words that the average destructive impact of a fire be minimum.

In the presence of an heterogeneous spatial probability density ρ of sparks, it is clear that the density r of firewalls should not be spatially uniform: more firewalls are needed in sensitive regions where the sparks are numerous. The density r of firewalls will thus not be constant after the optimization process but will adjust to the predefined distribution ρ of sparks. This spatial distribution ρ of sparks determines the probability p_m that a spark ignites a fire in a given domain m bounded by the fire walls: p_m is the sum of ρ over the cluster. Consider a patch of s_m trees whose boundary is a firewall. We define the average spark density on this cluster by $\rho_m = p_m/s_m$. ρ_m is thus the average probability per unit area that a spark falls in any tree of this patch, burning it to the ground. Conversely, the total probability for a fire to be ignited over the area s_m is $p_m = \rho_m s_m$. Since the typical linear size of the patch is $\sim s_m^{1/2}$ in two dimensions and $\sim s_m^{1/d}$ in the more general case of a d-dimensional space, the length of the surrounding firewill is proportional to $s_m^{(d-1)/d}$. Let us also denote c the cost per length (in 2D) or unit area (in higher dimensions) to construct the firewall. In units where the total area of the forest is one, the total yield is given by

$$L = 1 - \sum_{m=1}^{N} (\rho_m s_m) s_m - c \sum_{m=1}^{N} s_m^{(d-1)/d} , \qquad (14.155)$$

N is the total number of clusters defined by the firewalls, which is of course a function of the firewall configuration. Maximizing the yield L amounts to find the cluster sizes s_m such that $\partial L/\partial s_m = 0$, which gives (with the condition $\sum_m s_m = 1$)

$$\rho_m \sim 1/s_m^{1+1/d} . \qquad (14.156)$$

This result can be interpreted as the probability distribution of a fire of size s_m to occur. A subtlety has to be taken into account: since we deal with discrete sizes and because ρ_m is a power law of s_m, expression (14.156) is actually proportional to the probability for a fire to have a size between s_m and $s_m + \Delta s_m$, where Δs_m is proportional to s_m to ensure a correct correspondence between a discrete distribution of sizes and the continuous power law distribution. This gives the interpretation that expression (14.156) is proportional to the complementary cumulative distribution of fire sizes to be larger than s_m. In the notation of this book, this corresponds to $\mu = 1 + 1/d$. Newman et al. [676] uses a different argument to obtain the same result, based on a mixture of discrete and continuous notations.

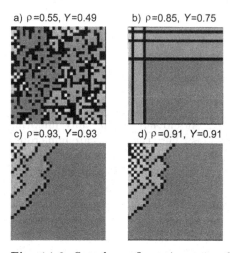

Fig. 14.6. Sample configurations on a 32×32 lattice for (**a**) the random case near the percolation threshold p_c, (**b**) a HOT grid, and HOT states obtained by evolution at (**c**) minimal loss, and (**d**) a somewhat lower density. Unoccupied sites are *black*, and clusters are *gray*, where *darker shades* indicate larger clusters. The designed systems are generated for an asymmetric spatial distribution of hitting probabilities with Gaussian tails, peaked at the *upper left corner* of the lattice. If a spark hits an unoccupied site, nothing burns. When the spark hits an occupied site the fire spreads throughout the associated cluster, defined to be the connected set of nearest-neighbor occupied sites. The yield Y is defined as the difference between tree density and average fire size. Reproduced from [140]

The optimization process with power law cost function provides robust performance despite the uncertainties quantified by the probabilities ρ_m's. In the forest fire example, the optimal distribution of spatial firewalls is the result of the interplay between our a priori knowledge of the uncertainty in the distribution of sparks and the cost resulting from fires. The solutions are robust with respect to the existence of uncertainties, i.e. to the fact that we do not know deterministically where sparks are going to ignite; we only know their probability distribution. In contrast, the spatial geometry of optimal firewalls is not robust with respect to an error in the quantification of the probabilities p_i's. It is not the uncertaintly that is dangerous but errors in quantifying this uncertainty: a different set of p_i's will lead to the same overall power law distribution (14.154), but with a very different spatial distribution of firewalls.

14.11.2 "Constrained Optimization with Limited Deviations" (COLD)

Newman et al. [676] have shown that the optimal solution (14.156) does not protect from rare and very large fires that may be disastrous. The reason is similar to the classic gambler's ruin, for which optimizing total return leads to ruin with probability one. To address this issue of the impact of the very large fires in the tail of their size distribution, as well as to cure the fragility with respect to mispecification of the probabilities p_i, Newman et al. [676] borrow the concept of "risk aversion" from economic theory. Technically, this amounts to changing the cost of a fire of size s_m from a value proportional to $1 - s_m$ as in (14.155) to a negative utility function $u(s_m)$ which decreases faster than linearly with s_m. A convenient mathematical form is $u(s_m) = (1 - s_m)^\alpha / \alpha$. The total function to maximize becomes

$$L(\alpha) = \sum_{m=1}^{N} (\rho_m s_m) u(s_m) - c \sum_{m=1}^{N} s_m^{(d-1)/d} . \tag{14.157}$$

For $\alpha < 1$ which expresses risk-aversion, the power law distribution (14.156) of fire sizes is found to be truncated at large s_m. For $\alpha < 1$, this distribution, which maximizes $L(\alpha < 1)$ does not maximize evidently $L(\alpha = 1)$ which is the real yield (in terms of trees): this means that one pays a cost for risk aversion in terms of yield. This concept of a compromise between return and risk is at the core of the theory of financial risks and of portfolio theory [605].

14.11.3 HOT versus Percolation

Finally, it is useful to contrast the organizations of trees obtained by the HOT optimization process and by the standard percolation described in Chap. 12. In the percolation model, one imagines that trees are planted randomly in space and, as a function of time, the density p of trees increases. As long as the density is smaller than the critical value p_c at which percolation occurs, there are only finite clusters of connected trees and any spark falling on a tree has a minor effect as it destroys only the cluster of trees connected to the ignited tree. For a density of trees above p_c, the trees form spontaneously a large cluster connecting the different borders of the system. This connecting cluster is often called the "infinite" percolation cluster. When a spark falls on one of the trees of this infinite percolating cluster, this has a significant effect because this cluster contains a finite fraction $P_\infty(p)$ of the total number of trees in the system. This fraction $P_\infty(p)$ plays the role of the order parameter of the percolation transition and is characterized by

$$P_\infty(p) = 0 , \qquad \text{for } p < p_c , \tag{14.158}$$

$$P_\infty(p) \propto (p - p_c)^\beta , \qquad \text{for } p \geq p_c , \tag{14.159}$$

where $0 < \beta < 1$ is the standard notation for the critical exponent of the order parameter of a critical transition. Thus, for $p > p_c$, there is a probability proportional to $P_\infty(p)$ that a spark will trigger a fire which will burn out the infinite percolation cluster, leading to a yield equal to $p - P_\infty(p)$ per unit surface. Since $0 < \beta < 1$, $p - P_\infty(p)$ has a cusp-like maximum at $p = p_c$. Thus, if trees are planted according to percolation theory and if the firewalls are just the boundaries of percolation clusters, we see that the maximum yield is obtained by the system functioning exactly at the critical point $p = p_c$. This suggests in passing another mechanism to attract the dynamics of the percolating system exactly to its critical point, in the sense of Sect. 15.4.2.

The optimal yield $p - P_\infty(p)$ of the percolation solution should be contrasted with the result of the HOT optimization process, which allows the system to function at a density of trees significantly higher than p_c by optimizing the firewalls in response to the density of sparks.

15. Self-Organized Criticality

15.1 What Is Self-Organized Criticality?

15.1.1 Introduction

The study of out-of-equilibrium dynamics (e.g. dynamical phase transitions) and of heterogeneous systems (e.g. spin-glasses) has progressively made popular the concept of complex systems and the importance of systemic approaches: systems with a large number of mutually interacting parts, exchanging energy, matter or information with their environment, self-organize their internal structure and their dynamics with novel and sometimes surprising macroscopic ("emergent") properties. The complex system approach, which involves "seeing" inter-connections and relationships i.e. the whole picture as well as the component parts, is nowadays pervasive in modern control of engineering devices and business management. It also plays an increasing role in most of the scientific disciplines, including biology (biological networks, ecology, evolution, origin of life, immunology, neurobiology, molecular biology, etc), geology (plate-tectonics, earthquakes and volcanoes, erosion and landscapes, climate and weather, environment, etc.), economy and social sciences (including cognition, distributed learning, interacting agents, etc.).

A central property of a complex system is the possible occurrence of coherent large-scale collective behaviors with a very rich structure, resulting from the repeated non-linear interactions among its constituents: the whole turns out to be much more than the sum of its parts. Punctuated dynamics seems to be an essential dynamical process for systems that evolve and become complex, with a specific behavior that is strongly contingent on its history. The punctuations correspond to rare and sudden transitions that occur over time intervals that are short compared to the characteristic time scales of their posterior evolution. Such large events express more than anything else the underlying "forces" usually hidden under almost perfect balance. They provide the potential for a better scientific understanding of complex systems. These crises have important societal impacts and range from large natural catastrophes such as earthquakes, volcanic eruptions, hurricanes and tornadoes, landslides, avalanches, lightning strikes, meteorite/asteroid impacts, catastrophic events of environmental degradation, to the failure of engineering structures, crashes in the stock market, social unrest leading to large-scale strikes and upheaval, economic drawdowns on national and global scales, regional power blackouts, traffic gridlocks, diseases and epidemics, etc. It is essential to realize that the long-term behavior of these complex systems is often controlled in large part by these rare events.

Self-organized criticality (SOC) views these large events as belonging to the natural non-linear organization of complex systems. Large events are seen to result from

- the long-range power law decay of spatial and temporal correlations and
- the heavy tail power law distributions of even sizes.

Such properties are also shared by equilibrium systems at a critical phase transition [44].[1] The burst events in the self-organized critical state are critical in the sense of a nuclear chain reaction process. In a supercritical system, a single local event, like the injection of a neutron, leads to an exponentially exploding process. A sub-critical process has exponentially decaying activity, always dying out. In the critical state, the activity is just able to continue indefinitely, with a power law distribution of stopping times, reflecting the power law correlations in the system, exactly like the branching process at its critical value discussed in Chap. 13.

Self-organized criticality describes complex systems that are situated at the delicately balanced edge between order and disorder in a self-organized critical state. Only at the critical state, does the compromise between order and fluctuations exist that can qualify as truly complex behavior. Due to very large correlations, the individual degrees of freedom cannot be isolated. The infinity of degrees of freedom interacting with one another cannot be reduced to a few, which is what makes critical systems complex.

In the literature, the term "self-organized criticality" has not always been used with the same meaning and, in some cases, it has been misused. To be useful, this concept must be specified accurately and related to a well-defined situation based on physical mechanisms rather than on observations. In particular, a system is not self-organized critical solely because it exhibits a power law distribution of event sizes, since many other mechanisms lead to such signatures as discussed in Chap. 14. Systems which have been proposed to result from a self-organized critical dynamics include fault networks and seismicity, river networks, propagation of forest fires and biological evolution processes [44, 463]. It is still a debated question whether SOC indeed applies to any of these systems and what are the specific underlying physical mechanisms.

---

[1] For a different view point combining SOC with the concept that the largest extreme events are potentially predictable outliers, see [435] and the contribution of the author in [666], [883] and [884]).

15.1.2 Definition

In the broadest sense, SOC refers to the spontaneous organization of a system driven from the outside into a globally stationary state, which is characterized by self-similar distributions of event sizes and fractal geometrical properties. This stationary state is dynamical in nature and is characterized by statistical fluctuations, which are refered generically to as "avalanches".

The term "self-organized criticality" contains two parts. The word "criticality" refers to the state of a system at a critical point at which the correlation length and the susceptibility become infinite in the infinite size limit (see Chap. 9). The label "self-organized" is often applied indiscriminately to pattern formation among many interacting elements. The concept is that the structuration, the patterns and large scale organization appear spontaneously. However there is some fuzziness in what is meant by "spontaneously". In an Ising system, the magnetization appears at the large scale from the interactions between the individual spins. Nevertheless, we do not refer to this situation as self-organized while it is in a sense. The notion of self-organization is thus relative to the absence of control of parameters that are considered artificial and thus depends to some degree on both the historical maturation of the scientific field and on the level of understanding of the underlying mechanisms. In SOC, there is a greater specificity as the emphasis is put on the mechanisms that maintain the system in a critical state.

Many different views have been expressed on SOC since the introduction of the sandpile model by Bak, Tang and Wiesenfeld in 1987 [46]. There is no consensus because the lack of a general understanding prevents the construction of a unifying framework. It is the opinion of the present author that the search for a degree of universality similar to the one found for thermal critical phase transitions is illusory and that the richness of out-of-equilibrium systems lies in the multiciplicity of mechanisms generating similar behaviors. Even for the same model and within the same mechanism, authors sometimes diverge with respect to the identification of the relevant variable or mechanism, which may reflect the fact that there are several possible descriptions of a self-organizing system (see below the parallel version of the Bak–Sneppen model for which we propose to see the sand flux h as an "order" parameter [894, 901] while Vespignani et al. use it as a control parameter [168, 169, 234, 655, 987–989]).

What may turn out to be "universal" is the resulting hierarchical organization, involving many scales, common to complex systems. In addition, all scales do not play the same role; a set of special discrete levels often play special roles in the global hierarchy. Examples are found in meteorology (dust devils, tornadoes, cyclones, large-scale weather systems), in the tectonic crust (joints, small faults, main faults, plate boundaries), in economy (traders, companies, countries, currency blocks), etc. In artificial intelligence, this is used in the development of hierarchical computational strategies. In Statistical Physics, hierarchies have been shown to appear dynamically due to a spontaneous breakdown of continuous scale invariance into discrete scale invariance [878].

15.2 Sandpile Models

15.2.1 Generalities

The concept of Self-Organized Criticality (SOC) was introduced by Bak, Tang and Wiesenfeld in 1987 [46] using the example of a sandpile. If a sandpile is formed on a horizontal circular base with any arbitrary initial distribution of sand grains, a sandpile of fixed conical shape (steady state) is formed by slowly adding sand grains one after another (external drive). In the steady state, the surface of the sandpile makes on the average a constant angle with the horizontal plane, known as the angle of repose. The addition of each sand grain results in some activity on the surface of the pile: an avalanche of sand mass follows, which propagates on the surface of the sandpile. In the stationary regime, avalanches are of many different sizes and Bak, Tang and Wiesenfeld argued that they would have a power law distribution. If one starts with an initial uncritical state, initially most of the avalanches are small, but the range of sizes of avalanches grows with time. After a long time, the system arrives at a critical state, in which the avalanches extend over all length and time scales [44, 227, 463].

Laboratory experiments on sandpiles, however, have not in general found evidence of criticality in sandpiles due to inertial and dilatational effects [661], except for small avalanches [409] or with elongated rice grains [313, 588] where these effects are minimized: indeed, small avalanches have small velocities and thus negligible kinetic energy and they activate only the first surface layer of the pile; elongated rice grains slip also essentially at the surface as a result of their anisotropy, thus minimizing the dilatational effects; they also build up scaffold-like structures which enhance the threshold nature of the dynamics.

Theoretically, a large number of discrete and continuous sandpile models have been studied. Among them, the Abelian sandpile model is the simplest and most popular [46, 226]. Other variants include Zhang's model which has modified rules for sandpile evolution [1057], a model for Abelian distributed processors and other stochastic rule models [227], the Eulerian Walkers model [749] and the Takayasu aggregation model [941, 942].

15.2.2 The Abelian Sandpile

In the abelian sandpile model, each lattice site is characterized by its height h. Starting from an arbitrary initial distribution of heights, grains are added one at a time at randomly selected sites n: $h_n \to h_n + 1$. The sand column at any arbitrary site i becomes unstable when h_i exceeds a threshold value h_c and topples to reduce its height to $h_i \to h_i - 2d$, where d is the space dimension of the lattice. The $2d$ grains lost for the site i are redistributed on the $2d$ neighbouring sites $\{j\}$ which gain a unit sand grain each: $h_j \rightarrow h_j + 1$. This toppling may make some of the neighbouring sites unstable. Consequently, these sites will topple themselves, possibly making further neighbors unstable. In this way, a cascade of topplings propagate, which finally terminates when all sites in the system become stable. Figure 15.1 shows a particular example of an avalanche on a square lattice of size 3×3 with open boundary conditions and with $h_c = 4$. When this avalanche has stopped, the next grain is added on a site chosen randomly. This condition is equivalent to assuming that the rate of adding sand is much slower than the natural rate of relaxation of the system. As we said above, the large separation of the driving and of the relaxation time scales is usually considered to be a defining characteristic of SOC. Finally, the system must be open to the outside, i.e. must dissipate energy or matter for instance. An outcoming flux of grains must balance the incoming flux of grains, for a stationary state to occur. Usually, the outcoming flux occurs on the boundary of the system: even if the number of grains is conserved inside the box, it loses some grains at the boundaries. Even in a very large box, the effect of the dissipating boundaries are essential: increasing the box size will have the effect of increasing the transient regime over which the SOC establishes itself; the SOC state is built from the long-range correlations that establish a delicate balance between internal avalanches and avalanches that touch the boundaries [623].

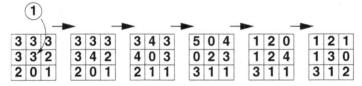

Fig. 15.1. An avalanche of the abelian sandpile model, generated on a 3×3 square lattice with open boundary conditions and $h_c = 4$. A sand grain is dropped on a stable configuration at the central site. The avalanche created has size $s = 6$, area $a = 6$, life-time $t = 4$ and radius $r = \sqrt{2}$. Reproduced from [596]

The simplicity of the abelian model is that the final stable height configuration of the system is independent of the sequence in which sand grains are added to the system to reach this stable configuration [226] (hence the name "abelian" refering to the mathematical property of commutativity). On a stable configuration $\mathcal{C}$, if two grains are added, first at i and then at j, the resulting stable configuration $\mathcal{C}'$ is exactly the same as in the case where the grains were added first at j and then at i. In other sandpile models, where the stability of a sand column depends on the local slope or the local curvature, the dynamics is not abelian, since toppling of one unstable site may convert another unstable site to a stable site. Many such rules have been studied in the literature [488, 595].

As shown in Fig. 15.1, an avalanche is a cascade of topplings of a number of sites created by the addition of a sand grain. The strength of an avalanche can be quantified in several ways:

1. size (s): the total number of topplings in the avalanche,
2. area (a): the number of distinct sites which toppled,
3. life-time (t): the duration of the avalanche and
4. radius (r): the maximum distance of a toppled site from the origin.

These four different quantities are not independent and are related to each other by scaling laws. Between any two such measures x, y belonging to the set $\{s, a, t, r\}$, one can define a mutual dependence by the scaling of the expectation of one quantity y as a function of the other x:

$$\langle y \rangle \sim x^{\gamma_{xy}} . \tag{15.1}$$

These exponents are related to one another, e.g.,

$$\gamma_{ts} = \gamma_{tr}\gamma_{rs} . \tag{15.2}$$

For the abelian sandpile model, it can be proven that the avalanche clusters cannot have any holes and in addition that $\gamma_{rs} = 2$ in two dimensions, i.e.

$$\langle s \rangle \sim r^2 . \tag{15.3}$$

It has also been shown that $\gamma_{rt} = 5/4$, i.e. [581]

$$\langle t \rangle \sim r^{5/4} . \tag{15.4}$$

A better way to estimate the exponents γ_{tx} is to average over the intermediate values of the size, area and radius at every intermediate time step during the growth of the avalanche [596]. One can calculate all the critical exponents for the directed version of the model in all dimensions. For the undirected case, the model is related to the $Q \to 0$ limit of the Potts model [226], where Q is the number of states that the Potts spin can take (see Chap. 11). Recall that the limit $Q \to 1$ maps the Potts model onto the percolation model as recalled in Chap. 12. This shows a deep relationship between certain SOC models and standard criticality [894, 901]. This mapping to the $Q \to 0$ limit of the Potts model enables exact calculation of some exponents in two dimensions, and there are some conjectures about others [226].

It was also shown that every recurrent configuration of the abelian model on an arbitrary lattice has a one-to-one correspondence to a random spanning tree graph on the same lattice. A spanning tree is a sub-graph of a lattice, having all sites and some bonds of that lattice. It has no loop and therefore, between any pair of sites, there exists a unique path through a sequence of bonds. There can be many possible spanning trees on a lattice. These trees have interesting statistics in a sample where they are equally likely. There are exactly the same number of distinct spanning trees as the number of recurrent Abelian sandpile model configurations on any arbitrary lattice [581].

Given a stable height configuration, there exists a unique prescription to obtain the equivalent Spanning tree. This is called the *Burning* method [581]. A fire front, initially at every site outside the boundary, gradually penetrates (burns) into the system using a deterministic rule. The paths of the fire front constitute the spanning tree. A fully burnt system is recurrent, otherwise it is transient.

Spanning trees exhibit remarkable properties. Let us randomly select such a tree and then randomly pick up one of the unoccupied bonds and occupy it. It thus leads to the formation of a loop of length ℓ. It has been shown that these loops have the length distribution $D(\ell) \sim \ell^{-1-\mu}$, with $\mu = 3/5$. Similarly, if a bond of a spanning tree is randomly selected and deleted, then the spanning tree is divided into two fragments. The sizes of the two generated fragments follow a probability distribution $D(a) \sim a^{-1-\mu}$ with $\mu = 3/8$ [597]. This is an illustration of the fact that geometrical objects defined from simple connectivity rules often embody a non-trivial scale invariant structure. A similar situation occurs in the percolation model close to the critical point (Chap. 12).

Quite generally, the finite size scaling form for the probability distribution function for any measure $x \in \{s, a, t, r\}$ is

$$P(x) \sim x^{-\tau_x} f_x \left(\frac{x}{L^{\sigma_x}} \right) . \tag{15.5}$$

The exponent σ_x determines the variation of the cut-off of the quantity x with the system size L. Scaling relations like $\gamma_{xy} = (\tau_x - 1)/(\tau_y - 1)$ connect any two measures. The scaling assumptions (15.5) for the avalanche sizes have not been demonstrated and may be open to doubt [488]. It has been argued that a multifractal distribution is better suited [217, 949] due to the dominance of rare outflowing avalanches. This seems to be due to the effect of rare large avalanches dissipating at the border which strongly influence the statistics.

Many different sandpile models have been studied. However, the precise classification of various models into different universality classes in terms of their critical exponents is not yet available. Exact values of all the critical exponents of the most widely studied Abelian model are still not known in two dimensions. Some effort has also been made towards the analytical calculation of avalanche size exponents [96, 533, 750]. Numerical studies for these exponents are found to give scattered values. Not much work has been done to study the version of sandpile models with threshold dynamics acting on their slopes or the curvature. Another still unsettled question is whether the conservation of the grain number in the toppling rules is a necessary condition to obtain a critical state. A difficulty that plagues this question is that extremely long transients occur during which numerical estimates give the impression of stationarity while, in reality, the system is still slowly evolving.

15.3 Threshold Dynamics

15.3.1 Generalization

An important sub-class of SOC is constituted by out-of-equilibrium systems driven with constant rate and made of many interactive components, which possess the following fundamental properties:

1. a highly nonlinear behavior, namely, essentially a threshold response,
2. a very slow driving rate,
3. a globally stationary regime, characterized by stationary statistical properties, and
4. power distributions of event sizes and fractal geometrical properties (including long range correlations).

The threshold plays a crucial role as a local rigidity which allows for a separation of time scales and, equally important, produces a large number of metastable states. The dynamics takes the system from one of these metastable states to another. It is believed that separation of time scale and metastability are essential for the existence of scale invariance in this class of systems. These concepts will be illustrated in details below with specific models.

The importance of the threshold dynamics (which is however not necessary to observe SOC in some systems, see below) is exemplified by the mapping of the stochastic critical forest-fire model [245] into a deterministic threshold model presenting the same statistical properties [846]. In the initial definition of the forest-fire model, no threshold appears explicitly and the separation of time scales is put in by hand by tuning the rates of two stochastic processes which act as driving forces for the model. The forest-fire [245] is defined on a d-dimensional square lattice. Empty sites are turned into "trees" with a probability p per site in every time step. A tree can catch fire stochastically when hit by "lightning", with probability f at each time step, or deterministically when a neighbouring site is on fire. The model is found to be critical in the limit $p \to 0$ together with $f/p \to 0$. This model is a generalization of a model first suggested by Bak, Chen and Tang [47] which is identical to the present model [245] except that it does not contain the stochastic ignition by lightning. The Bak, Chen and Tang system is not critical (in less than three dimensions). Thus the introduction of the stochastic lightning mechanism appeared to be necessary, at least in two dimensions, for the model to behave critically.

This model can be recast into a deterministic auto-ignition model. This model is identical to the model of [245], except that the spontaneous ignition probability f is replaced by an auto-ignition mechanism by which trees ignite automatically when their age T after inception reaches a value T_{max}. Choosing this value suitably with respect to p gives a system with exactly the same behavior and statistical properties as the stochastic forest fire model [245].

Thus, one stochastic driving process has been removed and a threshold introduced, while maintaining the SOC state; this model also displays explicitly the relationship between threshold dynamics and the separation of time scales so much needed for the SOC state.

The auto-ignition model can be turned into a completely deterministic critical model by eliminating the stochastic growth mechanism. In the deterministic model, called the regen FF in [846], each cell is given an integer parameter T which increases by one at each time step. If $T > 0$, the cell is said to be occupied, otherwise it is empty (or regenerating). The initial configuration is a random distribution of T-values and fires. Fires spread through nearest neighbours and the auto-ignition mechanism is again operative so that a tree catches fire when its T reaches T_{max}. However, in this model, when a tree catches fire, the result is a decrement of T_{regen} from its T-value. Note that when $T_{regen} < T_{max}$, a cell may still be occupied after it has been ignited. The parameters T_{max} and T_{regen} can be thought of as having a qualitatively reciprocal relationship with f and p respectively (in terms of the average "waiting time" for spontaneous ignition and tree regrowth), though this is less straightforward in the latter case because trees are not always burned down by fire. It is evident that T_{regen} also sets, and allows direct control of, the degree of dissipation of the T-parameter in the system. The forest-fire model, that was thought to belong to a different class, is now seen not fundamentally different from other threshold models discussed below.

Despite the many complexities concerning their initiation and propagation, real forest fires have been found to exhibit power-law frequency–area statistics over many orders of magnitude [582]. A simple forest fire model of the class discussed above exhibits similar behavior [582]. One practical implication of this result is that the frequency–area distribution of small and medium fires can be used to quantify the risk of large fires, as is routinely done for earthquakes. However, there are interesting differences in different parts of the world that we mention briefly. In many areas around the world, the dry season sees numerous large wildfires, with sometimes deaths of firefighters and people, the destruction of many structures and of large forests. It is widely accepted that livestock grazing, timber harvesting, and fire suppression over the past century have led to unnatural conditions – excessive biomass (too many trees without sufficient biodiversity and dead woody material) and altered species mix – in the pine forests of the West of the U.S.A., in the Mediterraneen countries and elsewhere. These conditions make the forests more susceptible to drought, insect and disease epidemics, and other forest-wide catastrophes and in particular large wildfires [356]. Interest in fuel management, to reduce fire control costs and damages, has thus been renewed with the numerous, destructive wildfires spread across the West of the U.S.A. The most-often used technique of fuel management is fire suppression. Recent reviews comparing Southern California on one hand, where management is active since 1900, and Baja California (north of Mexico) on the other hand where management is essentially absent ("let-burn" strategy) highlight a remarkable fact [628, 650]: only small and relatively moderate patches of fires occur in Baja California, compared to a wide distribution of fire sizes in Southern California including huge destructive fires. The selective elimination of small fires (those that can be controlled) in normal weather in Southern California restricts large fires to extreme weather episodes, a process that encourages broad-scale high spread rates and intensities. It is found that the danger of fire suppression is the inevitable development of coarse-scale bush fuel patchiness and large instance fires in contradistinction with the natural self-organization of small patchiness in left-burn areas. Thus, more work is needed before extrapolating naively the power-law frequency-area statistics, which may give a misleading sense of simplicity while in fact hiding the fact that it may result from mixing many different regions. The implications for policy management of forests as well as the lessons that can be learned for other complex systems, including stock markets, are discussed in [887].

15.3.2 Illustration of Self-Organized Criticality Within the Earth's Crust

It has been noticed early after the introduction of the concept of SOC by Bak, Tang and Wiesenfeld [46] that the earth's crust is one of the best systems that obeys the four conditions 1–4 stated above [45, 373, 860, 869, 893, 917] and has since then often been proposed as a paradigm of SOC (see [579, 869] for reviews).

- The threshold response can be associated with the stick–slip instability of solid friction or to a rupture threshold thought to characterize the behavior of a fault upon increasing applied stress.
- The slow driving rate is that of the slow tectonic deformations thought to be exerted at the borders of a given tectonic plate by the neighboring plates and at its base by the underlying lower crust and mantle. The large separation of timescales between the driving tectonic velocity ($\sim$ cm/yr) and the velocity of slip ($\sim$ m/s) makes the crust problem maybe the best natural example of self-organized criticality. It is important to realize that these two ingredients must come together.
  Let us imagine in contrast a system composed of elements with threshold dynamics, but which is driven at a finite rate compared to the typical time scale of its response. In the earthquake problem, this would correspond to plates moving at a velocity which is not much smaller than the rupture front velocity of brittle failure. In this case, the crust would rupture incessantly with earthquakes which could not be separated and which would create an average rapid plastic deformation flow of the crust. The frequency-size distribution would disappear and the system would not be in a SOC state

but rather in a kind of plastic turbulence. Artificial block–spring models of rapidly driven seismicity can be used to illustrate this idea [128, 141]. Alternatively, consider a system driven very slowly, with elements which do not possess a threshold but only a strong, continuous, nonlinear response without hysteresis or jump behavior. Due to the very slow driving at a given instant, only a small increment is applied to the system which responds adiabatically without any interesting behavior such as earthquakes with brittle rupture ("slow" earthquake regime? [567]).
It is thus the existence of the threshold that enables the system to accumulate and store the slowly increasing stress until the instability is reached and an earthquake is triggered. In turn, the slow tectonic driving allows for a response which is decoupled from the driving itself and which reflects the critical organization of the crust.

• The stationarity condition ensures that the system is not in a transient phase and distinguishes the long-term organization of faulting in the crust from, for instance, irreversible rupture of a sample in the laboratory and thus distinguishes SOC from the mechanism of avalanches associated with first-order transitions in disordered systems discussed in Chap. 14.

• The power laws and fractal properties reflect the notion of scale invariance; namely, measurements at one scale are related to measurements at another scale by a normalization involving a power of the ratio of the two scales. These properties are important and interesting because they characterize systems with many relevant scales and long-range interactions, which probably exist in the crust.

SOC is remarkable in the way it emerges. The spatial correlations between different parts of the system do not appear to be due to a progressive diffusion from a nucleus as in standard critical phase transitions but result from the repetitive action of rupture cascades. In other words, within the SOC hypothesis, different portions of the crust become correlated at long distances by the action of earthquakes which "transport" the stress field fluctuations in the different parts of the crust many times back and forth to finally organize the system. This physical picture is substantiated by various numerical and analytical studies of simplified models of the crust [45, 166, 184, 185, 907, 1057].
Actually, the existence of long-range height–height correlations in sandpile models are not necessary ingredients for the existence of a power law pdf of avalanche sizes. This can be seen from the fact that mean-field approaches to SOC [168, 169, 987] have no explicit reference to the correlation of the height or slope fields. In addition, the simple directed sandpile model in any dimension d studied by Dhar and Ramaswamy [228] has a power law pdf of avalanche sizes but no correlations in heights. Furthermore, a mean-field theory of a sandpile on a directed Bethe lattice does not give any height correlations, since the steady-state has a product measure [729].

15.4 Scenarios for Self-Organized Criticality

15.4.1 Generalities

Self-organized criticality was initially introduced with the hope of providing a universal and unifying mechanism for power laws and $1/f$ noise in nature. It is now clear that SOC may result from several distinct physical mechanisms. Similarly to the existence of many different mechanisms for the generation of power laws discussed in Chap. 14, the mechanism by which an open system self-organizes into a state with no characteristic scales is not unique. We are going to discuss several important mechanisms which have been studied in the literature. It is now understood qualitatively that there is a class of models exhibiting SOC as a result of the tendency for their elements to synchronize [105, 172, 182, 183, 338]. This tendency is however frustrated by constraints such as open boundary conditions [105, 172, 182, 183, 290, 338, 368, 457, 623, 857] and quenched disorder [908] which lead to a dynamical regime at the edge of synchronization, the SOC state. Another class, the so-called extremal models, exhibits SOC due to the competition between local strengthening and weakening due to interactions [713]. In a third class of models, SOC results from the tuning of the order parameter of a system exhibiting a genuine critical point to a vanishingly small, but positive value, thus ensuring that the corresponding control parameter lies exactly at its critical value for the underlying depinning transition [894, 901]. Another class consists of multiplicatively driven systems, such as spring–block models driven by temporally increasing spring coefficients [558, 559]. Due to its multiplicative driving, criticality occurs even with periodic boundary conditions via a coarsening process, similar to spinodal decomposition (see Chap. 9). The observed behavior should be relevant to a class of systems approaching equilibrium via a punctuated threshold dynamics. Conservation laws have also been conjectured to be essential for a class of SOC systems [378, 857]. In some systems [46, 139, 170, 488], the scale invariance can be shown to follow from a local conservation law (sand grains are conserved except at the boundaries of the pile) [581]. In this sense, the origins of long range correlations in SOC systems with conservation are well understood, though not all exponents have been calculated analytically. Let us also mention the intringuing suggestion by Obukhov [698] that SOC can be seen to result from non-linear interactions between Goldstone modes. Recall that Goltstone modes are the low-frequency and long-wavelength transverse fluctuations of a system that attempts to restore a broken symmetry such as a global translational or rotational invariance. As a consequence, the Goldstone modes usually exbibit scaling properties. If they interact non-linearly with scale free interactions, they can lead to self-similar spatial and temporal behavior. Let us finally mention the proposal that SOC results from a diffusive dynamics in which the diffusion coefficient exhibits a singular behavior $D \sim (h_c - h)^{-\phi}$. As a consequence of the acceleration of diffusion close to h_c, the field h is attracted to h_c and the diverging D leads to a broad distribution of fluctuation amplitudes [139]. Actually, one can also view the problem with the reverse logic: when a system exhibits SOC, its phenomenological description at large scales in terms of a diffusive field necessitates a singular behavior of the effective diffusion coefficient. This scenario thus has more descriptive than explanatory power.

15.4.2 Nonlinear Feedback of the "Order Parameter" onto the "Control Parameter"

Maybe the simplest idea to begin understanding SOC is through the process of artificially forcing standard critical transitions into a self-organized state [311, 872]. In Chap. 9, we have recalled that a critical phase transition is characterized by a so-called "order parameter" (OP), say the magnetization in a magnetic or Ising spin system, as a function of a "control parameter" (CP), the temperature. A critical phase transition is characterized by the existence of a critical value CP_c of the control parameter, such that the OP goes continuously from zero above CP_c to a non-zero value below CP_c. Right at CP_c, the system is critical, i.e., it exhibits fluctuations at all length scales which are reflected into a singular behavior of thermodynamic properties. One has thus to tune the CP close to CP_c to observe this self-similar structure.

At any given time, a system undergoing a critical phase transition has a fixed CP. In order to transform such a critical phase transition into a SOC state, the idea is to introduce a genuine dynamics on the CP and OP such that the OP exerts a non-linear feedback on the CP, this feedback being chosen such that the CP is attracted to the special critical value CP_c. How is it possible to recognize CP_c in a natural way and make it attractive? Indeed, CP_c is a number whose value depends on arbitrary units. One thus needs an absolute measure of the distance to the critical point. The trick is to realize that, at CP_c, the correlation length describing the typical spatial size of the fluctuations of the OP diverges. The susceptibility also diverges and thus offers another probe of the approach of the critical point. The idea is to imagine a feedback process of the OP on the CP whose amplitude is controlled by the spatial correlation length ξ or the susceptibility χ. This program has led to thought-experiments in which the correlation length is measured by some probing radiation by scattering methods or some electronic feedback using a microprocessor or analog device which push the temperature or analog control parameter to that value where the susceptibility or the correlation length is a maximum. The practical realization of the feedback thus corresponds to an optimization of the response of the system under the action of a probe or a disturbance. Possible implementations are for liquid–vapor and binary demixing critical points, the ^{4}He superfluid transition, magnetic systems, and superfluid transitions [311].

A clever independent implementation of this idea defines the "self-organized branching process" (SOBP) introduced by Zapperi et al. [1048]. In the mean-field description of the Bak–Tang–Wiesenfeld sandpile model, one neglects correlations, which implies that avalanches do not form loops and hence spread as a branching process. In the SOBP, an avalanche starts with a single active site, which then relaxes with probability p, leading to two new active sites. With probability $1 - p$, the initial site does not relax and the avalanche stops. If the avalanches does not stop, the procedure is repeated for the new active sites until no active sites remain. The parameter p is thus the probability that a site relaxes when it is triggered by an external input. For the SOBP, there is a critical value $p_c = 1/2$ such that for $p > p_c$ the probability to have an infinite avalanche is nonzero, while for $p < p_c$ all avalanches are finite. Thus, $p = p_c$ corresponds to the critical case, where avalanches are power law distributed. However, this fine-tuning of the control parameter p in the branching model cannot be the explanation of the occurrence of SOC, where the critical state is approached dynamically without the need to fine-tune any parameter. In order to resolve this paradox, Zapperi et al. introduce a feedback of the order parameter on the control parameter p. As shown in Sect. 15.4.4 and in Fig. 15.2, it is natural to think of the order parameter as the flux $J(t)$ of grains out of the sandpile, that is, the number of grains leaving the system at time step t. The idea of Zapperi et al. [1048] is that, if $J(t) = 0$, the control parameter p should increase as more grains are stored, making the system more prone to local instabilities. In contrast, if $J(t) > 1$, the system looses grains to the outside, make it more stable; this should lead to a smaller $p(t + 1)$. This concept is captured by the following equation:

$$p(t + 1) = p(t) + \frac{1 - J(p(t), t)}{N} \, , \tag{15.6}$$

where N is the total number of sites. The notation $J(p(t), t)$ reminds us that the flux of grains is also dictated by the instantaneous value of the control parameter. Using the branching model on a tree with n generations, we have $N = 2^{n+1} - 1$. The statistical average of the flux is simply $\langle J(p(t), t) \rangle = (2p)^n$ (one factor $2p$ per generation equal to the probability p to branch times the number 2 of new branches per generation as shown in Sect. 13.1). We can thus write $J(p(t), t) = (2p)^n + \eta(p, t)$, where the "noise" η describes the fluctuations around the average. Taking the continuous limit of (15.6) gives

$$\frac{dp}{dt} = \frac{1 - (2p)^n}{N} + \frac{\eta(p, t)}{N} \, . \tag{15.7}$$

By linearizing this expression, it is easy to see that the fixed point $p_c = 1/2$ is attractive and stable, thus rationalizing SOC in this SOBP as due to the feedback of the order parameter J onto the control parameter p. The distribution of avalanche sizes can then be calculated by standard techniques in branching processes using generating functions [404]. Not surprisingly, mean-field exponents are found, such as $\tau = 3/2$ for the pdf of avalanche sizes: $P(s) \sim 1/s^{\tau}$ [1048].

Several previous works [560, 774] have used a method to determine the critical percolation concentration which uses a similar feedback procedure (D. Stauffer, private communication): occupy a lattice with some probability p and check if it percolates. If yes, decrease p by δp, if not increase p by δp. Then, occupy the lattice again using the same sequence of random numbers and the new p. Then check and decrease δp by a factor 2, and change p again by the new δp. Thus, in a square site problem ($p_c = 0.593...$), one may obtain the sequence (starting with $p = 1/2, \delta p = 1/4$): $p = 1/2$, no; $p = 3/4$, yes; $p = 5/8$, yes; $p = 9/16$, no... This is an exponentially efficient iteration to determine p_c for one fixed configuration. The method has been called "Hunting the Lion" by A. Aharony: you move in the direction where you hear the lion roar, the lion being here the "infinite" percolating cluster.

Solomon et al. [859] have recently introduced a self-organized critical dynamics that converges to the percolation critical point, which is similar to the feedback method [311, 872]. The model intends to describe the organization of social imitative processes, such as in the decision to go see a new movie. The decision by a person i to go see the movie is supposed to depend on two variables: the quality q of the movie and the viewer's preference p_i; if $p_i \geq q$, the person goes to see the movie and does not go otherwise. In their Monte Carlo simulations of $L \times L$ square lattices, $p_c = 0.593$, and they restrict themselves to the simplest dynamics: the quality of the movie increases by δq if no cluster spans from top to bottom, while it decreases by δq otherwise. The viewer's preference p_i, initially distributed randomly between 0 and 1, changes by $\pm \delta p$ depending on whether agent i went to the movie or skipped it. For $\delta p = 0$, $\delta q > 0$, the quality q moves to the usual percolation threshold. For $\delta p > 0$, $\delta q = 0$, the p_i distribution drifts towards a single peak centered on the fixed q value, taken equal to 0.5 (no spanning cluster) or 0.593 (some spanning clusters). If both δp and δq are positive, p_i and q drift towards $p_c = 0.593$, even if the initial q was 0.5. Thus, generalizing invasion percolation, this dynamic percolator shows self-organized criticality: whereas in usual percolation, one has to put in 0.593 as the percolation threshold, the present social percolation model drifts towards this critical point automatically via the global feedback process. This model is offered as a possible explanation for the alternation of hits and failures rather than a featureless distribution of partial successes in certain markets such as toys, gadgets, movies industries, but also in the adoption of technological changes, of political and economical measures, and in the political arena.

15.4.3 Generic Scale Invariance

Hwa and Karder [443–445] and Grinstein et al. [376, 378] have used the field theoretical approach to model sandpiles and more generally a class of SOC systems that have eventually been coined as exhibiting "generic scale invariance". The philosophy of field theory is to search for the optimal representation of the system to be studied in which scale transformations used in the renormalization group approach (Chap. 11) become the simplest. Field theory makes strong use of the concept of "universality" according to which most microscopic details are not important in determining the large scale properties. From the very beginning, field theory gets rid of such details as the microscopic network structures, complicated interactions and so on, by keeping only a few relevant interactions. The price to pay is that, by construction, one looses all information on those ingredients that are not "universal" and the mathematical treatment that ensues is usually limited to some kind of perturbative approach in terms of one or a few interaction coupling terms.

In this spirit, the field theoretical approach to sandpiles consists in identifying the relevant coarse-grained dynamical parameter, namely the altitude $h(x, t)$ at point x and at time t. The simplest approach consists of modeling the coupling between neighboring sites by a diffusion process described by

$$\frac{\partial h}{\partial t} = -\frac{1}{\tau} h + \nu \nabla^2 h + \eta(x, t) \ , \tag{15.8}$$

where $\eta(x, t)$ is a Gaussian noise without correlations

$$\langle \eta(x, t)\eta(x', t')\rangle = 2D\delta(x - x')\delta(t - t') \ , \tag{15.9}$$

and the current is $\mathbf{j} = -\nu \nabla h$. The criticality in this system is reached when $1/\tau = 0$, i.e. when there is no local damping and the relaxation occurs globally by diffusing over the whole system. In this case, the susceptibility of the system in the Fourier space is defined by

$$\chi(k) \equiv \langle |\hat{h}(k, t)|^2\rangle \sim k^{-2} \ , \tag{15.10}$$

leading to the standard dependence $1/x^{d-2}$ of the Green function in d dimensions of a diffusive system. Simple diffusion with white noise driving is scale invariant.

A sandpile or any other system with a threshold relaxation has a highly non-linear behavior. In field theory, the idea is to replace the specific relaxation rules with thresholds by effective non-linear terms that are introduced in (15.8). The mathematical form of these terms are determined by principles of simplicity and symmetry. Hwa and Kardar [443–445] have shown that the leading non-linear term in increasing powers of the field $h(x, t)$ in an anisotropic sandpile presenting a prefered direction $\mathbf{e}$ for the sand flow is $\nabla_\parallel(h^2)$, which is the gradient of h^2 with respect to the direction parallel to the prefered flow. Generalizing (15.8) in the critical state $1/\tau = 0$ to account for anisotropy and to include the leading non-linear correction leads to

$$\frac{\partial h}{\partial t} = \nu_\parallel \nabla_\parallel^2 h + \nu_\perp \nabla_\perp^2 h - \frac{\lambda}{2}\nabla_\parallel(h^2) + \eta(x, t) \ , \tag{15.11}$$

corresponding to a flux current

$$\mathbf{j} = -\nu_\parallel \nabla_\parallel h - \nu_\perp \nabla_\perp^2 h + \frac{\lambda}{2}(h^2)\mathbf{e} \ . \tag{15.12}$$

Sornette and Virieux [917] have used this framework to obtain a relation between deformations at large time scales (Million years) and deformations at small time scales (seconds to hundred years) in the earth's crust, replacing h by the strain field. The first term of the r.h.s. of (15.11) corresponds to strain diffusion, the second one represents the non-linear rheological behavior of the lithosphere[2] and the noise term $\eta(x, t)$ accounts for the fact that tectonic deformations are locally irregular. The theory builds on the general conceptual framework that geological deformations and fault structures are the long term trace, or memory, of the cumulative short term fluctuations in earthquakes and perhaps other processes, and that the latter can be described as a "high frequency" noise for the former. This approach can be justified from the fact that the lithosphere is in a "critical" state of marginal mechanical stability, characterized by long-range correlations and power laws.

If one ignores the possible long-range noise correlation and its power law distribution, the Langevin equation (15.11) can be solved using the dynamical renormalization group and the general solution takes the homogeneous scaling form [443–445]

$$h\left(x_{\|}, x_{\perp}, t\right) = B^{\chi} h\left(x_{\|}/B, x_{\perp}/B^{\zeta}, t/B^{z}\right) , \tag{15.13}$$

where B is an arbitrary scaling factor. The exponents χ, ζ and z, which define how the field h scales with space and time, are given exactly by $z = 6/(7 - d)$ $(= 3/2$ in 3D), $\chi = (1 - d)/(7 - d)$ $(= -1/2$ in 3D) and $\zeta = 3/(7 - d)$ $(= 3/4$ in 3D).

The noise term $\eta(x, t)$ represents the sources of the field fluctuations around its long term trend which are not described by the continuous evolution. A Langevin equation indeed contains two main contributions: (1) a continuous one describing the long term average evolution and (2) short time fluctuations around the average evolution. For instance, the Langevin equation, describing the motion of a Brownian particle falling through a gas, contains both the average effect of the colliding molecules, i.e. a smooth drag term proportional to the particle velocity, but also the instantaneous effect of these collisions which create a fluctuating force at very small times comparable to the molecule–particle collision time and which is responsible for the erratic motion of the Brownian particle. In the case of tectonic deformations, the driving forces at the origin of plate motions are the analog of gravity. The lithosphere heterogeneity, fault roughness and the resulting stick–slip friction rheology are at the origin of the earthquakes which are the analog of the molecular collisions which create erratic motion and fluctuations. Averaged over long times, the particle appears to be falling through a viscous medium while fault motions and continental deformations appear to be controlled by a smooth friction. The noise term $\eta(x, t)$ thus describes the effects

---

[2] The lithosphere is the outmost 100 km of the crust and upper mantle which is relatively rigid (as seen from the propagation of seismic waves) and lies on top of a softer layer about 400 km thick called the asthenosphere.

of deformations at short times away from the long term trend. This noise term must thus describe the effect of all events on the fluctuations around the long term trend occurring at time scales much smaller than that at which the averaging over the strain is valid. Therefore, $\eta(x,t)$ must describe the effect of the various heterogeneous modes of deformations at short times, such as creep, silent and slow earthquakes, as well as genuine earthquakes of all sizes. Whereas the strength of the molecular collisions can be described in terms of a mean and standard deviation, earthquakes are known to follow a power law size distribution. Therefore, the noise $\eta(x,t)$ should be characterized by a probability distribution which is a power law given by the Gutenberg–Richter distribution.

By taking into account the power law nature of the noise term and assuming that the strain field is on average scale independent (i.e. that the exponent χ is zero), one obtains a self-consistent determination of the exponent b the Gutenberg–Richter distribution [917]: $b = 1$ for small earthquakes for which their energy or seismic moment is proportional to the third power of the linear size of the rupture (3D rupture propagation) and $b = 1.5$ for larger and great earthquakes for which their energy or seismic moment is proportional to the second power of the linear size of their rupture (2D rupture propagation). These predictions are consistent with recent observations for large ($M_W \geq 7$) earthquakes worldwide.

In the case where the noise term obeys exactly the conservation law corresponding to

$$\langle \eta(x,t)\eta(x',t') \rangle = 2 \left(D_\perp \nabla_\perp^2 + D_\parallel \nabla_\parallel^2 \right) \delta(x - x')\delta(t - t') , \qquad (15.14)$$

the susceptibility reads

$$\chi(k) = \frac{D_\perp k_\perp^2 + D_\parallel k_\parallel^2}{\nu |k|^2} , \qquad (15.15)$$

where $|k|^2 = k_\perp^2 + k_\parallel^2$. The Fourier transform of $\chi(k)$ gives the Green function [376, 378]

$$G(x) \equiv \langle h(x,t)h(0,t) \rangle . \qquad (15.16)$$

- In the anisotropic case, $D_\perp \neq D_\parallel$, detailed balance is violated and

$$G(x) \sim x^{-d} \qquad (15.17)$$

correspond to self-similarity and thus criticality [376, 378]. Recall that detailed balance is a sufficient (but not necessary) condition for an equilibrium (time independent) probability distribution P to exist. It reads $P(A,t)w(A \to B) = P(B,t)w(B \to A)$, where $w(A \to B)$ is the transition rate of the system from state A to state B. A clear exposition of the relationship between detailed balance and the diffusion and Fokker–Planck equations presented in Chap. 2 can be found in [398].

- In the isotropic case, $D_\perp = D_\parallel$, the next higher order term in (15.8) is necessary to obtain a meaningful description leading to

$$\frac{\partial h}{\partial t} = \nu_1 \nabla^2 h + \nu_2 (\nabla^2)^2 h + \eta(x,t) \ . \tag{15.18}$$

The corresponding Green function is

$$G(x) \sim e^{-x/\xi} \ , \qquad \text{where } \xi \propto (\nu_2/\nu_1)^{1/2} \ . \tag{15.19}$$

The system is no longer scale invariant as isotropy together with noise conservation leads to the introduction of a characteristic scale ξ.

To sum up, in the field theory, scale invariance results from the existence of conservation laws and may disappear or reappear depending on the interplay between anisotropy and conservation of the driving noise [376, 378]. Another interesting scenario involves the non-linear terms added in the diffusion equation which create a novel fixed point in the renormalization group flow [376].

The problem with this field theoretical approach is that the essential threshold dynamics is not captured by the "weak" and perturbative nonlinear terms added to the diffusive equation. In addition, avalanches, which constitute the hallmark of these systems, are not described and the origin of the self-organization is not really explained other than by the existence of a conservation law and its competition with anisotropy. As we said, this is due to the fact that the threshold dynamics is replaced by a "weak" perturbative nonlinear term. Furthermore, the driving occurs on a fast time scale (stochastic noise) in contrast with the very slow driving common to sandpile SOC models, whereas the order parameter exhibits slow diffusion-like relaxations similar to critical slowing down [428], in opposition with the fast relaxation induced by the avalanches. A physical system which exemplifies these features is provided by earthquakes which relax the stress accumulated over centuries during time scales of tens of seconds.

Recognizing the crucial role played by the threshold dynamics and the necessity to take it into account explicitly in a continuum formalism, several authors [52, 130, 230, 574, 874] have modeled the threshold nature of the dynamics by either a discontinuous or singular diffusion coefficient [52, 130, 574], or by series expansion of the heaviside function and its derivatives [230]. This kind of approach still contains an ad hoc discrete component and cannot thus be considered fully continuous. Furthermore, the very slow driving condition is rarely imposed except in [574]. A notable attempt to incorporate the threshold behavior in a continuous formalism has been done in [135] using the macroscopic phenomenological Coulomb solid friction law. It turns out that this law does not yield any SOC but only a large avalanche regime due to the linear growth of the state variable derived from the Coulomb law.

Gil and Sornette [338] have proposed a more microscopic and fundamental description which can both display SOC and the large avalanche regime, and furthermore provides a good model of solid friction [142]. This model provides a general correspondence between SOC and synchronization of threshold oscillators (see below). The basic idea is that the order parameter presents multistability and hysteresis captured by a subcritical bifurcation between two states. This ensures the threshold dynamics. In addition, it can be shown that a local hysteretic reponse qualifies as a microscopic model of solid friction [142], which is a key property of dry sand. Another important ingredient is to incorporate the dynamics of an order parameter (OP) and of the corresponding *control* parameter (CP) in order to understand why the CP self-organizes to a critical value. Within the sandpile picture, the CP $\partial h/\partial x$ is the slope of the sandpile, h being the local height, and the OP S is the state variable distinguishing between static grains ($S = 0$) and rolling grains ($S \neq 0$). Therefore, the sand flux is proportional to S. Coupling these two parameters is physically natural.

In the fully continuous Landau–Ginzburg sandpile model [338], SOC is identified as the regime where diffusive relaxation along the sandpile occurs faster than the instability growth rate transforming a static grain into a rolling one. The wide distribution of avalanches sizes can be seen to result from an effective negative diffusivity, favoring the unstability of small avalanches which are in competition with the tendency for synchronization leading to large avalanches. Indeed, in the other limit of slow diffusion, avalanches comparable to the system size become dominant but coexist with a power law pdf of avalanche sizes for small avalanches. The pdf is thus a power law for small avalanche sizes and develops a bump at large sizes corresponding to the characteristic avalanches sweeping the whole system. A qualitatively similar phenomenology has been found in sandpile models with stress-dependent stress drops [208].

15.4.4 Mapping onto a Critical Point

General Framework. Consider a "standard" *unstable* critical phase transition, such as the Ising ferromagnet or bond percolation. Here, a spin, up or down, is assigned to each site with an exchange coupling constant J. Furthermore, one defines two nearest neighbor sites to be connected with a probability $p = 1 - e^{-2J/k_B T}$ if both have spin up. For zero external field, this defines a critical temperature T_c or bond-density ρ_c below which the order parameter m_0, the magnetization or the probability of an infinite cluster, is zero above T_c and behaves as $m_0 \propto (T_c - T)^\beta$ below T_c. The transition is further characterized by a diverging correlation length $\xi \propto (T - T_c)^{-\nu}$ and susceptibility $\chi \propto (T - T_c)^{-\gamma}$ as T_c is approached, quantifying the spatial fluctuations of the order parameter.

Suppose that it turns out to be natural for the system under consideration that, instead of controlling T, the "operator" controls the order parameter m_0 and furthermore one takes the limiting case of fixing it to a positive but arbitrarily small value. The condition $m_0 \to 0^+$ is equivalent to $T \to T_c^-$.

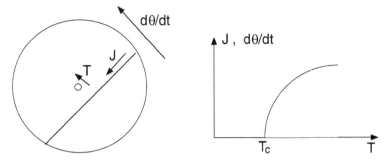

Fig. 15.2. Sandpile model in the rotating cylinder geometry, inspired from experiments and suitable for the mapping discussed in this section: the cylinder axis is horizontal and coincident with the rotation axis. The cylinder is partially filled with "sand" presenting an initially flat horizontal interface below the axis. Suppose that the axis of the cylinder is held to a fixed frame by a torsion spring on which one can exert a controlled torque T (CP). If $T = 0$, the cylinder takes the position such that the surface of the sand is horizontal, i.e. the rotation angle $\theta = 0$. If one starts to exert a non-vanishing T, the cylinder rotates up to an angle θ such that the torque exerted by the tilted sandpile balances exactly the applied T. Increasing T, one finally reaches a critical value T_c at which the sandpile reaches its slope θ_c of instability corresponding to the triggering of sand flow J, whose magnitude increases with $T > T_c$. In contrast, controlling the order parameter (the flux J of sand or the rotational velocity $d\theta/dt$) to a vanishingly small but non-zero value ensures the convergence to the SOC state

In other words, the system is at the critical value of the *unstable* critical point and must therefore exhibit fluctuations at all scales in its response. This is nothing but the hallmark of the underlying *unstable* critical point. This scenario applies most naturally to out-of-equilibrium driven systems.

In other words, some systems exhibiting SOC present a genuine critical transition when forced by a suitable control parameter, often in the form of a generalized force (torque for sandpile models, stress for earthquake models, force for depinning systems). Then, SOC appears as soon as one controls or drives the system via the order parameter of the critical transition (it turns out that in these systems, the order parameter is also the conjugate of the control parameter in the sense of the mechanical Hamilton–Jacobi equations). The order parameter hence in general takes the form of a velocity or flux. The condition that the driving is performed at $M \to 0^+$ illuminates the special role played by the constraint of a *very slow driving rate* common to all systems exhibiting SOC: this is the exact condition to control the order parameter at 0^+ which ensures the positioning at the exact critical value of the control parameter.

This general idea applies qualitatively and, in some cases, quantitatively to the sandpile models (Fig. 15.2), the earthquake models (Fig. 15.3), pinned–depinned lines or Charge-Density-Wave models, fractal growth processes and forest-fire models [42, 372, 894, 901] and has been shown experimentally to

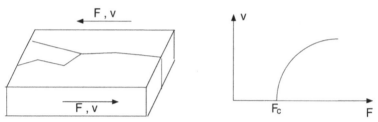

Fig. 15.3. Model of a tectonic plate on which a shear force F is imposed at its border. As the applied force F increases, the plate (which can contain pre-existing damage such as cracks and faults) starts to deform increasing the internal damage. For sufficiently low F, after some transient during which the system deforms and adjusts itself to the applied force, the system becomes static and nothing happens: the strain rate or velocity v of deformation becomes zero in the stationary state (here we are neglecting any additional creep or ductile behavior). As F increases, the transient becomes longer and longer since larger and larger plastic-like deformations will develop within the plate. There exists a critical *plasticity* threshold F_c at which the plate becomes globally "plastic", in the sense that it starts to flow with a non-zero strain rate $v \equiv d\epsilon/dt$ under fixed F. As F increases above F_c, the shear strain rate v increases. F is the control parameter and $v = d\epsilon/dt$ qualifies as the order parameter ($d\epsilon/dt = 0$ for $F < F_c$ and $d\epsilon/dt > 0$ for $F > F_c$). However, driving the plate at a constant and slow deformation rate v ensures SOC at F close to F_c

apply to the superfluid transition in Helium in the presence of a slow applied heat flux [636]. In this last example, when ^{4}He becomes superfluid, it can transport heat without any temperature gradient. The heat flux can thus be seen as a kind of order parameter directly related to the standard order parameter, namely the superfluid condensate wave function, which becomes non zero at the critical temperature (the control parameter). By establishing a small controlled heat flux Q, the system is driven by its order parameter and, close to the lambda point, the temperature gradient self-organizes itself to become independent of Q. Fulco et al. have implemented this mapping concept operationally by using a recursive relation [320]: once the order parameter is properly chosen, the recursion drives the physical system spontaneously to the value of the control parameter associated with such a choice; in particular, if the order parameter is set to a small value, the system is driven towards the critical point. Fulco et al. have shown that this method provides efficient estimations of critical properties, including critical points and exponents [320]. For Ising spin systems, and despite the small lattice sizes used, it yields critical temperatures and exponents in good agreement with those available in the literature.

The role of driving in sandpile models and numerical simulations at non-vanishing driving rates have been reported by Hwa and Kardar [445], Montakhad and Carlson [644] and Barrat et al. [59]. "Self-organized" criticality is recovered in the double limit of infinitely slow driving and small dissipation rates [59].

Extremal Version of the Bak–Sneppen Model as the Zero Temperature Limit of a Spin Glass Model. The one-dimensional version of the Bak–Sneppen model of evolution [48] is defined as follows: each of the N sites on a line array is filled with independent random numbers $p_i, i = 1, 2, \ldots, N$, drawn from a uniform distribution between $[0, 1]$. At each time step, the smallest p_i is selected and replaced by a new random number p'_i still in $[0, 1]$. At the same time, we update its two nearest-neighbors p_{i-1}, p_{i+1} with two new, independent random numbers. Periodic boundary conditions are enforced.

Note that the choice of a uniform distribution is by no way restrictive since what really matters is the non-decreasing character of the (cumulative) probability distribution function for any probability density. Bak and Sneppen [48] argued that their model is to be thought of as a model for darwinian biological evolution: each p_i represents the "fitness" of the species i, i.e. its adaptativity to changes and mutations. The species with the lowest p_i dies and its extinction affects its nearest neighbors p_{i-1}, p_{i+1} in the ecological nest, which must respond with an instantaneous mutation. Numerical evidence indicates that, in the long-time and large-N limits, the probability density of sites is strictly zero for all $p_i < p_c$ and constant above $p_c = 0.66702 \pm 0.00008$ [713]. In the context of statistical mechanics, a more appealing interpretation is to think of the Bak–Sneppen model as a model for self-organized depinning [388], where the interface undergoes a local rearrangement where the force $f_i = 1 - p_i$ is maximal. This was the original formulation of the Sneppen model [855].

In the Bak–Sneppen model, one usually defines avalanches in the following way: starting from a given $p_{\min}$ at time t, the duration t_0 of an avalanche is the minimum number of time steps t' required for $p_{\min}(t + t')$ to be greater than $p_{\min}(t)$. At the time $t+1$, among the three new random numbers, one at least can be smaller than $p_{\min}(t)$ and the smallest one of all three will initiate subsequent topplings which follow the same rule at each subsequent time step. The avalanche terminates when all sites are above $p_{\min}$. Then, an avalanche starts elsewhere, nucleating from the next minimum site, until a steady-state is reached where an infinite avalanche develops: the toppling of the minimal site will eventually trigger an infinite number of topplings among spatially connected sites in a system of infinite extent. The size of an avalanche is defined as the number of topplings. However, the avalanches in this extremal model are spatially but not necessarily *spatio-temporally* connected. This is due to the *extremal* nature of the dynamics in which an avalanche spreads out only from the minimal site and its two neighbors at all time steps, while other sites with fitnesses smaller than the $p_{\min}$ which initially triggered the avalanche are left unactivated, until eventually they get modified when the avalanche returns to them.

The Bak–Sneppen model has recently been shown [407] to implicitly assume that a single positive temperature-like parameter T is set arbitrarily close to zero. For finite T, the model is no longer critical. The underlying mechanism is the existence of a hierarchy of time scales that become separated as $T \to 0^+$, allowing extremal dynamics to be realised as the limiting case of a system with strictly local driving. To see this, we follow [407] and consider a system consisting of N elements, each of which is assigned a *barrier* E_i, $i = 1, \ldots, N$. The values of the barriers are drawn from the time-independent *prior distribution* $\rho(E)$, which is assumed to have no delta-function peaks so that there is a vanishing probability of two different elements having the same value of E.

The system evolves according to two rules. Firstly, each element becomes *activated* at a rate $e^{-E_i/T}$ where the constant parameter $T > 0$ has the same units as E. An activated site i is assigned a new barrier E_i drawn from $\rho(E)$, corresponding to a shift to a new metastable state with a new barrier height. Secondly, for every activated element, z other elements are chosen and also assigned new barrier values. The system behaviour changes qualitatively as the single parameter $T > 0$ is varied:

- $T \to 0^+$: In the limit of infinitesimal T, the activation rates $e^{-E_i/T}$ for different E_i diverge relative to each other. The element with the smallest barrier will become active on one timescale, the one with the second smallest barrier becomes active on another, much longer timescale, and so on. Thus, with probability one, the first element to become active will be that with the smallest barrier (which is always unique for a finite set of non-degenerate $\{E_i\}$). This is the way in which the Bak–Sneppen model is usually defined.

- *T small but finite:* The strict separation of time scales is lost for finite T and every element has a non-vanishing probability of being the first to become active, so the dynamics are no longer extremal. The model is not critical in this regime. This claim is supported by the results of numerical simulations [407] and the following mean field analysis.

For $N \to \infty$, the proportion $P(E,t)$ of elements with barriers in the range between E and $E + dE$ evolves according to the following continuous master equation:

$$\frac{\partial P(E,t)}{\partial t} = -\frac{e^{-E/T}}{\int_0^\infty e^{-E'/T} P(E',t)\, dE'} P(E,t)$$
$$-z\, P(E,t) + (z+1)\rho(E). \tag{15.20}$$

According to this equation, $P(E,t)$ decreases when an element changes its barrier value, which occurs either when it becomes active, or when it is selected as one of the z interacting elements. These two processes are described by the first and second terms on the right hand side of (15.20), respectively, where the prefactor to the first term is the probability for an element to be activated. Conservation of probability is ensured by the third term, which corresponds to the $z + 1$ new barriers drawn from $\rho(E)$. For the uniform $\rho(E)$ in the limit $t \to \infty$, one gets [407]

$$P(E,\infty) \approx \frac{z+1}{z} \left(1 + e^{-(E-E_c)/T} \right)^{-1}, \tag{15.21}$$

with $E_c = 1/(z + 1)$. As $T \to 0^+$, the exponential in (15.21) either blows up or decays depending on whether E is less than or greater than E_c, respectively. Thus $P(E, \infty)$ converges to the step function $(z + 1)/z \, \theta(E - E_c)$, in accord with the known solution of the mean field Bak–Sneppen model. However, there is no such discontinuity for finite T and $P(E, \infty)$ is smoothly varying for all $0 < E < 1$.

Mapping Directed Percolation onto the Parallel Bak–Sneppen Model. The parallel Bak–Sneppen model [42, 372, 894] is the same as the initial extremal version recalled above except that *all* sites which have their numbers below $p_{\min}$ (the value from which an avalanche started) and not solely the smallest one (which is of course smaller than $p_{\min}$) see their values and those of their neighbors replaced by new random numbers in $[0, 1]$. This model self-organizes to a critical point. To see this, let us introduce the *tuned* parallel Bak–Sneppen model, which consists in updating at a given time step *all* sites which are smaller than p and their (respective) left and right nearest-neighbors, where p is an externally controlled parameter chosen in $]0, 1[$. The updating is again done in parallel for all unstable sites, and any site neighbored by two toppling sites is updated once and only once. p plays the role of a control parameter. In this model, the system does not self-organize but rather ajusts itself in response to the imposed value p. In fact, this second model is indistinguishable from the directed percolation (DP) model discussed in Chap. 12, which exhibits a genuine critical transition at a specific value p_c.

Note first that, in contrast to the extremal Bak–Sneppen version, avalanches are both spatially and *spatio-temporally* connected, as in DP. It is natural to view the time-evolution as a two-dimensional lattice (i, t), $p_i(t)$ being the value of the fitness at (discrete) time t of the i-th site. To each site of the 2d-lattice, we associate a spin-like variable $n_{i,t}$ equal to 0 if $p_i(t) > p$, and $+1$ if $p_i(t) \leq p$. In words $n_{i,t} = 0$ if $p_i(t)$ is stable, and $+1$ if it is unstable. By the definition of the model, the value of $n_{i,t+1}$ just depends on the values of $n_{i-1,t}, n_{i,t}, n_{i+1,t}$: as soon as at least one among these three sites is unstable, the central site $p_i(t)$ will be updated. Since we redraw each site from an uniform distribution between 0 and 1, $p_i(t + 1)$ will be smaller than p and thus unstable with probability p, and stable with probability $1 - p$. This rule determines 14 out of the $2^4 = 16$ *local conditional probabilities* $P(n_{i,t+1}|n_{i-1,t}, n_{i,t}, n_{i+1,t})$, the last two probabilities being $P(0|0, 0, 0) = 1$ and of course $P(1|0, 0, 0) = 1 - P(0|0, 0, 0) = 0$: if a site and its two nearest-neighbors are stable, it will remain stable with probability 1. This last condition shows that the tuned parallel Bak–Sneppen model is not fully probabilistic: according to the conventional terminology, the phase formed of 0 spins is called an *absorbing* phase [505, 797]. The existence of an absorbing phase is a strong indication that the model should be in the DP universality class [369] (see Chap. 12). We are going to prove that this is indeed the case.

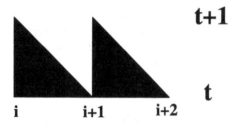

t+1

Fig. 15.4. 2d lattice on which the parallel Bak–Sneppen model is defined. By slightly bending the lattice, we get triangular plaquettes, although the connectivity is still that of a square lattice

- For the sake of pedagogy, we first consider a one-sided model where just one neighbor of an unstable site is updated at the next time step, the central site being also modified at the next time step. This is the topology of the original Sneppen model. By innocuously bending the lattice as shown in Fig. 15.4, we obtain triangular "plaquettes". Here, the local conditional probabilities of the tuned parallel Bak–Sneppen model read: $P(0|1,0) = P(0|0,1) = P(0|1,1) = 1 - p$ and $P(0|0,0) = 1$ (supplemented, of course, with $P(0|n_1,n_2) = 1 - P(1|n_1,n_2)$ for any n_1, n_2). These local conditional probabilities are defined on every other triangular "plaquette" – say the up-pointing ones if time is running upward as in Fig. 15.4, which connects spins between two successive time slices.
- Now imagine that we call 0 spins *dry* sites and 1 spins *wet* sites; the rules of the directed site–bond percolation model are that sites have probability p_s to be wet and that up-pointing bonds (the diagonal edges of a triangle) have probability p_b to conduct fluid. This is just a generalized mixed bond–site percolation problem (a particular case of the Domany–Kinzel automaton), where the local conditional probabilities are straightforward to write down [239, 337, 505]:

$$P(1|0,0) = 0$$

$$P(1|1,0) = P(1|0,1) = p_s p_b$$

$$P(1|1,1) = p_s[p_b{}^2 + 2p_b(1 - p_b)] = p_s p_b(2 - p_b) \ . \tag{15.22}$$

We can now readily proceed to the correspondence.

1. The two models (tuned parallel Bak–Sneppen model and directed percolation model) defined on the same lattice have the states and the same set of transitional probabilities $P(n_{i,t+1}|n_{i-1,t}, n_{i+1,t})$.
2. By identification, we obtain the correspondence between p for the tuned parallel Bak–Sneppen model and the p_s and p_b of the directed percolation model: $p = p_s p_b$ and $p = p_s p_b(2 - p_b)$ so that $p = p_s$ and $p_b = 1$.

The one-sided tuned parallel Bak–Sneppen is thus completely isomorphic to directed site percolation on a square lattice. The two models have a singular behavior at the critical value $p_c = 0.705489 \pm 0.000004$ [280] for the site directed percolation model on a 2d *square* lattice. The equivalence between the two models generalizes readily to the original (three sites) tuned parallel

BS model, i.e. one has: $P(1|1,0,0) = p_s p_b$, $P(1|1,1,0) = p_s[p_b^2 + 2p_b(1 - p_b)] = p_s p_b(2 - p_b)$ and $P(1|1,1,1) = p_s(p_b^3 + 3p_b^2(1 - p_b) + 3p_b(1 - p_b)^2) = p_s(1 - (1 - p_b)^3)$. Identifying with $P(1|1,0,0) = P(1|1,1,0) = P(1|1,1,1) = p$ gives again $p_s = p$ and $p_b = 1$.

From the definition of the tuned parallel Bak–Sneppen model, the control parameter p can be imposed to be larger than the DP threshold, corresponding to the depinned regime, where the fraction v of unstable or active sites (i.e. those with $p_i < p$) becomes non-zero. v is the *order parameter* for this depinning transition: $v \propto (p - p_c')^\beta$, where $\beta = 0.2764 \pm 0.0008$.

- *The Self-Oganized Parallel Bak–Sneppen Model Converges to the Critical Point of the Directed Percolation Model*

As already asserted, the DP critical state can be reached either by tuning a control parameter as in the tuned parallel BS model or by going to a very small rate of evolution, since all that really matters is how avalanches are defined. We now show that the parallel BS model converges indeed to the DP critical point [894].

Let us denote the gap by $p_{min}(t)$, i.e. the smallest number in the lattice at time t. There are two questions to address: (1) does $p_{min}(t)$ converge to a well-defined value p_∞ at long times? (2) if yes, what is this value p_∞?

Let us first assume that the answer to the first question is positive and then show that $p_\infty = p_c$ of directed percolation. Our demonstration is made by "ad absurdum" reasoning.

- Suppose that p_∞ is larger that the DP critical value p_c. Consider a time t_0 at which an avalanche starts with $p_c < p_{min}(t_0) < p_\infty$. Since the avalanche develops according to the parallel updating rule, nothing distinguishes its time evolution from the dynamics of the tuned parallel Bak–Sneppen model with p fixed to $p_{min}(t_0)$. However, since the tuned parallel Bak–Sneppen model is strictly equivalent to DP, the condition $p_c < p_{min}(t_0) = p$ implies that there is a finite probability that the avalanche is infinite. As a consequence, $p_{min}(t_0)$ cannot be less than p_∞, in contradiction with the hypothesis.

- Suppose conversely that $p_\infty < p_c$ of DP. Let us consider an avalanche starting at $p_{min}(t) = p_\infty$. Again, since the avalanche develops according to the parallel updating rule, nothing distinguishes its time evolution from the dynamics of the tuned parallel Bak–Sneppen model with p fixed to p_∞. Since p_∞ is supposed smaller than the critical value p_c of directed percolation, the activity must die after a finite number of time steps, in contradiction with the starting hypothesis $p_{min}(t) = p_\infty$, which is such that arbitrarily large avalanches can appear by the very definition of the critical point.

We are thus led to the conclusion that if p_∞ exists, it is equal to the critical threshold p_c of directed percolation.

We now return to the first question of the existence of the limit p_∞. We first note that the gap $p_{min}(t)$ is a monotonic non-decreasing function of time, by the very definition of the model. Being bounded by 1, $p_{\min}(t)$ either converges to a finite value p_∞ and within our demonstration, the final result follows, or $p_{\min}(t)$ could shift infinitely slowly upwards towards the limiting value 1 without definite convergence. However, the latter cannot be: indeed, suppose $p_{\min}(t)$ drifts above p_c; then, by the fact that an avalanche develops according to the parallel updating rule and has therefore the same time evolution as an avalanche of the tuned parallel Bak–Sneppen model and therefore of directed percolation, there is a finite probability that this avalanche is infinite and never stops. If it never stops, this means that the smallest fitness (also called "gap") is stuck for ever at this value $p_{\min}(t)$ at which an infinite avalanche has been triggered. This outcome can occur as soon as $p_{\min}(t)$ becomes larger than p_c, by the definition of p_c. This shows that $p_{\min}(t)$ cannot grow above p_c because, in an infinite system, as soon as there is a non-zero probability for an infinite avalanche to occur, it will occur and thus block forever the evolution of the gap. Is it possible for $p_{\min}(t)$ to drift very slowly without ever converging to p_∞? This is ruled out by the "gap" equation (15.28) discussed more extensively below, derived for the usual extremal Bak–Sneppen model [713], which also holds for the parallel Bak–Sneppen model. The gap equation quantifies the mechanism of approach to the self-organized critical attractor, whatever it might be. In words, the gap equation expresses the fact that $p_{\min}(t)$ increases by finite increments as long as it is smaller than p_∞ due to the finiteness of the avalanches. This is the last piece of reasoning needed to conclude that $p_{\min}(t)$ converges to a finite value less than or equal to p_c. With our previous "ad absurdum" reasoning, we conclude that the parallel BS model self-organizes spontaneously onto the critical point of directed percolation.

15.4.5 Mapping to Contact Processes

In a series of papers, Vespignani, Zapperi and co-authors [168, 169, 234, 655, 987–989] have developed a description of sandpiles and other SOC models such as the forest-fire model in terms of contact processes. A contact process is such that one state (the vacuum or inactive state) is absorbing, i.e. the system cannot escape spontaneously from it. An example is the directed percolation model discussed in Chap. 12. In order to obtain the mapping onto contact processes, they introduce additional control parameters, namely a non-vanishing driving rate h (incoming sand flux) and dissipation rate ϵ. The central idea is to distinguish three populations of sites.

- *stable* $[\sigma_i = s$ of coarse-grained concentration $\rho_s(r,t)]$: stable sites are those that do not relax (become active) if sand is added to them by external fields or interactions with active sites.
- *critical* $[\sigma_i = c$ of coarse-grained concentration $\rho_c(r,t)]$: critical sites become active with the addition of sand.

- *active* [$\sigma_i = a$ of coarse-grained concentration $\rho_a(r, t)$]: active sites are those transferring sand; they are interacting with other sites (usually nearest neighbors).

The approach of Vespignani, Zapperi and co-authors is reminiscent of the "mapping of SOC onto criticality" discussed in the previous section. However, there are important differences, in particular, their interpretation of the sand flux h as a control rather than an order parameter and the density $\rho_a(r, t)$ of active sites as the order parameter. The correspondence to contact processes to which the directed percolation model belongs make the two approaches even more similar. A value of Vespignani, Zapperi and co-authors' approach is that they have been able to develop further their theoretical description and obtain good predictions for the exponents.

In order to interpret the three populations of sites, we note that SOC indeed refers generally to systems in which the only state that generates dynamical evolution is the active one; i.e. stable and critical sites can change their state only because of external fields or by interacting with an active nearest neighbor. Therefore, SOC models correspond to three state cellular automata. This description is only approximate, since a certain amount of information is lost in grouping together stable sites. For instance, in the Bak, Tang and Wiesenfeld model, one observes several energy levels which pertain to a stable site. The three state description is exact for instance in the forest-fire model [245].

Coarse-graining the dynamical evolution rules, Vespignani, Zapperi and co-authors obtain the following set of mean-field equations for $\rho_s(r, t), \rho_c(r, t)$ and $\rho_a(r, t)$ by neglecting higher orders in h and ρ_a

$$\frac{\partial}{\partial t}\rho_a(t) = -\rho_a(t) + h\rho_c(t) + (g - \epsilon)\rho_c(t)\rho_a(t) + \mathcal{O}(h\rho_a, \rho_a^2) \ , \quad (15.23)$$

$$\frac{\partial}{\partial t}\rho_s(t) = \rho_a(t) - uh\rho_s(t) + u(g - \epsilon)\rho_s(t)\rho_a(t) + \mathcal{O}(h\rho_a, \rho_a^2) \ , \quad (15.24)$$

where g is the number of sites involved in the dynamical relaxation process ($= 2d$ in the Bak, Tang and Wiesenfeld model), and u ($= 1$ for three level models) takes into account the presence of several height levels of the sandpile and can be determined self-consistently in multilevel models by using the conservation of sand. ρ_c is determined by

$$\rho_c = 1 - \rho_a - \rho_s \ . \quad (15.25)$$

The mean-field character of these equations stems from the fact that there is no space dependence, i.e. the concentrations are assumed to be homogeneous. The mapping onto contact processes is now clear by inspection of (15.23) which shows that, in the absence of the external field h, a state with no active site $\rho_a = 0$ remains stable for ever.

These equations have been used to study the subcritical regime as well as the critical regime when both h and ϵ go to zero with $h/\epsilon \to 0$ at the same time, such that the density of active sites also vanishes in this limit, ensuring criticality. Scaling laws relating spreading critical exponents and avalanche exponents can thus be obtained in these kinds of systems with absorbing states [655]. This allows one to improve the precision of exponents of some well-known models of contact processes, such as directed percolation and dynamical percolation in different spatial dimensions.

Dickman et al. [232] propose the following recipe to obtain SOC. Starting from a system with a continuous absorbing state phase transition at a critical value ρ_c of a density ρ, interpret this density ρ_c as the global value of a local dynamical variable conserved by the dynamics. To this conservation constrain, add two processes: (1) one for increasing the density in infinitesimal steps $\rho \to \rho + d\rho$ when the local dynamics reaches an absorbing configuration; (2) the second process decreases the density at an infinitesimal rate while the system is active. The basic ingredients of this recipe [232] are an absorbing state phase transition and a method for forcing the model to reach its critical point by adding or removing particles when the system is frozen or active.

15.4.6 Critical Desynchronization

Out-of-equilibrium driven systems with threshold dynamics exhibit a strong tendency for synchronization [417, 935, 936]. After some transient regime, collective synchronization is characterized by coherent oscillatory activity of the set of coupled oscillators. This effect has attracted much interest in biology for the study of large scale rhythms in populations of interacting elements [936]. The south–eastern fireflies, where thousands of individuals gathered on trees flash together in unison, is the most cited example [930]. Other examples are the rhythmic activity of cells of the heart pacemaker and the emergence of synchronization in globally coupled Integrate and Fire (IF) oscillators. Synchronization occurs under broad conditions on the properties of the oscillators and is stable against a frozen disorder of the oscillator properties [75].

Beside synchronization, another form of collective organized behavior is known to occur in large assemblies of elements with pulse interactions: Self-Organized Criticality. In some of these models, the external drive acts globally and continuously on all the lattice sites [700], until one of them reaches the threshold, in which case it relaxes to zero: each site is therefore a *relaxation oscillator*. These are the stick–slip-like models which may or may not be deterministic. When the driving is performed at each time step by the increment of a unique site, sites are *not* oscillators: this case covers the sandpile-like models [46].

Consider the mean field model of N relaxation oscillators O_i with state variable $E_i = E(t) \in [0, E_c = 1]$, monotonically increasing in time with period 1. The interaction between these relaxation oscillators is that, when

$E_i \geq 1$, it relaxes to zero and increments all the other oscillators by a pulse α/N:

$$E_i \geq 1 \Rightarrow \begin{cases} E_i & \to 0 \\ E_{i \neq j} & \to E_j + \alpha/N \end{cases} \tag{15.26}$$

where $\alpha \in [0, 1]$ is a dissipation parameter. The model is taken into the slow drive limit by assuming that any avalanche of firings, i.e. any succession of firings triggered by the relaxation of one oscillator, is instantaneous. Immediately after their relaxation, oscillators within the same avalanche are not incremented by the pulses resulting from the successive relaxations in the avalanche.

The theorem of Mirollo and Strogatz [630], that applies to this model, states that, for a convex function $E_i = E(t)$, the system synchronizes completely but for an exceptional set of initial conditions (technically of zero Lebesgue measure). In practice, for all physically interesting situations, the system can be shown to synchronize even for a linear and a range of concave functions $E(t)$ [105]. To show this, we follow [105].

The set of ordered distinct values $E_1^{(j)} < E_2^{(j)} < \ldots < E_{m_j}^{(j)} = 1$ of the state variables present in the system just before the $(j + 1)$-th avalanche defines a configuration. To each $E_i^{(j)}$ corresponds a group of $N_i^{(j)}$ oscillators at this value $(\sum_{i=1}^{m_j} N_i^{(j)} = N)$. The time necessary for all the m_j groups to avalanche exactly once is called a cycle. To trace the evolution of the system, it is useful to follow, cycle after cycle, the gaps $s_i^{(j)} = E_{i+1}^{(j)} - E_i^{(j)}$ between the values of successive groups. If one of these gaps $s_i^{(j)}$ becomes smaller than the value $\alpha N_{i+1}^{(j)}/N$ of the pulse of the $(i + 1)$-th group, then the (i)-th group gets absorbed by the $(i+1)$-th group. For a linear $E(t)$, an elementary calculation shows that the gap between a group $(i + 1)$ and a smaller one (i) is reduced during one cycle by $\delta s_i = |\alpha(N_i - N_{i+1})/N|$. Therefore large groups unavoidably absorb the smaller ones that follow them and become larger and larger: it is this positive feedback that leads to synchronization. Introducing frozen disorder in the natural relaxation frequencies in general yields complete or partial synchronization depending on the couplings.

As mentioned above, besides synchronization or quasi-synchronization, lattice models of pulse-coupled oscillators may display SOC, which appears when a system is perturbed, which otherwise should synchronize totally or partially [172], or which should be periodic [368]. The most striking example is the Olami–Feder–Christensen (OFC) model [700] which consists of oscillators E_i on a square lattice, that relax to zero when they exceed a given threshold E_c, thus incrementing their nearest neighbors by a pulse which is α ($\alpha \leq 1/4$) times their value:

$$E_i \geq E_c \Rightarrow \begin{cases} E_i & \to 0 \\ E_{nn} & \to E_{nn} + \alpha E_i \, . \end{cases} \tag{15.27}$$

With open boundary conditions, this model seems to show SOC, while the Feder and Feder (FF) model [290], which is identical except for the increment, which is a constant that can be seen as the mean of αE_i (pulse of the FF model $= \Delta = \overline{\alpha E_i}$), shows partial synchronization [172]. This means that the randomness of the initial conditions, which is dynamically eliminated in the FF model while it is maintained via the increment in the OFC model, changes the behavior of the system from partial synchronization to apparent SOC. Furthermore, different kinds of perturbations, incompatible with a periodic behavior, change periodically ordered states for SOC. If, for instance, a random noise is added to the increment in the FF model, synchronization disappears and the system becomes apparently SOC [172]. On the other hand, if instead of open boundary conditions, periodic conditions are used, SOC disappears in favor of a periodic state, the period being the number of sites of the lattice. It seems to reappear if one inhomogeneity is introduced on the lattice [368].

A random neighbor version of the OFC model has been studied analytically [154]: avalanches are found with finite size for all values of α up to the conserving limit α_c equal to the inverse of the number of neighbors ($1/4$ in the 2D square lattice). However, their mean size $\langle s \rangle$ grows exponentially fast as $\exp[C/(\alpha_c - \alpha)]$, for $\alpha \to \alpha_c$, so fast that $\langle s \rangle$ becomes rapidly larger than the system size, giving the false impression of a diverging quantity at $\alpha < \alpha_c$. This might explain the false appearance of SOC in some numerical models [568] and serve as a cautionary note about similar numerical evidence obtained for other claimed SOC models.

The importance of heterogeneities in desynchronizing otherwise synchronized oscillators of relaxation has also been exempliflied in a simple model of self-organization of faults and earthquakes [184, 185, 627, 907, 908]. The model describes brittle ruptures and slip events in a continental plate and its spontaneous organization by repeated earthquakes in terms of coarse-grained properties of the mechanical plate. It simulates anti-plane shear (i.e. scalar) deformation of a thin plate with inhomogeneous elastic properties subjected to a constant applied strain rate at its borders, mimicking the effect of neighboring plates. Rupture occurs when the local stress reaches a threshold value. Broken elements are instantaneously healed and retain the original material properties, enabling the occurrence of recurrent earthquakes. The most startling feature of this model is that ruptures become strongly correlated in space and time leading to the spontaneous development of multifractal structures [184] that gradually accumulate large displacements. The formation of the structures and the temporal variation of rupture activity is due to a complex interplay between the random structure, long range elastic interactions and the threshold nature of rupture physics. The spontaneous formation of fractal fault structures by repeated earthquakes is mirrored at short-times by the spatio-temporal chaotic dynamics of earthquakes, well-described by a Gutenberg–Richter power law. The fault structures can be understood as pure geometrical objects [627], namely minimal manifolds which in two-dimensions correspond to the random directed polymer (RDP) problem. This mapping allows one to use the results of many studies on the RDP in the field of statistical physics, where it is an exact result that the minimal random manifolds in 2D systems are self-affine with a roughness exponent $2/3$ [388] (see Fig. 5.15).

The competition between the disordering effect of frozen heterogeneity and the ordering effect of elastic coupling leads to a phase diagram in the space $(\beta, \Delta\sigma)$, where β is a measure of the coupling between faults (proportional to the stress drop associated with a rupture) and $\Delta\sigma$ quantifies the amount of quenched disorder. A synchronized regime is found for small disorder or large coupling and coexists with a SOC regime for larger disorder or smaller couplings. This suggests to view SOC as a critical "desynchronization", i.e. a regime where the system is marginally desynchronized. In other words, in these classes of SOC, the system attempts again and again to synchronize, leading to large avalanches but never completely succeeds due to the desynchronization resulting from the competing factors, such a heterogeneity, boundary conditions, etc. The self-similar distribution of avalanche sizes reflects the failure of the system to achieve large synchronized avalanches in the system.

Middleton and Tang [623] confirm this picture by studying the SOC model without conservation [700]. They find that the homogeneous system with periodic boundary condition is periodic and neutrally stable. A change to open boundaries results in the invasion of the interior by a "self-organized" region spreading from the boundaries. The mechanism for the self-organization is closely related to the synchronization or phase-locking of the individual elements with each other.

15.4.7 Extremal Dynamics

Definition. Inspired by the metaphoric picture of biological evolution proposed by S. Wright (for a review, see [1027]), who introduced the idea of species evolving over a rugged fitness landscape with random mutations and relative selection towards higher fitness, Bak and Sneppen introduced a simple model of extremal dynamics [48] that we have already mentioned above. Its generalization to M traits (the case $M = 1$ is the Bak–Sneppen model) is defined as follows [96, 714]: a species is represented by a single site on a d-dimensional lattice. Each species possesses a collection of M traits represented by a set of M numbers in the unit interval. The dynamics consists of mutating, at every time step, the smallest number λ in the entire system by replacing it by a new (possibly smaller) number that is randomly drawn from a uniform distribution in the unit interval. Choosing the smallest random number mimics the Darwinian principle that the least fit species mutates [197]. The dynamical impact of this event on neighboring species is simulated by also replacing one of the M numbers on each neighboring site with a new random number. Which one of the M numbers is selected for such an update is determined at random. The interaction between the fitnesses of species leads to a chain reaction of coevolution. Such a chain reaction leads to avalanches: an avalanche starts at a novel increase of the minimum fitness λ and stops when the next one starts, i.e. when the minimum λ first becomes greater than the starting value.

This concept has been found to provide a common language for describing such diverse physical situations such as roughening of a crack front in fracture [822], wetting front motion on heterogeneous surfaces [277], the dynamics of a ferromagnetic domain wall driven by an external magnetic field [1047], motion of vortices in type-II superconductors (for a review, see T. Giamarchi and P. LeDoussal in [1036]), fluid invasion in porous media [793], solid friction [945] or, as we said, biological evolution [713].

In these different systems, the dissipative behavior of the system is explained by the competition between an elastic restoring force and a nonlinear, randomly distributed, time-independent, pinning force. In the case of the spreading of a partially wetting liquid for example, the *pinning* force is due to surface chemical heterogeneities or roughness, and the elastic restoring force is a result of the surface tension at the liquid/vapor interface. For strong pinning, the wetting front displays local instabilities that force it to advance quasistatically in a jerky motion with jumps punctuating phases of stress build up [474]. In the stationary regime, the main contribution to the global displacement is from jumps of local parts of the chain resulting from these instabilities [945]. To describe this sort of evolution, Tanguy et al. [945] have proposed an *extremal* model: only the site closest to its instability threshold advances. After a jump, the instability thresholds of all the sites are modified by the (elastic) couplings between sites. More precisely, in their model, the interface of size L is defined on a discrete lattice (x, h). Initially the front $h(x) = 0$, and the pinning forces $f_p(x, t = 0) = f_0(x, h = 0)$ are assigned independently from a flat distribution $f_0(x, y)$. The site x_0 subjected to the minimum pinning force (and hence closest to its instability threshold) advances first, thus $h(x_0) \to h(x_0) + \Delta h$. At this new position, a new random pinning force is encountered $f_p(x_0, t + \delta t) = f_0(x_0, h(x_0) + \Delta h)$. The external loading F on the system, and interactions along the front, produce a local driving force on each site x proportional to $f(x, t) = F \int G(x - y) h(y, t) \, dy$ where the kernel $G(x) \propto |x|^{-1-b}$ accounts for long range interactions mediated by the medium. The loading F is then adjusted so that only one site depins $f(x, t) = f_p(x, t)$; the others remain trapped since $f(y, t) \le f_p(y, t)$ for $y \ne x$. The dynamics of advancing the minimum site and readjusting the others is continued indefinitely.

In this spirit, a wide class of extremal models have already been studied extensively by Paczuski et al. [713]. These models include the Bak–Sneppen evolution model [48] and the Sneppen Interface Model [855]. All these models try to explain driven motion under strong pinning by means of a discrete,

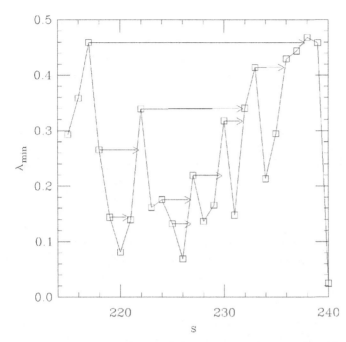

Fig. 15.6. Plot of a part of the sequence of minimal random numbers $\lambda_{\min}(s)$ chosen for an update at time s in a λ_c avalanche for $M = \infty$. The durations of a hierarchy of λ avalanches is indicated by *forward arrows*, where $\lambda = \lambda_{\min}(s)$. Longer avalanches with larger values of λ contain many shorter avalanches which have to finish before the longer avalanche can terminate. Note that an update punctuates any λ avalanche with $\lambda \leq \lambda_{\min}(s)$. Taken from [96]

for spatial and temporal correlations. In the continuum limit, they showed that the avalanche dynamics can be described in terms of a Schrödinger equation with a nonlocal potential. Its solution yields a non-Gaussian tail in the distribution with an avalanche dimension $D = 4$ ($r \sim t^{1/D}$), signaling subdiffusive behavior for the spread of activity.

Consider first the probability $P_\lambda(r, s)$ that the λ avalanche dies precisely after s updates and does *not* affect a particular site r away from the origin of the avalanche. The quantities $P_\lambda(r, s)$ and $F_\lambda(r, s)$ are related by

$$P_\lambda(r, s) = P_\lambda(r = \infty, s) - F_\lambda(r, s) , \qquad (15.29)$$

where $P_\lambda(r = \infty, s) = P_\lambda(s)$. Since an avalanche begins with a single active barrier at $r = 0$ and $s = 0$, then $P_\lambda(r = 0, s) \equiv 0$ for all $s \geq 0$, and $P_\lambda(r, s = 0) \equiv 0$ for all r. The remaining properties of a λ avalanche can be deduced from the properties of avalanches that ensue after the first update. It will terminate with probability $(1 - \lambda)^2$ after the first update when the update does not produce any new active barriers. Thus, $P_\lambda(r, s = 1) \equiv (1 - \lambda)^2$. For avalanches surviving until $s \geq 2$, one finds for $r \geq 1$ [96]

$$P_\lambda(r, s) = \lambda(1 - \lambda) \left[P_\lambda(r - 1, s - 1) + P_\lambda(r + 1, s - 1) \right]$$

$$+\lambda^2 \sum_{s'=0}^{s-1} P_\lambda(r - 1, s') P_\lambda(r + 1, s - 1 - s') . \tag{15.30}$$

The first update may create exactly one new active barrier with probability $\lambda(1 - \lambda)$ either to the left or to the right of the origin (i. e. one step towards or away from the chosen site of distance r). In this case, the properties of the original avalanche of duration s are related to the properties of an avalanche of duration $s - 1$ with regard to a site of distance $r - 1$ or $r + 1$, respectively. Furthermore, the first update may create two new active barriers with probability λ^2 to the left and the right of the origin. Then, the properties of the original avalanche of duration s are related to the properties of all combinations of two avalanches of combined duration $s - 1$. Both of these avalanches evolve in a *statistically independent* manner for $M = \infty$. Since only one of these avalanches can be updated at each time step, their combined duration has to add up to $s - 1$ for this process to contribute to the avalanche of duration s. For any such combination, the probability not to affect the chosen site of distance r from the origin is given simply given by the product of the probabilities for the two ensuing avalanches not to affect a chosen site of distance $r - 1$ or $r + 1$, respectively.

The equation governing $F_\lambda(r, s)$ is obtained by inserting (15.29) into (15.30). It is $F_\lambda(0, s) \equiv P_\lambda(s)$, $F_\lambda(r, 0) \equiv 0$, $F_\lambda(r \geq 1, s = 1) = 0$, and for $s \geq 1$, $r \geq 1$,

$$F_\lambda(r, s + 1) = \lambda(1 - \lambda) \left[F_\lambda(r - 1, s) + F_\lambda(r + 1, s) \right]$$

$$+\lambda^2 \sum_{s'=0}^{s} P_\lambda(s - s') \left[F_\lambda(r - 1, s') + F_\lambda(r + 1, s') \right]$$

$$-\lambda^2 \sum_{s'=0}^{s} F_\lambda(r - 1, s') F_\lambda(r + 1, s - s') . \tag{15.31}$$

Focusing on the spatiotemporal correlations at the critical point λ_c, for sufficiently large values of r and s, $F_\lambda(r, s)$ goes to 0 for $r \to \infty$ sufficiently fast such that we can neglect the nonlinear term in (15.31). The continuum limit yields

$$\frac{\partial F(r, s)}{\partial s} \sim \frac{1}{2} \nabla_r^2 F(r, s) + \frac{1}{2} \int_0^s V(s - s') F(r, s') \, ds' , \tag{15.32}$$

which is a "Schrödinger" equation in imaginary time, with a nonlocal memory kernel $V(s) = P(s) - 2\delta(s)$, where $P(s) \sim t^{-\alpha}$ with $\alpha = 3/2$ [96] is the avalanche size distribution and $\delta(s)$ is the usual Dirac δ function. Note that it is the statistical independence of the avalanches that gives $V(s)$ in terms

of the probability distribution of avalanche sizes. The memory in the system
is characterized solely in terms of this distribution. The dependence $P(s) \sim t^{-3/2}$ results from the fact that the length of an avalanche is the same as the
first return to the origin of a random walk [96].

Using the Laplace transform, $\tilde{F}(r, y) = \int_0^\infty dt\, e^{-yt} F(r, t)$, (15.32) turns
into an ordinary second-order differential equation in r,

$$\nabla_r^2 \tilde{F}(r, y) \sim \left[2y - \tilde{V}(y) \right] \tilde{F}(r, y) \,, \tag{15.33}$$

where $\tilde{V}(y)$ is the Laplace transform of $V(t)$; $\tilde{V}(y) \sim y^{\alpha-1}$ for small y. In
the limit of large times, the effect of the history-dependent potential dominates over the time derivative in (15.32), signaling the deviation from simple
diffusive behavior. The solution of (15.33) which decreases for large r is

$$\tilde{F}(r, y) \sim \exp\left(-A\, r y^{(\alpha-1)/2} \right) , \tag{15.34}$$

where A is a constant. The inverse Laplace transform can be estimated by
the saddle-point approximation and finally yields

$$F(r, t) \sim \sqrt{\frac{24}{\pi}} t^{-3/2} \left(\frac{r^4}{t} \right)^{1/3} e^{-(3/4)(r^4/t)^{1/3}} \quad (r^4 \gg t \gg 1) \,. \tag{15.35}$$

This anomalous tail in the probability distribution for the activity in SOC
is also characteristic of glassy systems. For example, the directed polymer in
a random media, which contains many features of frustrated random systems,
also has a non-Gaussian tail for $G(x, t)$ [388]. While this model is inherently
dynamical with no quenched disorder, the difference between frozen disorder
and dynamically generated ones has been shown to be less than thought
previously: both may lead to glassy dynamics with long-range memory and
aging [113, 603].

Linear Fractional Stable Motion for Extremal Dynamics. We stressed
above in the definition of extremal dynamics that all the information in this
sort of dynamics is contained in the "activity" map, defined as a space–time
plot of where the front is active at every instant of time. Previous studies
regarding extremal models [557, 945] have shown that most of the relevant
information is actually contained in the probability density function (pdf) of
the activity map. In the stationary regime, assuming that the activity was
located at x_0 at time t_0, the probability that it is at x at time t is:

$$p(|x - x_0|, t - t_0) = (t - t_0)^{-1/z} \phi\left(\frac{|x - x_0|}{(t - t_0)^{1/z}} \right) \tag{15.36}$$

with

$$\phi(r) \propto \begin{cases} r^{-1-\alpha} & \text{for} \quad r \gg 1 \\ r^0 & \text{for} \quad r \ll 1 \,. \end{cases} \tag{15.37}$$

While the exponent α controls the asymptotic behavior of the time-independent function ϕ, z controls the propagation of the activity along the system as a function of time and is therefore known as the "dynamical exponent." The above distribution is self-affine and therefore its temporal evolution is completely defined through the exponents z and α.

The Linear Fractional Stable Motion [810] has been used by Krishnamurthy et al. [532] to combine the effect of long range time correlations with a fat tail distribution of jump sizes. Let x denote a process generated in the following manner:

$$x(t) = \sum f(t, u)\eta(u)\,, \tag{15.38}$$

where $\eta(u)$ is an uncorrelated noise with a symmetric distribution $p(\eta = x) \sim |x|^{-1-\alpha}$. For stationary processes, it is natural to assume $f(t, u) = f(t - u)$. The definition above implies basically that x consists of a sum of uncorrelated Lévy jumps weighted in time by $f(t - u)$. This weight function therefore controls the time dependence of the statistical properties of x. It is easy to show that the sum in (15.38) can be performed much as for independant Lévy flights and the random variable $X = x(t)$ (given that $X = 0$ at $t = 0$) has a probability density function

$$p(X, t) = \int \exp(ikX) \exp(-\sigma_t^\alpha |k|^\alpha)\, \mathrm{d}k/2\pi\,. \tag{15.39}$$

where $(\sigma_t)^\alpha = \sum_{u=0}^{t} |f(t - u)|^\alpha$. For the Linear Fractional Stable Motion [810], the function $f(t, u)$ reads:

$$f(t, u) = (t - u)^d\,, \tag{15.40}$$

where the parameter d satisfies $-1/\alpha < d < 1 - 1/\alpha$. The Linear Fractional Stable Motion [810] is thus a self-similar process with stationary increments. One can define a Hurst exponent H describing the self-similarity of this process through the definition

$$\langle |x(t) - x(t')|^q \rangle^{1/q} \approx |t - t'|^H \quad \text{for } q < \alpha\,. \tag{15.41}$$

H accounts for possible temporal statistical dependence between jumps. For the process (15.38) giving (15.39), $H = d + 1/\alpha$ and [532]

$$\sigma_t^\alpha = \sigma_1^\alpha |t|^{\alpha H}\,. \tag{15.42}$$

The Linear Fractional Stable Motion offers the following prediction for the exponents α and z defined in the scaling form (15.36) obtained for an extremal model. From (15.39) and (15.42), the pdf $p(|x - x_0|, t - t_0)$ satisfies the scaling form (15.36) with $z = 1/H$. This prediction is verified with high accuracy in numerical simulations [532].

The analytical expression (15.39) for the space–time plot of the activity provides new predictions, as shown by Krishnamurthy et al. [532]. Consider the two-point function $P(l_1, l_2)$, which is the probability of having a jump l_1 and l_2 at two consecutive instants. Using (15.39), the expression of this function in Fourier space is just $\sim \int \exp(ik(l_1 + l_2)) \exp(-|2|^{\alpha H}|k|^\alpha)\, \mathrm{d}k$. In real space, this corresponds to the function $1/(l_1 + l_2)^{\alpha+1}$, asymptotically.

The conditional probability $P(l_2|l_1)$ – the probability of having a jump of length l_2 at time $t = 2$ given that there was a jump l_1 at time $t = 1$ – can be calculated using $P(l_2|l_1) = P(l_1, l_2)/P(l_1)$. This gives

$$P(l_2|l_1) \sim 1/(1 + l_2/l_1)^{\alpha+1} , \qquad (15.43)$$

confirming the numerical evidence presented in [530, 531]. This reasoning can be generalized to the full n-point function $P(l_1, l_2, l_3, \dots, l_n)$ (discussed in [530, 531]) of having a jump of length l_1 at time $t = 1$, a jump of length l_2 at time $t = 2$ and so on till $t = n$. Systematic expansions can be obtained for the conditional probabilities just as for the two-point function and can also be verified numerically on the models.

Another interesting consequence of the definition of the Linear Fractional Stable Motion for extremal dynamics is the following equation for the front propagation under this dynamics. Defining the height $h(X, t)$ of an interface at time t at a spatial location X as simply the accumulated activity there up to time t, we get

$$h(X, t) = \Delta h \int_0^t \delta(X - x(t')) \, dt'. \qquad (15.44)$$

Let us perform the scale transformation commonly used for self-affine surfaces: $X \to bX$, $t \to b^z t$ and $h \to b^\chi h$, where χ is the so-called *roughness exponent* for the height. A power counting on both sides of this equation gives the relation $\chi = z - 1$, known to hold for extremal dynamics [713, 945].

15.4.8 Dynamical System Theory of Self-Organized Criticality

Blanchard, Cessac and Krüger [94, 151] have developed a dynamical system theory for a certain class of SOC models (like the Zhang's model [735, 1057]), for which the whole SOC dynamics can either be described in terms of Iterated Function Systems, or as a piecewise hyperbolic dynamical system of skew-product type where one coordinate encodes the sequence of activations. The product involves activation (corresponding to a a kind of Bernoulli shift map) and relaxation (leading to to contractive maps).

Following [95], let us first recall briefly the rules of Zhang's model, defined in a connected graph Λ, with nearest neighbors edges. The boundary of Λ is the set of points in the set complementary to Λ at distance 1 from Λ. N is the cardinality of Λ. Each site $i \in \Lambda$ is characterized by its "energy" X_i, which is a non-negative real number. The "state" of the network is completely defined by the configuration of energies. Let E_c be a real, positive number, called the critical energy, and $\mathcal{M} = [0, E_c[^N$. A configuration is "stable" if it is in $\mathcal{M}$ and "unstable" or "overcritical" otherwise. If the configuration is stable, then we choose a site i at random with probability $1/N$, and add to it the energy δX. Since the physically relevant parameter is the local rigidity $E_c/\delta X$ [735], one can investigate the cases where E_c varies, and where δX is a constant. If a site i is overcritical $(X_i \geq E_c)$, it loses a part of its energy in equal parts to its $2d$ neighbours. Namely, we fix a parameter $\epsilon \in [0, 1[$ such that, after relaxation of the site i, the remaining energy of i is ϵX_i, while the $2d$ neighbours receive the energy $(1 - \epsilon)X_i/2d$. In the original Zhang's model [1057], ϵ was taken to be zero. If several nodes are simulaneously overcritical, the local distribution rules are additively superposed, i.e. the time evolution of the system is synchronous. The sites of the boundary have always zero energy (dissipation at the boundaries). The succession of updating leading from an unstable configuration to a stable one is an *avalanche*. Because of the dissipation at the boundaries, all avalanches are *finite*. The structure of an avalanche can be encoded by the sequence of overcritical sites $A = \{A_i\}_{0 \leq i}$ where $A_0 = \{a\}$, the activated site, and $A_i = \{j \in \Lambda | X_j \geq E_c$ in the i-th step of avalanche$\}, i > 0$. The addition of energy is *adiabatic*. When an avalanche occurs, one waits until it stops before adding a new energy quantum. Further activations eventually generate a new avalanche, but, because of the adiabatic rule, each new avalanche starts from *only one* overcritical site.

Since the avalanche after activation of site a maps overcritical to stable configurations, one can view this process as a mapping from $\mathcal{M} \to \mathcal{M}$ where one includes the process of activation of site a. One can hence associate a map T_a with the activation at vertex a. This map usually has singularities and therefore different domains of continuity denoted by M_a^k where k runs through a finite set depending on a. Call $T_a^k = T_a|_{\mathcal{M}_a^k}$. The dynamical system approach is especially suited to study the properties of the family of mappings $\{T_a^k\}$ and to link these properties to the asymptotic behavior of the SOC model.

Several deep results from the theory of hyperbolic dynamical systems can then be used in this framework, having interesting implications on the SOC dynamics, provided one makes some natural assumption (like ergodicity). With this dynamical approach, a precise definition of the SOC attractor can be obtained. Within this approach, one can readily show that the SOC attractor has a fractal structure for low values of the critical energy. The dynamical system approach defines the structure of the dynamical SOC attractor in a natural way, which is nothing but the natural invariant measure. The Lyapunov exponents, the geometric structure of the support of the invariant measure (Hausdorff dimensions), and the system size are related to the probability distribution of the avalanche size via the Ledrappier–Young formula [552] which relates the Kolmogorov–Sinai entropy to the Lyapunov exponents: if h_μ is the Kolmogorov–Sinai entropy of the invariant measure defined by the map dynamics and λ_i^+'s are the positive Lyapunov exponents, for ergodic measures the Ledrappier–Young formula is [552]

$$h_\mu = \sum_i \lambda_i^+ \sigma_i \,, \qquad (15.45)$$

where σ_i is the transverse dimension of the measure on the sub-manifold i defined such that the contraction (resp. the expansion) on this sub-manifold is governed by the Lyapunov exponent λ_i. By using the ergodic theorem, on gets the log-average volume contraction which is also the sum of Lyapunov exponents and thus relates the local volume contraction to the average avalanche size. It connects therefore microscopic dynamical quantities (Lyapunov exponents) to a macroscopic observable (average avalanche size). In particular it allows to establish a link between the Lyapunov spectrum and the critical exponents of the avalanche size distribution. This relationship reads

$$\sum_i \lambda_i^- = (\ln \epsilon) \sum_{s=1}^{S_N} s P_N(s) = (\ln \epsilon)\,\bar{s}\,, \qquad (15.46)$$

where $\bar{s}$ is the average avalanche size and s_N the maximal avalanche size.

Blanchard, Cessac and Krüger [94, 151] have noticed that the dynamics of the Zhang's model is essentially equivalent to a graph probabilistic Iterated Function System (IFS) [58, 287], namely, a set of quasi-contractions F_i randomly composed along a Markov graph admitting a unique invariant measure μ^*. Note that IFS are usually defined for true contractions, however, in the present case, any finite composition along the graph is a contraction, hence the classical theory of graph directed Iterated Functions Systems applies and allows one to obtain interesting results with respect to the geometrical structure of the invariant set. Note that this connection to IFS which are not everywhere contracting provides a bridge with the multiplicative random maps discussed in Chap. 14, which have been shown to produce power law distributions as a result of intermittent amplification.

The dynamical system approach also predicts an interesting phase transition which is similar to the synchronized–SOC transition found numerically as a function of the ratio disorder over coupling strength between threshold oscillators of relaxation [908]. The gist of the argument is as follows [95]. Domains of continuity $\mathcal{M}_a^k$ are bounded by hyperplanes, which are moving when E_c varies. In general, a small variation in E_c does not lead to structural changes in the dynamics, if all these hyperplanes are intersecting the interior of $\mathcal{M}$. In this case, the structure of the transition graph is not modified. Moreover, the corresponding mapping T_a^k does not change under this motion. More precisely, changes in E_c just change the shape of $\mathcal{M}_a^k$ but not the matrix of the mapping T_a^k. However, for some E_c values, some hyperplanes have intersection only with the boundary of $\mathcal{M}_a^k$. This implies that a small change in E_c can push these hyperplanes outside $\mathcal{M}$. Hence the corresponding transition graph changes in structure. Since the asymptotic dynamics and therefore the invariant distribution is dependent on the graph structure, one expects changes in the SOC picture when crossing these *critical* E_c values. It is easy to see in particular that the limiting cases $E_c \to \infty$ and $E_c \to 0$ are completely different. For $E_c \to \infty$, relaxation events are more and more seldom. One obtains a kind of frozen state where the energy increases (on average) monotonously with some rare (but large) avalanches. Moreover, the asymptotic energy distribution is sensitive to the initial conditions (loss of ergodicity). Furthermore, the attractor has a large Haussdorf dimension. On the other hand, for $E_c \to 0$, each activation generates a very large avalanche (that has to reflect many times on the boundary before it has lost enough energy to stop). This implies larger and larger contraction, and therefore the sum of Lyapunov exponents decreases to $-\infty$. As a corollary of Ledrappier–Young formula, the partial fractal dimensions have to go to zero in order to maintain the product equal to $\ln N$.

Explicit formula for the transport modes that appear as diffusion modes in a landscape where the metric is given by the density of active sites have been established [152]. Finite size scaling can then be used for the Lyapunov spectrum which allows one to relate the scaling exponents to the scaling of quantities such as avalanche size, duration and density of active sites. The generating functions of probability distributions in the generalized Zhang model have also been found to exhibit a Lee–Yang phenomenon [1033]. Namely, their zeros pinch the real axis at $z = 1$, as the system size goes to infinity. This establishes an additional link between the classical theory of critical phenomena and SOC. A scaling theory of the Lee–Yang zeros has been proposed in this setting [153], which allows one to calculate the critical exponents. This approach also reveals artificial biases in the distributions obtained by numerical simulations.

15.5 Tests of Self-Organized Criticality in Complex Systems: the Example of the Earth's Crust

What are the possible observable consequences of the SOC hypothesis in the Earth's crust? This discussion is taken from [373].

A hypothesis cannot be tested by the empirical evidence that served to shape it. We thus exclude the Gutenberg–Richter power law as well as the fractal geometrical structure of the earthquake epicenters and fault patterns. In the present discussion, we address two novel properties/predictions that derive naturally from detailed numerical studies of simplified models of the crust.

The first one is the most obvious for geologists but we nevertheless address it as some confusion might exist, probably seeded by the statistical physics community. It concerns the localization of earthquake activity on faults. When elasticity is correctly incorporated, SOC (defined by the four conditions given in the definition at the beginning of this Chapter) is found to coexist (and is, in fact, deeply intertwinned) with a spontaneous organization of a fault structure on which the earthquake activity is clustered [184, 185, 408, 576, 627, 907, 915]. SOC is thus not synonymous with a diffuse "avalanche" activity covering uniformly all the available space, as extrapolations from sandpile models would imply [45, 166, 173, 700].

The incorporation of elasticity in models of SOC, in fact, leads to an enrichment of the concept, since fault structures are found to be geometrical objects themselves subjected to the self-organizing principles and which can be perceived as self-organized structures solving a global optimization problem as defined by Crutchfield and Mitchell [192]. The most interesting aspect of SOC is probably in its prediction that the stress field exhibits long-range spatial correlations as well as important amplitude fluctuations. The exact solution of a simple SOC model [226, 228, 580] has shown that the spatial stress-stress correlation of the stress fluctuations around the average stress is long range and decays as a power law. A similar conclusion on the related strain field fluctuations has been obtained within a general mathematical formalism [917]. The conclusion we can draw from the understanding brought by these conceptual models is that the stress fluctuations not only reflect but also constitute an active and essential component of the organizing principle leading to SOC.

A substantial fraction of the crust is close to rupture instability. Together with the localization of seismicity on faults, this leads to the conclusion that a significant fraction of the crust is susceptible to rupture, while presently being quiescent. The quantitative determination of the susceptible fraction is dependent on the specificity of the model [735, 907, 1057] and cannot thus be ascertained with precision for the crust. What is important, however, is that the susceptible part of the crust can be activated with relatively small perturbations or by modification of the overall driving conditions. This remark leads to a straighforward interpretation of induced seismicity by human activity [373].

If a finite fraction of the crust is susceptible and can easily be brought to an unstable state, not all the crust is in such a marginal stability state. In fact, the complementary finite fraction of the crust is relatively stable and resistant to perturbations. The assertion often found in the SOC literature that "the crust is almost everywhere on the verge of rupture" is simply wrong, as found in the simplified SOC models. For instance, numerical simulations show that in discrete models made of interacting blocks carrying a continuous scalar stress variable, the average stress is around 0.6 times the threshold stress at rupture [735]. In these models, the crust is far, on the average, from rupture. However, it exhibits strong fluctuations such that a subset of space is very close to rupture as already pointed out. The average is thus a poor representation of the large variability of the stress amplitudes in the crust. In Chap. 17, we show in fact that the distribution of the stress induced by a random distribution of dislocation sources is a stable Lévy law (Chap. 4). The power law tail is in practice rounded off by the strength of the rocks.

16. Introduction to the Physics of Random Systems

16.1 Generalities

Quenched (frozen) randomness may bring qualitative novel behaviors. Disorder has traditionally been addressed as a perturbation, a nuisance, leading to distorsions without much surprise. The methods known as homogeneization theory [80] and effective medium theories [171] exemplify this by starting from the concept that an heterogeneous medium can be replaced for all its relevant observables by an effective and equivalent homogeneous medium.

However, a large amount of work has shown that there are many situations where this conclusion is incorrect. A first milestone can be traced back to Anderson who showed in 1958 that sufficiently strong quenched disorder can trap or localize waves in the multiple scattering regime [25, 27, 877]. In this goal, he recognized that a suitable representation of the variability of measurable quantities is not provided by the variance but by the full probability distribution. Other important discoveries are the breaking of ergodicity (the average over an ensemble of realizations may not be the same as the average over time evolution) and the rugged-multi-valley structure of the energy landscape of Ising models, in other words optimization problems in the presence of quenched disorder (or competing or frustrated constraints) have exponentially many solutions (in the number of degrees of freedom) with about the same cost functions [622]. The two behaviors turn out to be intimately linked: the existence of many competing minima of the energy of a system leads to the trapping of its dynamics over long times in certain narrow regions of phase space and prevents the exploration of all possible configurations, hence breaking ergodicity.

The present chapter is necessarily naive in its presentation as the field has developed tremendously in the last thirty years. Here, we survey rather superficially some key concepts and results with the view point of providing hints to the reader that the novel concepts developed in this field are worth studying and will certainly be useful for applications to many other fields which are still in their infancy.

The paradigms of quenched disorder systems are the spin glass systems [91]. The term "spin glass" derives from condensed matter applications in which ferromagnetic systems (iron for instance) are corrupted by non-ferromagnetic impurities (gold atoms!). For sufficiently strong concentrations of impurities, the system looses its ferromagnatic properties even at very low temperature and enters a novel "spin glass" phase. The two main physical ingredients controlling this regime are the existence of "frustration" and frozen disorder. Frustration is the concept that not all constraints can be satisfied simultaneously. For instance, consider a triangle whose nodes each carry a spin. If the coupling constants K defined in (9.1) or (9.4) are negative (antiferromagnetic) such that two neighboring spins "prefer" to anti-align, it is clear that at least one pair is not in its optimal configuration, as shown in Fig. 16.1. Such frustration occurs even in absence of disorder. It is at the origin of the coexistence of several states with the same energy as shown in Fig. 16.1. The addition of frozen (also refered to as "quenched") disorder often multiplies the number of states with equivalent energy.

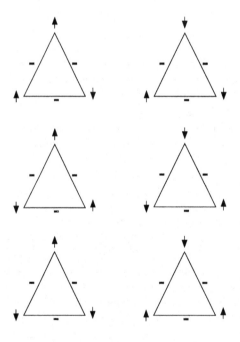

Fig. 16.1. Illustration of the concept of "frustration": each spin at the nodes of a triangle can only take two states, up or down. The − signs on the bonds indicate that the energy of a configuration of two neighboring spins is −1 if they are anti-parallel and +1 if they are parallel. This corresponds to an anti-ferromagnetic coupling. In contrast to the ferromagnetic case where the configuration with minimum energy is only two-fold degenerate (all spins up or down), there are six states with minimum energy equal to −1 that are shown in the figure

Spin glasses constitute a well-studied but still only partially understood paradigm of the physics of systems with frozen disorder. In the context of thermodynamics systems, their specific behavior that makes them apart from other systems is as follows.

- If we cool the system below some temperature (the so-called glass transition temperature T_g), its energy and other thermodynamic variables depends on the cooling rate in a significant way as well as on the previous thermal history of the system.
- No thermodynamic anomaly is observed: the entropy (extrapolated at very slow cooling) is a linear function of the temperature in the region where

such an extrapolation is possible. For a finite value of the cooling rate, the specific heat is nearly discontinuous. Data are consistent with the possibility that the true equilibrium value of the specific heat is also discontinuous at a temperature T_c lower than T_g.

• The relaxation time τ (and quantities related to it, e.g. the viscosity) diverges at low temperature. In many glasses (the fragile ones), the experimental data can be fitted by the so-called Vogel–Fulcher law [319, 995]:

$$\tau = \tau_0 e^{\beta B(T)} \tag{16.1}$$

$$B(T) \propto (T - T_c)^{-\beta} , \tag{16.2}$$

where τ_0 is a typical microscopic time, T_c is near the value at which we could guess the presence of a discontinuity in the specific heat and the exponent β is of order 1.

The mechanism of combination of exponentials for power laws studied in Chap. 14 provides a simple and intuitive understanding of the glass transition [109]. Indeed, if the distribution of energy barriers between states is an exponential with average x_0, the trapping or waiting time τ to cross over a given energy barrier is given by the exponential Arrhenius factor as in (14.37) where y is replaced by τ and X is the temperature T. As shown in Chap. 14, the combination of these two exponentials give a power law distribution for the trapping times with exponent $\mu = T/x_0$. For $T \le x_0$, $\mu \le 1$ for which the average trapping time $\langle \tau \rangle$ is infinite. The value $T_g = x_0$ thus corresponds to the glass transition temperature. The mechanism of the glass transition is the occurrence of arbitrary large trapping times (responsible for the divergence of $\langle \tau \rangle$) in a complex hierarchical system of energy barriers.

We can imagine many applications to natural systems of this rich phenomenology. As the physics of spin glasses is still in its infancy and under active construction, the concepts and methods have not yet really penetrated many other fields, except for biology (neural networks [418], theory of the brain, physics of protein folding) and computation of (or in) complex systems [622]. For instance, sequential reordering dynamics in the reordering of proteins after a "proteinquake" follows a hierarchy of protein substates that are arranged in a Bethe tree manner [308–310].

From a theoretical point of view, the mean field theory of spin glasses has been developed to a high degree of sophistication [604, 622], using the so-called replica symmetry breaking approach. Technically, the problem results from the fact that precise theoretical predictions can only be performed for observables such as the free energy that are averaged over many realizations of the disorder, i.e. over many equivalent system realizations. A way to perform this is to use the replica method which involves an analytic continuation from integer to non integer values of the number of replicas of the physical system under study. The breaking of the symmetry between replicas refers to the fact that the true energy minima of the system are found when one breaks the equivalence between all replicas [719–721] and organize them in a specific hierarchical (ultrametric) structure [759]. The concept of replica symmetry breaking has become a crucial tool in the study of frustrated disordered systems and there are no doubts that it describes correctly what happens in infinite range models. On the contrary, its correctness in the case of short range models remains controversial.

An alternative tool is the so-called cavity approach [622], in which one starts by assuming that, in a finite volume, the decomposition in pure (minimum energy) states is possible and has some suitable properties [604]. Then, one compares a system containing N spins to a system with $N + 1$ spins: in the limit of N large, the probability $\mathcal{P}_{N+1}[w]$ that the system be in a given state w can be written in explicit form as function of $\mathcal{P}_N[w]$. Symbolically, one gets

$$\mathcal{P}_{N+1} = \mathcal{R}[\mathcal{P}_N] \ . \tag{16.3}$$

The probability $\mathcal{P}_\infty$, which is at the heart of the replica approach, can be obtained by solving the fixed point equation

$$\mathcal{P}_\infty = \mathcal{R}[\mathcal{P}_\infty] \ . \tag{16.4}$$

The probability distribution (embedded with an ultrametric structure) which was found by using replica theory turns out to be a solution of this fixed point equation. Alas, it is not known if it is the only solution.

"Chaoticity" has been found to be a very important property of spin glasses and more generally of frustrated disordered systems. It amounts to say that if one considers a finite system and adds to the total Hamiltonian a perturbation $\delta\mathcal{H}$ such that

$$1 \ll \delta\mathcal{H} \ll N \ , \tag{16.5}$$

the unperturbed and the perturbed system are as different as possible. Chaoticity can also be formulated by saying that the states of the perturbed systems have minimal overlap with the states of the system in absence of the perturbation. Examples are found with respect to a random energy perturbation, a change in the application of the external (magnetic) field or of the temperature or when changing the number of spins (known in this case as "chaotic volume dependence"). A simple case of chaotic dependence on the volume is the one of an Ising ferromagnet in presence of a random symmetric distribution of magnetic field at low temperature. The total magnetization is well approximated by

$$\mathrm{sign}\left(\sum_{i=1,N} h_i \right)$$

(where the h_i are the random fields) and this quantity changes in a random way when N goes to infinity. There is no central limit theory for such quantity and usual notions discussed in Chap. 2 has to be revised.

16.2 The Random Energy Model

The random energy model (REM) [218] is the simplest model illustrating this kind of phenomenology. Many of the main features of the model are present in more sophisticated versions, if the appropriate modifications are done. The idea is that randomness with frustration will create many states with complicated and almost random energies. It is thus natural to take the limit where these energies can be completely random.

Specifically, the model is defined as follows. There are N Ising spins (σ_i, $i = 1, N$) which may take values ± 1; the total number of configurations is equal to $M \equiv 2^N$ and they can be identified by a label k in the interval $1, ..., M$. The spins here encode a boolan parameterization of the system and can be thought to describe systems other than magnetic. For instance, active and inactive, intact or ruptured, etc.

Usually, one writes an explicit expression for the energies E_k of a system as function of the specific spin configuration as in (9.1) or (9.4). On the contrary, in the REM, one assumes that the values of the E_k are random, with a probability distribution $n(E)$ taken Gaussian:

$$n(E) = \frac{1}{\sqrt{\pi J\, N}}\, \mathrm{e}^{-E^2/JN} \ . \tag{16.6}$$

J plays the role of an interaction strength and the normalization with a variance proportional to N ensures, as we will see, the correct extensive dependence with the size N of the system. $n(E)$ is also called the average density of states.

Based on pure statistical considerations, we have seen in Chap. 7 that the probability that the system is found in a state of energy E, given a constraint on the average energy of the system, is given by the Boltzmann factor (3.45)

$$P(E) = n(E)\frac{\mathrm{e}^{-\beta E}}{Z(\beta)}, \tag{16.7}$$

where the partition function is simply given by

$$Z(\beta) = \sum_{k=1,M} \exp(-\beta E_k) = \int \mathrm{d}E\, \rho(E) \exp(-\beta E), \tag{16.8}$$

$$\rho(E) \equiv \sum_{k=1,M} \delta(E - E_k) \ . \tag{16.9}$$

$\rho(E)\, \mathrm{d}E$ is the number of levels between E and $E + \mathrm{d}E$ such that $\langle \rho(E) \rangle = n(E)$. The inverse "temperature" β is the Lagrange multiplier ensuring the validity of the constraint (see Chaps. 3 and 7). We have seen that all quantities of interest can be derived from the knowledge of the partition function $Z(\beta)$. The value of the partition function and of the free energy density $f = -\ln Z/N\beta$ depends on all the values of the energies E_k. $Z(\beta)$ is also nothing but the characteristic function of the distribution of the energies.

Notice that, as the distribution of the energies is Gaussian, the distribution of the factors $\exp(-\beta E_k)$ is log-normal with a tail which becomes heavier and heavier as β increases (the "temperature" decreases). Since $Z(\beta)$ is the sum of log-normal variables, as β becomes large, the extreme fluctuations of the log-normal variables will dominate the sum. The problem is thus related to the extreme value distributions discussed in Chaps. 1 and 3 (see also [114]).

This density of states $\rho(E)$ satisfies

$$\begin{cases} \langle \rho(E) \rangle \sim \exp\left\{ N \left[\log 2 - (E/N)^2 \right] \right\} \\ \langle n(E)^2 \rangle - \langle n(E) \rangle^2 \sim \langle n(E) \rangle \;, \end{cases} \tag{16.10}$$

so that two different energy regions can be identified:

1. If $|E| < N\epsilon_c$, with $\epsilon_c = (\log 2)^{1/2}$, then the number of states $\mathcal{N}(E)$ is much greater than one and its fluctuations are small, so for a typical sample we will have

$$\mathcal{N}^{\text{typ}}(E) = \langle \rho(E) \rangle + \eta_E \langle \rho(E) \rangle^{1/2} \;, \tag{16.11}$$

where η_E is a random number of order 1.

2. On the other hand, if $|E| > N\epsilon_c$, then the number of states is exponentially small meaning that a typical sample will have no levels in this region

$$\mathcal{N}^{\text{typ}}(E) = 0 \;. \tag{16.12}$$

The partition function for a typical sample will then be

$$\mathcal{Z}^{\text{typ}} = A + B \;, \tag{16.13}$$

with

$$A = \sum_{|E| < N\epsilon_c} \langle \rho(E) \rangle \mathrm{e}^{-\beta E} \tag{16.14}$$

and

$$B = \sum_{|E| < N\epsilon_c} \eta_E \langle \rho(E) \rangle^{1/2} \mathrm{e}^{-\beta E} \;. \tag{16.15}$$

The calculation of A may be done by the steepest descent method

$$A \sim 2^N \int_{-\epsilon_c}^{+\epsilon_c} \mathrm{d}e \, \exp\{-N(e^2 + \beta e)\} \;. \tag{16.16}$$

When the integration limits lay on opposite sides of the saddle point $e = -\beta/2$, we can write

$$\int_{-\epsilon_c}^{+\epsilon_c} = \int_{-\epsilon_c}^{-\infty} + \int_{-\infty}^{+\infty} + \int_{+\infty}^{+\epsilon_c} \tag{16.17}$$

and A has contributions both from the saddle point and from the integration limits for large N. This happens if $|\beta| < \beta_c$ where

$$\beta_c = 2\epsilon_c = \frac{J}{2}(\log 2)^{1/2} \tag{16.18}$$

and we get

$$A \sim \exp(N\beta\epsilon_c) + \exp\left\{N[\log 2 + (\beta/2)^2]\right\} , \tag{16.19}$$

whereas, if $|\beta| > \beta_c$, the two steepest descent paths run from the integration limits down to infinity on the same side of the saddle point so it does not contribute and we have

$$\int_{-\epsilon_c}^{+\epsilon_c} = \int_{-\epsilon_c}^{\pm\infty} + \int_{\pm\infty}^{+\epsilon_c} \tag{16.20}$$

$$A \sim \exp(N\beta\epsilon_c) . \tag{16.21}$$

The contribution B of the fluctuations can be calculated for large N. Owing to the randomness of η_E, B is not the integral of an analytic function so the steepest descent method is no longer applicable. On the other hand, the η_E are uncorrelated for different values of E so the term with the largest modulus will dominate the sum in (16.15). We can then estimate the modulus of B as

$$|B| \sim \max_{-N\epsilon_c \le E \le N\epsilon_c} \left| \langle n(E) \rangle^{1/2} e^{-\beta E} \right| . \tag{16.22}$$

It is easily seen that when $|\beta| < \beta_c/2$, the maximum is in between the limits so

$$|B| \sim \exp\left[\frac{N}{2}(\log 2 + \beta^2)\right] , \tag{16.23}$$

while if $|\beta| > \beta_c/2$, one of the limits dominates and

$$|B| \sim \exp[N|\beta_1|\epsilon_c)] . \tag{16.24}$$

For the free energy, one thus finds

$$\frac{F}{N} = -\frac{\ln 2}{\beta} - \frac{J^2}{4}\beta , \qquad \text{for } \beta < \beta_c , \tag{16.25}$$

$$\frac{F}{N} = -J (\ln 2)^{1/2} , \qquad \text{for } \beta > \beta_c . \tag{16.26}$$

The average energy is given by

$$\langle E \rangle = -\frac{d \ln Z}{d\beta} = \frac{d(\beta F)}{d\beta} , \tag{16.27}$$

i.e.

$$\langle E \rangle = -\frac{J^2}{2}\, \beta \;, \qquad \text{for } \beta < \beta_{\mathrm{c}} \;, \tag{16.28}$$

and

$$\langle E \rangle = -J\, (\ln 2)^{1/2}\, \beta \;, \qquad \text{for } \beta > \beta_{\mathrm{c}} \;. \tag{16.29}$$

The entropy is proportional to $-\mathrm{d}F/\mathrm{d}\beta$ and is positive at high temperature $1/\beta$, decreases with $1/\beta$ (decreasing temperature) and vanishes exactly at $\beta = \beta_{\mathrm{c}}$ (glass transition) at which a small number of states dominate the behavior. It follows that, in the high temperature region, an exponentially large number of configurations contributes to the partition function (usual case) while, in the low temperature region, the probability is concentrated on a finite number of configurations.

As soon as we enter in the low temperature (large β) region, the probability of finding two equal configurations is not zero. The transition is quite strange from the thermodynamic point of view.

- It looks like a critical (second-order) transition because there is no latent heat. It is characterized by a jump in the specific heat (which decreases going toward low temperatures).
- It looks like a first order transition. There are no divergent susceptibilities coming from above or below (which within mean field theory should imply no divergent correlation length).

These strange characteristics can be summarized by saying that the transition is of order one and half, because it shares some characteristics with both the first order and the second order (critical) transitions. The thermodynamic behavior of real glasses near T_{c} is very similar to the order one and half transition of REM. This behavior is typical of the mean field approximation to glassy systems.

The random energy model assumes that the energy is completely upset by a single spin flip, i.e. it assumes the property of chaoticity to hold strongly. Refinements of the REM can be developed in terms of the so-called p-spins models [334, 379], in which the energies of nearby configurations are also nearby. Energy density (as function of the configurations) is not a continuous function in the REM while it is continuous in the p-spins models, in the topology induced by the distance. In this new case, some of the essential properties of the REM are valid but new features are present.

The Hamiltonian one considers in p-spins models [334, 379] depends on some control variables J, which have a Gaussian distribution and play the same role of the random energies of the REM, and on the spin variables σ. For $p = 1, 2, 3$ the Hamiltonian is respectively

$$H_J^1(\sigma) = \sum_{i=1,N} J_i \sigma_i \;, \tag{16.30}$$

$$H_J^2(\sigma) = \sum_{i,k=1,N}' J_{i,k}\sigma_i\sigma_k \ , \tag{16.31}$$

$$H_J^3(\sigma) = \sum_{i,k,l=1,N}' J_{i,k,l}\sigma_i\sigma_k\sigma_l \ ,$$

where the primed sum indicates that all the indices are different. The variables J must have a variance of $O(N^{(1-p)/2})$ in order to have a non trivial thermodynamical limit.

It is possible to prove by an explicit computation that, if we send first $N \to \infty$ and later $p \to \infty$, one recover the REM. Indeed the energy differences corresponding to one spin flip are of order p for large p (they are of order N in the REM), so that in the limit $p \to \infty$ the energies in nearby configurations become uncorrelated and the REM is recovered.

16.3 Non-Self-Averaging Properties

16.3.1 Definitions

The question of self-averaging arises in disordered models when one considers the statistical fluctuations of a given thermodynamic or statistical extensive property X of the model. Suppose we have a certain system of size N (under "size" can be considered the linear dimension of the system as well as the number of parts forming it, or equally a measure of its phase space, such as the number of its possible configurations) in which disorder is represented by some quenched random variables. For finite N, X is sample-dependent, in the sense that to each sample of the system, namely to each realization of the quenched disorder, corresponds a unique value of X, and the ensemble average $\langle X \rangle$ of X is obtained by averaging over all possible realizations of the quenched disorder. The sample-to-sample fluctuations of X are described by the normalized variance

$$\mathcal{D}_N(X) = \frac{\langle X^2 \rangle - \langle X \rangle^2}{\langle X \rangle^2} \ , \tag{16.32}$$

which clearly depends on N and is non-zero for finite N. If $\mathcal{D}_N(X) \to 0$ in the thermodynamic limit $N \to \infty$ then X is said to be *self-averaging* and a sufficiently large sample is a good representative of the whole ensemble. But if $\mathcal{D}_N(X)$ tends to a finite positive value, then X remains sample-dependent even in the thermodynamic limit, and an evaluation of X made on an arbitrarily large sample is not significant of the value of X on other samples. If this is the case, X is said to be *non self-averaging*.

The appearance of non self-averaging effects in the low temperature phase has been the most interesting outcome of the replica approach to spin glasses [622]. It has later been recognized that such effects are present in a large class of even simpler models ranging from condensed matter theory to population biology [419], to dynamical systems' theory [222] and mathematics as well (see [219] for a review of some of them). In the low temperature spin glasses, phase space can be thought of as if it was decomposed into infinitely many pure states α, the weights W_α of which remain sample dependent even in the thermodynamic limit [219]. This results in non-vanishing sample-to-sample fluctuations of the weights W_α even in the thermodynamic limit, so that any sample, no matter how large, is never a good representative of the whole ensemble. Furthermore, one may find finite-weighted pure states in each sample, something which sounds strange since the normalizing condition

$$\sum_\alpha W_\alpha = 1 \tag{16.33}$$

must always be satisfied. The same holds for many other models. In some cases, the expression obtained for the fluctuations of the weights coincides with that obtained for spin glasses in particular limits (as pointed out in [219]), so that it looks as if the spin glass problem, at least in its mean field version, belongs to a larger class of problems, and it would be interesting to develop a more general theory to treat them.

By now, the standard method to detect non self-averaging effects (i.e. breaking of ergodicity) is to study the quantity

$$Y = \sum_\alpha W_\alpha^2 \ . \tag{16.34}$$

It is possible to show (see [219] for applications) that, if both the ensemble average $\langle Y \rangle$, and the variance $\mathrm{var}(Y) = \langle Y^2 \rangle - \langle Y \rangle^2$ of Y are non zero in the thermodynamic limit so that the probability density $\Pi(Y)$ remains "broad" when the system's size goes to infinity, then Y and consequently the weights W_α are non self-averaging. $\langle Y \rangle$ is the average of Y over all possible samples, that is over all possible realizations of disorder, represented by a number of quenched random variables. For the REM, one finds

$$\langle Y^2 \rangle = \frac{\langle Y \rangle + 2\langle Y \rangle^2}{3} \ , \tag{16.35}$$

$$\langle Y^3 \rangle = \frac{\langle Y \rangle + \langle Y \rangle^2}{3} \ , \tag{16.36}$$

$$\langle Y^4 \rangle = \frac{2\langle Y \rangle + 3\langle Y \rangle^2 + \langle Y \rangle^3}{6} \ . \tag{16.37}$$

Similar results hold for boolean networks and random map models [222] and for simple models of evolution [224]. These results derive from the fact that, in many disordered models, in the thermodynamic limit the system's phase space appears broken into infinite basins: (i) such breaking is sample dependent; (ii) for each breaking, there are finite sized basins; (iii) the sizes of the basins are non self-averaging quantities.

16.3.2 Fragmentation Models

Broken objects are perhaps the most intuitive and simplest models showing the same non-self-averaging behavior. Consider fragmenting a given object of size 1 into an infinite number of pieces of sizes W_α according to a given breaking process. A sample corresponds to a specific rupture, hence to particular values of a set of quenched random variables on which the process depends. For some processes, one finds that the sizes of the pieces lack of self-averaging, that is they remain sample dependent despite of the fact that the number of pieces is infinite and that $\sum_\alpha W_\alpha = 1$. This is the case of Derrida and Flyvbjerg's randomly broken object [221], where the breaking process depends on an infinite number of quenched random variables. Not all breaking processes lead to non self-averaging effects. Suppose for example that the object is broken uniformly into N basins of size $W_s \sim 1/N$ each; then $Y \sim 1/N$, and both W_s and Y would go to zero in the $N \to \infty$ limit.

Kolmogorov's Log-Normal Fragmentation Model. Before giving examples where lack of self-averaging occurs, one cannot speak of fragmentation without refering to Kolmogorov's famous paper on the log-normal distribution law of particle sizes in fragmentation processes [520], which we now briefly summarize. I acknowledge exchanges with V.F. Pisarenko on this problem.

Let $N(r, t)$ be the number of particles with sizes $\rho \leq r$ at times $t = 0, 1, 2, ...$, for $0 \leq r < +\infty$. Let p_n be the probability for the generation of n particle-progenies from a given particle-progenitor in the unit time interval $[t, t + 1[$. Let $F_n(a_1, ..., a_n) = P\{k_1 \leq a_1, ..., k_n \leq a_n\}$ be the conditional distribution function of the ratios $k_i = r_i/r$ of children particle sizes to the mother size (under condition of fixed n). No mass conservation is imposed. The only condition is that children have sizes not exceeding the mother's size, that is, $0 \leq k_1 \leq ... \leq k_n \leq 1$. In addition, the degenerate case when all children have the mother's size r with probability one is excluded. Finally, let $Q(k)$ be the expectation of the number of children particles of sizes $\rho \leq kr$, where r is the mother's size. Since k measures the ratio of a child's size to the mother's size, it does not depend on the later. Thus, it is important to realize that the model expresses a condition of self-similarity. The additional assumptions of the model are:

1. Particles act independently.
2. Both p_n and F_n do not depend on the absolute size of the mother particle and on its past history;

3. $Q(1) > 1$. This last condition means that there are more than one child generated by a given mother particle, on average. This ensures that, as $t \to +\infty$, infinitely many particles are created.
4. The following integral is finite:

$$\int_0^1 |\ln k|^3 \, dQ(k) < +\infty \ . \tag{16.38}$$

This last condition is not very restrictive and is similar to the assumption in the Lyapunov's proof of the Central Limit Theorem (CLT) for the sum of random variables in which the finiteness of the third statistical moment of addends is sufficient (but not necessary) for the validity of the CLT (see Chap 2).

Let us note

$$A = \frac{1}{Q(1)} \int_0^1 \ln(k) \, dQ(k) \ ;$$

$$B = \frac{1}{Q(1)} \int_0^1 (\ln(k) - A)^2 \, dQ(k) \ . \tag{16.39}$$

Then, the fundamental result of Kolmogorov, which launched the interest in the log-normal distribution for fragmentation, is the following. Under the above conditions and, if B is strickly positive (that is, the degenerate case when all children have the same size is excluded), and calling $N(t)$ the number of all particles at time t, then the random ratio $N(e^x, t)/N(t)$ quantifying the relative number of particles with sizes $\leq e^x$ tends in probability at large times to the integral

$$\frac{N(e^x, t)}{N(t)} \stackrel{d}{=} \frac{1}{\sqrt{2\pi B t}} \int_{-\infty}^x du \, \exp\left[-\frac{(u - At)^2}{2B^2 t}\right] \ . \tag{16.40}$$

The symbol $\stackrel{d}{=}$ means that the cumulative distribution of $N(e^x, t)/N(t)$ is equal to the right-hand-side of (16.40). Notice that the asymptotic result (16.40) is similar to a central limit theorem on the logarithm of the particle size.

It is important to stress that $F_n(a_1, ..., a_n)$ does not appear in (16.40): the limit distribution of the fragment sizes at long times does not depend neither on the mother size nor on the specific past history of the fragmentation process. This is true because of the self-similarity of the fragmentation process outlined above. Kolmogorov's theorem is very general indeed and covers many fragmentation schemes and branching processres schemes with all possible types of fragmentation (branching).

An important question is how useful if the asymptotic result (16.40) in practical cases. For instance, let us consider the simple case where only fragments of sizes larger than δr occur, such that $Q(k < \delta) = 0; Q(k \geq D) =$ constant, with $Q(k)$ varying continuously between $k = \delta$ and D. Then, the convergence to the log-normal law (in its central part) is very fast. After about 7–10 generations (7–10 random factors), the distribution of fragment size is well-approximated by the log-normal law. Away from the central part of the distribution, large deviations occur along the lines discussed in Chap. 3 but which are more difficult to predict for such classes of multiplicative processes, in contrast with the case of additive processes for which the situation is clear [268, 812]. Let us also mention in this vein that Molchan has recently stressed the important role of the log-normal distribution in multiplicative (Mandelbrot) cascades see [639], proposing a novel view of the emergence of multifractality.

The Simplest Fragmentation Model. In the following example, we follow De Martino [207] and consider an object of size 1 and break it according to the following process: given a real number $p \in [0, 1]$, we tear the object in two pieces of sizes $1 - p$ and p respectively; then we do the same thing with the piece of size p, obtaining two pieces of sizes $(1 - p)p$ and p^2 respectively, plus the one of size $1 - p$ obtained at the first breaking step. If we repeat the same procedure with the pieces of sizes p^2, p^3, ... that are obtained at the second, third, ... breaking steps, keeping the pieces of sizes $(1-p)p$, $(1-p)p^2$, ... obtained at the same steps, we finally have a set of pieces of sizes

$$
\begin{aligned}
W_1 &= 1 - p \\
W_2 &= (1 - p)p \\
&\cdots \\
W_s &= (1 - p)p^{s-1} \\
&\cdots ,
\end{aligned}
\tag{16.41}
$$

where W_s denotes the size of the piece kept at the s-th step. Clearly, $\sum_s W_s = \sum_{s=1}^{\infty}(1 - p)p^{s-1} = 1$. Note that the sizes of the resulting pieces form a geometric sequence. Now suppose that p is a random variable with probability density $\rho(p)$. In this case, we can imagine that a breaking sample is produced by choosing a random value of p from $\rho(p)$ and fix it during each fragmentation history, so that averaging over all samples means simply averaging over all possible values of p.

Let again $Y = \sum_s W_s^2$. The value of Y for a single sample is given by

$$
Y = \sum_s (1 - p)^2 p^{2(s-1)} = \frac{(1 - p)^2}{1 - p^2} = \frac{1 - p}{1 + p} ,
\tag{16.42}
$$

and the ensemble average of Y is simply

$$
\langle Y \rangle = \int_0^1 \frac{1 - p}{1 + p} \rho(p) \, \mathrm{d}p .
\tag{16.43}
$$

When ρ is uniform, we obtain the value

$$
\langle Y \rangle = \log 4 - 1 \simeq 0.386 \ldots .
\tag{16.44}
$$

The probability density $\Pi(Y)$ over the samples may be calculated from the relation $\Pi(Y)dY = \rho(p)\,dp$ expressing the conservation of probability. From (16.42), we get $p = (1+Y)^{-1}(1-Y)$ and thus

$$\Pi(Y) = \left|\frac{dp}{dY}\right|\rho\left(\frac{1-Y}{1+Y}\right) ,\tag{16.45}$$

namely

$$\Pi(Y) = \frac{2}{(1+Y)^2}\rho\left(\frac{1-Y}{1+Y}\right) .\tag{16.46}$$

Y's ensemble average is given by $\langle Y \rangle = \int_0^1 Y\Pi(Y)\,dY$ and one can verify that, for a uniform ρ, the value (16.44) is recovered using

$$\int \frac{2Y}{(1+Y)^2}dY = 2\left(\log|1+Y| + \frac{1}{1+Y}\right) + \mathrm{const} .\tag{16.47}$$

The second moment $\langle Y^2 \rangle = \int_0^1 Y^2\Pi(Y)\,dY$ may also be calculated easily for a uniform ρ. Using

$$\int \frac{2Y^2}{(1+Y)^2}dY$$
$$= 2\left[-\frac{Y^2}{1+Y} + 2\left(1+Y - \log|1+Y|\right)\right] + \mathrm{const} ,\tag{16.48}$$

we obtain $\langle Y^2 \rangle = 3 - \log 16 \simeq .227\ldots \neq \langle Y \rangle^2$. The variance of Y is finally given by

$$\mathrm{Var}(Y) = \langle Y^2 \rangle - \langle Y \rangle^2 \simeq 0.078\ldots .\tag{16.49}$$

The fact that Y has a non-zero variance in the thermodynamic limit which, for this model, is represented by the infinite number of pieces in which the object is broken, proves that Y, and consequently the sizes W_s, lack of self-averaging for a geometrically broken object.

Another Recursive Fragmentation Model. We follow Krapivsky et al. [529] and define the following recursive fragmentation process. Starting with the unit interval, a break point l is chosen in $[0,1]$ with a probability density $p_l(l)$. Then, with probability p, the interval is divided into two fragments of length ratios l and $1-l$, while with probability $q = 1-p$, the interval becomes "frozen" and never fragmented again. If the interval is fragmented, we recursively apply the above fragmentation procedure to both of the resulting fragments.

The probability density $P(x)$ of fragments of length x averaged over many different realizations (corresponding to a mean field theory) is given by

$$P(x) = q\delta(x-1) + 2p\int_0^1 dl\, p_l(l)\int_x^1 dy\, P(y)\delta(ly-x) .\tag{16.50}$$

The first term corresponds to no fragmentation occurring. The second term indicates that a fragment of size x can only be obtained from a larger fragment $y \geq x$ under the application of the breaking process with length ratio l. The factor 2 accounts for the creation of two segments per fragmentation step. Integrating over y to get rid of the delta function, we get

$$P(x) = q\delta(x - 1) + 2p \int_0^1 dl \, \frac{p_l(l)}{l} P\left(\frac{x}{l}\right) . \tag{16.51}$$

Equation (16.51) can be solved by introducing the Mellin transform

$$\hat{P}(s) = \int_0^{+\infty} dx \, x^{s-1} P(x) . \tag{16.52}$$

Equations (16.51) and (16.52) yield

$$\hat{P}(s) = \frac{q}{1 - 2p\hat{p}_l(s)} , \tag{16.53}$$

where $\hat{p}_l(s)$ is the Mellin transform of $p_l(l)$. Simple poles of $\hat{P}(s)$, i.e. simple solutions of

$$2p\hat{p}_l(s) = 1 , \tag{16.54}$$

give power laws by the inverse Mellin transform. This can be seen directly by assuming a power law solution and replacing in (16.51). This retrieves the equation (16.54). Note that, in general, there will be solutions in the variable s of (16.54) with imaginary parts. For instance, take $p(l) = \delta(l - l_0)$, corresponding to an exact self-similar fragmentation process. Then, $\hat{p}(s) = l_0^{s-1}$ and the solutions of (16.54) are

$$s_n^* = -\frac{\ln(2p/l_0)}{\ln l_0} + i\frac{2\pi n}{\ln l_0} . \tag{16.55}$$

The complex exponents s_n^* signal the existence of log-periodic corrections to the pure power laws and reflect a discrete scale invariance created by the multiplicative fragmentation process [712] with the reduction factor l_0. This kind of phenomena has been discussed in Chap. 5. The existence of these log-periodic corrections are robust with respect to the presence of disorder [476, 712], i.e. remain for a large set of choice for $p(l)$ which can become broad.

However, when the disorder becomes too large, as for a uniform $p_l(l) = 1$, the distribution of fragment sizes becomes a pure power law, as seen from the existence of a single simple pole:

$$\hat{P}(s) = q\left[1 + \frac{2p}{s - 2p}\right] . \tag{16.56}$$

The inverse Mellin transform of (16.56) gives

$$P(x) = q\left[\delta(x - 1) + 2px^{-2p}\right] . \tag{16.57}$$

Apart from the obvious δ-function corresponding to the process where fragmentation stopped at the first step, the average length density distribution is a purely algebraic function. In particular, the fragment distribution diverges algebraically in the limit of small fragments. The mechanism for this power law distribution results essentially from the multiplicative structure of the fragmentation model together with the fact that fragments can become unbreakable. This mechanism is similar to that discussed in Chap. 14, in relation to multiplicative processes repelled from the origin. This is confirmed by the fact that (16.51) has exactly the same structure as (14.125) obtained for multiplicative models with repulsion from the origin.

The recursive fragmentation process can be generalized to higher dimensions. In two dimensions, we start with the unit square, choose a point (x_1, x_2) with a uniform probability density, and divide, with probability p, the original square into four rectangles of sizes $x_1 \times x_2$, $x_1 \times (1 - x_2)$, $(1 - x_1) \times x_2$, and $(1 - x_1) \times (1 - x_2)$. With probability q, the square becomes frozen and we never again attempt to fragment it. The process is repeated recursively whenever a new fragment is produced. This model of fragmentation also leads to a power law distribution of fragment sizes, as measured by their surface (volume in the higher dimensional case).

The recursive fragmentation processes described above thus exhibit a number of features that arise in other complex and disordered systems, such as non-self-averaging behavior and the existence of an infinite number of singularities in the distribution of the largest fragment. These features indicate that even in the "thermodynamic limit", sample to sample fluctuations remain, and that knowledge of first order averages may not be sufficient for characterizing the system.

17. Randomness
and Long-Range Laplacian Interactions

17.1 Lévy Distributions from Random Distributions of Sources with Long-Range Interactions

We now present another aspect of quenched randomness that occurs in relation to long-range interactions. Our exposition is non-rigorous and emphasizes the underlying physical mechanisms. For a rigorous presentation, we refer to V. Zolotarev on "Model of Point Sources of Influence" (see [1066], Chap. 1, Sect. 1.1). V. Zolotarev provides in particular a general formula (number (1.1.9) of his book) for the characteristic function of random fields generated by randomly distributed (Poissonian) point sources. The corresponding distribution is always an infinitely divisible distribution. There are numerous examples of such fields with power-like tails (gravitational fields of stars, temperature distribution in nuclear reactors, strain fields in crystals generated by random imperfections, magnetic fields generated by elementary magnets, some problems in communication theory, and so on). We also mention the problem of the sum of impulsive sources in signal analysis: many of the natural and man-made impulsive interferences may be considered as the results of a large number of spatially and temporally distributed sources that produce random noise sources of short duration. Nikias and Shao [685] present a derivation of symmetric stable Lévy distributions to model the first-order statistics of such impulsive noise. The stable Lévy distributions can be derived from the standard filtered-impulse mechanism of the noise process under the condition that the spatial and temporal distributions of noise sources as well as the propagation properties are scale invariant and characterized by power laws [321, 624].

17.1.1 Holtsmark's Gravitational Force Distribution

A classic problem in astronomy is as follows: consider a universe of identical pointwise stars of unit mass distributed at random leading to an average uniform density ρ for the universe. Take a point at random in the universe: you will measure a gravitational pull. At another point, the pull will probably be different. Dropping measurements at random, what is the distribution of the components of the gravitational forces created by the random distribution of stars?

If the density was not concentrated at stars but uniformely distributed in gas clouds evenly spread within the universe, the gravitational force will be exactly zero in an infinite universe due to the exact cancellation of the pulling force exerted by all pairs of points symmetric to each other with respect to the measuring point. A non-trivial phenomenon appears solely due to heterogeneity as we show below. Thus, replacing a random system by an average homogeneous system is very misleading. This is an example where homogeneization or effective medium theories do not apply.

This problem has been solved first by the astronomer Holtsmark. Let us give two derivations that illuminate the relationship between this problem and the stable Lévy laws studied in Chap. 4 and with the statistics of large fluctuations described in Chap. 3.

First Intuitive Derivation. If the density is ρ and the stars have unit mass, then the typical distance between stars is $r^* \sim (3/4\pi\rho)^{1/3}$. However, if we drop a measurement point at random, it will sometimes fall very close to a star and will thus feel a strong attraction. Due to the uncorrelated randomness of the star position, the probability, that the closest star is at a distance between r and $r + dr$ to the measurement point taken at the origin, is

$$P(r)dr \sim \rho r^{d-1} dr \qquad \text{for } r \text{ small .} \tag{17.1}$$

The factor r^{d-1} stems from the spherical geometry of finding a star at a distance r in a space of dimension d. For $d = 3$, $r^{d-1} = r^2$ falls to zero quite rapidly for $r < r^*$. According to the law of gravitation, the corresponding pull due to a neighboring star at a distance r is

$$F \sim r^{-2} , \tag{17.2}$$

for all components of the force. Assuming that the contribution of the other stars that are further away do not contribute significantly compared to the closest star, we get the distribution $P(F) dF$ of the forces F. It is such that $P(F) dF = P(r) dr$, which yields

$$P(F) dF \sim \frac{dF}{F^{1+\mu}} , \tag{17.3}$$

for F large, with $\mu = d/2 = 3/2$ in our 3D space. This simple reasoning retrieves the result by Holtsmark that the distribution of the forces in a random universe is very broad, with *no* variance since $\mu < 2$ (see Chap. 4). This derivation uses the mechanism of power law or inversion change of variable (14.10) presented in Chap. 14 to generate a power law distribution.

Lévy Distributions. Let us now discuss the assumption that the contribution of the other stars that are further away do not contribute significantly compared to the closest star. This will lead us to an argument borrowed from Feller [293]. We consider a ball of radius r centered at the origin (measurement point) which contains N pointwise stars of unit mass placed independently and randomly. Call X_i the x-component of the gravitational force created by the i-th star at the origin. The same reasoning applies to the other force components along y and z. Then, the total force along x is

$$S_N = \sum_{i=1} X_i \; . \tag{17.4}$$

We are interested in the limit $r \to \infty$, $N \to \infty$ and

$$\frac{4\pi}{3} r^3 N^{-1} \to \frac{1}{\rho} \; , \tag{17.5}$$

where ρ is the average large scale star density. Then, the distribution of S_N tends to the symmetric stable Lévy distribution with characteristic exponent $\mu = 3/2$. To see this result, let us consider the density ρ as a free parameter. Consider two clusters of stars with density ρ_1 and ρ_2 respectively and the corresponding force $S_N^{(1)}$ and $S_N^{(2)}$ that they exert at the origin along x. Because the distribution of stars in each cluster is random and uncorrelated, these two clusters may be combined to form a new cluster of density $\rho_1 + \rho_2$ which exerts a total force along x equal to $S_T = S_N^{(1)} + S_N^{(2)}$. In probabilistic terms, the sum of two independent variables $S_N^{(1)}$ and $S_N^{(2)}$ should have the same distribution as S_T. This requires that the distribution of S_N in (17.4) be one of the stable distributions, either the Gaussian law or one of the Lévy laws. To determine which of these to choose, consider a change of density from 1 to ρ. Since density is mass per unit volume, this change of density amounts to a change of length from 1 to $\rho^{-1/3}$. As the gravitational force varies inversely with the square of the distance, we see that S_N must have the same distribution as $N^{2/3} S_1$, as seen from (17.5). In other words, $S_N / N^{2/3}$ must have the same distribution as S_1. From the scaling law (4.24) that characterizes Lévy laws, this implies that the distribution of S_N in the limit of large N is the Lévy distribution with exponent $\mu = 3/2$.

Specifically, the probability density of the force F is given by

$$P(F) = \frac{H(\beta)}{F_0} \tag{17.6}$$

where

$$F_0 = (4/15)^{2/3} (2\pi G M) \rho^{2/3} \tag{17.7}$$

is the normalizing force (ρ is the average density of sources, M is the mass of each point source and G is the gravitational constant), $\beta = F/F_0$ is a dimensionless force and

$$H(\beta) = \frac{2}{\pi \beta} \int_0^\infty \mathrm{d}x \; \exp[-(x/\beta)^{3/2}] \, x \, \sin(x) \; . \tag{17.8}$$

The main result is that, in the thermodynamic limit ($V \to \infty$ with ρ constant), the force distribution has a finite first moment and an infinite variance. As we already stressed, this divergence is due to the possibility of being arbitrarily close to a source. The approximate solution given by the nearest neighbor approximation (i.e. by considering only the effect of the nearest neighbor source) and the exact Holtsmark's distribution (17.8) agrees over most of the range of F as shown in Fig. 17.1. The region where they differ mostly is when $F \to 0$. This is due to the fact that a weak force arises when there is an almost perfect cancellation of the pulls from all the stars: this case involves a collective effect for which the nearest neighbor approximation fails.

This result (17.6)–(17.8) extends and makes more precise the previous asymptotic result (17.3) which is valid only in the tail of the distribution for very large forces. The Lévy stable distribution describes the full distribution of the gravitational pulls, in the suitable limit of an infinite volume V and an infinite number N of stars with a finite ratio N/V.

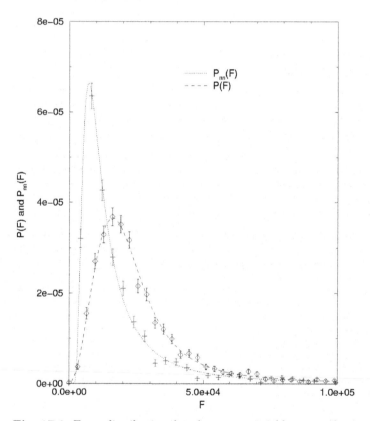

Fig. 17.1. Force distribution due the nearest neighbor contribution (*crosses*) and due to all the field sources (*diamonds*) for an homogeneous sample with $\rho = N/V = 2.39 \times 10^4$ in three dimensions. The *dotted line* represents the force distribution $W(F)$ computed by the nearest neighbor approximation, while the *dashed line* is the Holtsmark's distribution (17.6). The agreement is quite satisfactory at strong field while there is a deviation at weak forces F. Reproduced from [323]

The gravitational force distribution arising from a fractal set of sources must take into account the fluctuations in the position of the sources [323]. In the case of real structures in finite samples, an important role is found to be played by morphological properties and finite size effects. For fractal dimensions d_f smaller than $d - 1$ where d is the space dimension, the convergence of the net gravitational force is assured by the fast decaying of the density, while for fractal dimension $d_f > d - 1$, the morphological properties of the structure determine the eventual convergence of the force as a function of distance.

The probability density of the absolute value F (generalized Holtsmark's distribution) of the field intensity is equal to

$$H(\beta, d_f) = \frac{2}{\pi\beta} \int_0^\infty dx \, \exp[-(x/\beta)^{d_f/2}] \, x \, \sin(x) \ . \tag{17.9}$$

In this case $F_0 = (4/15)^{2/d_f}(2\pi GM)(d_f B/(4\pi))^{2/d_f}$, where B is a constant characterizing the average mass in the unitary sphere and $\beta = F/F_0$. The main change due to the fractal structure is that the scaling exponent in (17.9) is $d_f/2$ rather than $3/2$. In this case, the tail of the probability density has a slower decay than in the homogeneous case shown in Fig. 17.1. This means that the variance of the force is larger for $d_f < 3$ than for the $d = 3$. The case $d_f < 2$ is rather well described by (17.9): this is not the case for $d_f > 2$ where the n-th point correlations must be taken into account in the case of real structures [323]. An important limit is the strong field one ($F \to \infty$). In this case, it is possible to show that the force distribution of (17.9) can be reduced to the one derived under the nearest neighbor approximation:

$$P_{nn}(F) \, dF$$

$$= \frac{d_f B}{2}(GM)^{d_f/2} F^{-(d_f+2)/2} \exp\left[-B(GM)^{d_f/2}F^{-d_f/2}\right] dF \ . \tag{17.10}$$

The nearest neighbor approximation is good for $F \gg F_1$, where $F_1 = (B)^{2/d_f}$. GM. In the fractal case, as in the homogeneous one, the divergences of the force moments (the second for $d_f \le 3$ and the first for $d_f \le 2$) are due only to the fact that a source field can be arbitrarily close to the measurement point.

17.1.2 Generalization to Other Fields (Electric, Elastic, Hydrodynamics)

Consider more generally a field f that depends on the distance r from a pointwise source as (17.2) but with an exponent s different from 2:

$$f \sim r^{-s} \ . \tag{17.11}$$

Let us again assume that the sources are randomly and uniformly distributed in space. Then, the probability, that the distance from a given point to the closest source is r, is again given by (17.1) for small r. Using the conservation of the probability under a change of variable $P(f)\,df = P(r)\,dr$, we have

$$P(f)\,df \sim \frac{df}{f^{1+\mu}} \qquad \text{for } f \text{ large}, \qquad \text{with } \mu = \frac{d}{s}. \qquad (17.12)$$

This result (17.12) describes the tail of the distribution for large forces controlled by the closest approach to a source. The same reasoning as for the gravitational force above shows that the full distribution is the symmetric stable Lévy law with characteristic exponent $\mu = d/s$, when one takes the limit of an infinite number of sources in an infinite volume with finite density of sources.

This derivation applies to many situations as soon as the forces depend on r as in (17.11) for small r, because randomness and defects are ubiquitous. Let us mention the case of electrostatics and the electric field distribution from random dipole sources as in electrorheological fluids [389, 390, 392], interactions between vortex lines in superconductors and in fluids [629, 919] and strain and stress fields in solids with impurities or defects [18, 528]. In this last application, there are many cases to consider that give different stable distributions.

- For straight parallel dislocations in a solid, we have $s = 1$ and $d = 2$ leading to the stable Lévy law with characteristic exponent $\mu = 2$.
- For small dislocation loops or elastic point defects, we have $s = 3$ and $d = 3$, leading to the stable Lévy law with characteristic exponent $\mu = 1$. This situation applies to relatively small earthquakes, for which the rupture mechanism is equivalent to the introduction of a dislocation loop in the crust [10, 885]. Zolotarev [1067] has shown that, for the spatially uniform distribution of defects (faults) in an elastic medium, the stress distribution is the stable Lévy law with $\mu = 1$, called the Cauchy law. The characteristic function for the random stress distribution can be written

$$\ln \hat{P}(k) = \int_0^\infty dr\, r^2 \left[e^{ik\sigma/r^3} - 1 \right], \qquad (17.13)$$

where σ is the normalized (for $r = 1$) stress Green function of a defect. The r^{-3} term expresses the decay with distance of the stress due to a defect in 3D.

- Kagan [490] has shown that, assuming that earthquakes form a fractal set with dimension $d_f < 3$, the stress distribution should follow a stable law with an exponent $\mu = d_f/3$ less than 1. This result is simply obtained from our derivation given above, by replacing d by d_f in (17.12). This derives from the fact that d must be replaced by d_f in (17.1) as a result of the definition of a fractal set with (capacity) dimension d_f (see Chap. 5). The interesting and challenging consequence is that we cannot talk anymore about an average stress or a stress correlation function: these quantities are not defined for stable distributions with $\mu \leq 1$, as we have seen in Chap. 4

(see also [685, 810]). In such case, instead of averages and correlations, we should use quantiles and codifferences. Moreover, stress is a tensor, not a scalar, with enormous complications ensuing.

- For parallel dislocation dipoles, we have $s = 3$ and $d = 2$ leading to the stable Lévy law with characteristic exponent $\mu = 2/3$.

17.2 Long-Range Field Fluctuations Due to Irregular Arrays of Sources at Boundaries

17.2.1 Problem and Main Results

"Self-screening" of periodic assemblies of elements whose individual influence is long-range is well-known. Consider for instance a periodic column or plane of electric dipoles. Even though the field of each individual dipole decays as $1/(\text{distance})^3$, the total field decays exponentially. This is due to the almost exact cancellation between the angular structure of the dipole fields. This effect is also well-known for periodic arrays of dislocations [548] and for periodic boundary conditions with applications to gravity and thermal anomalies [970].

An interesting question, relevant to real-life situations, is the robustness of this cancellation in the presence of disorder. The main results are as follows.

- For weak disorder around an average periodicity, the ensemble average field still decays exponentially. However, the standard deviation σ^2 or second moment of the field, which quantifies the amplitude of its fluctuations, decays only algebraically as $(\text{distance})^{-3}$:

$$\sigma \sim z^{-3/2} . \tag{17.14}$$

This slow power law decay (17.14) of the field fluctuations is due to the breakdown of the almost exact cancellation of the sources at all multipole orders. Because of the disorder, Fourier components of arbitrarily large wavelengths in the power spectrum of the source array appear and are responsible for the slow decay of the field fluctuations. Since small random fluctuations around a periodic modulation are always present in nature, the slow decay as $(\text{distance})^{-3/2}$ of the typical field fluctuations must be ubiquitous in nature. For instance, it probably explains the experimental observations of attraction and dissolution of precipitates by migrating grain boundaries [107]. In electro- and magneto-rheological fluids [389, 390, 392], chains are found to interact more strongly that predicted from a naive prediction based on a regular periodic necklace structure. Similarly, intermediate wavelength magnetic and gravity anomalies measured at various heights above the ocean crust may sometimes arise from the random structure of geological polarity reversals. Another interesting domain of application is the problem of sedimentation of suspensions and permeability of porous media. One is interested for instance in characterizing the sedimentation of a cloud of heavy particles in a fluid. The fluctuations of particle speeds around the average speed of the sedimenting cloud involves taking correctly into account the renormalization and screening effects of all other particles on a given test particle, a problem which is still unsolved in general. The formalism presented below allows one to obtain the velocity fluctuations once the configuration of the particles is known.

• For strong disorder (to be made precise below), expression (17.14) is modified into

$$\sigma \sim z^{-1/2} \ . \tag{17.15}$$

• One needs also to account for other intermediate disorder strengths and for the various types of constitutive elements, including dislocations, dipoles, quadrupoles or more complex entities.

17.2.2 Calculation Methods

Three different mathematical methods have been used to obtain these results:

• a perturbation approach in conjunction with the Poisson summation formula [106, 801, 873, 1051–1053],
• brute force calculations of the standard deviation of the field expressed as an infinite series [389, 390, 392, 669–671],
• a spectral method [801, 873].

All of them are consistent in the weak disorder regime and confirm the validity of (17.14). In the strong disorder regime, the first method in terms of the Poisson summation formula does not hold since it relies upon a perturbative expansion of the disorder around the perfect periodic structure. The two other methods are in principle applicable to arbitrary situations, the spectral method being however the most general and powerful.

In what follows, we show how expression (17.14) is obtained and specify the meaning of "weak" disorder. Then, we present the very general form of the solution of a general problem within the spectral method framework [801, 873]. This allows us to recover simply all previous results including the result (17.15) for the large disorder regime and furthermore to derive a series of results for various types of constitutive elements. The value of the exponent of the power law controlling the decay of the variance σ^2 of the field, which determine the "universality class", is controlled by two properties:

1. the power spectrum of the disorder at low wavenumber,
2. the local structure of the constitutive elements determining their multipolar order.

Derivation of the Power Law Decay and Meaning of "Weak Disorder". Consider the simple problem of a Laplacian field V obeying Laplace's equation

$$\Delta V = 0 \tag{17.16}$$

in a semi-infinite medium bounded by a frontier, with imposed boundary values or sources at the boundary. We take the boundary to be the plane $(0x, 0y)$. The semi-infinite medium extends from $z = 0$ to $+\infty$. An alternate distribution of sources $(+)$ and $(-)$ of strength $\pm S_0$ are arranged spatially in the plane $(0x, 0y)$. We study first a two dimensional version of the problem and shall return later to the three-dimensional case. The set of sources $(+)$ and $(-)$ are assumed to be spatially disordered around an average periodic modulation:

$$(+) \quad x_n^+ = n\lambda + d_n^+ \qquad (-) \quad x_n^- = (2n+1)\lambda/2 + d_n^- \ , \tag{17.17}$$

where d_n^+ and d_n^+ are random variables which are independent from site to site with zero average

$$\langle d_n^{\pm} \rangle = 0 \tag{17.18}$$

and a variance

$$\langle d_n^{\pm} d_m^{\pm} \rangle = 0 \quad \text{if} \ \ n \neq m \ \ \text{and} \ \ = \langle [d_n^{\pm}]^2 \rangle \quad \text{for} \ \ n = m \ . \tag{17.19}$$

λ is the average period of the modulation. We assume furthermore that $V(x, z \to \infty) \to 0$. The problem is thus to solve the Poisson equation

$$\Delta V = -4\pi S_0 \delta(z) \sum_{n=-\infty}^{+\infty} [\delta(x - x_n^+) - \delta(x - x_n^-)] \ , \tag{17.20}$$

where δ is the Dirac's δ function. This formulation encaptures different boundary conditions, for instance the cases when the distribution of sources is replaced by the knowledge of the field V at the boundary. This mathematical formulation, which has the advantage of simplicity, already describes various problems of gravity, magnetic, temperature and resistivity anomalies and will constitute the backbone of our simple derivation.

When the disorder is absent ($d_n^{\pm} = 0$ for all n), we replace the set of sources by the boundary condition obtained by taking the first term in its Fourier series expansion:

$$V(x, z = 0) = \lambda S_0 \cos\left(\frac{2\pi x}{\lambda}\right) \ . \tag{17.21}$$

The method of separation of variables

$$V(x, z) = Z(z) \cos\left(\frac{2\pi x}{\lambda}\right) \tag{17.22}$$

satisfies automatically the boundary condition. Substitution in (17.16) yields

$$\frac{d^2 Z}{dz^2} - \frac{4\pi^2}{\lambda^2} Z = 0 \tag{17.23}$$

whose solution obeying $V(x, z \to \infty) \to 0$ is

$$V(x, z) = \lambda S_0 \cos(2\pi x/\lambda) \; e^{-2\pi z/\lambda} . \tag{17.24}$$

We thus retrieves the classical result [970] that the field disturbance introduced by the boundary condition (17.21) decays exponentially with depth in a distance proportional to the horizontal wavelength λ.

When the disorder is present, we use the Green function method. In 2D, the solution of the equation

$$\Delta V = -4\pi\delta(x - x_0)\delta(z - z_0) \qquad \text{with} \;\; V(x, z \to \infty) \to 0 \tag{17.25}$$

is [651]

$$G(x, z) = -\ln[(x - x_0)^2 + (z - z_0)^2] . \tag{17.26}$$

Using this expression for the Green function, the general solution of (17.20) reads formally

$$-\frac{V(x, z)}{S_0} = \sum_{n=-\infty}^{+\infty} f^+(n + d_n^+/\lambda) - \sum_{n=-\infty}^{+\infty} f^-(n + d_n^-/\lambda) , \tag{17.27}$$

where

$$f^+(t) = \ln[(x - t\lambda)^2 + z^2] \qquad f^-(t) = \ln[(x - (2t+1)\lambda/2)^2 + z^2] . \tag{17.28}$$

Note that for $d_n^\pm = 0$ for all n, the series can be summed up using the theory of analytical functions ([651] Tome II, p. 1236) and yields

$$-\frac{V(x, z)}{S_0} = |\ln[\tan(2\pi(x + iz)/\lambda)]|^2 \tag{17.29}$$

which reduces to (17.24) for large $z \gg \lambda$.

In order to estimate $V(x, z)$ given by (17.27), we expand each term $f^\pm(n + d_n^\pm/\lambda)$ up to second order in $d_n^\pm/\lambda$, since the disorder is assumed to be small:

$$f^\pm(n + d_n^\pm/\lambda) = f^\pm(n) + \frac{d_n^\pm}{\lambda} \frac{df^\pm(n)}{dn}$$

$$+ \frac{1}{2}\left(\frac{d_n^\pm}{\lambda}\right)^2 \frac{d^2 f^\pm(n)}{dn^2} + \dots \tag{17.30}$$

The average $\langle V(x, z) \rangle$ over the disorder reduces to

$$-\frac{\langle V(x, z) \rangle}{S_0} = \sum_{n=-\infty}^{+\infty} f(n) + (1/2) \left\langle \left(\frac{d}{\lambda}\right)^2 \right\rangle \sum_{n=-\infty}^{+\infty} \frac{d^2 f(n)}{dn^2} + \dots \tag{17.31}$$

with

$$f(n) = f^+(n) - f^-(n) .$$ (17.32)

To obtain (17.31), we have used the conditions (17.18) and (17.19). Using Poisson's summation rule which reads

$$\sum_{n=-\infty}^{+\infty} f(n) = \sum_{k=-\infty}^{+\infty} \int_{-\infty}^{+\infty} f(t)e^{2i\pi kt} \, dt ,$$ (17.33)

we separate the $k = 0$ contribution which allows one to write expression (17.31) as

$$-\frac{\langle V(x,z) \rangle}{S_0} = \int_{n=-\infty}^{+\infty} dt \left[f(t) + \left\langle \left(\frac{d}{\lambda}\right)^2 \right\rangle \frac{d^2 f(t)}{dt^2} + \dots \right]$$

$$+2 \sum_{k=-\infty}^{+\infty} \int_{-\infty}^{+\infty} dt \left[f(t) + \left\langle \left(\frac{d}{\lambda}\right)^2 \right\rangle \frac{d^2 f(t)}{dt^2} + \dots \right] \cos(2\pi kt) .$$ (17.34)

Averaging the Laplacian field amounts essentially to recovering the periodic case with a small perturbation proportional to the second moment $\langle (d/\lambda)^2 \rangle$ of the disorder. For $f(t)$ given by (17.32) with (17.28), the first integral in the r.h.s. of (17.34) is identically zero since $f^+(t)$ and $f^-(t)$ cancel exactly. The leading behavior of $V(x,z)$ is thus given by the first ($k = 1$) term in (17.34) which can be shown, after some tedious calculations, to recover expression (17.24). The fact that the zero wavevector $k = 0$ contribution vanishes is the mathematical expression of the mutual screening at all multipole orders of the field produced by each source in the periodic case. Thus, averaging the Laplacian field would seem to imply that the effect of fluctuations are negligible.

In fact, the effect of fluctuations are quantified by the second moment σ^2 of $V(x,z)$ defined by

$$\sigma^2 \equiv \frac{\langle [V(x,z)]^2 \rangle - \langle V(x,z) \rangle^2}{S_0^2}$$

$$= \left\langle \left(\frac{d}{\lambda}\right)^2 \right\rangle \sum_{n=-\infty}^{+\infty} \left(\frac{df(n)}{dn}\right)^2 + \dots$$ (17.35)

Applying again Poisson's summation rule to the sum in (17.35), we find that the $k = 0$ term is now proportional to $\langle (d/\lambda)^2 \rangle \int_{-\infty}^{+\infty} [df(t)/dt]^2 + \dots$ which is no longer vanishing. This term is at the origin of the long-range power law decay (17.14) as found from a direct calculation:

$$\sigma \sim C \left\langle \left(\frac{d}{\lambda}\right)^2 \right\rangle^{1/2} \left(\frac{z}{\lambda}\right)^{-3/2} + \mathcal{O}(z^{-2}) ,$$ (17.36)

where C is a numerical factor of order unity. The strength of the disorder affects only the prefactor of the power law decay and not its power law dependence.

General Spectral Formulation and Solution of the Random Source Problem. We now consider a general problem, electric, elastic or other, characterized by the value of the Green function $G(x - x', z)$ giving the field at position (x, z) created by a unit element placed at $(x', 0)$. The Green function $G(x - x', z)$ fully characterizes the given problem. The element can be a single source, a dipole, a multipole, or a dislocation among others. The distribution of element sources in the plane $z = 0$ is described by the function $r(x')$. We do not need to specify if it is periodic, weakly random or otherwise. Then, the formal solution of the field is given by

$$V(x, z) = \int r(x')G(x - x', z) \, dx' \; . \tag{17.37}$$

This expression consists of a convolution between the distribution of sources and the Green function which depends on the depth z of the source layer beneath the observation plane. Convolution in the space domain becomes a multiplication in the Fourier domain, so that we have

$$\hat{V}(\mathbf{k}, z) = \hat{r}(\mathbf{k})\hat{G}(\mathbf{k}, z) \; , \tag{17.38}$$

where $\mathbf{k}$ is the wavenumber in the x-plane. Note that the Fourier transform $\hat{G}(\mathbf{k}, z)$ of the Green function is nothing but the field V created by a periodic array of elements with periods $2\pi/k_x$ and $2\pi/k_y$ respectively in the x and y directions. For suitable elements, such as dipoles, dislocations, etc., $G(\mathbf{k}, z)$ decays exponentially with z as

$$\hat{G}(\mathbf{k}, z) \sim e^{-kz} \; , \tag{17.39}$$

where $k = |\mathbf{k}|$, thus recovering the general self-screening property of periodic systems of dipoles or dislocations discussed above. By inverse Fourier transform, expression (17.38) transforms into

$$V(x, z) = \mathcal{R} \int \hat{r}(\mathbf{k})\hat{G}(\mathbf{k}, z)e^{i\mathbf{k}x} \, d\mathbf{k} \; , \tag{17.40}$$

where $\mathcal{R}$ means that we take the real part of the expression on the r.h.s.

The ensemble average field and the second moment of the field are given respectively by

$$\langle V(x, z) \rangle = \mathcal{R} \int \langle \hat{r}(\mathbf{k}) \rangle \hat{G}(\mathbf{k}, z)e^{i\mathbf{k}x} \, d\mathbf{k} \tag{17.41}$$

and

$$\sigma^2 = \mathcal{R} \int \int \langle \hat{r}(\mathbf{k})\hat{r}^*(\mathbf{k}') \rangle \hat{G}(\mathbf{k}, z)\hat{G}^*(\mathbf{k}', z)e^{i(\mathbf{k}-\mathbf{k}')x} \, d\mathbf{k} \, d\mathbf{k}' \; , \tag{17.42}$$

where the symbol $*$ stands for the complex conjugate. Consideration of these two equations allows us to recover previous results with ease.

- For periodic systems, $\langle r(\mathbf{k}) \rangle = r_0 \delta(\mathbf{k} - \mathbf{k}_0)$, which together with (17.39), yields the exponential decay

$$V(x, z) >\sim e^{-k_0 z} \cos(k_0 x) , \qquad (17.43)$$

characteristic of the self-screening property. Note that we have excluded the case where the spatial average yields a non-zero contribution at $k = 0$, implying a non-vanishing source density in the continuous limit, i.e. a non-vanishig total charge. Our discussion applies only to the contributions at finite wavevectors, the $k = 0$ contribution being easily taken into account in the continuous limit.
- For weak disorder such that the power spectrum of the distribution is given by

$$\langle \hat{r}(\mathbf{k}) \hat{r}^*(\mathbf{k}') \rangle = r_0^2 k^2 \delta(\mathbf{k} - \mathbf{k}') , \qquad (17.44)$$

we obtain (17.14) at distances $z \gg k_0^{-1}$. The z-dependence of the field and of its fluctuations at large distances z is controlled by the behavior of these integrals in the small wavevector domain. Note that the x-dependence disappears in the ensemble averaging due to the assumed absence of correlations of the density of sources, in other words due to the property of average translational invariance (average uniform spatially distributed disorder) in an infinite system.
- These considerations allow us to give the general rules controlling the decay of the standard deviation σ of the fluctuations of the field. If a given physical problem is such that the power spectrum of the disorder is given by

$$\langle \hat{r}(\mathbf{k}) \ \hat{r}^*(\mathbf{k}') \rangle \sim k^a \ \delta(\mathbf{k} - \mathbf{k}') , \qquad (17.45)$$

and the field created by a periodic array of elements of wavevector $\mathbf{k}$ is given by

$$\hat{G}(\mathbf{k}, z) \sim k^b z^c e^{-kz} , \qquad (17.46)$$

where a, b and c are in general positive but possibly negative exponents, then from expression (17.42), we immediatly get from power counting

$$\sigma^2 \sim z^{-\alpha} , \qquad \text{where } \alpha = 1 + a + 2(b - c) . \qquad (17.47)$$

The universality class describing the fluctuation of the field is determined entirely by three exponent a, b and c, defined by

1. the spectral content of the disorder at low wavenumber given by expression (17.45) which gives the value of exponent a,
2. the structure of the constitutive elements (which controls the order of the multipole) and the nature of the physical problem, which determines the kernel structure given by (17.46) and which gives the exponents b and c.

It is clear that the characteristic decay length of a source of spectral component with wavenumber k is of the order of k^{-1}. Thus, the power law decay of σ^2 can be traced back to the existence of very low wavenumbers in the power spectrum of the disorder. Disorder destroys the exact cancellation of all multipoles and introduces a continuous spectrum down to $k \to 0$. A given value of α does not need to correspond to the same physics, since there are many combinations of a, b and c for a given value of the decay exponent α. For instance, the case $\alpha = 3$ is obtained for $a = 2$ and $b = c = 0$, which corresponds to a system of planar dipoles in electrostatics. The same value $\alpha = 3$ is also obtained for a planar lattice of dislocations in the elasticity problem for which $a = 2$ and $b = c = 1$. More generally, we expect that the decay exponent α may be modified either due to a change of the spectral content of the disorder or due to the multipole nature of the constitutive elements in the source array.

For very large disorder, where each constitutive element could take any position within the plane, the power spectrum of the disorder has a white noise structure corresponding to the exponent $a = 0$. In the two previous examples, this changes the decay exponent to $\alpha = 1$.

A fractal disordered source network is characterized by a non-integer power law spectrum with exponent $a = 2 - \beta$ with $0 \leq \beta \leq 2$. The corresponding decay exponent is $\alpha = (3 - \beta)/2$ in the case of simple elements ($b = c$).

The domain of validity of these results is typically for distances z larger than the average period in the weak disorder case around an average periodicity. In the strong disorder case, it is valid in an ensemble average sense over many disorder configurations.

Distribution of the Field Strengths at Fixed Distance z from the Sources and Relation to Holtsmark's Distribution. We have derived the dependence of the average and variance of the field as a function of the distance z to the sources. Can we say more and obtain the full distribution of $V(x, z)$ at fixed distance z? The question is also motivated by the analogy between this problem and Holtsmark's distribution.

All the information is contained in the expression (17.40) together with (17.45) and (17.46). Expression (17.45) tells us that $\hat{r}(\mathbf{k})$ is a delta-correlated noise with variance proportional to k^a, i.e. its distribution is a Gaussian with a variance proportional to k^a. Equation (17.40) shows that $V(x, z)$ is a linear sum of random Gaussian variables. From the central limit theorem discussed in Chap. 2, $V(x, z)$ is therefore also a Gaussian variable with variance equal to the sum of the variance of the individual variables. Its variance is thus given by (17.42) and (17.47). Thus, in contrast to Holtsmark's power law distribution, the distribution of the field created by a random array of sources at a boundary is Gaussian! The reason for the difference is clear: the point of measurement (x, z) is at a finite bounded distance from all sources, with a minimum distance z. There is thus no possibility for the occasional close approach to a source, which is the phenomenon at the origin of Holtsmark's power law distribution.

Let us finally give the spatial correlation along x of the field $V(x, z)$ at fixed distance z:

$$\langle V(x, z)V(x', z)\rangle = \mathcal{R} \iint \langle \hat{r}(\mathbf{k})\hat{r}^*(\mathbf{k}')\rangle \hat{G}(\mathbf{k}, z)\hat{G}^*(\mathbf{k}', z)e^{\mathrm{i}\mathbf{k}x - \mathrm{i}\mathbf{k}'x'} \, \mathrm{d}\mathbf{k} \, \mathrm{d}\mathbf{k}$$

$$= \mathcal{R} \int \mathrm{d}k \, k^{2b+a} z^{2c} e^{-2kz} e^{\mathrm{i}k(x-x')}$$

$$\sim \mathcal{R} \frac{z^{2c}}{[2z - \mathrm{i}k(x - x')]^{1+2b+a}} \, . \tag{17.48}$$

For instance, for $b = c = 0$, we get

$$\langle V(x, z)V(x', z)\rangle \sim \frac{1}{8z^3} \frac{1 - 3((x - x')/2z)^2}{[1 + ((x - x')/2z)^2]^3} \, . \tag{17.49}$$

The correlation function at fixed z along the direction parallel to the source boundary is thus long-range with a power law decay $1/((x - x')/2z)^4$ at large $x - x'$ with a characteristic range proportional to z. It is interesting to note that the correlation changes sign for $|x - x'| > (2/\sqrt{3})z$ showing an anti-correlation at large horizontal distances.

17.2.3 Applications

The above calculations suggest that the effect of heterogeneity and disorder at scales that are unresolved may contribute a significant distorsion to large scale averages and thus may be misinterpreted as an anomaly which does not exist.

• The question we ask is the following: what would be the amplitude A_{per} of a periodic modulation of period λ that would give the same field amplitude as the one created by a disorder with amplitude r_0 as defined in expression (17.44)? The answer is given by comparing the solution (17.43) to the standard deviation amplitude (17.36) due to the noise:

$$A_{\mathrm{per}}e^{-2\pi z/\lambda} \approx r_0 \left(\frac{\lambda}{z}\right)^{3/2} \, . \tag{17.50}$$

The exponential term in the l.h.s. provides a strong amplification. Suppose for instance that $z = 3\lambda$. This yields $A_{\mathrm{per}} \approx 3 \times 10^7 r_0$. Thus, even a very small noise amplitude may lead to a strong apparent anomaly. There is no real surprise: this stems from the exponential filter of the Green function (17.46), which makes the problem ill-conditionned with respect to noise. It is notorious that the inversion problem of getting the source from the field measurement is ill-conditioned. This can be seen to result from the form of the eigenfunctions and eigenvalues of the Laplace transform and the similar dilationally invariant Fredholm integral equation [614, 709].

• Another question also relevant to the spurious anomaly problem is what would be the amplitude A_{ano} of a localized source perturbation that could give the same field amplitude at a distance z as the one created by a disorder with amplitude r_0 as defined in expression (17.44)? In 2D, the field created by a localized source is controlled by the Green function (17.26), which also gives the field created by an extended line charge in 3D. A dipole along z at $(x = x_0, z = 0)$ of unit charge and distance l between the positive and negative charges gives the following field

$$l \frac{\partial G(x,z)}{\partial z} = -\frac{l}{2\pi} \frac{z}{(x - x_0)^2 + z^2} . \tag{17.51}$$

For the worst case $x = x_0$, this dipole field decays as $(1/2\pi)(l/z)$ which must be compared to the standard deviation decay $(d/\lambda)(\lambda/z)^{3/2}$ created by a random noise decorating the periodic array of sources. We have used the parameterization $r_0 = d/\lambda$ in (17.50). This yields

$$l = 2\pi \sqrt{\frac{\lambda}{z}} \, d . \tag{17.52}$$

The equivalent localized dipole source is thus at worst of the order of magnitude at the amplitude d of the disorder and becomes weaker and weaker for larger distances $z \gg \lambda$. Thus, this is not a strong effect. If we consider the same question for a localized quadrupole source, the effect is stronger and actually increases with z since the corresponding Green function decays as z^{-2}, faster than $z^{-3/2}$.

Flow Induced by Randomly Vibrating Boundaries. Let us consider the case of a planar boundary that is being vibrated parallel to its own plane. For an infinite solid boundary located at $z = 0$, and an incompressible fluid that occupies the region $z > 0$, the Navier–Stokes equation and boundary conditions are

$$\partial_t u = \nu \partial_z^2 u , \tag{17.53}$$

$$u(0,t) = u_0(t) , \quad u(\infty, t) < \infty , \tag{17.54}$$

where z is the coordinate normal to the boundary, $u(z,t)$ is the x component of the velocity, and $u_0(t)$ is the prescribed velocity of the boundary. The solution for harmonic vibration $u_0(t) = u_0 \cos(\Omega t)$ was given by Stokes [932]. It is a transversal wave that propagates into the bulk fluid with an exponentially decaying amplitude

$$u(z,t) = u_0 e^{-z/\delta_{\text{S}}} \cos(\Omega t - z/\delta_{\text{S}}) , \tag{17.55}$$

where $\delta_{\text{S}} = (2\nu/\Omega)^{1/2}$ is the Stokes layer thickness.

Volfson and Vinals [999] have derived the solution of this problem for the case where the boundary motion is white noise. They introduce the retarded, infinite space Green's function corresponding to (17.53), with boundary conditions (17.54),

$$G(z,t|z',t') = \frac{1}{[4\pi\nu(t-t')]^{1/2}} \left(e^{-(z-z')^2/4\nu(t-t')} - e^{-(z+z')^2/4\nu(t-t')} \right) ,$$

for $t > t'$ and $G(z,t|z',t') = 0$ for $t < t'$. If the fluid is initially quiescent, $u(z,0) = 0$, one finds

$$u(z,t) = \nu \int_0^t dt' \, u_0(t') \, (\partial_{z'} G)_{z'=0} , \qquad (17.56)$$

with

$$(\partial_{z'} G)_{z'=0} = \frac{z}{[4\pi\nu^3(t-t')^3]^{1/2}} \, e^{-z^2/4\nu(t-t')} . \qquad (17.57)$$

Equations (17.56–17.57) determine the transient behavior for any given $u_0(t)$.

If $u_0(t)$ is a Gaussian white noise process, the ensemble average of (17.56) yields $\langle u(z,t) \rangle = 0$. The corresponding equation for the variance reads

$$\langle u^2(z,t) \rangle = 2D\nu^2 \int_0^t dt' \, [(\partial_{z'} G)_{z'=0}]^2 = \frac{2D\nu}{\pi z^2} \left(1 + \frac{z^2}{2\nu t} \right) e^{-z^2/2\nu t} .$$

The variance of the induced fluid velocity propagates into the fluid diffusively. Saturation occurs for $t \gg z^2/2\nu$, at which point the variance does not decay exponentially far away from the wall, but rather as a power law.

$$\langle u^2(z,\infty) \rangle = \frac{2D\nu}{\pi z^2}. \qquad (17.58)$$

Long-range random flows can thus be induced by local boundary vibration. The mechanism is the same as discussed previously in this chapter, namely the fact that white noise is the superposition of all possible monochromatic vibrations: the power law decay is the superposition of all the exponential decays with the continuous spectrum of Stokes layer thicknesses $\delta_S = (2\nu/\Omega)^{1/2}$ corresponding to all possible angular frequencies Ω. Volfson and Vinals [999] show in addition that a breakdown from planarity leads to a continuous component, i.e. a steady streaming of the ensemble average. These results have applications for instance for experiments carried out under low level random acceleration fields that are typical of microgravity environments. Another possible application is to oceanic currents induced by turbulent winds.

Satellite Gravity Altimetry. Radar altimeter measurements of the marine geoid gives the possibility of uncovering the gravity field over all the ocean basins. The combination of two high-density data sets obtained by ERS-1 during its geodetic mapping phase (April 1994–March 1995) and the Geosat altimeter data from the U.S. Navy provided the first global view of all the ocean basins at a wavelength resolution of 20–30 km. There are many physical limitations of satellite altimetry that limit the resolution to wavelengths of about 10 km and to amplitudes of 10 mGal [394, 1032]. The report of the National Research Council [665] explores the scientific questions that could be addressed with a better global gravity field, and in particular the new access to dynamical properties of earth processes, as in ocean dynamics, continental water variation, sea-level rise and glaciology, solid-earth processes and the dynamic atmosphere.

Can satellite-derived gravity measurements be improved using more measurements or better processing? How well does satellite-derived gravity compare with more accurate local surveys, especially near land? Because the range in crustal density contrasts is fairly limited (500 kg/m^3), the effects are generally below the resolution limit of standard gravity measurements. The effect of disorder is also a factor that has to be taken into account.

In a series of papers, Maus and Dimri [608–610] have performed calculations of the theoretical power spectrum of the three-dimensional potential field caused by an arbitrary three-dimensional source distribution, with applications to gravity and magnetic data. These calculations are similar to those presented above with the spectral method. Starting from a scale-invariant source density with spectrum $\propto k^{-\beta}$, they find that the power spectrum of the gravity and magnetic field is anisotropic and that a specific scaling exponent exists only for lower-dimensional cross-sections of the fields. The scaling exponent of the density distribution and of the gravity field are related by

$$\beta_{\text{dens}}^{\text{3D}} = \beta_{\text{dens}}^{\text{2D}} + 1 = \beta_{\text{dens}}^{\text{1D}} + 2 \tag{17.59}$$

$$= \beta_{\text{field}}^{xy} - 1 = \beta_{\text{field}}^{x} = \beta_{\text{field}}^{y} = \beta_{\text{field}}^{z} \ . \tag{17.60}$$

The relationship between the scaling exponents of the susceptibility distribution and the magnetic field reduced to the pole can be stated as

$$\beta_{\text{susc}}^{\text{3D}} = \beta_{\text{susc}}^{\text{2D}} + 1 = \beta_{\text{susc}}^{\text{1D}} + 2 = \beta_{\text{field}}^{xy} + 1 \tag{17.61}$$

$$= \beta_{\text{field}}^{x} + 2 = \beta_{\text{field}}^{y} + 2 = \beta_{\text{field}}^{z} + 2 \ . \tag{17.62}$$

In [609, 610], Maus and Dimri propose a method for inverting the power spectrum of gravity and magnetic data and they test it on aeromagnetic and borewell data from the German Continental Deep Drilling Project. They model the sources of the potential field by a random function with scaling properties with spectrum $\propto k^{-\beta}$, defined on a half-space with its top at a specified depth beneath the observation plane. Comparing the theoretical power spectrum for this model with the power spectrum of the measured data, they obtain the best values for the depth to source and the scaling exponent as a global minimum of the misfit function. The most important result is that the low-wavenumber part of the power spectrum can be dominated by the scaling properties of the source distribution and not by the depth to some kind of reference level. It is found that the scaling exponent of the field varies with the type of surface geology and Maus and Dimri conjecture that the scaling exponent could be used to identify different types of geological structures.

Magnetic Field of the Earth. In 1981, Harrison and Carle [405] and Shure and Parker [843] presented two opposing interpretations of the power spectrum of the magnetic field of the earth. The spectra they discussed were calculated from data collected along three long ship traverses in the Pacific and Atlantic Oceans. Harrison and Carle argued that the observed spectra contained additional intermediate wavelength (400–4000 km) energy not predicted by a crustal model that only contains the sea-floor spreading magnetization polarity reversal pattern. Shure and Parker argued that this additional intermediate wavelength energy is an artifact of the fact that the ship profiles provide only a 1-D sample whereas the profiles may well have traversed 2-D structures obliquely. LaBreque et al. [545, 546] tried to avoid the problems identified by Shure and Parker [843] by using data from many profiles with different orientations. They showed that intermediate wavelength energy does exist and they suggested that additional crustal sources needed to be considered to explain the spectra as Harrison and Carle had proposed in 1981. From the above analysis, the random component in the structure of a source layer can be important in the interpretation of structures. It can lead to the appearance or enhancement of longer wavelengths in the field. This may lead to either an error in the location or in the strength of the source. See [138, 724].

References

1. Abramowitz, E. and Stegun, I.A. (1972) *Handbook of Mathematical Functions* (Dover Publications, New York).
2. Adamic, L.A. and Huberman, B.A. (1999) The nature of markets in the WWW, preprint http://www.parc.xerox.com/istl/groups/iea/www/novelty.html
3. Adler, A., Feldman, R. and Taqqu, M.S., Eds. (1998) *A Practical Guide to Heavytails: Statistical Techniques for Analyzing Heavy Tailed Distributions* (Birkhauser, Boston).
4. D'Agostini, G. (1995) Probability and Measurement Uncertainty in Physics – a Bayesian Primer, preprint http://xxx.lanl.gov/abs/hep-ph/9512295.
5. Aharony, A. (1983) "Tricritical phenomena", in *Critical Phenomena*, ed. by Hahne, F.J., *Lecture Notes in Physics*, Vol. 186 (Springer, Berlin, Heidelberg) p. 209.
6. Aharony, A. (1989) Critical properties of random and constrained dipolar magnets, *Phys. Rev. B* **12**, 1049–1056.
7. Aharony, A. and Feder, J., eds. (1989) *Fractals in Physics, Physica D* **38**, nos. 1–3 (North Holland, Amsterdam).
8. Aitcheson, J. and Brown, J.A.C. (1957) *The Log-Normal Distribution* (Cambridge University Press, London, England).
9. Aki, K. (1965) Maximum likelihood estimate of b in the formula $\log N = a - bm$ and its confidence limits, *Bull. Earthquake Res. Inst. Tokyo Univ.* **43**, 237–239.
10. Aki, K. and Richards, P.G. (1980) *Quantitative Seismology* (Freeman, San Francisco).
11. Albano, E.V. (1992) Critical exponents for the irreversible surface reaction $A + B \to AB$ with B desorption on homogeneous and fractal media, *Phys. Rev. Lett.* **69**, 656–659.
12. Albert, R., Jeong, H. and Barabasi, A.L. (1999) The diameter of the world wide web, *Nature* **401**, 130–131.
13. Alemany, P.A. and Zanette, D.H. (1994) Fractal random walks from a variational formalism for Tsallis entropies, *Phys. Rev. E* **49**, R956–R958.
14. Allègre, C.J., Le Mouel, J.L. and Provost, A. (1982) Scaling rules in rock fracture and possible implications for earthquake predictions, *Nature* **297**, 47–49.
15. Alexander, F.J. and Eyink, G.L. (1997) Rayleigh–Ritz calculation of effective potential far from equilibrium, *Phys. Rev. Lett.* **78**, 1–4.
16. Alstrom, P. (1988), Mean-field exponents for self-organized critical phenomena, *Phys. Rev. A* **38**, 4905–4906.
17. Alstrom, P. (1990) Self-organized criticality and fractal growth, *Phys. Rev. A* **41**, 7049–7052.

18. Altarelli, M., Núñez-Regueiro, M.D. and Papoular, M. (1995) Coexistence of two length scales in X-ray and neutron critical scattering: a theoretical interpretation, *Phys. Rev. Lett.* **74**, 3840–3843.
19. Amit, D.J. (1984) *Field Theory, the Renormalization Group, and Critical Phenomena*, 2nd ed. (World Scientific, Singapore).
20. An, L.-J. and Sammis, C.G. (1994) Particle size distribution of cataclastic fault materials from southern California: a 3-D study, *Pure and Applied Geophysics* **143**, 203–227.
21. Andersen, J.V., Gluzman, S., and Sornette, D. (2000) Fundamental Framework for Technical Analysis, *Eur. Phys. J.* B **14**, 579–601.
22. Andersen, J.V., Sornette, D. and Leung, K.-T. (1997) Tri-critical behavior in rupture induced by disorder, *Phys. Rev. Lett.* **78**, 2140–2143.
23. Anderson, D.L., The San Andreas fault (1971) *Sci. Am.* **225**, 52–68.
24. Anderson, D.L. (2002) How many plates? *Geology* **30**, 411–414.
25. Anderson, P.W. (1958) Absence of diffusion in certain random lattices, *Phys. Rev.* **109**, 1492–1505.
26. Anderson, P.W. (1972) More is different (Broken symmetry and the nature of the hierarchical structure of science), *Science* **177**, 393–396.
27. Anderson, P.W. (1978) in Ill-Condensed Matter III, Les Houches, edited by Balian, R., Maynard, R. and Toulouse, G. (North Holland, Amsterdam and World Scientific, Singapore).
28. Andreotti, B. and Douady, S. (1999) On probability distribution functions in turbulence. Part 1. A regularisation method to improve the estimate of a PDF from an experimental histogram, *Physica D* **132**, 111–132.
29. Anifrani, J.-C., Le Floc'h, C., Sornette, D. and Souillard, B. (1995) Universal log-periodic correction to renormalization group scaling for rupture stress prediction from acoustic emissions, *J. Phys. I France* **5**, 631–638.
30. Anselmet, F., Gagne, Y., Hopfinger, E.J. and Antonia, R.A. (1984) High-order velocity structure functions in turbulent shear flows, *J. Fluid Mech.* **140**, 63–89.
31. Arnéodo, A., Bacry, E., Graves, P.V. and Muzy, J.-F. (1995) Characterizing long-range correlations in DNA sequences from wavelet analysis, *Phys. Rev. Lett.* **74**, 3293–3296.
32. Arnéodo, A., Argoul, F., Bacry, E., Elezgaray, J. and Muzy, J.-F. (1995) *Ondelettess, multifractales et turbulences* (Diderot Editeur, Arts et Sciences, Paris).
33. Arnéodo, A., Bouchaud, J.-P., Cont, R., Muzy, J.-F., Potters, M. and Sornette, D. (1996) Comment on "Turbulent cascades in foreign exchange markets", (reply to Ghashghaie, S., Breymann, W., Peinke, J., Talkner, P. and Dodge, Y. Nature **381** 767 (1996).), unpublished preprint at http://xxx.lanl.gov/abs/cond-mat/9607120.
34. Arnéodo, A., Muzy, J.-F. and Sornette, D. (1998) "Direct" causal cascade in the stock market, *Eur. Phys. J.* B **2**, 277–282.
35. Arnol'd, V. (1988) *Geometrical Methods in the Theory of Ordinary Differential Equations*, 2nd ed. (Springer, New York).
36. Arous, G.B., Bovier, A., and Gayrard, V. (2002) Aging in the Random Energy Model, *Phys. Rev. Lett.* **88**, 0872012.
37. Asmussen, S. and Klüppelberg, C. (1996) Large deviations results for subexponential tails with applications to insurance risks, *Stochast. Proc. and Their Appl.* **64**, 103–125.
38. Auerbach, F. (1913) Das Gesetz der Bevlkerungskonzentration, *Petermanns Geographische Mitteilungen* **LIX**, 73–76.

39. Aukrust, T., Browne, D.A. and Webman, I. (1989) Kinetic phase transition in a one-component irreversible reaction model, *Europhys. Lett.* **10**, 249–255.

40. Aukrust, T., Browne, D.A. and Webman, I. (1990) Critical behavior of an autocatalytic reaction model, *Phys. Rev. A* **41**, 5294–5301.

41. Bacry, E., Delour, J. and Muzy, J.-F. (2001) Multifractal random walk, *Phys. Rev. E* **64**, 026103.

42. Bagnoli, F., Palmerini, P. and Rechtman, R. (1997) Algorithmic mapping from criticality to self-organized criticality, *Phys. Rev. E* **55**, 3970–3976.

43. Baillie, R.T. (1996) Long memory processes and fractional integration in econometrics, *Journal of Econometrics* **73**, 5–59.

44. Bak, P. (1996) *How Nature Works: the Science of Self-organized Criticality* (Copernicus, New York).

45. Bak, P. and Tang, C. (1989) Earthquakes as a self-organized critical phenomenon, *J. Geophys. Res.* **94**, 15,635–15,637.

46. Bak, P., Tang, C. and Weisenfeld, K. (1987) Self-organized criticality: An explanation of $1/f$ noise, *Phys. Rev. A* **38**, 364–374.

47. Bak, P., Chen, K. and Tang, C. (1990) A forest-fire model and some thought on turbulence, *Phys. Lett. A* **147**, 297–300.

48. Bak, P. and Sneppen, K. (1993) Punctuated equilibriun and criticality in a simple model of evolution, *Phys. Rev. Lett.* **71**, 4083–4086.

49. Baker, G.A. Jr. and Graves-Moris P. (1996) *Padé-Approximants* (Cambridge University Press, Cambridge).

50. Balasubramanian, V. (1997) Statistical inference, Occam's razor, and statistical mechanics on the space of probability distributions, *Neural Computation* **9**, 349–368.

51. Banik, S.K., Chaudhuri, J.R. and Ray, D.S. (2000) The generalized Kramers' theory for nonequilibrium open systems, *J. Chem. Phys.* **112**, 8330–8337

52. Bántay, P. and Jánosi, I.M. (1992) Avalanche dynamics from anomalous diffusion, *Phys. Rev. Lett.* **68**, 2058–2061.

53. Barber, M.N. and Ninham, B.W. (1970) *Random walks and restricted walks* (Gordon and Breach, New York).

54. Barenblatt, G.I. (1987) *Dimensional Analysis* (Gordon and Breach, New York).

55. Barenblatt, G.I. (1996) *Scaling, Self-similarity, and Intermediate Asymptotics* (Cambridge University Press, New York).

56. Barenblatt, G.I. and Goldenfeld, N. (1995) Does fully developed turbulence exist? Reynolds number independence versus asymptotic covariance, *Phys. Fluids* **7**, 3078–3082.

57. Barkai, E., Metzler, R. and Klafter, J. (2000) From continuous time random walks to the fractional Fokker–Planck equation, *Phys. Rev. E* **61**, 132–138.

58. Barnsley, M. (1988) *Fractals everywhere* (Academic Press, Boston).

59. Barrat, A., Vespignani, A. and Zapperi, S. (1999) Fluctuations and correlations in sandpile models, *Phys. Rev. Lett.* **83**, 1962–1965.

60. Barton, C.C. and La Pointe P.R., eds. (1995) *Fractals in the Earth Sciences* (Plenum Press, New York and London).

61. Barton, C.C. and La Pointe P.R., eds. (1995) *Fractals in petroleum geology and earth processes* (Plenum Press, New York and London).

62. Beck, C. (2001) Dynamical foundations of nonextensive statistical mechanics, *Phys. Rev. Lett.* **87**, 180601.

63. Beck, C. (2002) Generalized statistical mechanics and fully developed turbulence, *Physica A* **306**, 189–198.

64. Beck, C. (2002) Nonextensive methods in turbulence and particle physics, *Physica A* **305**, 209–217.

65. Beck, C. (2002) Non-additivity of Tsallis entropies and fluctuations of temperature, *Europhys. Lett.* **57**, 329–333.
66. Beck, C. (2003) Superstatistics in hydrodynamic turbulence (http://arXiv. org/abs/physics/0303061)
67. Beck, C. (2003) Superstatistics: Theory and Applications (http://arXiv.org/ abs/cond-mat/0303288)
68. Beck, C. and Cohen, E.G.D. (2003) Superstatistics, *Physica* **322A**, 267–275.
69. Bender, C. and Orszag, S.A. (1978) *Advanced Mathematical Methods for Scientists and Engineers* (McGraw-Hill, New York).
70. Beney, P., Droz, M. and Frachebourg, L. (1990) On the critical behavior of cellular automata models of non-equilibrium phase transition, *J. Phys. A* **23**, 3353–3359.
71. Benson, D., Wheatcraft, S. and Meerschaert, M. (2000) Application of a fractional advection-dispersion equation, *Water Resour. Res.* **36**, 1403–1412.
72. Benson, D., Wheatcraft, S. and Meerschaert, M. (2000) The fractional-order governing equation of Lévy motion, *Water Resour. Res.* **36**, 1413–1424.
73. Benson, D., Wheatcraft, S. and Meerschaert, M. (2001) Fractional dispersion, Lévy motion and the MADE tracer tests, *Transport in Porous Media* **42**, 211–240.
74. Benzi, R., Ciliberto, S., Tripiccione, R., Baudet, C. et al. (1993) Extended self-similarity in turbulent flows, *Phys. Rev. E* **48**, R29–R32.
75. Beran, J. (1994) *Statistics for Long-Memory Processes* (Chapman & Hall, New York).
76. Berezin, A.A. (1996) More spirited debate on Physics, Parapsychology and paradigms, *Physics Today* **April**, 15–15.
77. Bergé, P., Pomeau, Y. and Vidal, C. (1984) *Order within Chaos* (Wiley, New York).
78. Berger, A.S. and Berger, J. (1991) *The encyclopedia of Parapsychology and Psychical Research.* 1st ed. (Paragon House, New York).
79. Bergman, B. (1986) Estimation of Weibull parameters using a weight function, *Journal of Materials Science Letters* **5**, 611–614.
80. Bergman et al. (1985) *Les méthodes de l'homogeneisation : théorie et applications en physique* (Paris : Eyrolles, Series title: Collection de la Direction des études et recherches d'Electricité de France 57).
81. Berkowitz, B., and Scher, H. (1997) Anomalous transport in random fracture networks, *Phys. Rev. Lett.* **79**, 4038–4041.
82. Berkowitz, B., and Scher, H. (1998) Theory of anomalous chemical transport in random fracture networks, *Phys. Rev. E* **57**, 5858–5869.
83. Bernardo, J.M. and Smith, A.F.M. (1994) *Bayesian Theory* (Wiley, Chichester).
84. Bernasconi, J., Schneider, W.R. and Wyss, W. (1980) Diffusion and hopping conductivity in disordered one-dimensional lattice systems, *Zeitschrift fur Physik B* **37**, 175–184.
85. Berry, M.V. (1982) Universal power-law tails for singularity-dominated strong fluctuations, *J. Phys. A* **15**, 2735–2749.
86. Bertsekas, D.P. (1982) *Constrained optimization and Lagrange multiplier methods* (Academic Press, New York).
87. Bialek, W., Callan, C.G. and Strong, S.P. (1996) Field Theories for Learning Probability Distributions, *Phys. Rev. Lett.* **77**, 4693–4697.
88. Biham, O., Malcai, O., Lidar, D.A. and Avnir, D. (1998) Is nature fractal? – Response, *Science* **279**, 785–786.
89. Biham, O., Malcai, O., Lidar, D.A. and Avnir, D. (1998) Fractality in nature – Response, *Science* **279**, 1615–1616.

90. Binder, K. (1981) Finite size scaling analysis of Ising model block distribution functions, *Zeitschrift fr Physik B* **43**, 119–140.

91. Binder, K. and Young, A.P. (1986) Spin glasses: experimental facts, theoretical concepts, and open questions, *Reviews of Modern Physics* **58**, 801–976.

92. Bird, P. (2003) An updated digital model of plate boundaries, *Geochemistry, Geophysics, Geosystems* **4**(3), 1027, doi:10.1029/2001GC000252.

93. Bird, P. and Rosenstock, R.W. (1984) Kinematics of present crust and mantle flow in southern California, *Geol. Soc. Am. Bull.* **95**, 946–957.

94. Blanchard, Ph., Cessac, B. and Krüger, T. (1997) A dynamical systems approach for SOC models of Zhang's type, *J. Stat. Phys.* **88**, 307–318.

95. Blanchard, Ph., Cessac, B. and Krüger, T. (2000) What can one learn about Self-Organized Criticality from Dynamical Systems theory? *J. Stat. Phys.* **98**, 375–404

96. Boettcher, S. and Paczuski, M. (1996) Ultrametricity and memory in a solvable model of self-organized criticality, *Phys. Rev. E* **54**, 1082–1095.

97. Bogolubov, N.N. and Shirkov, D.V. (1983) *Quantum Fields* (Benjamin, London).

98. Bohr, T., van Hecke, M., Mikkelsen, R. and Ipsen, M. (2001) Breakdown of universality in transitions to spatiotemporal chaos, *Phys. Rev. Lett.* **86**, 5482–5485.

99. Bollobas, B., Riordan, O., Spencer, J. and Tusnady, G. (2001) The degree sequence of a scale-free random process, *Random Structures and Algorithms* **18**(3), 279–290.

100. Bonabeau, E., Dagorn, L. and Freon, P. (1999) Scaling in animal group-size distributions, *Proc. Nat. Acad. Sci, USA* **96**, 4472–4477.

101. Bookstein, A. (1990) Informetric distributions, Part I: Unified overview, *Journal of the American Society for Information Science* **41**(5), 368–375.

102. Borkowski, J. (1984) *The structure of Turbulence in the Surface Layer of the Atmosphere* (Warszawa-Lodz, Panstwowe Wydawn. Nauk.).

103. Borovkov, A.A. (1998) *Mathematical Statistics* (Overseas Publishers Association, Gordon and Breach Science Publishers, Amsterdam).

104. Bovier, A., Kurkova, I. and Löwe, M. (2002) Fluctuations of the free energy in the REM and the p-spin SK models, *Ann. Probab.* **30**, 605–651.

105. Bottani, S. (1995) Pulse-coupled relaxation oscillators – From biological synchronization to self-organized criticality, *Phys. Rev. Lett.* **74**, 4189–4192.

106. Bouchaud, E. and Bouchaud, J.-P. (1992) Long-range stress field produced by a planar assembly of edge dislocations – the effect of disorder, *Philos. Mag. Lett.* **65**, 339–343.

107. Bouchaud, E., Bouchaud, J.-P., Naka, S., Lepasset, G. and Octor, H. (1992) Dissolution of precipitates by migrating grain boundaries, *Acta metall. mater.* **40**, 3451–3458.

108. Bouchaud, E., Lapasset, G., Planès, J. and Naveos, S. (1993) Statistics of branched fracture surfaces, *Phys. Rev. B* **48**, 2917–2928.

109. Bouchaud, J.-P. (1992) Weak ergodicity breaking and aging in disordered systems, *J. Physique (Paris)* **2**, 1705–1713.

110. Bouchaud, J.P., Bouchaud, E., Lapasset, G. and Planès, J. (1993) Models of fractal cracks, *Phys. Rev. Lett.* **71**, 2240–2243.

111. Bouchaud, J.-P. and Georges, A. (1990) Anomalous diffusion in disordered media: Statistical mechanisms, models and physical applications, *Physics Reports* **195**, 127–293.

112. Bouchaud, J.-P., Georges, A., Koplik, J., Provata, A. and Redner, S. (1990) Superdiffusion in random velocity fields, *Phys. Rev. Lett.* **64**, 2503–2506.

113. Bouchaud, J.-P. and Mézard, M. (1994) Self-induced quenched disorder – A model for the glass transition, *J. Phys. I (Paris)* **4**, 1109–1114.

114. Bouchaud, J.P. and Mézard, M. (1997) Universality classes for extreme-value statistics, *J. Phys. A Math. Gen.* **30**, 7997–8015.

115. Bouchaud, J.-P., Sornette, D., Walter, C., and Aguilar, J.-P. (1998) Taming large events: Optimal portfolio theory for strongly fluctuating assets, *International Journal of Theoretical and Applied Finance* **1**, 25–41.

116. Bowman, D.D., Ouillon, G., Sammis, C.G., Sornette, A. and Sornette, D. (1998) An Observational test of the critical earthquake concept, *J. Geophys. Res.* **103**, 24359–24372.

117. Boyanovsky, D. and Cardy, J.L. (1982) Critical behavior of m-component magnets with correlated impurities, *Phys. Rev. B* **26**, 154–170.

118. Boyanovsky, D., de Vega, H.J. and Holman, R. (1999) Non-equilibrium phase transitions in condensed matter and cosmology: spinodal decomposition, condensates and defects, Lectures delivered at the NATO Advanced Study Institute: Topological Defects and the Non-Equilibrium Dynamics of Symmetry Breaking Phase Transitions.

119. Boyer, D., Tarjus, G., and Viot, P. (1995) Shattering transition in a multivariate fragmentation model, *Phys. Rev. E* **51**, 1043–1046.

120. Bradley, C.E., and Price, T. (1992) Graduating sample data using generalized Weibull functions, *Applied Mathematics and Computation* **50**, 115–144.

121. Bradley, R.M. and Wu, K. (1994) Dynamic fuse model for electromigration failure of polycrystalline metal films, *Phys. Rev. E* **50**, R631–R634.

122. Bradley, R.M. and Wu, K. (1994) Crack propagation in a dynamic fuse mode of electromigration, *J. Phys. A* **27**, 327–333.

123. Bray, A.J. (1994) Theory of phase-ordering kinetics, *Advances in Physics* **43**, 357–459.

124. Bray, A.J. and Moore, M.A. (1987) Chaotic nature of the spin-glass phase, *Phys. Rev. Lett.* **58**, 57–60.

125. Breiman, L. (1965) On some limit theorems similar to the arc-sin law, *Theory of Probability and its Applications* **10**, 323–329.

126. Brewer, J.W. and Smith, M.K., eds. (1981) *Emmy Noether: a tribute to her life and work* (New York : M. Dekker).

127. Bruce, A.D. (1995) Critical finite-size scaling of the free energy *J. Phys. A* **28**, 3345–3349 (1995).

128. Burridge, R. and Knopoff, L. (1967) Model and theoretical seismicity, *Bull. Seismo. Soc. Am.* **57**, 341–371.

129. Cacciari, M., Mazzanti, G. and Montanari, G.C. (1996) Comparison of Maximum Likelihood unbiasing methods for the estimation of the Weibull parameters, *IEEE Transactions on Dielectrics and Electric Insulation* **3**, 18–27.

130. Cafiero, R., Loreto, V., Pietronero, L., Vespignani, A. and Zapperi, S. (1995) Local rigidity and self-organized criticality for avalanches, *Europhys.Lett.* **29**, 111–116.

131. Callen, H.B. (1985) *Thermodynamics and an introduction to thermostatistics.* 2nd ed. (Wiley, New York).

132. Callen, E. and Shapero, D. (1974) A theory of social imitation, *Physics Today* **July**, 23–28.

133. Cannelli, G., Cantelli, R. and Cordero, F. (1993) Self-organized criticality of the fracture processes associated with hydrogen precipitation in Niobium by acoustic emission, *Phys. Rev. Lett.* **70**, 3923–3926.

134. Caponeri, M. and Ciliberto, S. (1992) Thermodynamics aspects of the transition to spatio-temporal chaos, *Physica D* **58**, 365–383.

135. Caponeri, C.A., Douady, S., Fauve, S. and Laroche, C. (1995) in "Mobile Particulate systems", *Nato Asi Series* **287**, 19 (Ed. E. Guazzelli and Oger, L., Kluwer).

136. Cardy, J., and Sugar, R.L. (1980) Directed percolation and Reggeon field theory, *J. Phys. A* **13**, L423–L427.

137. Cardy, J. (1999) Critical exponents near a random fractal boundary, *J. Phys. A* **32**, L177–L182.

138. Caress, D.W. and Parker, R.L. (1989) Spectral interpolation and downward continuation of marine magnetic and anomaly data, *J. Geophys. Res.* **94**, 17393–17407.

139. Carlson, J.M., Chayes, J.T., Grannan, E.R. and Swindle, G.H. (1990) Self-organized criticality and singular diffusion, *Phys. Rev. Lett.* **65**, 2547–2550.

140. Carlson, J.M. and Doyle, J. (1999) Highly optimized tolerance: A mechanism for power laws in designed systems, *Phys. Rev. E* **60**, 1412–1427.

141. Carlson, J.M., Langer, J.S., Shaw, B. and Tang, C. (1991) Intrinsic properties of the Burridge–Knopoff model of a fault, *Phys. Rev. A* **44**, 884–897.

142. Caroli, C. and Nozières, P. (1995) in "The physics of friction", Peisson, B. ed.

143. Carter, N.L. and Tsenn, M.C. (1987) Flow properties of continental lithosphere, *Tectonophysics* **136**, 27–63.

144. Castaing, B., Gagne, Y. and Hopfinger, E. (1990) Velocity probability density functions of high Reynolds number turbulence, *Physica D* **46**, 177.

145. Castaing, B. (1996) The temperature of turbulent flows, *J. Phys. II France* **6**, 105–114.

146. Castaing, B. (1997), pp. 225–234 in [251].

147. Cardy, J.L., ed. (1988) *Finite-size Scaling* (North-Holland, Amsterdam).

148. Castillo, E. (1988) *Extreme Value Theory in Engineering* (Academic Press, Boston).

149. Caves, C.M. (2000) Predicting future duration from present age: a critical assessment, *Contemporary Physics* **41**, 143–153.

150. Caves, C.M., Fuchs, C.A. and Schack, R. (2002) Quantum probabilities as Bayesian probabilities, *Phys. Rev.* **65**, 022305, DOI: 10.1103/PhysRevA.65.022305.

151. Cessac, B., Blanchard, Ph. and Krüger, T. (1998) A dynamical system approach to Self-Organized Criticality, *Mathematical results in Statistical Mechanics*, Marseille (Word Scientific Singapore).

152. Cessac, B., Blanchard, P. and Kruger, T. (2001) Lyapunov exponents and transport in the Zhang model of self-organized criticality, *Phys. Rev. E* **64**, 016133.

153. Cessac, B. and Meunier, J.-L. (2002) Anomalous scaling and Lee–Yang zeros in self-organized criticality, *Phys. Rev. E* **65**, 036131.

154. Chabanol, M.-L and Hakim, V. (1997) Analysis of a dissipative model of self-organized criticality with random neighbors, *Phys. Rev. E* **56**, R2343–R2346.

155. Chabanol, M.-L., Hakim, V. and Rappel, W.-J. (1997) Collective chaos and noise in the globally coupled complex Ginzburg-Landau equation, *Physica D* **103**, 273–293.

156. Chaitin, G.J. (1987) *Algorithmic Information Theory* (Cambridge University Press, Cambridge and New York).

157. Chandrasekhar, B.K. (1997) Estimation of Weibull parameter with a modified weight function, *Journal of Materials Research* **12**, 2638–2642.

158. Charmet, J.C., Roux, S. and Guyon, E. eds. (1990) *Disorder and Fracture*. NATO ASI Series B: Physics Vol. 235 (Plenum Press, New York and London).
159. Chaté, H. and Manneville, P. (1989) Role of defects in the transition to turbulence via spatiotemporal intermittency, *Physica D* **37**, 33–41.
160. Chaté, H. and Manneville, P. (1990) Criticality in cellular automata, *Physica D* **45**, 122–135.
161. Chauvin, R. (1985) *Parapsychology: When the Irrational Rejoins Science* (Jefferson, N.C., McFarland).
162. Chavanis, P.-H. (2002) Effective velocity created by a point vortex in two-dimensional hydrodynamics, *Phys. Rev. E* **65**, 056302.
163. Chavanis, P.-H. and Sire, C. (2000) The statistics of velocity fluctuations arising from a random distribution of point vortices: the speed of fluctuations and the diffusion coefficient, *Phys. Rev. E* **62**, 490–506.
164. Chaves, A.S. (1998) A fractional diffusion equation to describe Lévy flight, *Phys. Lett. A* **239**, 13–16.
165. Chen, J.-H. and Lubensky, T.C. (1977) Mean field and ε-expansion study of spin glasses, *Phys. Rev. B* **16**, 2106–2114.
166. Chen, K., Bak, P. and Okubov, S.P. (1991) Self-organized criticality in crack-propagation model of earthquakes, *Phys. Rev. A* **43**, 625–630.
167. Cheng, Z., and Redner, S. (1988) Scaling theory of fragmentation, *Phys. Rev. Lett.* **60**, 2450–2453.
168. Chessa, A., Marinari, E., Vespignani, A. and Zapperi, S. (1998) Mean-field behavior of the sandpile model below the upper critical dimension, *Phys. Rev. E* **57**, R6241–R6244.
169. Chessa, A., Stanley, H.E., Vespignani, A. and Zapperi, S. (1999) Universality in sandpiles, *Phys. Rev. E* **59**, R12–R15.
170. Chhabra, A., Feigenbaum, M., Kadanoff, L., Kolan, A. and Procaccia, I. (1993) Sandpiles, avalanches and the statistical mechanics of non-equilibrium stationary states, *Phys. Rev. E* **47**, 3099–3121.
171. Choy, T.C. (1999) *Effective medium theory : principles and applications* (New York: Oxford University Press).
172. Christensen, K. (1992) Self-organization in Models of Sandpiles, Earthquakes and Flashing Fireflies (PhD Thesis, Oslo University).
173. Christensen, K. and Olami, Z. (1992) Variation of the Gutenberg–Richter b values and nontrivial temporal correlations in a spring-block model for earthquakes, *J. Geophys. Res.* **97**, 8729–8735.
174. Clark, R.M., Cox, S.J.D. and Laslett, G.M. (1999) Generalizations of power-law distributions applicable to sampled fault-trace length: model choice, parameter estimation and caveats, *Geophys. J. Int.* **136**, 357–372.
175. Cline, D.B.H. and Resnick, S.J. (1992) Multivariate subexponential distributions, *Stochastic Processes and their Applications* **42**, 49–72.
176. Cohen, A.C., Maximum likelihood estimation in the Weibull distribution based on complete and on censored samples, *Technometrics* **7**, 579–588.
177. Coleman, P.H. and Pietronero, L. (1992) The fractal structure of the Universe, *Physics Reports* **213**, 311–389.
178. Compte, A. (1996) Stochastic foundations of fractional dynamics, *Phys. Rev. E* **53**, 4191–4193.
179. Compte, A. (1997) Continuous time random walks on moving fluids, *Phys. Rev. E* **55**, 6821–6831.
180. Coniglio, A. and Klein, W. (1980) Clusters and Ising critical droplets: a renormalisation group approach, *J. Phys. A* **13**, 2775–2780.

181. Cornelius, R.R. and Voight, B. (1995) Graphical and PC-Software analysis of volcano eruption precursors according to the materials failure forecast method (FFM), *J. Volcanology and Geothermal Research* **64**, 295–320.
182. Corral, A., Pérez, C.J. and Díaz-Guilera, A. (1997) Self-organized criticality induced by diversity, *Phys. Rev. Lett.* **78**, 1492–1495.
183. Corral, A., Pérez, C.J., Díaz-Guilera, A. and Arenas, A. (1995) Self-organized criticality and synchronization in a lattice model of integrate-and-fire oscillators, *Phys. Rev. Lett.* **74**, 118–121.
184. Cowie, P., Sornette, D. and Vanneste, C. (1995) Multifractal scaling properties of a growing fault population, *Geophys. J. Int.* **122**, 457–469.
185. Cowie, P., Vanneste, C. and Sornette, D. (1993) Statistical physics model for the spatio-temporal evolution of faults, *J. Geophys. Res.* **98**, 21809–21821.
186. Cramer, H. (1946) *Mathematical Methods of Statistics* (Princeton University Press, Princeton).
187. Crane, T. (1996) Soul-searching – Human nature and supernatural belief – N. Humphrey, *Nature* **379**, 685–685.
188. Crave, A. and Davy, P. (1997) Scaling relationships and channel networks from two large-magnitude watersheds in French Brittany, *Tectonophysics* **269**, 91–111.
189. Crisanti, A., Jensen, M.H., Vulpiani, A. and Paladin G. (1992) Strongly intermittent chaos and scaling in an earthquake model, *Phys. Rev. A* **46**, R7363–R7366.
190. Cross, M.C. and Hohenberg, P.C. (1993) Pattern formation outside of equilibrium, *Rev. Mod. Phys.* **65**, 851–1112.
191. Cross, M.C. and Hohenberg, P.C. (1994) Spatio-temporal chaos, *Science* **263**(N5153), 1569–1570.
192. Crutchfield, J.P. and Mitchell, M. (1995) The evolution of emergent computation, *Proc. Nat. Acad. Sci. U.S.A.* **92**, 10742–10746.
193. D'Agostini, G. (1995) *Probability and Measurement Uncertainty in Physics – a Bayesian Primer*, Notes based on lectures to the DESY summer students (Hamburg, Germany), preprint hep-ph/9512295.
194. Daguier, P., Henaux, S., Bouchaud, E. and Creuzet, F. (1996) Quantitative analysis of a fracture surface by atomic force microscopy, *Phys. Rev. E* **53**, 5637–5642.
195. Dahmen, K.A. and Sethna, J.P. (1993) Hysteresis loop critical exponents in $6 - \epsilon$ dimensions, *Phys. Rev. Lett.* **71**, 3222–3225.
196. Dahmen, K.A. and Sethna, J.P. (1996) Hysteresis, avalanches and disorder-induced critical scaling – A renormalization group approach, *Phys. Rev. B* **53**, 14872–14905.
197. Darwin, C. (1910) *The Origin of Species by Means of Natural Selection, 6th ed.* (John Murray, London).
198. Davy, P. (1993) On the frequency-length distribution of the San-Andreas fault system, *J. Geophys. Res.* **98**, 12141–12151.
199. Davy, P. and Cobbold, P.R. (1988) Indentation tectonics in nature and experiments. 1. Experiments scaled for gravity, *Bull. Geol. Inst. Uppsala, N.S.* **14**, 129–141.
200. Davy, P. and Cobbold, P.R. (1991) Experiments on shortening of four-layer model of the continental lithosphere, *Tectonophysics* **188**, 1–25.
201. Davy, P., Sornette, A. and Sornette D. (1990) Some consequences of a proposed fractal nature of continental faulting, *Nature* **348**, 56–58.
202. Davy, P., Sornette, A. and Sornette D. (1992) Experimental discovery of scaling laws relating fractal dimensions and the length distribution exponent of fault systems, *Geophys. Res. Lett.* **19**, 361–364.

203. de Arcangelis, L., Hansen, A., Herrmann, H.J. and Roux, S. (1989) Scaling laws in fracture, *Phys. Rev. B* **40**, 877–880.

204. de Arcangelis, L., Redner, S. and Herrmann, H.J. (1985) A random fuse model for breaking processes, *J. Phys. Lett. France* **46**, L585–L590.

205. de Calan, C., Luck, J.-M., Nieuwenhuizen, T.M. and Petritis, D. (1985) *J. Phys. A* **18**, 501–523.

206. de Finetti, B. (1990) *Theory of Probability* (Wiley, New York).

207. De Martino, A. (1998) The geometrically broken object, preprint cond-mat/9805204

208. de Sousa Vieira, M. (2002) Breakdown of self-organized criticality in sandpiles, *Phys. Rev. E* **66**, 51306–51315.

209. Deemer, W.L. and Votaw, D. F. (1955) Estimation of parameters of truncated or censored exponential distributions, *Ann. Math. Stat.* **26**, 498–504.

210. DeMets, C., Gordon, R.G., Argus, D.F. and Stein, S. (1990) Current plate motions, *Geophys. J. Int.* **101**, 425–478.

211. DeMets, C., Gordon, R.G., Argus, D.F. and Stein, S. (1994) Effect of recent revisions to the geomagnetic reversal time scale on estimate of current plate motions, *Geophys. Res. Lett.* **21**, 2191–2194.

212. de Finetti, B. (1974) *Theory of Probability* (Wiley, New York).

213. de Gennes, P.-G. (1979) *Scaling Concepts in Polymer Physics* (Cornell University Press, Ithaca, N.Y.).

214. de Haan, L. and Resnick, S. (1996) Second-order regular variation and rates of convergence in extreme-value theory, *The Annals of Probability* **24**, 97–124.

215. de Haan, L., Resnick, S.I., Rootzén, H. and de Vries, C.G. (1989) Extremal behaviour of solutions to a stochastic difference equation with applications to ARCH processes, *Stochastic Processes and Applics.* **32**, 213–224.

216. Dekkers, A.L. and De Haan, L. (1989) On the estimation of the extreme-value index and large quantile estimation, *The Annals of Statistics* **17**, 1795–1832.

217. DeMenech, M., Stella, A.L. and Tebaldi, C. (1998) Rare events and breakdown of simple scaling in the Abelian sandpile model, *Phys. Rev. E* **58**, R2677–R2680.

218. Derrida, B. (1981) Random-energy model: an exactly solvable model of disordered systems, *Phys. Rev. B* **24**, 2613–2626.

219. Derrida, B. (1997) From random walks to spin glasses, *Physica D* **107**, 186–198.

220. Derrida, B., Eckmann, J.P. and Erzan, A. (1983) Renormalization group trajectories with periodic and aperiodic orbits, *J. Phys. A* **16**, 893–906.

221. Derrida, B. and Flyvbjerg, H. (1987) Statistical properties of randomly broken objects and of multivalley structures in disordered systems, *J. Phys. A: Math. Gen.* **20**, 5273–5288.

222. Derrida, B. and Flyvbjerg, H. (1987) The random map model: a disordered model with deterministic dynamics, *J. Physique (Paris)* **48**, 971–978.

223. Derrida, B., Itzykson, C. and Luck, J.M. (1984) Oscillatory critical amplitudes in hierarchical models, *Commun. Math. Phys.* **94**, 115–132.

224. Derrida, B. and Peliti, L. (1991) Evolution in a flat fitness landscape, *Bull. Math. Biol.* **53**, 355–382.

225. Deutscher, G., Zallen, R. and Adler, J. eds. (1983) *Percolation Structures and Processes*. The Israel Physical Society, Bristol: A. Hilger; Jerusalem: Israel Physical Society in association with The American Institute of Physics, New York.

226. Dhar, D. (1990) Self-organized critical state of sandpile automaton models, *Phys. Rev. Lett.* **64**, 1613–1616.

227. Dhar, D. (1999) The Abelian sandpile and related models, *Physica A* **263**, 4–25.

228. Dhar, D. and Ramaswamy, R. (1989) Exactly solved model of self-organized critical phenomena, *Phys. Rev. Lett.* **63**, 1659–1662.

229. Dhar, D., Shukla, P. and Sethna, J.P. (1997) Zero-temperature hysteresis in the random-field Ising model on a Bethe lattice, *J. Phys. A* **30**, 5259–5267.

230. Díaz-Guilera, A. (1992) Noise and dynamics of self-organized critical phenomena, *Phys. Rev. A* **45**, 8551–558.

231. Dickman, R. (1990) Nonequilibrium critical behavior of the triplet annihilation model, *Phys. Rev. A* **42**, 6985–6990.

232. Dickman, R., Munoz, M.A., Vespignani, A. and Zapperi, S. (2000) Paths to self-organized criticality, *Brazilian J. Phys.* **30**, 27–41.

233. Dickman, R. and Tomé, T. (1991) First-order phase transition in a one-dimensional nonequilibrium model, *Phys. Rev. A* **44**, 4833–4838.

234. Dickman, R., Vespignani, A. and Zapperi, S. (1998) Self-organized criticality as an absorbing-state phase transition, *Phys. Rev. E* **57**, 5095–5105.

235. Dieterich, J.H. (1992) Earthquake nucleation on faults with rate- and state-dependent strength, *Tectonophysics* **211**, 115–134.

236. Diodati, P., Marchesoni, F. and Piazza, S. (1991) Acoustic emission from volcanic rocks – An example of self-organized criticality, *Phys. Rev. Lett.* **67**, 2239–2243.

237. Dodds, P.S. and Rothman, D. (2000) Scaling, universality, and geomorphology, *Annual Review of Earth and Planetary Sciences* **28**, 571–610.

238. Dodds, P.S. and Rothman, D. (1999) Unified view of scaling laws for river networks, *Phys. Rev. E* **59**, 4865–4877.

239. Domany, E. and Kinzel, W. (1984) Equivalence of cellular automata to Ising models and direct percolation, *Phys. Rev. Lett.* **53**, 311–314.

240. Doob, J. (1941) Probability as Measure, *Ann. Math. Stat.* **12**, 206.

241. Dorogovtsev, S.N., Mendes, J.F.F. and Samukhin, A.N. (2000) Structure of growing networks with preferential linking, *Phys. Rev. Lett.* **21**, 4633–4636.

242. Dorogovtsev, S.N. and Mendes, J.F.F. (2001) Effect of the accelerating growth of communications networks on their structure, *Phys. Rev. E* **63**, 025101.

243. Dowling, J.P. (1995) Parapsychological Review A?, *Physics Today* **July**, 78–78.

244. Dowling, J.P. (1996) More spirited debate on Physics, Parapsychology and paradigms – reply, *Physics Today* **April**, 81–81.

245. Drössel, B. and Schwabl, F. (1992) Self-organized critical forest-fire model, *Phys. Rev. Lett.* **69**, 1629–1632.

246. Dubey, S.D., *Monte Carlo study of the moment and maximum likelihood estimators of Weibull parameters*, technical report.
    Dubey, S., "Hyper-efficient Estimator of the Location Parameter of the Weibull Laws," Naval Research Logistics Quarterly, 13, 253–263 (1966).

247. Dubrulle, B. (1996) Anomalous scaling and generic structure function in Turbulence, *J. Phys. France II* **6**, 1825–1840.

248. Dubrulle, B. and Graner, F. (1996) Possible statistics of scale invariant systems, *J. Phys. II* **6**, 797–816.

249. Dubrulle, B. and Graner, F. (1996) Scale invariance and scaling exponents in fully developed turbulence, *J. Phys. II* **6**, 817–824.

250. Dubrulle, B. and Graner, F. (1997) Analogy between scale symmetry and relativistic mechanics. II. Electric analog of turbulence, *Phys. Rev. E* **56**, 6435–6442.

251. Dubrulle, B., Graner, F. and Sornette, D. eds. (1997) *Scale Invariance and Beyond* (EDP Sciences and Springer, Berlin).

252. DuMouchel, W.H. (1983) Estimating the stable index α in order to measure tail thickness: a critique, *The Annals of Statistics* **11**, 1019–1031.

253. Duplantier, B. (1988) Random walks and quantum gravity in two dimensions, *Phys. Rev. Lett.* **81**, 5489–5492

254. Duplantier, B. (1999) Harmonic measure exponents for two-dimensional percolation, *Phys. Rev. Lett.* **82**, 3940–3943.

255. Duplantier, B. (2000) Conformally Invariant Fractals and Potential Theory, *Phys. Rev. Lett.* **84**, 1363–1367.

256. Duxbury, P.M., Beale, P.D. and Leath, P.L. (1986) Size effects of electrical breakdown in quenched random media, *Phys. Rev. Lett.* **57**, 1052–1055.

257. Duxbury, P.M., Leath, P.L. and Beale, P.D. (1987) Breakdown properties of quenched random systems: the random-fuse network, *Phys. Rev. B* **36**, 367–380.

258. Dziewonski, A.M., Ekstrom, G. and Salganik, M.P. (1993) Centroid-moment tensor solutions for January–March, 1992, *Phys. Earth Planet. Inter.* **77**, 143–150.

259. Earman, J. (1992) *Bayes or Bust? A Critical Examination of Bayesian Confirmation Theory* (MIT Press, Cambridge, MA).

260. Eberlein, E. and Taqqu, M.S. eds. (1986) *Dependence in Probability and Statistics: a Survey of Recent Results, Oberwolfach, 1985* (Birkhauser, Boston).

261. Eckmann, J.-P. and Wittwer, P. (1986) Multiplicative and Additive Renormalization, in Osterwalder, K. and Stora, R. eds., *Critical Phenomena, Random Systems, Gauge Theories.* Les Houches, Session XLIII, 1984, Elsevier Science Publishers B.V., 455–465.

262. Economou, E.N. (1983) *Green's Functions in Quantum Physics*, 2nd ed. (Springer, Berlin and New York).

263. Edgar, G.A., ed. (1993) *Classics on Fractals* (Addison-Wesley Publishing Company, Reading, Massachusetts).

264. Efron, B. (1982) The jackknife, the bootstrap and other resampling plans, *SIAM monograph* **38**, CBMS-NSF.

265. Egolf, D.A. (2000) Equilibrium regained: from nonequilibrium chaos to statistical mechanics, *Science* **287**, 101–104.

266. Eguiluz, V.M., Ospeck, M., Choe, Y., Hudspeth, A.J. and Magnasco, M.O. (2000) Essential nonlinearities in hearing, *Phys. Rev. Lett.* **84**, 5232–5235.

267. Eistein, B. and Sobel, M. (1953) *J. Amer. Statistical Association* **48**, 486–502.

268. Ellis, R.S. (1984) Large deviations for a general class of random vectors, *Ann. Probab.* **12**, 1–12.

269. Elshamy, M. (1992) *Bivariate Extreme Value Distributions.* National Aeronautics and Space Administration, Office of Management, Scientific and Technical Information Program; [Springfield, Va.: For sale by the National Technical Information Service, Washington, DC.

270. Emak, D.L. and McCammon, J.A. (1978) *J. Chem. Phys.* **69**, 1352.

271. Embrechts, P., Subexponential distribution functions and their applications: a review, *Proc. 7th Conf. Probability Theory (Brasov, Romania, 1985)*, 125–136.

272. Embrechts, P. and Goldie, C.M. (1980) On closure and factorization properties of subexponential and related distributions, *J. Austral. Math. Soc. (Series A)* **29**, 243–256.

273. Embrechts, P., Goldie, C.M. and Veravereke, N. (1979) Subexponentiality and infinite divisibility, *Z. Wahrscheinlichkeitstheorie verw. Gebiete* **49**, 335–347.

274. Embrechts, P., Kluppelberg, C. and Mikosch, T. (1997) *Modelling Extremal Events for Insurance and Finance* (Springer, New York).

275. Embrechts, P., and Omey, E. (1984) A property of longtailed distributions, *J. Appl. Prob.* **21**, 80–87.

276. Erdélyi, A., ed. (1954) *Tables of Integral Transforms* (Bateman Manuscript Project, Vol. I, McGraw-Hill, New York).

277. Ertaş, D. and Kardar, M. (1994) Critical dynamics of contact line depinning, *Phys. Rev. E* **49**, R2532–R2535.

278. Erzan, A. (1997) Finite q-differences and the discrete renormalization group, *Phys. Lett. A* **225**, 235–238.

279. Erzan, A. and Eckmann, J.-P. (1997) q-analysis of Fractal Sets, *Phys. Rev. Lett.* **78**, 3245–3248.

280. Essam, J., Guttmann, A. and De 'Bell, K. (1988) On two-dimensional directed percolation, *J. Phys. A* **31**, 3815–3832.

281. Estoup, J.B. (1916) Gammes Stenographiques, *Institut Stenographique de France*, Paris.

282. Eyink, G.L. (1996) Action principle in nonequilibrium statistical dynamics, *Phys. Rev.* **54**, 3419–3435.

283. Eyink, G.L. (1997) Fluctuations in the irreversible decay of turbulent energy, *Phys. Rev. E* **56**, 5413–5422.

284. Eyink, G.L. (1998) Action principle in statistical dynamics, *Progress of Theoretical Physics Supplement* **130**, 77–86.

285. Fabiani, J.-L. and Theys, J. (1987) *La Société Vulnérable, Evaluer et Maîtriser les Risques* (Presses de l'Ecole Normale Supérieure, Paris).

286. Falconer, K. (1990) *Fractal Geometry, Mathematical Foundations and Applications.* Wiley, New York.

287. Falconer K. "Techniques in fractal geometry", John Wiley - sons, 1997.

288. Fama, E. (1965) *Management Science* **11**, 404–419.

289. Farnum, N.R., and Booth, P. (1997) Uniqueness of maximum likelihood estimators of the two-parameter Weibull distribution, *IEEE Transactions on Reliability* **46**, 523–525.

290. Feder, H.J.S. and Feder, J. (1991) Self-organized criticality in a stick-slip process, *Phys. Rev. Lett.* **66**, 2669–2672.

291. Feder, J. (1988) *Fractals* (Plenum Press, New York).

292. Feller, W. (1957) *Introduction to Probability Theory and Its Applications.* vol. I (Wiley, New York).

293. Feller, W. (1971) *An Introduction to Probability Theory and its Applications.* vol. II (Wiley, New York).

294. Feng, S., Halperin, B.I. and Sen, P.N. (1987) Transport properties of continuum systems near the percolation threshold, *Phys. Rev. B* **35**, 197–214.

295. Fermi, E. (1949) On the origin of cosmic radiation, *Phys. Rev.* **75**, 1169–1174.

296. Fermi, E., Pasta, I.R. and Ulam, S.M. (1955) Los Alamos National Laboratory Report No. LA-1940, (unpublished).

297. Fernandez, R., Frolich, J. and Sokal, A.D. (1992) *Random Walks, Critical Phenomena, and Triviality in Quantum Field Theory* (Springer, Berlin, New York).

298. Ferrari, R., Manfroi, A.J. and Young, W.R. (2001) Strongly and weakly self-similar diffusion, *Physica D* **154**, 111–137.

299. Feynman, R.P. (1989) *The Feynman Lectures on Physics.* vol. 1, Chapter 56: Probability (Addison-Wesley, Redwood City, California).
300. Feigenbaum, J.A., and Freund, P.G.O. (1996) Discrete scale invariance in stock markets before crashes, *Int. J. Mod. Phys.* **10**, 3737–3745.
301. Fineberg, J. and Marder, M. (1999) Instability in dynamic fracture, *Physics Reports* **313**, 2–108.
302. Fisher, M.E. (1998) Renormalization group theory – Its basis and formulation in statistical physics, *Reviews of Modern Physics* **70**, 653–681.
303. Fisher, M.E. and Privman, V. (1985) First-order transitions breaking $O(n)$ symmetry: finite-size scaling, *Phys. Rev. B* **32**, 447–64.
304. Fisher, R.A. (1973) *Statistical Methods and Scientific Inference*, [3d ed., revised and enlarged] (Hafner Press, New York).
305. Fisher, R.A., and Tippett, L.H.C. (1928) Limiting forms of the frequency distribution of the largest or smallest member of a sample, *Proceedings of the Cambridge Philosophical Society* **XXIV**, 180–190.
306. Flandrin, P. (1999) *Time-Frequency/Time-Scale Analysis* (Academic Press, San Diego).
307. Fox, C. (1961) The G and H-functions as symmetrical Fourier kernels, *Trans. Amer. Math. Soc.* **98**, 395–429.
308. Frauenfelder, H., Sligar, S.G. and Wolynes, P.G. (1991) The energy landscape and motions of proteins, it Science **254**, 1598–1603.
309. Frauenfelder, H. and Wolynes, P.G. (1994) Biomolecules – Where the physics of complexity and simplicity meet, *Physics Today* **47**, 58–64.
310. Frauenfelder, H., Wolynes, P.G. and Austin, R.H. (1999) Biological physics, *Rev. Mod. Phys.* **71**, S419-S430.
311. Fraysse, N., Sornette, A. and Sornette, D. (1993) Critical transitions made self-organized: proposed experiments, *J. Phys. I France* **3**, 1377–1386.
312. Freidlin, M.I. and Wentzell, A.D. (1984) *Random Perturbations of Dynamical Systems* (Springer, New York).
313. Frette, V., Christensen, K., Malte-Sorensen, A., Feder, J., Josang, T. and Meakin, P. (1996) Avalanche dynamics in a pile of rice, *Nature* **379**, 49–52.
314. Friedrich, R., Siegert, S., Peinke, J., Luck, S., Siefert, M., Lindemann, M., Raethjen, J., Deuschl, G., Pfister, G. (2000) Extracting model equations from experimental data, *Physics Letters A* **271**, 217–222.
315. Frisch, H. (1988) A Cauchy integral equation method for analytic solutions of half-space convolution equations, *J. Quant. Spectrosc. Radiat. Transfer* **39**, 149–162.
316. Frisch, U. (1995) *Turbulence, The Legacy of A.N. Kolmogorov* (Cambridge University Press, Cambridge).
317. Frisch, U., Bec, J., and Villone, B. (2001) Singularities and the distribution of density in the Burgers/adhesion model, *Physica D* **152–153**, 620–635.
318. Frisch, U. and Sornette, D. (1997) Extreme deviations and applications, *J. Phys. I France* **7**, 1155–1171.
319. Fulcher, G.S. (1925) *J. Am. Ceram. Soc.* **6**, 339.
320. Fulco, U.L., Nobre, F.D., da Silva, L.R., Lucena, L.S. and Viswanathan, G.M. (2000) Efficient search method for obtaining critical properties, *Physica A* **284**, 223–230.
321. Furutsu, K. and Ishida, T. (1961) On the theory of amplitude distribution of impulsive random noise, *J. Appl. Phys.* **32**(7), 1206–1221.
322. Gabaix, X. (1999) Zipf's law for cities: an explanation, *Quarterly Journal of Economics* **114**, 739–767.
323. Gabrielli, A., Labini, F.S. and Pellegrini, S. (1999) Gravitational force distribution in fractal structures, *Europhys. Lett.* **46**, 127–133.

324. Gabrielov, A., Newman, W.I. and Knopoff, L. (1994) Lattice models of Fracture: Sensitivity to the Local Dynamics, *Phys. Rev. E* **50**, 188–197.
325. Gaillard-Groleas, G., Lagier, M. and Sornette, D. (1990) Critical behaviour in piezoelectric ceramics, *Phys. Rev. Lett.* **64**, 1577–1580.
326. Galambos, J. (1987) *The Asymptotic Theory of Extreme Order Statistics* 2nd ed. (R.E. Krieger Pub. Co., Malabar, Fla).
327. Galambos, J., Lechner, J. and Simiu, E. eds. (1994) *Extreme value theory and applications: Proceedings of the Conference on Extreme Value Theory and Applications, Gaithersburg, Maryland, 1993* (Kluwer Academic, Boston).
328. Gallavotti, G. (2000) Non-equilibrium in statistical and fluid mechanics – Ensembles and their equivalence, Entropy driven intermittency, *J. Math. Phys.* **41**, 4061–4081.
329. Gallavotti, G. (2000) *Statistical Mechanics. A Short Treatise* (Springer, Berlin, Heidelberg)
330. Gallavotti, G. (2002) *Foundations of Fluid Dynamics* (Springer, Berlin, Heidelberg).
331. Gallavotti, G. (2003) Nonequilibrium thermodynamics? (http://arXiv.org/abs/cond-mat/0301172)
332. Gallavotti, G. and Cohen, E.G.D. (1995) Dynamical ensembles in nonequilibrium statistical mechanics, *Phys. Rev. Lett.* **74**, 2694–2697.
333. Garcimartin, A., Guarino, A., Bellon, L. and Ciliberto, S. (1997) Statistical properties of fracture precursors, *Phys. Rev. Lett.* **79**, 3202–3205.
334. Gardner, E. (1985) Spin Glasses with p-Spin Interactions, *Nucl. Phys. B* **257**, 747–765.
335. Geilikman, M. B., Pisarenko, V.F. and Golubeva, T.V. (1990) Multifractal Patterns of Seismicity, *Earth and Planetary Science Letters* **99**, 127–138.
336. Gelfand, I.M., Guberman, Sh.A., Keilis-Borok, V.I., Knopoff., L., Press, F., Ranzman, E.Ya., Rotwain, I.M. and Sadovsky, A.M. (1976) Pattern recognition applied to earthquake epicenters in California, *Physics of the Earth and Planetary Interiors* **11**, 227–283.
337. Georges, A. and Le Doussal, P. (1989) From equilibrium spin models to probabilistic cellular automata, *J. Stat. Phys.* **54**, 1011–1064.
338. Gil, L. and Sornette, D. (1996) Landau–Ginzburg theory of self-organized criticality, *Phys. Rev. Lett.* **76**, 3991–3994.
339. Gilabert, A., Sornette, A., Benayad, M., Sornette, D. and Vanneste, C. (1990) Conductivity and rupture in crack-deteriorated systems, *J. Phys. France* **51**, 247–257.
340. Gilabert, A., Vanneste, C., Sornette, D. and Guyon, E. (1987) The random fuse network as a model of rupture in a disordered medium, *J. Phys. France* **48**, 763–770.
341. Gilman, J.J. (1996) Mechanochemistry, *Science* **274**, 65–65.
342. Gluzman, S. and Sornette, D. (2002) Classification of possible finite-time singularities by functional renormalization, art. no. 016134, *Phys. Rev. E* **6601** N1 PT2:U315-U328.
343. Gluzman, S. and Sornette, D. (2002) Log-periodic route to fractal functions, art. no. 036142, *Phys. Rev. E* **6503** N3 PT2A:U418-U436.
344. Gluzman, S. and Yukalov, V.I. (1997) Algebraic self-similar renormalization in the theory of critical phenomena, *Phys. Rev. E* **55**, 3983–3999.
345. Gluzman S. and Yukalov, V.I. (1998) Unified approach to crossover phenomena, *Phys. Rev. E* **58**, 4197–4209.
346. Gluzman, S., Sornette, D. and Yukalov, V.I. (2002) Reconstructing generalized exponential laws by self-similar exponential approximants, in press

in *International Journal of Modern Physics C* (http://arXiv.org/abs/cond-mat/0204326)

347. Gluzman, S., Yukalov, V.I. and Sornette, D. (2002) Self-similar factor approximants, *Phys. Rev. E* **67** (2), art. 026109, DOI: 10.1103/PhysRevE.67.026109.

348. Gluzman, S. and Sornette, D. (2003) Self-similar approximants of the permeability in heterogeneous porous media from moment equation expansions, submitted to *Transport in Porous Media* (http://arXiv.org/abs/cond-mat/0211565)

349. Gnedenko, B. (1943) Sur la distribution limite du terme maximum d'une série aléatoire, *Ann. Math.* **44**, 423–453.

350. Gnedenko, B.V. and Kolmogorov, A.N. (1954) *Limit Distributions for Sum of Independent Random Variables* (Addison Wesley, Reading MA).

351. Goldenfeld, N. (1992) *Lectures on Phase Transitions and the Renormalization Group* (Addison-Wesley, Advanced Book Program, Reading, Mass.).

352. Goldie, C.M. (1991) Implicit renewal theory and tails of solutions of random equations, *Annals of Applied Probability* **1**, 126–166.

353. Goldie, C.M. and Resnick, S. (1988) Subexponential distribution tails and point processes, *Cumm. Statist.-Stochastic Models* **4**, 361–372.

354. Gómez, J.B., Iñiguez, D.I. and Pacheco, A.F. (1993) Solvable fracture model with local load transfer, *Phys. Rev. Lett.* **71**, 380–383.

355. Gorshkov, A.I., Kossobokov, V.G., Rantsman, E.Ya. and Soloviev, A.A. (2001) Recognition of earthquake prone areas: validity of results obtained from 1972 to 2000, *Computational Seismology* **32**, 48–57.

356. Gorte, R.W. (1995) Forest Fires and Forest Health, Congressional Research Service Report, The Committee for the National Institute for the Environment, 1725 K Street, NW, Suite 212, Washington, D.C. 20006.

357. Gott, J.R. (1993) Implications of the Copernican principle for our future prospects, *Nature* **363**, 315–319.

358. Gould, S.J. (1996) *Full House* (Harmony Books, New York).

359. Gouriéroux, C. and Monfort, A. (1994) Testing non nested hypothesis, *Handbook of Econometrics* **4**, 2585–2637.

360. Graham, R. and Tel, T. (1985) Weak-noise limit of Fokker–Planck models and nondifferentiable potentials for dissipative dynamical systems, *Phys. Rev. A* **31**, 1109–1122.

361. Graham, R. and Tel, T. (1986) Nonequilibrium potential for coexisting attractors, *Phys. Rev. A* **33**, 1322–1337.

362. Graham, R. and Tel, T. (1987) Nonequilibrium potentials for local codimension-2 bifurcations of dissipative flows, *Phys. Rev. A* **35**, 1328–1349.

363. Graner, F. and Dubrulle, B. (1997) Analogy between scale symmetry and relativistic mechanics. I. Lagrangian formalism, *Phys. Rev. E* **56**, 6427–6434.

364. Grassberger, P. (1982) On phase transitions in Schlogl's second model, *Z. Phys. B* **47**, 365–374.

365. Grassberger, P. (1989) Noise-induced escape from attractors, *J. Phys. A* **22**, 3283–3290.

366. Grassberger, P. (1989) Some further results on a kinetic critical phenomenon, *J. Phys. A* **22**, L1103–L1107.

367. Grassberger, P. (1992) Numerical studies of critical percolation in 3-dimensions, *J. Phys. A* **25**, 5867–5888.

368. Grassberger, P. (1994) Efficient large-scale simulations of a uniformly driven system, *Phys. Rev. E.* **49**, 2436–2444.

369. Grassberger, P. (1995) Are damage spreading transitions generically in the universality class of directed percolation? *J. Stat. Phys.* **79**, 13–23.

370. Grassberger, P., Krause, F. and von der Twer, T. (1984) A new type of kinetic critical phenomenon J. Phys. A **17**, L105–L107.

371. Grassberger, P. and Sundermeyer, K. (1978) Phys. Lett. B **77**, 220.

372. Grassberger, P. and Zhang, Y.-C. (1996) Self-organized formulation of standard percolation phenomena, Physica A **224**, 169–179.

373. Grasso, J.R. and Sornette, D. (1998) Testing self-organized criticality by induced seismicity, J. Geophys. Res. **103**, 29965–29987.

374. Gratzer, W. (1996) An encyclopedia of claims, frauds and hoaxes of the occult and supernatural, J. Randi, Nature **379**, 782–783.

375. Grimmett, G. (1989) Percolation (Springer, New York).

376. Grinstein, G., Jayaprakash, C. and Socolar, J.E.S. (1993) Scale invariance of non-conserved quantities in driven systems, Phys. Rev. E **48**, R643–R646.

377. Grinstein, G., Lai, Z.-W. and Browne, D.A. (1989) Critical phenomena in a nonequilibrium model of heterogeneous catalysis, Phys. Rev. A **40**, 4820–4823.

378. Grinstein, G., Lee, D.-H. and Sachdev, S. (1990) Conservation laws, anisotropy and self-organized criticality in noisy non-equilibrium systems, Phys. Rev. Lett. **64**, 1927–1930.

379. Gross, D.J. and Mezard, M. (1984) The Simplest Spin Glass, Nucl. Phys. B **240**, 431–452.

380. Gumbel, E.J. (1954) Statistical Theory of Extreme Values and Some Practical Applications; a Series of Lectures (U. S. Govt. Print. Office, Washington).

381. Gumbel, E.J. (1958) Statistics of Extremes (Columbia University Press, New York).

382. Guyon, E., Nadal, J.-P. and Pomeau, Y. eds. (1988) Disorder and Mixing: Convection, Diffusion, and Reaction in Random Materials and Processes (Kluwer Academic Publishers, Boston).

383. Haan, C.T. and Beer, C.E. (1967) Determination of maximum likelihood estimators for the three parameter Weibull distribution, Iowa State Journal of Science **42**, 37–42.

384. Haken H. (1983) Synergetics, an Introduction 3rd ed. (Springer, Berlin).

385. Haken H. (1987) Advanced Synergetics. 2nd ed. (Springer, Berlin).

386. Hallinan, A.J. (1993) A review of the Weibull distribution, Journal of Quality Technology **25**, 85–93.

387. Halperin, B.I., Feng, S. and Sen, P.N. (1985) Differences between lattice and continuum percolation transport exponents, Phys. Rev. Letts. **.54**, 2391–2394.

388. Halpin-Healy, T. and Zhang, Y.-C. (1995) Kinetic roughening phenomena, stochastic growth, directed polymers and all that – Aspects of multidisciplinary statistical mechanics, Physics Reports **254**, 215–415.

389. Halsey, T.C. (1992) Electrorheological fluids, Science **258**, N5083, 761–766.

390. Halsey, T.C. (1993) Electrorheological fluids structure and dynamics, Adv. Materials **5**, 711–718.

391. Halsey, T.C., Jensen, M.H., Kadanoff, L.P., Procaccia, I, Shraiman, B.I. (1986) Fractal measures and their singularities: the characterization of strange sets, Phys. Rev. A **33**, 1141–1151.

392. Halsey, T.C. and Toor, W. (1990) Fluctuation-induced coupling between defect lines or particle chains, J. Stat. Phys. **61**, 1257–1281.

393. Hamburger, D., Biham, O and Avnir, D. (1996) Apparent fractality emerging from models of random distributions, Phys. Rev. E **53**, 3342–3358.

394. Hammer, P.T.C., Hildebrand, J.A. and Parker, R.L. (1991) Gravity inversion using seminorm minimization – density modeling of Jasper Seamount, Geophysics **56**, 68–79.

395. Hamilton, J.D. (1994) *Time Series Analysis* (Princeton University Press, Princeton, New Jersey).
396. Hänggi, P. and Hung, P. (1995) Colored noise in dynamical systems, *Advances in Chemical Physics* **89**, 239–326.
397. Hänggi, P., Talkner, P. and Borkovec, M. (1990) Reaction-rate theory: fifty years after Kramers, *Reviews of Modern Physics* **62**, 251–341.
398. Hänggi, P. and Thomas, H. (1982) Stochastic processes: time evolution, symmetries and linear response, *Physics Reports* **88**, 207–319.
399. Hansel, D. and Sompolinsky, H. (1993) Solvable model of spatio-temporal chaos, *Phys. Rev. Lett.* **71**, 2710–2713.
400. Hansen, A. and Hemmer, P.C. (1994) Burst avalanches in bundles of fibers – Local versus global load-sharing, *Phys. Lett. A* **184**, 394–396.
401. Hansen, A., Hinrichsen, E. and Roux, S. (1991) Scale-invariant disorder in fracture and related breakdown phenomena, *Phys. Rev. B* **43**, 665–678.
402. Hardy, G.H. (1916) Weierstrass's non-differentiable function, *Trans. Amer. Math. Soc.* **17**, 301–325.
403. Harlow, D.G. and Phoenix, S.L. (1981) Probability distribution for the strength of composite materials II: a convergent series of bounds, *Int. J. Fract.* **17**, 601–630.
404. Harris, T.E. (1963) *The theory of branching processes* (Springer, Berlin).
405. Harrison, C.G.A., and Carle, H. M. (1981) Intermediate wavelength magnetic anomalies over Ocean Basins, *J. Geophys. Res.* **86**, 11585–11599.
406. Hawking, S.W. (1988) *A Brief History of Time: from the Big Bang to Black Holes* (Bantam Books, Toronto).
407. Head, D.A. (2000) Temperature scaling, glassiness and stationarity in the Bak–Sneppen model, *European Physical Journal B* **17**, 289–294. Sep.
408. Heimpel, M. and Olson, P. (1996) A seismodynamical model of lithospheric deformation – Development of continental and oceanic rift networks, *J. Geophys. Res.* **101**, 16155–16176.
409. Held, G.A., Solina, D.H., Keane, D.T., Haag, W.J., Horn, P.M. and Grinstein, G. (1990) Experimental study of critical-mass fluctuations in an evolving sandpile, *Phys. Rev. Lett.* **65**, 1120–1123.
410. Helmstetter, A. (2003) Is earthquake triggering driven by small earthquakes?, submitted to *Phys. Rev. Lett.* (http://arxiv.org/abs/physics/0210056)
411. Helmstetter, A. and Sornette, D. (2002) Sub-critical and super-critical regimes in epidemic models of earthquake aftershocks, *J. Geophys. Res.* **107** (B10), 2237, doi:10.1029/2001JB001580.
412. Helmstetter, A and Sornette, D. (2002) Diffusion of epicenters of earthquake aftershocks, Omori's law, and generalized continuous-time random walk models, *Phys. Rev. E* **6606**, DOI: 10.1103/PhysRevE.66.061104.
413. Helmstetter, A. and Sornette, D. (2003) Foreshocks and cascades of triggered seismicity, submitted to *J. Geophys. Res.* (http://arXiv.org/abs/physics/0210130)
414. Helmstetter, A., Sornette, D. and Grasso, J.-R. (2003) Mainshocks are Aftershocks of conditional foreshocks: how do foreshock statistical properties emerge from aftershock laws, *J. Geophys. Res.* **108** (B10), 2046, doi:10.1029/2002JB001991.
415. Hemmer, P.C. and Hansen, A. (1992) The distribution of simultaneous fiber failures in fiber bundles, *J. Appl. Mech.* **59**, 909–914.
416. Herrmann, H.J. and Roux, S. eds. (1990) *Statistical models for the fracture of disordered media* (Elsevier, Amsterdam).

417. Herz, A.V.M. and Hopfield, J.J. (1995) Earthquake cycles and neural reverberations – Collective oscillations in systems with pulse-coupled threshold elements, *Phys. Rev. Lett.* **75**, 1222–1225.

418. Hopfield, J.J. (1999) Brain, neural networks, and computation, *Rev. Mod. Phys.* **71**, S431-S437.

419. Higgs, P. (1995) Frequency distributions in population genetics parallel those in statistical physics, *Phys. Rev. E* **51**, 95–101.

420. Hilfer, R. (1997) Fractional derivatives in static and dynamic scaling, in [251], 53–62.

421. Hilfer, R., ed. (2000) *Applications of Fractional Calculus in Physics* (World Scientific, Singapore; River Edge, NJ).

422. Hill, B.M. (1975) A simple general approach to inference about the tail of a distribution, *Ann. Statistics* **3**, 1163–1174.

423. Hill, B.M. (1994) Bayesian forecasting of extreme values in an exchangeable sequence, *J. Res. Natl. Inst. Stand. Technol.* **99**, 521–538.

424. Hill, C.T. (2000) Teaching Symmetry in the Introductory Physics Curriculum, http://www.emmynoether.com/

425. Hill, T.P. (1998) The first digit phenomenon, *American Scientist* **86**, 358–363.

426. Hinrichsen, H. (2000) On possible experimental realizations of directed percolation, *Brazilian J. Phys.* **30**, 69–82.

427. Hirst, W. (1994) *Fractal Landscapes: from the Real World* (Cornerhouse, Manchester).

428. Hohenberg, P.C. and Halperin, B.I. (1977) Theory of dynamical critical phenomena, *Rev. Mod. Phys.* **49**, 435–479.

429. Hohenberg, P.C. and Swift, J.B. (1992) Effects of additive noise at the onset of Rayleigh–Bénard convection, *Phys. Rev. A* **46**, 4773–4785.

430. Holy, T.E. (1997) Analysis of data from continuous probability distributions, *Phys. Rev. Lett.* **79**, 3545–3548.

431. Houle, P.A. and Sethna, J.P. (1996) Acoustic emission from crumpling paper, *Phys. Rev. E* **54**, 278–283.

432. Hu, C.K., Chen, C.N. and Wu, F.Y. (1995) Histogram Monte-Carlo position-space renormalization group – Applications to the site percolation, *J. Stat. Phys.* **82**, 1199–1206.

433. Hu, T.-Y. and Lau, K.-S. (1993) Fractal dimensions and singularities of the Weierstrass-type functions, *Trans. Am. Math. Soc.* **335**, 649–665.

434. Huang, Y., Ouillon, G., Saleur, H. and Sornette, D. (1997) Spontaneous generation of discrete scale invariance in growth models, *Phys. Rev. E* **55**, 6433–6447.

435. Huang, Y., Saleur, H., Sammis, C.G. and Sornette, D. (1998) Precursors, aftershocks, criticality and self-organized criticality, *Europhys. Lett.* **41**, 43–48.

436. Huang, Y., Johansen, A., Lee, M.W., Saleur, S. and Sornette, D. (2000) Artifactual log-periodicity in finite-size data: Relevance for Earthquake Aftershocks, *J. Geophys. Res.* **105**, 25451–25471.

437. Huberman, B.A., Pirolli, P.L.T., Pitkow, J.E. and Lukose, R.M. (1998) Strong regularities in World Wide Web surfing, *Science* **280**, 95–97.

438. Huberman, B.A. and Adamic, L.A. (1999) Internet – Growth dynamics of the World-Wide Web, *Nature* **401**, 131–131.

439. Huberman, B.A. and Adamic, L.A. (2000) The nature of markets in the World Wide Web, *Quarterly Journal of Economic Commerce* **1**, 5–12.

440. Hughes, B.D., Shlesinger, M.F., and Montroll, E.W. (1981) Random walks with self-similar clusters, *Proc. Natl. Acad. Sci. U.S.A.* **78**. 3287–3291.

441. Hurst, H.E. (1951) Long term storage capacity of reservoirs, *Transactions of the American Society of Civil Engineers* **116**, 770–808.
442. Huse, D. A. and Henley, C. L. (1985) Pinning and roughening of domain walls in Ising systems due to random impurities, *Phys. Rev. Lett.* **54**, 2708–2711.
443. Hwa, T. and Kardar, M. (1989) Fractals and self-organized criticality in dissipative dynamics, *Physica D* **38**, 198–202.
444. Hwa, T. and Kardar, M. (1989) Dissipative transport in open systems: an investigation of self-organized criticality, *Phys. Rev. Lett.* **62**, 1813–16.
445. Hwa, T. and Kardar, M. (1992) Avalanches, Hydrodynamics and discharge events in models of sandpiles, *Phys. Rev. A* **45**, 7002–7023.
446. Ibragimov, I.A., and Linnik, Yu.V. (1971) *Independent and Stationary Sequences of Random Variables*. Ed. by Kingman, J.F.C. (Wolters-Noordhoff, Groningen).
447. Ijjasz-Vasquez, E.J., Bras, R.L., Rodriguez-Iturbe, I., Rigon, R. and others (1993) Are river basins optimal channel networks, *Advances in Water Ressources* **16**, 69–79.
448. International Organization for Standardization – ISO (1993) *Guide to the Expression of Uncertainty in Measurement*, Geneva, Switzerland.
449. Isichenko, M.B. (1992) Percolation, Statistical topography and transport in random media, *Rev. Mod. Phys.* **64**, 961–1043.
450. Jackson, D.D. et al. (1995) Seismic hazards in Southern California: probable earthquakes, 1994 to 2024, *Bull. Seism. Soc. Am.* **85**, 379–439.
451. Jackson, F.H. (1904) A generalization of the functions $\Gamma(n)$ and x^n, *Proc. Roy. Soc. London* **74**, 64–72.
452. Jackson, F.H. (1910) On $q$definite integrals, *Quart. J. Pure Appl. Math.* **41**, 193–203.
453. Jackson, F.H. (1951) *Quart. J. Math. Oxford Ser. 2* **1**.
454. Jacquelin, J. (1996) Inference of sampling on Weibull parameter estimation, *IEEE Transactions on Dielectrics and Electrical Insulation* **3**, 809–816.
455. Jan, N., Moseley, L., Ray, T. and Stauffer, D. (1999) Is the Fossil Record Indicative of a Critical System? *Advances in Complex Systems* **12**, 137–141.
456. Janicki, A. and Weron, A. (1994) *Simulation and Chaotic Behavior of α-stable stochastic processes* (Marcel Dekker, New York).
457. Jánosi, I.M. and Kertész, J. (1993) Self-organized criticality with and without conservation, *Physica A* **200**, 179–188.
458. Jansen, D.W. and de Vries, C.G. (1991) On the frequency of large stock returns: putting booms and busts into perspective, *The Review of Economics and Statistics* **18**, 18–24.
459. Janssen, H.K. (1981) On the nonequilibrium phase transition in reaction-diffusion systems with an absorbing stationary state, *Z. Phys. B* **42**, 151–154.
460. Jaynes, E.T. (1968) Prior probabilities, *IEEE Trans. Systems Sci. Cyb.* **4**, 227.
461. Jaynes, E.T. (1989) Concentration of distributions at entropy maxima, in: Jaynes, E.T. *Papers on Probability, Statistics and Statistical Mechanics*. ed. by Rosenkrantz, R.D. (Kluwer Academic, Dordrecht).
462. Jeffreys, H. (1961) *Theory of Probability*, 3rd ed. (Oxford University Press).
463. Jensen, H.J. (1998) *Self-Organized Criticality* (Cambridge university Press, Cambridge).
464. Jensen, I. (1992) Ph.-D. thesis, Aarhus Univ., Denmark.
465. Jensen, I. (1993) Critical behavior of branching annihilating random walks with an odd number of offsprings, *Phys. Rev. E* **47**, R1–R4.

466. Jensen, I. (1993) Conservation laws and universality in branching annihilating random walks, *J. Phys. A* **26**, 3921–3930.

467. Jensen, I. (1993) Critical behavior of the pair contact process, *Phys. Rev. Lett.* **70**, 1465–1468.

468. Jensen, I. (1994) Critical behavior of nonequilibrium models with infinitely many absorbing states, *Int. J. Mod. Phys. B* **8**, 3299–3311.

469. Jensen, I. (1994) Critical behavior of a surface reaction model with infinitely many absorbing states, *J. Phys. A* **27**, L61–L68.

470. Jensen, I., Fogedby, H.C. and Dickman, R. (1990) Critical exponents for an irreversible surface reaction model, *Phys. Rev. A* **41**, 3411–3414.

471. Jensen, I. and Dickman, R. (1993) Nonequilibrium phase transitions in systems with infinitely many absorbing states, *Phys. Rev. E* **48**, 1710–1725.

472. Jensen, I., Fogedby, H.C. and Dickman, R. (1990) Critical exponents for an irreversible surface reaction model, *Phys. Rev. A* **41**, 3411–3414.

473. Jevtić, M.M. (1995) Noise as a diagnostic and prediction tool in reliability physics, *Microelectron. Reliab.* **35**, 455–477.

474. Joanny, J.F. and de Gennes, P.G. (1984) A model for contact angle hysteresis, *J. Chem. Phys.* **81**, 552–562.

475. Jögi, P. and Sornette, D. (1998) Self-organized critical random directed polymers, *Phys. Rev. E* **57**, 6931–6943.

476. Jögi, P., Sornette, D. and Blank, M. (1998) Fine structure and complex exponents in power law distributions from random maps, *Phys. Rev. E* **57**, 120–134.

477. Johansen, A., Sornette, D., Wakita, G., Tsunogai, U., Newman, W.I. and Saleur, H. (1996) Discrete scaling in earthquake precursory phenomena: evidence in the Kobe earthquake, Japan, *J. Phys. I France* **6**, 1391–1402.

478. Johansen, A. and Sornette, D. (1998) Evidence of discrete scale invariance by canonical averaging, *Int. J. Mod. Phys. C* **9**, 433–447.

479. Johansen, A. and Sornette, D. (1999) Critical crashes, *Risk* **12**, 91–94.

480. Johansen, A. and Sornette, D. (2000) Critical ruptures, *Eur. Phys. J. B* **18**, 163–181.

481. Johansen, A., Sornette, D. and Ledoit, O. (1999) Predicting Financial Crashes using discrete scale invariance, *Journal of Risk* **1**, 5–32.

482. Johansen, A., Ledoit, O. and Sornette, D. (2000) Crashes as critical points, *Int. J. Theor. Applied Finance* **3**, 219–255.

483. Johansen, A., Sornette, D. and Hansen, A.E. (2000) Punctuated vortex coalescence and discrete scale invariance in two-dimensional turbulence, *Physica D* **138**, 302–315

484. Johns, M.V., and Lieberman, G.J. (1966) An exact asymptotically efficient confidence bound for reliability in the case of the Weibull distribution, *Technometrics* **78**, 135–175.

485. Johnson, N.L., Kotz, S. and Balakrishnan, N. (1995) *Continuous Univariate Distributions*, vol. 2 (Wiley, New York)

486. Jona-Lasinio, G. (1975) The renormalization group: a probabilistic view, *Nuovo Cimento* **26B**, 99.

487. Jona-Lasinio, G. (2001) Renormalization group and probability theory, *Physics Reports* **352**, 439–458.

488. Kadanoff, L.P., Nagel, S.R., Wu, L. and Zhou, S. (1989) Scaling and universality in avalanches, *Phys. Rev. A* **39**, 6524–6537.

489. Kagan, Y.Y. (1990) Random stress and earthquake statistics: Spatial dependence, *Geophys. J. Int.* **102**, 573–583.

490. Kagan, Y. Y. (1994) Distribution of incremental static stress caused by earthquakes, *Nonlinear Processes in Geophysics* **1**, 172–181.

491. Kagan, Y.Y. (1997) Seismic moment-frequency relation for shallow earthquakes: Regional comparison, *J. Geophys. Res.* **102**, 2835–2852.

492. Kagan, Y.Y. and Knopoff, L. (1980) Spatial distribution of earthquakes: the two-point correlation function, *Geophys. J. Roy. Astron. Soc.* **62**, 303–320.

493. Kahng, B., Batrouni, G.G., Redner, S., de Arcangelis, L. and Herrmann, H. (1988) Electrical breakdown in a fuse network with random, continuously distributed breaking strengths, *Phys. Rev.* B **37**, 7625–7637.

494. Kaplan, J.L., Mallet-Paret, J., and Yorke, J.A. (1984) The Lyapunov dimension of a nowhere differentiable attracting torus, *Ergod. Th. Dynam. Sys.* **4**, 261–281.

495. Kardar, M., Parisi, G. and Zhang, Y.-C. (1986) Dynamic scaling of growing interfaces, *Phys. Rev. Lett.* **56**, 889–892.

496. Kardar, M. (1987) Replica Bethe ansatz studies of two-dimensional interfaces with quenched random impurities, *Nucl. Phys.* B **290**, 582–602.

497. Kasahara, K. (1981) *Earthquake Mechanics* (Cambridge University Press, Cambridge, UK), Chap. 7, Section 7.1.2.

498. Kass, R.E. and Raftery, A.E. (1995) Bayes factors, *Journal of the American Statistical Association* **90**, 773–795.

499. Kautz, R.L. (1988) Thermally induced escape: the principle of minimum available noise energy, *Phys. Rev.* A **38**, 2066–2080.

500. Kenney, J.F. and Keeping, E.S (1951) Moment-generating and characteristic functions, some examples of moment-generating functions, and uniqueness theorem for characteristic functions. Sections 4.6–4.8 in *Mathematics of Statistics*, Pt. 2, 2nd ed. (Van Nostrand, Princeton, NJ) 72–77.

501. Kesten, H. (1973) Random difference equations and renewal theory for products of random matrices, *Acta Math.* **131**, 207–248.

502. Khanin, W.E., Mazel, A. and Sinai, Ya.G. (1997) Probability distribution functions for the random forced Burgers equation, *Phys. Rev. Lett.* **78**, 1904–1907.

503. Khmelnitskii, D.E. (1978) Impurity effect on the phase transition at $T = 0$ in magnets. Critical oscillations in corrections to the scaling laws, *Phys. Lett.* A **67**, 59–60.

504. Kinnison, R.R. (1985) *Applied extreme value statistics* (Battelle Press, Columbus, Ohio, Distributed by Macmillan, New York).

505. Kinzel, W. (1985) Learning and pattern recognition in spin glass models, *Zeitschrift fr Physik* B **60**, 205–213.

506. Kirby, S.H. and Kronenberg, A.K. (1987) Rheology of the lithosphere: selected topics, *Reviews of Geophysics* **25**, 1–1244.

507. Kirchner, J.W., Feng, X. and Neal, C. (2000) Fractal stream chemistry and its implications for contaminant transport in catchments, *Nature* **403**, 524–527.

508. Kirkpatrick, S. (1973) Percolation and conduction, *Rev. Mod. Phys.* **45**, 574–588.

509. Klafter, J., Shlesinger, M.F. and Zumofen, G. (1996) Beyond Brownian motion, *Physics Today* **49**, 33–39.

510. Klein, W. and Coniglio, A. (1981) Thermal phase transitions at the percolation threshold, *Phys. Lett.* A **84**, 83–84.

511. Klein, W., Rundle, J.B. and Ferguson, C.D. (1997) Scaling and nucleation in models of earthquake faults, *Phys. Rev. Lett.* **78**, 3793–3796.

512. Kliman, G. and Stein, J. (1992) Methods of motor current signature analysis, *Elec. Mach. Power Syst.* **20**, 463–474.

513. Klinger, M.I. (1988) Glassy disordered systems: topology, atomic dynamics and localized electron states, *Phys. Rep.* **165**, 275–397.

514. Kloster, M., Hansen, A and Hemmer, P.C. (1997) Burst avalanches in solvable models of fibrous materials, *Phys. Rev. E* **56**, 2615–2625.
515. Klüpperberg, C. (1988) Subexponential distributions and integrated tails, *J. Appl. Prob.* **25**, 132–141.
516. Klüpperberg, C. (1989) Subexponential distributions and characterizations of related classes, *Probab. Th. Rel. Fields* **82**, 259–269.
517. Knopoff, L. (1996) The organization of seismicity on fault networks, *Proc. Nat. Acad. Sci. U.S.A.* **93**, 3830–3837.
518. Knopoff, L. and Kagan, Y. Y. (1977) Analysis of the theory of extremes as applied to earthquake problems, *J. Geophys. Res.* **82**, 5647–5657.
519. Kolmogorov, A.N. (1933) *Grundberiffe der Warscheinlichkeitsrechnung* (Springer, Berlin).
520. Kolmogorov, A.N. (1941) On log-normal distribution law of particle sizes in fragmentation process, *Doklady AN USSR* **31**, 99–101 (in Russian).
521. Kolmogorov, A.N. (1965) *Theory of Probability* in *Mathematics, its Content, Methods and Meaning*, Chapter 9 (Academy of Sciences of the USSR, Moscow), p. 270 (in Russian).
522. Kolmogorov, A.N. (1965) Three approaches to the notion of "information quantity," *Problems of Transmission of Information* **1**, 3–11 (in Russian).
523. Kolmogorov, A.N. (1969) On the logical foundation of the theory of information and theory of probability, *Problems of Transmission of Information* **5**, 3–7 (in Russian).
524. Koopman, B. (1940) The axioms and algebra of intuitive probability, *Ann. Math. Stat.* **41**, 269–292.
525. Koplik, J., Redner, S. and Hinch, E.J. (1995) Universal and non-universal first-passage properties of planar multipole flows, *Phys. Rev. Lett.* **74**, 82–85.
526. Koplik, J., Redner, S. and Hinch, E.J. (1995) Tracer dispersion in planar multipole flows, *Phys. Rev. E* **50**, 4650–4671.
527. Koponen, I. (1995) Analytic approach to the problem of convergence of truncated Lévy flights towards the Gaussian stochastic process, *Phys. Rev. E* **52**, 1197–1199.
528. Korzhenevskii, A.L., Herrmanns, K. and Heuer, H.-O. (1999) Inhomogeneous ground state and the coexistence of two length scales near phase transitions in real solids, *Europhys. Lett.* **45**, 195–200.
529. Krapivsky, P.L., Grosse, I. and Ben-Naim, E. (1999) Scale Invariance and Lack of Self-Averaging in Fragmentation, *Phys. Rev. E* **61**, R993–R996
530. Krishnamurthy, S. and Barma, M. (1996) Active-site motion and pattern formation in self-organized interface depinning, *Phys. Rev. Lett.* **76**, 423–426.
531. Krishnamurthy, S. and Barma, M. (1998) Pattern formation in interface depinning and other models: erratically moving spatial structures, *Phys. Rev. E* **57**, 2949–2964.
532. Krishnamurthy S., Tanguy, A., Abry, P. and Roux, S. (2000) A stochastic description of extremal dynamics, *Europhysics Letters* **51**, 1–7.
533. Ktitarev, D.V. and Priezzhev, V.B. (1998) Expansion and contraction of avalanches in the two-dimensional Abelian sandpile, *Phys. Rev. E* **58**, 2883–2888.
534. Kubo, R. (1959) *Lectures in Theoretical Physics* **1**, 120, W. Brittin, ed. (Intercience, New York).
535. Kullback, S. (1958). *Information Theory and Statistics* (John Wiley, New York) Chapter 1.
536. Kulldorff, G. (1962) *Contributions to the Theory of Estimation from Grouped and Partially Grouped Samples* (Wiley, New York).

537. Kullmann, L. and Kertesz, J. (2001) Preferential growth: exact solution of the time-dependent distributions, *Phys. Rev. E* **63**, 051112.
538. Kullmann, L. and Kertesz, J. (2001) Preferential growth: solution and application to modeling stock market, *Physica A* **299**, 121–126.
539. Kundagrami, A., Dasgupta, C., Punyindu, P. and DasSarma, S. (1998) Extended self-similarity in kinetic surface roughening, *Phys. Rev. E* **57**, R3703–R3706.
540. Kuo, C.C. and Phoenix, S.L. (1987) *J. Appl. Prob.* **24**, 137–159.
541. Kumamoto, H. and Henley, E.J. (1996) *Probabilistic Risk Assessment and Management for Engineers and Scientists*. 2nd ed. (IEEE PRESS Marketing, New York).
542. Kumar, R., Raghavan, P., Rajagopalan, S., Sivakumar, D., Tomkins, A., and Upfal, E. (2000) Stochastic models for the Web graph, In *Proceedings of the 41st Annual Symposium on Foundations of Computer Science*, 57–65.
543. Kurchan, J. (2000) Emergence of macroscopic temperatures in systems that are not thermodynamical microscopically: towards a thermodynamical description of slow granular rheology, *Journal of Physics: Condensed Matter* **12**, 6611–6617.
544. La Barbera, P. and Roth, G. (1994) Invariance and scaling properties in the distributions of contributing area and energy in drainage basins, *Hydrol. Proc.* **8**, 123–125.
545. LaBreque, J.L., Cande, S.C. and Jarrard, R.D. (1985) Intermediate-wavelength magnetic anomaly field of the North Pacific and possible source distributions, *J. Geophys. Res.* **90**, 2549–2564.
546. LaBreque, J.L. and Raymond, C.A. (1985) Sea floor spreading anomalies in the Magsat field of the North Atlantic, *J. Geophys. Res.* **90**, 2565–2575.
547. Laherrère, J. and Sornette, D. (1998) Stretched exponential distributions in Nature and Economy: "Fat tails" with characteristic scales, *Eur. Phys. J. B* **2**, 525–539.
548. Landau, L. and Lifshitz, E. (1970) *Theory of Elasticity*. 2d English ed. (Pergamon Press, Oxford, New York).
549. Lanford, O.E. (1973) Entropy and equilibrium states in classical mechanics, in Statistical Mechanics and Mathematical Problems, ed. A. Lenard, Springer, Berlin, *Lect. Notes Phys.* **20**, 1–113.
550. Langer, J.S. in 'Solids Far from Equilibrium', Ed. C. Godrèche, (Cambridge Univ. Press 1992); J.S. Langer in 'Far from Equilibrium Phase Transitions', Ed. L. Garrido, (Springer, 1988); J.S. Langer in 'Fluctuations, Instabilities and Phase Transitions', Ed. T. Riste, Nato Advanced Study Institute, Geilo Norway, 1975 (Plenum, 1975).
551. Laslett, G.M. (1982) Censoring and edge effects in areal and line transect sampling of rock joint traces, *Math. Geol., Vol.* **14**, 125–140, 1982.
552. Ledrappier F., Young L. S., "The metric entropy for diffeomorphisms", Annals of Math. 122, (1985) 509–574
553. Lee, M.W., Unstable fault interactions and earthquake self-organization, PhD Thesis, UCLA (1999).
554. Lee, M.W., Sornette, D. and Knopoff, L. (1999) Persistence and Quiescence of Seismicity on Fault Systems, *Phys. Rev. Lett.* **83**, 4219–4222.
555. Lee, M.W. and Sornette, D. (2000) Novel mechanism for discrete scale invariance in sandpile models, *Eur. Phys. J. B* **15**, 193–197.
556. Lekkerkerker, H.N.W. and Boon, J.-P., Brownian motion near hydrodynamic-regime transition, *Phys. Rev. Lett.* **36**, 724–725.
557. Leschhorn, H. and Tang, L.-H. (1994) Avalanches and correlations in driven interface depinning, *Phys. Rev. E* **49**, 1238–1245.

558. Leung, K.T., Andersen, J.V. and Sornette, D. (1998) Self-organized criticality in stick-slip models with periodic boundaries, *Phys. Rev. Lett.* **80**, 1916–1919.

559. Leung, K.T., Andersen, J.V. and Sornette, D. (1998) Self-organized criticality in an isotropically driven model approaching equilibrium, *Physica* **254**, 85–96.

560. Levinshtein et al. (1976) *Soviet Phys. JETP* **42**, 197.

561. Levy, M. and Solomon, S. (1996) Power laws are logarithmic Boltzmann laws, *Int. J. Mod. Phys. C* **7**, 595–601.

562. Li, W., $1/f$-noise, http://linkage.rockefeller.edu/wli/1fnoise/ and references therein (site checked July 3, 2000).

563. Liebovitch, L.S. et al. (1999) Nonlinear properties of cardiac rhythm abnormalities, *Phys. Rev. E* **59**, 3312–3319.

564. Liebowitz, H., ed. (1984) *Fracture*, Academic, New York, Vols. I–VII.

565. Liggett, T.M. (1985) *Interacting Particle Systems* (Springer, New York).

566. Liggett, T.M. (1997) Stochastic models of interacting systems. *The Annals of Probability* **25**, 1–29.

567. Linde, A.T. et al., A slow earthquake sequence on the San Andreas fault, *Nature* **383**, 65–68.

568. Lise, S. and Jensen, H.J. (1996) Transitions in non-conserving models of self-organized criticality, *Phys. Rev. Lett.* **76**, 2326–2329.

569. Livi, R., Pettini, M., Ruffo, S., Sparpaglione, M. and Vulpiani, A. (1985) Equipartition threshold in nonlinear large Hamiltonian systems: the Fermi–Pasta–Ulam model, *Phys. Rev. A* **31**, 1039–1045.

570. Livi, R., Pettini, M., Ruffo, S. and Vulpiani, A. (1985) Further results on the equipartition threshold in large nonlinear Hamiltonian systems, *Phys. Rev. A* **31**, 2740–2743.

571. Lotka, A.J. (1926) The frequency distribution of scientific productivity, *Journal of the Washington Academy of Sciences* **16**, 317–323.

572. Lomnitz-Adler, J., Knopoff, L. and Martinez-Mekler, G. (1992) Avalanches and epidemic models of fracturing in earthquakes, *Phys. Rev. A* **45**, 2211–2221.

573. Loretan, M. and Phillips, P.C.B. (1994) Testing the covariance stationarity of heavy-tailed time series, *J. Empirical Finance* **1**, 211–248.

574. Lu, E.T. (1995) Avalanches in continuum driven dissipative systems, *Phys. Rev. Lett.* **74**, 2511–2514.

575. Lukacs, E. (1983) *Developments in Characteristic Function Theory* (McMillan, New York).

576. Lyakhovsky, V., BenZion, Y. and Agnon, A., Distributed damage, faulting, and friction, *J. Geophys. Res.* **102**, 27635–27649.

577. Machta, J. and Guyer, R.A. (1987) Largest current in a random resistor network, *Phys. Rev. B* **36**, 2142–2146.

578. Mahmound, H., Smythe, R., and Szymanski, J. (1993) On the structure of plane-oriented recursive trees and their branches, *Random Structures and Algorithms* **3**, 255–266.

579. Main, I.G. (1996) Statistical physics, seismogenesis, and seismic hazard, *Rev. Geophys.* **34**, 433–462.

580. Majumdar, S.N. and Dhar, D. (1991) Height correlations in the Abelian sandpile model, *J. Phys. A Math. Gen.* **24**, L357–L362.

581. Majumdar, S.N. and Dhar, D. (1992) Equivalence between the abelian sandpile model and the $Q \to 0$ limit of the Potts model, *Physica A* **185**, 129–145.

582. Malamud, B.D., Morein, G. and Turcotte, D.L. (1998) Forest fires – an example of self-organized critical behavior, *Science* **281**, 1840–1842.

583. Malcai, O., Lidar, D.A., Biham, O. and Avnir, D. (1997) Scaling range and cutoffs in empirical fractals, *Phys. Rev. E* **56**, 2817–2828.

584. Malevergne, Y., Pisarenko, V.F. and Sornette, D. (2003) Empirical distributions of log-returns: between the stretched exponential and the power law? submitted to *Quantitative Finance* (http://arXiv.org/abs/physics/0305089)

585. Malin, S. (1996) More spirited debate on Physics, Parapsychology and paradigms, *Physics Today* **April**, 15–15.

586. Mallat, S. (1997) *A Wavelet Tour of Signal Processing* (Academic Press, San Diego).

587. Maloy, K.J., Hansen, A., Hinrichsen, E.L. and Roux, S. (1992) Experimental measurements of the roughness of brittle cracks, *Phys. Rev. Lett.* **68**, 213–216.

588. Malte-Sorensen, A., Feder, J., Christensen, K., Frette, V., Josang, T. and Meakin, P. (1999) Surface fluctuations and correlations in a pile of rice, *Phys. Rev. Lett.* **83**, 764–767.

589. Mandelbrot, B. (1953) An Information Theory of the Statistical Structure of Languages, Proceedings of the Symposium on Applications of Communication Theory, London, Betterworths, London , pp. 486–502.

590. Mandelbrot, B.B. (1963) The variation of certain speculative prices, *Journal of Business* **36**, 394–419.

591. Mandelbrot, B.B. (1967) How long is the coast of Britain? Statistical self-similarity and fractional dimension, *Science* **155**, 636–638.

592. Mandelbrot, B.B. (1982) *The Fractal Geometry of Nature* (W.H. Freeman, San Francisco).

593. Mandelbrot, B.B. (1998) Is nature fractal? *Science* **279**, 783–784.

594. Mandelbrot, B.B. and Van Nessm, J.W. (1968) Fractional Brownian motions, fractional noise and applications, *Society for Industrial and Applied Mathematics Review* **10**, 422–437.

595. Manna, S.S. (1991) Critical exponents of the sandpile models in two dimensions, *Physica A* **179**, 249–268.

596. Manna, S.S. (1999) Sandpile models of self-organized criticality, *Current Science* **77**, 388.

597. Manna, S.S., Dhar, D., and Majumdar, S.N. (1992) Spanning trees in two dimensions, *Phys. Rev. A* **46**, R4471–R4474.

598. Mantegna, R. and Stanley, H.E. (1994) Stochastic process with ultraslow convergence to a Gaussian – the truncated Lévy flight, *Phys. Rev. Lett.* **73**, 2946–2949.

599. Mantegna, R. and Stanley, H.E. (1997) Scaling behaviour in the dynamics of an economic index, *Nature* **376**, N6535, 46–49.

600. Mantegna, R. and Stanley, H.E. (1997) Econophysics: Scaling and its breakdown in finance, *J. Stat. Phys.* **89**, 469–479.

601. Mantegna, R.N., Buldyrev, S.V., Goldberger, A.L., Halvin, S. and Stanley, H.E. (1995) Systematic analysis of coding and non-coding sequences using methods of statistical linguistics, *Phys. Rev. E* **52**, 2939–2950.

602. Marder, M. and Fineberg, J. (1996) How things break, *Physics Today* **49**, 24–29.

603. Marinari, E., Parisi, G. and Ritort, F. (1994) Replica field theory for deterministic models. 2-A non-random spin glass with glassy behavior, *J. Phys. A* **27**, 7647–7668.

604. Marinari, E., Parisi, G., Ricci-Tersenghi, F., Ruiz-Lorenzo, J.J. and Zuliani, F. (2000) Replica Symmetry Breaking in Short Range Spin Glasses: A Review of the Theoretical Foundations and of the Numerical Evidence, *J. Stat. Phys.* **98**, 973–1047.

605. Markowitz, H. (1959) *Portfolio Selection: Efficient Diversification of Investment* (Wiley, New York).
606. Maslov, D.L. (1993) Absence of self-averaging in shattering fragmentation processes, *Phys. Rev. Lett.* **71**, 1268–1271.
607. Mathai, A.M. and Saxena, R.K. (1978) *The H-function with applications in statistics and other disciplines* (Wiley Eastern Limited, New Delhi).
608. Maus, S. and Dimri, V.P. (1994) Scaling properties of potential fields due to scaling sources, *Geophys. Res. Lett.* **21**, 891–894.
609. Maus, S. and Dimri, V.P. (1995) Potential field power spectrum inversion for scaling geology, *J. Geophys. Res.* **100**, 12605–12616.
610. Maus, S. and Dimri, V.P. (1996) Depth estimation from the scaling power spectrum of potential fields, *Geophys. J. Int.* **124** 113–120.
611. McCartney, L.N. and Smith, R.L. (1993) Statistical theory of the strength of fiber bundles, Transactions of the ASME, *J. Appl. Mech.* **105**, 601–608.
612. McKay, D.S., Gibson, E.K., Thomaskeprta, K.L., Vali, H. et al. (1996) Search for past life on Mars – Possible relic biogenic activity in Martian meteorite ALH84001, *Science* **273**, 924–930.
613. McKay, S.R., Berker, A.N. and Kirkpatrick, S. (1982) Spin-glass behavior in frustrated Ising models with chaotic renormalization group trajectories, *Phys. Rev. Lett.* **48**, 767–770.
614. McWhirter, J.G. and Pike, E.R. (1978) On the numerical inversion of the Laplace transform and similar Fredholm integral equations of the first kind, *J. Phys. A* **11**, 1729–1745.
615. Meakin, P. (1998) *Fractals, Scaling, and Growth far from Equilibrium* (Cambridge University Press, Cambridge, U.K.; New York).
616. Mehta, A. and Edwards, S.F. (1989) Statistical mechanics of powder mixtures, *Physica A* **157**, 1091–1100.
617. Melnikov, V.I. (1991) The Kramers problem – 50 years of development, *Physics Reports* **209**, 1–71.
618. Mendes, J.F.F., Dickman, R., Henkel, M. and Marques, M. C. (1994) Generalized scaling for models with multiple absorbing states, *J. Phys. A* **27**, 3019–3028.
619. Metzler, R. and Klafter, J. (2000) The random walk's guide to anomalous diffusion: a fractional dynamics approach, *Physics Reports* **339**, 1–77.
620. Metzler, R., Klafter, J. and Jortner, J. (1999) Hierarchies and logarithmic oscillations in the temporal relaxation patterns of proteins and other complex systems, *Proc. Natl. Acad. Sci. USA* **96**, 11085–11089.
621. Metzler, R., Klafter, J. and Sokolov, I. (1998) Anomalous transport in external fields: continuous time random walks and fractional diffusion equations extended, *Phys. Rev. E* **58**, 1621–1633.
622. Mézard, M., Parisi, G. and Virasoro, M. (1987) *Spin Glass Theory and Beyond* (World Scientific, Singapore).
623. Middleton, A.A. and Tang, C. (1995) Self-organized criticality in nonconserved systems, *Phys. Rev. Lett.* **74**, 742–745.
624. Middleton, D. (1977) Statistical-physical models of electromagnetic interference, *IEEE Trans. Electromagn. Compat.* **EMC-19**(3), 106–127.
625. Miller, G.A. (1957) Some effects of intermittent silence *American Journal of Psychology* **70**, 311–314.
626. Miller, J. and Huse, D.A. (1993) Macroscopic equilibrium from microscopic irreversibility in a chaotic coupled-map lattice, *Phys. Rev. E* **48**, 2528–2535.
627. Miltenberger, P., Sornette, D. and Vanneste, Fault self-organization as optimal random paths selected by critical spatio-temporal dynamics of earthquakes, *Phys. Rev. Lett.* **71**, 3604–3607.

628. Minnich, R.A. and Chou, Y.H. (1997) Wildland fire patch dynamics in the chaparral of southern California and northern Baja California, *International Journal of Wildland Fire* **7**, 221–248.

629. Mints, R.G., Snapiro, I.B. and Brandt, E.H. (1997) Long-range interaction of fluctuating vortices with a parallel surface in layered superconductors, *Phys. Rev. B* **55**, 8466–8472.

630. Mirollo, R.E. and Strogatz, S.H. (1990) Synchronization of pulse-coupled biological oscillators, *SIAM J. Appl. Math.* **50**, 1645–1662.

631. Mises, R. (1941) On the foundations of probability, *Ann. Math. Stat.* **12**, 191.

632. Mittag-Leffler, M.G. (1903) Sur la nouvelle fonction, *Comptes Rendus Acad. Sci. Paris* **137**, 554–558.

633. Mittag-Leffler, M.G. (1905) Sur la representation analytique d'une branche uniforme d'une fonction monogene, *Acta Math.* **29**, 101–181.

634. Mittnik, S. and Rachev, S.T. (1993) Reply to comments on 'Modelling asset returns with alternative stable distributions' and some extensions, *Econometric Reviews* **12**, 347–389.

635. Mitzenmacher, M. (2003) A brief history of generative models for power law and lognormal distributions, *Internet Mathematics* **1**, to appear. (http://www.eecs.harvard.edu/~michaelm/NEWWORK/papers.html).

636. Moeur, W.A., Day, P.K., Liu, F.C., Boyd, S.T.P., Adriaans, M.J. and Duncan, R.V. (1997) Observation of self-organized criticality near the superfluid transition in He-4, *Phys. Rev. Lett.* **78**, 2421–2424.

637. Mogi K. (1969) Some features of recent seismic activity in and near Japan 2: activity before and after great earthquakes, *Bull. Eq. Res. Inst. Tokyo Univ.* **47**, 395–417.

638. Mogi, K. (1995) Earthquake prediction research in Japan, *J. Phys. Earth* **43**, 533–561.

639. Molchan, G.M. (1996) Scaling exponents and multifractal dimensions for independent random cascades, *Commun. Math. Phys.* **179**, 681–702.

640. Mollison, D. (1977) *J. Roy. Statist. Soc. B* **39**, 283.

641. Molnar, P. and Ramirez, J.A. (1998) Energy dissipation theories and optimal channel characteristics of river networks, *Water Ressources Research* **34**, 1809–1818.

642. Monette, L. and Klein, W. (1992) Spinodal decomposition as a coalescence process, *Phys. Rev. Letts.* **68**, 2336–2339.

643. Monette, L. (1994) Spinodal decomposition, *Int. J. Mod. Phys. B* **11**, 1417–1527.

644. Montakhab, A. and Carlson, J.M. (1998) Avalanches, transport, and local equilibrium in self-organized criticality, *Phys. Rev. E* **58**, 5608–5619.

645. Montroll, E.W. (1981) On the entropy function in sociotechnical systems, *Proc. Natl. Acad. Sci. USA* **78**, 7839–7843.

646. Montroll, E.W. and Shlesinger, M.F. (1982) On 1/f noise and other distributions with long tails, *Proc. Nat. Acad. Sci. USA* **79**, 3380–3383.

647. Montroll, E.W. and Shlesinger, M.F. (1983) Maximum entropy formalism, fractals, scaling phenomena, and 1/f noise: a tale of tails, *J. Stat. Phys.* **32**, 209–230.

648. Montroll, E.W. and Shlesinger, M.F. (1984) in *Studies in Statistical Mechanics*, ed. by Lebowitz, J. and Montroll, E. (North-Holland, Amsterdam), Vol. 11, p. 1.

649. Montroll, E.W. and West, B.J. (1979) On an enriched collection of stochastic processes, in *Fluctuation Phenomena*, Eds. E.W. Montroll and Lebowitz, J.L. (Elsevier Science Publishers B.V., North Holland, Amsterdam).

650. Moreno, J.M., ed. (1998) *Large Forest Fires* (Backhuys, Leiden).

651. Morse, P.M. and Feshbach, H. (1953) *Methods in Theoretical Physics* (McGraw-Hill, New York).
652. Moshe, M. (1978) *Phys. Rep. C* **37**, 255.
653. Mosteller, F., Rourke, R. and Thomas, G.B. (1961) *Probability: a First Course* (Addison Wesley, London).
654. Moura, A. and Yukalov, V.I. (2002) Self-similar extrapolation for the law of acoustic emission before failure of heterogeneous materials, *International Journal of Fracture* **115** (1), L3–L8.
655. Munoz, M.A., Dickman, R., Vespignani, A. and Zapperi, S. (1999) Avalanche and spreading exponents in systems with absorbing states, *Phys. Rev. E* **59**, 6175–6179.
656. Murphree, E.S. (1989) Some new results on the subexponential class, *J. Appl. Prob.* **27**, 892–897.
657. Murray, J.B., Voight, B. and Glot, J.P. (1994) Slope movement crisis on the east flank of Mt Etna Volcano – Models for eruption triggering and forecasting, *Engineering Geology* **38**, 245–259.
658. Muzy, J.-F., Delour, J. and Bacry, E. (2000) Modelling fluctuations of financial time series: from cascade process to stochastic volatility model, *Eur. Phys. J. B* **17**, 537–548.
659. Muzy, J.-F., Sornette, D., Delour, J. and Arneodo, A. (2001) Multifractal returns and hierarchical portfolio theory, *Quantitative Finance* **1**, 131–148.
660. Nagaev, S.V. (1963) On the asymptotic behavior of one-sided large deviation probabilities, *Doklady Akademii Nauk SSSR* **143**(2), 280 (in Russian). translated into English in *Theory of Probability and Its Applications (USA)* **26**(2), 362–366 (1981).
661. Nagel, S.R. (1992) Instabilities in a sandpile, *Rev. Mod. Phys.* **64**, 321–325.
662. Narkunskaya, G.S. (1988) Phase transition in a hierarchical model, *Computational Seismology* **21**, 25–31 (English translation (1989), 59–66.
663. Narkunskaya, G.S. and Shnirman, M.G. (1990) Hierarchical model of defect development and seismicity, *Phys. Earth Planet. Inter.* **61**, 29–35.
664. National Materials Advisory Board Commission on Engineering and Technical Systems, *"Aging of U.S. Air Force Aircraft"* (1997) FINAL REPORT from the Committee on Aging of U.S. Air Force Aircraft, National Research Council, Publication NMAB-488-2 (National Academy Press, Washington, D.C.).
665. National Research Council (1997) *Satellite Gravity and the Geosphere: Contribution to the Study of the Solid Earth and its Fluid Earth.* 112 pp., (National Academy Press, Washington, D.C.).
666. Nature debate, April 1999, "Is the reliable prediction of individual earthquakes a realistic scientific goal?", http://helix.nature.com/debates/earthquake/
667. Nauenberg, M. (1975) Scaling representations for critical phenomena, *J. Phys. A* **8**, 925.
668. Naylor, M.A., Mandl, G. and Sijpesteijn, C.H.K. (1986) Fault geometries in basement-induced wrench faulting under different initial stress states, *J. Struct. Geol.* **8**, 737–752.
669. Nazarov, A.A., Romanov, A.E. and Valiev, R.Z. (1993) On the structure, stress fields and energy of nonequilibrium grain boundaries, *Acta metall. Mater.* **41**, 1033–1040.
670. Nazarov, A.A. and Romanov, A.E. (1993) Stress fields of disordered dislocation arrays – finite walls, *Phil. Mag. Lett.* **68**, 297–301.

671. Nazarov, A.A., Romanov, A.E. and Baudelet, B. (1993) Long-range stress fields of disordered dislocation arrays: two types of disorder, and two decaying laws, *Phil. Mag. Lett.* **68**, 303–307.

672. Newell, G.F. (1959) Platoon formation in traffic, *Operations Research* **7**, 589–598.

673. Newman, M.E.J. (1996) Self-organized criticality, evolution and the fossil extinction record, *Proc. R. Soc. London B* **263**, 1605–1610.

674. Newman, M.E.J. (1997) Evidence for self-organized criticality in evolution, *Physica D* **107**, 292–296.

675. Newman, M.E.J. (1997) A model of mass extinction, *J. Theor. Biol.* **189**(3), 235–252.

676. Newman, M.E.J., Girvan and Farmer, J.D. (2002) Optimal design, robustness, and risk aversion *Phys. Rev. Lett.* **89**, DOI: 10.1103/PhysRevLett.89.028301.

677. Newman, M.E.J. and Sneppen, K. (1996) Avalanches, scaling and coherent noise, *Phys. Rev. E* **54**, 6226–6231.

678. Newman, W.I. and Turcotte, D.L. (1990) Cascade model for fluvial geomorphology, *Geophys. J. Int.* **100**, 433–439.

679. Newman, W.I. and Gabrielov, A.M. (1991) Failure of hierarchical distribution of fibre bundles.I, *Int. J. Fract.* **50**, 1–14.

680. Newman, W.I., Gabrielov, A., Durand, T., Phoenix, S.L. and Turcotte, D. (1994) An exact renormalization model for earthquakes and material failure, *Physica D* **77**, 200–216.

681. Newman, W.I., Turcotte, D. and Gabrielov, A. (1995) Log-periodic behavior of a hierarchical failure model with applications to precursory seismic activation, *Phys. Rev. E* **52**, 4827–4835.

682. Neyman, J. (1952) *Lectures and Conferences on Mathematical Statistics* (Graduate School, U.S. Department of Agriculture, Washington, D.C.).

683. Niemeijer, T. and van Leeuwen, J.M.J. (1976) Renormalization theory for Ising-like spin systems, in Phase Transitions and Critical Phenomena, Vol. 6, C. Domb and Green, M.S. (eds.), Academic Press, London, pp. 425–507.

684. Nieuwenhuizen, T.M. and Van Rossum, M.C.W. (1991) Universal fluctuations in a simple disordered system, *Phys. Lett. A* **160**, 461–464.

685. Nikias, C.L., and Shao, M. (1995) *Signal Processing with Alpha-Stable Distributions and Applications* (Wiley, New York).

686. Nishenko, S.P. and Barton, C.C. (1993) Scaling laws for natural disasters: application of fractal statistics to life and economic loss data (abstract), *Geol. Soc. Amer., Abstracts with Programs* **25**, 412.

687. Normand, C., Pomeau, Y. and Velarde, M. (1977) Convective instability: a physicist's approach, *Rev. Mod. Phys.* **49**, 581–624.

688. Normand, J.M. and Herrmann, H.J. (1995) Precise determination of the conductivity exponent of 3D percolation using Percola, *Int. J. Mod. Phys. C* **6**, 813–817.

689. Nottale, L. (1993) *Fractal Space–Time and Microphysics: Towards a Theory of Scale Relativity* (World Scientific, Singapore).

690. Nottale, L. (1996) Scale relativity and fractal space–time – applications to quantum physics, cosmology and chaotic systems, *Chaos Solitons & Fractals* **7**, 877–938.

691. Nottale, L. (1997) Scale-relativity and quantization of the universe. 1. Theoretical framework, *Astronomy & Astrophysics* **327**, 867–889.

692. Nottale, L. (1999) The scale-relativity program, *Chaos Solitons & Fractals* **10**, 459–468.

693. Nottale, L. (2001) Scale relativity and gauge invariance, *Chaos Solitons & Fractals* **12**, 1577–1583.
694. Nottale, L. (2003) Scale-relativistic cosmology, *Chaos Solitons & Fractals* **16**, 539–564.
695. Noullez, A., Wallace, G., Lempert, W., Miles, R.B., and Frisch, U. (1997) Transverse velocity increments in turbulent flow using the RELIEF technique, *J. Fluid Mech.* **339**, 287–307.
696. Novikov, E.A. (1966), *Dokl. Akad. Nauk SSSR* **168/6**, 1279.
697. Novikov, E.A. (1990), The effect of intermittency on statistical characteristics of turbulence and scale similarity of breakdown coefficients, *Phys. Fluids A* **2**, 814–820.
698. Obukhov, S.P. (1990) Self-organized criticality – Goldstone modes and their interactions, *Phys. Rev. Lett.* **65**, 1395–1398.
699. Ogata, Y. and Katsura, K. (1993) Analysis of temporal and spatial heterogeneity of magnitude frequency distribution inferred from earthquake catalogues, *Geophys. J. Int.* **113**, 727–738.
700. Olami, Z., Feder, H.J.S. and Christensen, K. (1992) Self-organized criticality in a continuous, non-conservative cellular automaton modelling of earthquakes, *Phys. Rev. Lett.* **68**, 1244–1247.
701. Onsager, L. (1931) *Phys. Rev.* **37**, 405.
702. Onsager, L. (1931) *Phys. Rev.* **38**, 2265.
703. Onsager, L. (1944) *Phys. Rev.* **65**, 117.
704. Onsager, L. and Machlup, S. (1953) (1931) *Phys. Rev.* **91**, 1505.
705. Ord, A. and Hobbs, B.E. (1989) The strength of the continental crust, detachment zones and the development of plastic instabilities, *Tectonophysics* **158**, 269–289.
706. Orlean, A. (1989) Mimetic contagion and speculative bubbles, *Theory and Decision* **27**, 63–92.
707. Orlean, A. (1991) Disorder on the stock market, *La Recherche* **22**, 668–672.
708. Osen, L.M. (1974) *Women in mathematics* Cambridge, Mass., MIT Press).
709. Ostrowsky, N., Sornette, D., Parker, P., and Pike, E.R. (1981) Exponential sampling method for light scattering polydispersity analysis, *Optica Acta* **28**, 1059–1070.
710. Ouillon, G. and Sornette, D. (1996) Unbiased multifractal analysis: application to fault patterns, *Geophys. Res. Lett.* **23**, 3409–3412.
711. Ouillon, G., Castaing, C. and Sornette, D. (1996) Hierarchical scaling of faulting, *J. Geophys. Res.* **101**, 5477–5487.
712. Ouillon, G., Sornette, D., Genter, A. and Castaing, C. (1996) The imaginary part of the joint spacing distribution, *J. Phys. I France* **6**, 1127–1139.
713. Paczuski, M., Maslov, S. and Bak, P. (1996) Avalanche dynamics in evolution, growth and depinning models, *Phys. Rev. E* **53**, 414–443.
714. Paczuski, M. and Boettcher, S. (1997) Avalanches and waves in the Abelian sandpile model, *Phys. Rev. E* **56**, R3745–R3748.
715. Paling, J. and Paling, S. (1993) *Up to the Armpits in Alligators* (The Environmental Institute, Gainesville, Florida).
716. Palmer, R.G., Stein, D.L., Abrahams, E. and Anderson, P.W. (1984) Models of hierarchically constrained dynamics for glassy relaxation, *Phys. Rev. Lett.* **53**, 958–961.
717. Panchang, V.G. and Gupta, R.C. (1989) On the determination of three-parameter Weibull Maximum Likelihood estimators, *Communications in Statistics – Simulation and Computation* **18**, 1037–1057.
718. Pareto, V. (1896–1965) *Cours d'Economie Politique*. Reprinted as a volume of *Oeuvres Complètes* (Droz, Geneva).

719. Parisi, G. (1980) A sequence of approximated solutions to the S-K model for spin glasses, *J. Phys. A: Math. Gen.* **13**, L115–L121 (1980).
720. Parisi, G. (1980) The order parameter for spin glasses: a function on the interval 0-1, *J. Phys. A: Math. Gen.* **13**, 1101–1112.
721. Parisi, G. (1980) Magnetic properties of spin glasses in a new mean field theory, *J. Phys. A: Math. Gen.* **13**, 1887–1895.
722. Parisi, G. and Frisch, U. (1985) On the singularity structure of fully-developed turbulence, in *Turbulence and Predictability in Geophysical Fluid Dynamics, Proceedings International School Enrico Fermi, 1983, Varenna, Italy*, eds. Ghil, M., Benzi, R. and Parisi, G. (North Holland, Amsterdam).
723. Park, H., Köhler, J., Kim, I.-M., Ben-Avraham, D. and Redner, S. (1993) Excluded volume effects in heterogeneous catalysis – Reactions between dollars and dimes, *J. Phys. A* **26**, 2071–2079.
724. Parker, R.L. and O Brien, M.S. (1997) Spectral analysis of vector magnetic field profiles, *J. Geophys. Res.* **102**, 24815–24824.
725. Paul, G. (1999) Coefficient scaling, *Phys. Rev. E* **59**, 4847–4856.
726. Peebles, P.J.E. (1980) *The Large Scale Structures of the Universe* (Princeton University Press, Princeton, NY).
727. Pelletier, J.D. (1997) Kardar-Parisi-Zhang scaling of the height of the convective boundary layer and fractal structure of Cumulus cloud fields, *Phys. Rev. Lett.* **78**, 2672–2675.
728. Peltzer, G. and Tapponnier, P. (1988) Formation and evolution of strike-slip faults, rifts, and basins during the India-Asia collision: an experimental approach, *J. Geophys. Res.* **93**, 15085–15117.
729. Peng, G. (1992) Self-organized critical state in a directed sandpile automaton on Bethe lattices: equivalence to site percolation, *J. Phys. A: Math. Gen.* **25**, 5279–5282.
730. Percovic, O., Dahmen, K. and Sethna, J. P. (1995) Avalanches, Barkhausen noise and plain old criticality, *Phys. Rev. Lett.* **75**, 4528–4531.
731. Perković, O. and Sethna, J.P. (1997) Improved magnetic information storage using return-point memory, *J. Appl. Phys.* **81**, 1590–1597.
732. Perković, O., Dahmen, K.A. and Sethna, J.P. (1999) Disorder-induced critical phenomena in hysteresis: Numerical scaling in three and higher dimensions, *Phys. Rev. B* **59**, 6106–6119.
733. Phoenix, S.L. and Smith, R.L. (1983) A comparison of probabilistic techniques for the strength of fibrous materials under local load-sharing among fibers, *Int. J. Solids and Structures* **19**, 479–496.
734. Phoenix, S.L. and Smith, R.L. (1989) The Strength Distribution and Size Effect in a Prototypical Model for Percolation Breakdown in Materials, *Technical Report* **89–43** (Mathematical Sciences Institute, Cornell University).
735. Pietronero, L., Tartaglia, P. and Zhang, Y.-C. (1991) Theoretical studies of self-organized criticality, *Physica A* **173**, 22–44.
736. Piggot, A.R. (1997) Fractal relations for the diameter and trace length of disc-shaped fractures, *J. Geophys. Res.* **102**, 18121–18125.
737. Pisarenko, V.F. (1998) Non-linear growth of cumulative flood losses with time, *Hydrological Processes* **12**, 461–470.
738. Pisarenko, V.F. and Sornette, D. (2003) Characterization of the frequency of extreme events by the generalized pareto distribution, in press in *Pure and Applied Geophysics* (http://arXiv.org/abs/cond-mat/0011168)
739. Pisarenko, V.F. and Sornette, D. (2003) Rigorous statistical detection and characterization of a deviation from the Gutenberg–Richter distribution above magnitude 8 in subduction zones, in press in *Pure and Applied Geophysics* (http://arXiv.org/abs/cond-mat/0201552)

740. Pitman, E.J.G. (1980) Subexponential distribution functions, *J. Austral. Math. Soc. (Series A)* **29**, 337–347.

741. Pollock, A.A. (1981) Acoustic Emission Amplitude Distributions, *International Advances in Non-Destructive Testing* **7**, 215–239.

742. Pollock, A.A. (1989) Acoustic Emission Inspection, in *Metal Handbook Ninth Edition* **17**, Non-destructive evaluation and quality control, ASM INTERNATIONAL, 278–294.

743. Pomeau, Y. (1982) Symétrie des fluctuations dans le renversement du temps (Symmetry of fluctuations under time reversal), *J. Physique* **43**, 859–867.

744. Pomeau, Y. (1986) Front motion, metastability and subcritical bifurcations in hydrodynamics, *Physica D* **23**, 3–11.

745. Pomeau, Y., Pumir, A. and Young, W. (1989) Anomalous diffusion of tracer in convection rolls, *Phys. Fluids A* **1**, 462–469.

746. Prato, D. and Tsallis, C. (1999) Nonextensive foundation of Levy distributions, *Phys. Rev. E* **60**, 2398–2401.

747. Press, S.J. (1989) *Bayesian Statistics: Principles, Models, and Applications* (Wiley, New York).

748. Preston, E. and Kraus T.L. (1993) In Andrew's Path: a Historical Report on FAA's Response to and Recovery from Hurricane Andrew, [Washington, D.C.]: U.S. Dept. of Transportation, Federal Aviation Administration, History Staff, Office of Public Affairs.

749. Priezzhev, V.B., Dhar, A., Krishnamurthy, S. and Dhar, D. (1996) Eulerian walkers as a model of self-organized criticality, *Phys. Rev. Lett.* **77**, 5079–5082.

750. Priezzhev, V.B., Ktitarev, D.V. and Ivashkevitch, E.V. (1996) Formation of avalanches and critical exponents in an abelian sandpile model, *Phys. Rev. Lett.* **76**, 2093–2096.

751. Privman, V. and Fisher, M.E. (1983) Finite-size effects at first-order transitions, *J. Stat. Phys.* **33**, 385–417.

752. Privman, V. ed. (1990) *Finite Size Scaling and Numerical Simulation of Statistical Systems* (World Scientific, Singapore; Teaneck, NJ).

753. Prosen, T. and Campbell, D.K. (2000) Momentum Conservation Implies Anomalous Energy Transport in 1D Classical Lattices, *Phys. Rev. Letts.* **84**, 2857–2860.

754. Pumir, A. (1985) Statistical properties of an equation describing fluid interfaces, *J. Physique* **46**, 511–522.

755. Pury, P.A. (1990) Asymmetry and convergence in the central limit theorem: an approach for physicists, *Am. J. Phys.* **58**, 62–67.

756. QueirosConde, D. (1997) Geometrical extended self-similarity and intermittency in diffusion-limited aggregates, *Phys. Rev. Lett.* **78**, 4426–4429.

757. Qiao, H.Z., and Tsokos, C.P. (1994) Parameter estimation of the Weibull probability distribution, *Mathematics and Computers in Simulation* **37**, 47–55.

758. Ramakrishna Rao, K., ed. (1984) *The Basic Experiments in Parapsychology* (McFarland, Jefferson, N.C.).

759. Rammal, R., Toulouse, G. and Virasoro, M.A. (1986) Ultrametricity for physicists, *Rev. Mod. Phys.* **58**, 765–788.

760. Ramsey, J.B and Rothman, P. (1996) Time irreversibility and business cycle asymmetry, *Journal of Money, Credit & Banking* **28**, 1–21.

761. Rao, C. (1965) *Linear statistical inference and its applications* (Wiley, New York), Chapter 6, Section 6e.3.

762. Rau, J. (1997) Statistical Mechanics in a Nutshell, *preprint physics/9805024*, Lecture Notes (Part I of a course on "Transport Theory" taught at Dresden University of Technology, Spring 1997).
763. Ray, T.S. and Klein, W. (1990) Nucleation near the spinodal in long-range Ising models, *J. Stat. Phys.* **61**, 891–902.
764. Redner, S. (1990) Random multiplicative processes: An elementary tutorial, *Am. J. Phys.* **58**, 267–273.
765. Redner, S. (1990) Fragmentation, in [416].
766. Reed, W.J. (2001) The Pareto, Zipf and other power laws *Economics Letters* **74**, 15–19.
767. Reed, W.J. (2003) The Pareto law of incomes – an explanation and an extension, *Physica A* **319**, 469–486.
768. Reed, W.J. and Jorgensen, M. (2002) The double Pareto-lognormal distribution – A new parametric model for size distribution, Submitted to *Commun. Stat. Theory & Methods* Available at http://www.math.uvic.ca/faculty/reed/index.html.
769. Reichhardt, T. (1996) Rocket failure leads to grounding of small US satellites, *Nature* **384**, 99.
770. Reif, S. (1965) *Fundamentals of Statistical and Thermal Physics* (McGraw-Hill Book Company, New York).
771. Resnick, S.I. (1987) *Extreme Value, Regular Variations and Point Processes* (Springer, New York).
772. Reynolds, P.J., Klein, W. and Stanley, H.E. (1977) Renormalization Group for Site and Bond Percolation, *J. Phys. C* **10**, L167–L172.
773. Reynolds, P.J., Stanley, H.E. and Klein, W. (1978) Percolation by Position-Space Renormalization Group with Large Cells, *Physica A* **11**, L199–L207.
774. Reynolds, P.J., Stanley, H.E. and Klein, W. (1980) Large-Cell Monte Carlo Renormalization Group for Percolation, *Phys. Rev. B* **21**, 1223–1245.
775. Rice, J.R. (1988) *Transactions of the ASME* **55**, 98–103.
776. Rice, J.R., Suo, Z. and Wang, J.-S. (1990) Mechanics and thermodynamics of brittle interface failure in bimaterial systems, in Metal-Ceramic Interfaces (eds. Ruhle, M., Evans, A.G., Ashby, M.F. ad J.P. Hirth), Acta-Scripta Metallurgica Proceedings Series, vol. 4 (Pergamon Press), pp. 269–294
777. Richardson, L.F. (1926) Atmospheric diffusion on a distance–neighbour graph, *Proc. Roy. Soc. London A* **110**, 709.
778. Richardson, L.F. (1961) The problem of contiguity: an appendix of statistics of deadly quarrels, *General Systems Yearbook* **6**, 139–187.
779. Rigon, R., Rinaldo, A., Rodriguez-Iturbe, I., Bras, R.L. and others (1993) Optimal channel networks – A framework for the study of river basin morphology, *Water Ressources Research* **29**, 1635–1646.
780. Risken, H. (1989) *The Fokker-Planck Equation: Methods of Solution and Applications.* 2nd ed. (Springer, Berlin; New York).
781. Riste, T. and Sherrington, D. eds. (1991) *Spontaneous Formation of Space-Time Structures and Criticality.* Proc. NATO ASI, Geilo, Norway (Kluwer, Dordrecht).
782. Robin, T. and Souillard, B. (1989) Long wavelength properties of granular metal insulator films – A microscopic approach, *Physica A* **157**, 285–292.
783. Robin, T. and Souillard, B. (1993) Electromagnetic properties of fractal aggregates, *Europhys. Lett.* **21**, 273–278.
784. Robin, T. and Souillard, B. (1993) Concentration-dependent exponent for AC hopping conductivity and permittivity, *Europhys. Lett.* **22**, 729–734.
785. Rocco, A. and West, B.J. (1999) Fractional Calculus and the Evolution of Fractal Phenomena, *Physica A* **265**, 535–546

786. Rodkin, M.V. and Pisarenko, V.F. (2000) Earthquake losses and casualties: a statistical analysis, in *Problems in Dynamics and Seismicity of the Earth*: Coll. Sci. Proc. Moscow, *Computational Seismology* **31**, 242–272 (in Russian).

787. Rodriguez-Iturbe, I., Rinaldo, A., Rigon, R., Bras, R.L. and others (1992) Fractal structures at least energy patterns – The case of river networks, *Geophys. Res. Lett.* **19**, 889–892.

788. Rodriguez-Iturbe, I. and Rinaldo, A. (1997) *Fractal River Basins: Chance and Self-organization*, New York: Cambridge University Press.

789. Roehner, B. and Winiwarter, P. (1985) Aggregation of independent paretian random variables, *Adv. Appl. Prob.* **17**, 465–469.

790. Romeo, M., Da Costa, V. and Bardou, F. (2002) Broad distribution effects in sums of lognormal random variables (http://arXiv.org/abs/physics/0211065)

791. Ross, R. (1996) Bias and standard deviation due to Weibull parameter estimation for small data sets, *IEEE Transactions on Dielectrics and Electrical insulation* **3**, 28–42.

792. Roux, S. and Hansen, A. (1990) Multifractals, in [158].

793. Roux, S. and Hansen, A. (1994) Interface roughening and pinning, *J. Phys I France* **4**, 515–538.

794. Rowe, W.D. (1977) *An Anatomy of Risk* (Wiley, New York).

795. Ruelle, D. (1995) *Turbulence, Strange Attractors and Chaos* (World Scientific, New York).

796. Ruelle, D. (1999) Smooth Dynamics and New Theoretical Ideas in Nonequilibrium Statistical Mechanics, *J. Stat. Phys.* **95**, 393–468.

797. Ruján, P. (1987) Cellular automata and statistical mechanical models, *J. Stat. Phys.* **49**, 139–222.

798. Rundle, J.B., Klein, W., Gross, S. and Turcotte, D.L. (1995) Boltzmann fluctuations in numerical simulations of non-equilibrium threshold systems, *Phys. Rev. Lett.* **75**, 1658–1661.

799. Rundle, J.B., Klein, W., Gross, S. and Turcotte, D.L. (1997) Non-Boltzmann fluctuations in numerical simulations of non-equilibrium lattice threshold systems – Reply, *Phys. Rev. Lett.* **78**, 3798–3798.

800. Rundle, J.B., Gross, S., Klein, W., Ferguson, C. and Turcotte, D. (1997) The statistical mechanics of earthquakes, *Tectonophysics* **277**, 147–164.

801. Saada, G. and Sornette, D. (1995) Long-range stress field fluctuations induced by random dislocation arrays: a unified spectral approach, *Acta Metallurgica Material* **43**, 313–318.

802. Sahimi, M. (1993) Flow phenomena in rocks: from continuum models to fractals, percolation, cellular automata and simulated annealing, *Rev. Mod. Phys.* **65**, 1393–1534.

803. Sahimi, M. (1994) *Applications of Percolation Theory* (Taylor and Francis, London).

804. Sahimi, M. and Arbabi, S. (1992) Percolation and fracture in disordered solids and granular media: approach to a fixed point, *Phys. Rev. Lett.* **68**, 608–611.

805. Sahimi, M. and Arbabi, S. (1996) Scaling laws for fracture of heterogeneous materials and rocks, *Phys. Rev. Lett.* **77**, 3689–3692.

806. Saichev, A.I. and Zaslavsky, G.M. (1997) Fractional kinetic equations: solutions and applications, *Chaos* **7**, 753–764.

807. Saleur, H. and Sornette, D. (1996) Complex exponents and log-periodic corrections in frustrated systems, *J. Phys. I France* **6**, 327–355.

808. Saleur, H., Sammis, C.G. and Sornette, D. (1996) Discrete scale invariance, complex fractal dimensions and log-periodic corrections in earthquakes, *J. Geophys. Res.* **101**, 17661–17677.
809. Saleur, H., Sammis, C.G. and Sornette, D. (1996) Renormalization group theory of earthquakes, *Nonlinear Processes in Geophysics* **3**, 102–109.
810. Samorodnitsky, G. and Taqqu, M.S. (1994) *Stable Non-Gaussian Random Processes: Stochastic Models with Infinite Variance* (Chapman & Hall, New York).
811. Samuelson, P.A. (1977) St-Petersburg paradox, defanged, dissected, and historically described, *J. Economic Literature* **15**, 24–55.
812. Saulis, L. and Statulevicius, V.A. (1991) *Limit Theorems for Large Deviations* (Kluwer Academic, Dordrecht, Boston).
813. Savage, L.J. (1962) *The Foundations of statistical inference, a discussion* (Methuen, London).
814. Savage, L.J. (1972) *The Foundations of Statistics* (Dover, New York).
815. Savage, S.B. (1994) *Advances of Applied Mechanics* **24**, 289.
816. Scher, H., Margolin, G., Metzler, R., Klafter, J. and Berkowitz, B. (2002) The dynamical foundation of fractal stream chemistry: the origin of extremely long retention times, *Geophys. Res. Lett.* **29**, 1–4.
817. Scher, H. and Montroll, E.W. (1975) Anomalous transit-time dispersion in amorphous solids, *Phys. Rev. B* **12**, 2455–2477.
818. Schertzer, D. and Lovejoy, S. (1987) Physical modeling and analysis of rain and clouds by anisotropic scaling multiplicative processes, *J. Geophys. Res.* **92**, 9693–9714.
819. Schertzer, D. and Lovejoy, S. (1992) Hard and soft multifractal processes, *Physica A* **185**, 187–194.
820. Schmidt, H. and Stapp, H. (1993) Psychokinetic with prerecorded random events and the effects of the pre-observation, *Journal of Parapsychology* **57**, 331–349.
821. Schmidt, H. (1993) Observation of a psychokinetic effect under highly controlled conditions, *Journal of Parapsychology* **57**, 351–372.
822. Schmittbuhl, J., Roux, S., Vilotte, J.-P. and Maloy, K.J. (1995) Interfacial crack pinning: effect of nonlocal interactions, *Phys. Rev. Lett.* **74**, 1787–1790.
823. Schneider, W.R. (1986) in S. Albeverio, Casati, G., Merlini, D. (Eds.), *Stochastic Processes in Classical and Quantum Systems*, Lecture Notes in Physics vol. 262 (Springer, Berlin).
824. Schneider, W.R. and Wyss, W. (1989) Fractional diffusion and wave equations, *Journal of Mathematical Physics* **30**, 134–44.
825. Schoen, R., Habetler, T., Kamran, F. and Bartheld, R. (1995) Motor bearing damage detection using stator current monitoring, *IEEE Trans. Ind. Applicat.* **31**, 1274–1279.
826. Scholz, C.H. (1990) *The Mechanics of Earthquakes and Faulting* (Cambridge Univ. Press, New York).
827. Scholz, C.H. and Mandelbrot B.B., eds. (1989) *Fractals in Geophysics* (Birkhuser, Basel).
828. Schumer, R., Benson, D., Meerschaert, M. and Wheatcraft, S. (2001) Eulerian derivation of the fractional advection–dispersion equation, *Journal of Contaminant Hydrology* **38**, 69–88.
829. Searles, D.J. and Denis J.E. (1999) The Fluctuation Theorem for Stochastic Systems, *Phys. Rev. E* **60**, 159–164.
830. Seddon, G. (1991) The nature of Nature (the natural and the supernatural in Western thought), *Westerly* **36**, 7–14.

831. Seki, T., and Yokoyama, S. (1996) Robust parameter-estimation using the Bootstrap method for the two-parametr Weibull distribution, *IEEE Transactions on Reliability* **45**, 34–41.
832. Sen, P.K. (1973) *J. Appl. Prob.* **10**, 586.
833. Sen, P.K. (1973) *Ann. Stat.* **1**, 526.
834. Sethna, J.P., Dahmen, K., Kartha, S., Krumhansl, J. A., Roberts, B.W. and Shore, J.D. (1993) Hysteresis and hierarchies-Dynamics of disorder-driven 1st-order phase transitions, *Phys. Rev. Lett.* **70**, 3347–3350.
835. Sethna, J.P., Perkovic, O. and Dahmen, K. (1997) pp. 87–97, in [251].
836. She, Z.S. and Jackson, E. (1993) Constrained Euler system for Navier–Stokes turbulence, *Phys. Rev. Lett.* **70**, 1255–1258.
837. She, Z.-S. and Leveque, E. (1994) Universal scaling laws in fully developed turbulence, *Phys. Rev. Lett.* **72**, 336–339.
838. She, Z.-S. and Waymire, E.C. (1995) Quantized energy cascade and log-poisson statistics in fully developed turbulence, *Phys. Rev. Lett.* **74**, 262–265.
839. Shlesinger, M.F. and West, B.J. (1991) Complex fractal dimension of the broncial tree, *Phys. Rev. Lett.* **67**, 2106–2108.
840. Shlesinger, M.F., Zaslavsky, G.M. and Klafter, J. (1993) Strange kinetics, *Nature* **363**, 31–37.
841. Shnirman, M.G. (1987) Dynamical hierarchical model of defect formation, *Computational Seismology* **20**, 87–95; English translation, 85–92 (1988).
842. Shnirman, M.G. and Blanter, E.M. (1998) Self-organized criticality in a mixed hierarchical system, *Phys. Rev. Lett.* **81**, 5445–5448.
843. Shure, L. and Parker, R.L. (1981) An alternative explanation for intermediate wavelength magnetic anomalies, *J. Geophys. Res.* **86**, 11600–11608.
844. Simon, H.A. (1955) On a class of skew distribution functions, *Biometrika* **42**(3/4), 425–440.
845. Sinclair, K. and Ball, R.C. (1996) Mechanism for global optimization of river networks from local erosion rules, *Phys. Rev. Lett.* **76**, 3360–3363.
846. Sinha-Ray, P. and Jensen, H.J. (2000) Forest-fire models as a bridge between different paradigms in self-organized criticality, *Physical Review E.* **62**, 3215–3218.
847. Singh, A.N. (1953) The theory and construction of non-differentiable functions, in *Squaring the Circle and Other Monographs*, ed. by Hobson, E.W. (Chelsea Publishing Company, New York).
848. Smalley, R.F., Turcotte, D.L. and Solla, S.A. (1985) A renormalization group approach to the stick-slip behavior of faults, *J. Geophys. Research* **90**, 1894–1900.
849. Smalley, R.F., Chatelain, J.-L., Turcotte, D.L. and Prevot, R., A fractal approach to the clustering of earthquakes: applications to the seismicity of the New Hebrides, *Bull. Soc. Seism. Soc. Am.* **77**, 1368–1381.
850. Smelyanskiy, V.N., Dykman, M.I., Rabitz, H. and Vugmeister, B.E. (1997) Fluctuations, Escape, and Nucleation in Driven Systems: Logarithmic Susceptibility, *Phys. Rev. Lett.* **79**, 3113–3116.
851. Smith, R.L., (1983) Limit theorems and approximations for the reliability of load-sharing systems, *Advances in Applied Probability* **15**, 304–330.
852. Smith, R.L. and Phoenix, S.L. (1981) Asymptotic distributions for the failure of fibrous materials under series-parallel structure and equal load-sharing, *Transactions of the ASME, Journal of Applied Mechanics* **48**, 75–82.
853. Smith, R.L. (1991) Weibull regression models for reliability data, *Reliability Engineering & System Safely* **34**, 55–77.

854. Smythe, R. and Mahmound, H. (1995) A survey of recursive trees, *Theoretical Probability and Mathematical Statistics* **51**, 1–27.
855. Sneppen, K. (1992) Self-organized pinning and interface growth in a random medium, *Phys. Rev. Lett.* **69**, 3539–3542.
856. Sneppen, K. and Newman, M.E.J. (1997) Coherent noise, scale invariance and intermittency in large systems, *Physica D* **110**, 209–222.
857. Socolar, J.E.S., Grinstein, G. and Jayaprakash, C. (1993) On self-organized criticality in non-conservative systems, *Phys. Rev. E* **47**, 2366–2376.
858. Solomon, S. and Levy, M. (1996) Spontaneous scaling emergence in generic stochastic systems, *Int. J. Mod. Phys. C* **7**, 745–751.
859. Solomon, S., Weisbuch, G., de Arcangelis, L., Jan, N. and Stauffer, D. (1999) Social Percolation Models, *Physica A* **277**, 239–247
860. Sornette, A. and Sornette, D. (1989) Self-organized criticality and earthquakes, *Europhys. Lett.* **9**, 197–202.
861. Sornette, A. and Sornette, D. (1990) Earthquake rupture as a critical point: Consequences for telluric precursors, *Tectonophysics* **179**, 327–334.
862. Sornette, A., Davy, P. and Sornette, D. (1990) Growth of fractal fault patterns, *Phys. Rev. Lett.* **65**, 2266–2269.
863. Sornette, A., Davy, P. and Sornette, D. (1993) Fault growth in brittle–ductile experiments and the mechanics of continental collisions, *J. Geophys. Res.* **98**, 12111–12139.
864. Sornette, A. and Sornette, D. (1999) Renormalization of earthquake aftershocks, *Geophys. Res. Lett.* **26**, 1981–1984.
865. Sornette, D. (1988) Critical transport and failure exponents in continuum crack percolation, *J. Physique (Paris)* **49**, 1365–1377.
866. Sornette, D. (1989) Failure thresholds in hierarchical and euclidian space by real space renormalization group, *J. Phys. France* **50**, 745–755.
867. Sornette, D. (1989) Elasticity and failure of a set of elements loaded in parallel, *J. Phys. A* **22**, L243–L250.
868. Sornette, D. (1989) Acoustic waves in random media: I Weak disorder regime, *Acustica* **67**, 199–215.
869. Sornette, D. (1991) Self-organized criticality in plate tectonics in [781], 57–106.
870. Sornette, D. (1992) Mean-field solution of a block-spring model of earthquakes, *J. Phys. I France* **2**, 2089–2096.
871. Sornette, D. (1992) $z^{-3/2}$ powerlaw decay of Laplacian fields induced by disorder: consequences for the inverse problem, *Geophys. Res. Lett.* **12**, 2377–2380.
872. Sornette, D. (1992) How to make critical phase transitions self-organized: a dynamical system feedback mechanism for Self-Organized Criticality, *J. Phys. I France* **2**, 2065–2073.
873. Sornette, D. (1993) Decay of long-range field fluctuations induced by random boundaries: a unified spectral approach, *J. Phys. I France* **3**, 2161–2170.
874. Sornette, D. (1993) Les Phénomènes Critiques Auto-Organisés, *Images de la Physique* (édition du CNRS, Paris).
875. Sornette, D. (1994) Sweeping of an instability: an alternative to self-organized criticality to get powerlaws without parameter tuning, *J. Phys. I France* **4**, 209–221.
876. Sornette, D. (1994) Power laws without parameter tuning: An alternative to self-organized criticality, *Phys. Rev. Lett.* **72**, 2306–2306.
877. Sornette, D. (1996) Anderson localization and quantum chaos in acoustics, *Physica B* **220**, 320–323.

878. Sornette, D. (1998) Discrete scale invariance and complex dimensions, *Phys. Reports* **297**, 239–270.
879. Sornette, D. (1998) Linear stochastic dynamics with nonlinear fractal properties, *Physica A* **250**, 295–314.
880. Sornette, D. (1998) Multiplicative processes and power laws, *Phys. Rev. E* **57**, 4811–4813.
881. Sornette, D. (1998) Discrete scale invariance in turbulence? Proceedings of the *Seventh European Turbulence Conference (ETC-7)*, Frisch, U., ed. (Kluwer Academic, Boston) (cond-mat/9802121).
882. Sornette, D. (1998) Large deviations and portfolio optimization, *Physica A* **256**, 251–283.
883. Sornette, D. (1998) Scientific prediction of catastrophes: a new approach, essay selected as one of the ten finalist to the James S. McDonnell Centennial Fellowships, document available at http://www.ess.ucla.edu/facpages/sornette.html
884. Sornette, D. (1999) Using complexity to comprehend catastrophe, Millennium Issue of *Physics World* **12**, 57–57.
885. Sornette, D. (1999) Earthquakes: from chemical alteration to mechanical rupture, *Physics Reports* **313**, 238–292.
886. Sornette, D. (2002) Predictability of catastrophic events: material rupture, earthquakes, turbulence, financial crashes and human birth, *Proceedings of the National Academy of Sciences USA* **99**, 2522–2529.
887. Sornette, D. (2003) *Why Stock Markets Crash: Critical Events in Complex Financial Systems* (Princeton University Press, Princeton, NJ)
888. Sornette, D. (2003) Critical market crashes, *Physics Reports* **378**, 1–98.
889. Sornette, D. and Andersen, J.-V. (1998) Scaling with respect to disorder in time-to-failure, *Eur. Phys. J. B* **1**, 353–357.
890. Sornette, D., Carbone, D., Ferré, F., Vauge, C. and Papiernik, E. (1995) Modèle mathématique de la parturition humaine: implications pour le diagnostic prenatal, *Médecine/Science* **11**, 1150–1153.
891. Sornette, D. and Cont, R. (1997) Convergent multiplicative processes repelled from zero: power laws and truncated power laws, *J. Phys. I France* **7**, 431–444.
892. Sornette, D. and Davy, P. (1991) Fault growth model and the universal fault length distribution, *Geophys. Res. Lett.* **18**, 1079–1081.
893. Sornette, D., Davy, P. and Sornette, A. (1990) Structuration of the lithosphere in plate tectonics as a self-organized critical phenomenon, *J. Geophys. Res.* **95**, 17353–17361.
894. Sornette, D. and Dornic, I. (1996) Parallel Bak–Sneppen model and directed percolation, *Phys. Rev. E* **54**, 3334–3338.
895. Sornette, D., Ferré, F. and Papiernik, E. (1994) Mathematical model of human gestation and parturition: implications for early diagnostic of prematurity and post-maturity, *Int. J. Bifurcation and Chaos* **4**, 693–699.
896. Sornette, D., Gilabert, A. and Vanneste, C. (1988) Rupture in random media, Proceeding of the (18–30 July 1988) Cargèse summer school on *Random Fluctuations and Pattern Growth: Experiments and Models*, NATO ASI Series, eds. Stanley, H.E. and Ostrowsky, N. **157** (Kluwer Academic Publisher, Boston).
897. Sornette, D. and Helmstetter, A. (2003) Endogeneous versus exogeneous shocks in systems with memory, *Physica A* **318** (3–4), 577–591.
898. Sornette, D. and Johansen, A. (1997) Large financial crashes, *Physica A* **245**, 411–422.

899. Sornette, D., Johansen, A., Arnéodo, A., Muzy, J.-F. and Saleur, H. (1996) Complex fractal dimensions describe the internal hierarchical structure of DLA, *Phys. Rev. Lett.* **76**, 251–254.

900. Sornette, D., Johansen, A. and Bouchaud, J.-P. (1996) Stock market crashes, precursors and replicas, *J. Phys. I France* **6**, 167–175.

901. Sornette, D., Johansen, A. and Dornic, I. (1995) Mapping self-organized criticality onto criticality, *J. Phys. I France* **5**, 325–335.

902. Sornette, D. and Knopoff, L. (1997) The paradox of the expected time until the next earthquake, *Bull. Seismol. Soc. Am.* **87**, 789–798.

903. Sornette, D., Knopoff, L., Kagan, Y.Y. and Vanneste, C. (1996) Rank-ordering statistics of extreme events: application to the distribution of large earthquakes, *J. Geophys. Res.* **101**, 13883–13893.

904. Sornette, D., Lagier, M., Roux, S. and Hansen, A. (1989) Critical piezoelectricity in percolation, *J. Phys. France* **50**, 2201–2216.

905. Sornette, D., Leung, K.-T. and Andersen, J. V. (1998) Conditions for abrupt failure in the democratic fiber bundle model, *Phys. Rev. Lett.* **80**, 3158–3158.

906. Sornette, D., Malevergne, Y. and Muzy, J.-F. (2003) What causes crashes? *Risk* **16** (2), 67–71. (http://arXiv.org/abs/cond-mat/0204626)

907. Sornette, D., Miltenberger, P. and Vanneste, C. (1994) Statistical physics of fault patterns self-organized by repeated earthquakes, *Pure and Applied Geophysics* **142**, 491–527.

908. Sornette, D., Miltenberger, P. and Vanneste, C. (1995) Statistical physics of fault patterns self-organized by repeated earthquakes: synchronization versus self-organized criticality, in *Recent Progresses in Statistical Mechanics and Quantum Field Theory*, Proceedings of the conference "Statistical Mechanics and Quantum Field Theory", USC, Los Angeles, May 16–21, eds. Bouwknegt, P., Fendley, P., Minahan, J., Nemeschansky, D., Pilch, K., Saleur, H. and Warner, N. (World Scientific, Singapore), pp. 313–332.

909. Sornette, D. and Pisarenko, V.F. (2003) Fractal plate tectonics, *Geophys. Res. Lett.* **30**(3), 1105, doi:10.1029/2002GL015043.

910. Sornette, D. and Sammis, C.G. (1995) Complex critical exponents from renormalization group theory of earthquakes: implications for earthquake predictions, *J. Phys. I France* **5**, 607–619.

911. Sornette, D., Simonetti, P. and Andersen, J.V. (2000) ϕ^q-field theory for portfolio optimization: "fat tails" and non-linear correlations, *Physics Reports* **335** (2), 19–92.

912. Sornette, D. and Sornette, A. (1999) General theory of the modified Gutenberg–Richter law for large seismic moments, *Bull. Seism. Soc. Am.* **89**, 1121–1130.

913. Sornette, D. and Vanneste, C. (1992) Dynamics and memory effects in rupture of thermal fuse networks, *Phys. Rev. Lett.* **68**, 612–615.

914. Sornette, D. and Vanneste, C. (1994) Dendrites and fronts in a model of dynamical rupture with damage, *Phys. Rev. E* **50**, 4327–4345.

915. Sornette, D. and Vanneste, C. (1996) Fault self-organization by repeated earthquakes in a quasi-static antiplane crack model, *Nonlinear Processes Geophys.* **3**, 1–12.

916. Sornette, D., Vanneste, C. and Knopoff, L. (1992) Statistical model of earthquake foreshocks, *Phys. Rev. A* **45**, 8351–8357.

917. Sornette, D. and Virieux, J. (1992) A theory linking large time tectonics and short time deformations of the lithosphere, *Nature* **357**, 401–403.

918. Sornette, D. and Zhang, Y.-C. (1993) Non-linear langevin models of geomorphic erosion processes, *Geophys. J. Int.* **113**, 382–386.

919. Sow, C.H., Harada, K., Tonomura, A., Crabtree, G., and others, Measurement of the vortex pair interaction potential in a type-II superconductor, *Phys. Rev. Lett.* **80**, 2693–2696.
920. Srivasta, H. M., Gupta, K.C. and Goyal, S.P. (1982) *The H-Function of One and Two Variables with Applications* (South Asian Publishers, New Delhi).
921. Srivastava, H.M. and Kashyap, B.R.K. (1982) *Special Functions in Queuing Theory and Related Stochastic Processes* (Academic Press, New York).
922. Stanley, H.E. (1987) *Introduction to Phase Transitions and Critical Phenomena* (Oxford University Press, New York).
923. Stanley, H.E. and Ostrowsky, N. eds. (1986) *On Growth and Form* (Martinus Nijhoff, Kluwer Academic Publishers, Boston).
924. Stanley, H.E. and Ostrowsky, N. eds. (1988) *Random Fluctuations and Pattern Growth: Experiments and Models, NATO ASI Series* **157** (Kluwer Academic Publisher, Boston).
925. Stapp, H. (1994) Theoretical model of a purported empirical violation of the predictions of quantum mechanics, *Phys. Rev. A* **50**, 18–22.
926. Stapp, H.P. (1995) Parapsychological Review A? – Reply, *Physics Today* **July** 78–79;
927. Stauffer, D. (1997) Minireview: New results for old percolation, *Physica A* **242**, 1–7.
928. Stauffer, D. and Aharony, A. (1994) *Introduction to Percolation Theory.* 2nd ed. (Taylor & Francis, London; Bristol, PA).
929. Stauffer, D. (1998) Monte Carlo investigation of rare magnetization fluctuations in Ising models, *Int. J. Mod. Phys. C* **9**, 625–631.
930. Stewart, I. (1999) The synchronicity of firefly flashing, *Scientific American* **280**, 104–106.
931. Stokes, D.M. (1996) Leaps of faith – Science, miracles and the search for supernatural consolation – N. Humphrey, *J. Am. Soc. for Psychical Research* **90**, 241–246.
932. Stokes, G.G. (1851), *Trans. Camb. Phil. Soc.* **9**, 8, Mathematical and Physical Papers **3**,1.
933. Stratonovich, R.L. (1967) *Topics in the Theory of Random Noise.* Vol. I and II (Gordon and Breach, New York).
934. Stratonovich, R.L. (1992) *Nonlinear Nonequilibrium Thermodynamics I: Linear and Nonlinear Fluctuation-Dissipation Theorems* (Springer, Berlin; New York).
935. Strogatz, S.H. (1993) Norbert Wiener's Brain waves, *Lecture Notes in Biomathematics* **100** (Springer, Berlin).
936. Strogatz, S.H. and Steward, I. (1993) Coupled oscillators and biological synchronization, *Scientific American* **269**, 102–109.
937. Stuart, A. and Ord, J.K. (1994) *Kendall's Advanced Theory of Statistics.* 6th ed. (Edward Arnold, London and Halsted Press, New York).
938. Suh, M.W., Battacharyya, B.B. and Grandage, A. (1970) *J. Appl. Prob.* **7**, 712.
939. Sun, T., Meakin, P. and Jossang, T. (1995) Minimum energy dissipation river networks with fractal boundaries, *Phys. Rev. E* **51**, 5353–5359.
940. Swift, J.B., Babcock, K.L., and Hohenberg, P.C. (1994) Effect of thermal noise in Taylor-Couette flow with corotation and axial through-flow, *Physica A* **204**, 625–649.
941. Takayasu, H. (1989) Steady-state distribution of generalized aggregation system with injection, *Phys. Rev. Lett.* **63**, 2563–2565.
942. Takayasu, H., Nishikawa, I. and Tasaki, H. (1988) Power-law mass distribution of aggregation systems with injection, *Phys. Rev. A* **37**, 3110–3117.

943. Takayasu, H. and Tretyakov, A.Yu. (1992) Extinction, survival and dynamical phase transition of branching, *Phys. Rev. Lett.* **68**, 3060–3063.

944. Takayasu, H., Sato, A.-H. and Takayasu, M. (1997) Stable infinite variance fluctuations in randomly amplified Langevin systems, *Phys. Rev. Lett.* **79**, 966–969.

945. Tanguy, A., Gounelle, M. and Roux, S. (1998) From individual to collective pinning: effect of long-range elastic interactions, *Phys. Rev. E* **58**, 1577–1590.

946. Taqqu, M.S., Teverovsky, V. and Willinger, W. (1995) Estimators for longrange dependence – An empirical study. *Fractals* **3**, 785–798.

947. Taqqu, M.S., Teverovsky, V. and Willinger, W. (1997) Is network traffic selfsimilar or multifractal? *Fractals* **5**, 63–73.

948. Tchéou, J.-M. and Brachet, M.E. (1996) Multifractal scaling of probability density function – A tool for turbulent data analysis, *J. Phys. II France* **6**, 937–943.

949. Tebaldi, C., Menech, M.D. and Stella, A.L. (1999) Multifractal scaling in the Bak-Tang-Wiesenfeld Sandpile and edge events, *Phys. Rev. Letts.* **83**, 3952–3955.

950. Tel, T., Graham, R. and Hu, G. (1989) Nonequilibrium potentials and their power-series expansions, *Phys. Rev. A* **40**, 4065–4071.

951. Teugels, J.L. (1974) The subexponential class of probability distributions, *Theory Probab. Appl.* **19**, 821–822.

952. Teugels, J.L. (1975) The class of subexponential distributions, *The Annals of Probability* **3**, 1000–1011.

953. Theiler, J., Eubank, S., Longtin, A., Galdrikian, B. and Farmer, D. (1992) Testing for nonlinearity in time series – The method of surrogate data, *Physica D* **58**, 77–94.

954. Thom, R. (1972) *Stabilité Structurelle et Morphogénèse* (Benjamin, New York).

955. Thomae, J. (1869) Beitrge zur Theorie der durch die Heinesche Reihe., *J. reine angew. Math.* **70**, 258281.

956. Thoman, D.R., Bain, L.J. and Antle, C.E. (1969) Inferences on the parameter of the Weibull distribution, *Technometrics* **11**, 445–460.

957. Thoman, D.R., Bain, L.J. and Antle, C.E. (1970) Maximum likelihood estimation, exact confidence intervals for reliability, and tolerance limits in the Weibull distribution, *Technometrics* **12**, 363–371.

958. Thomas, G.M., Gerth, R., Velasco, T. and Rabelo, L.C. (1995) Using realcoded genetic algorithms for Weibull parameter estimation, *Computer & Industrial Engineering* **29**, 377–381.

959. Throburn, W.M. (1915) Occam's razor, *Mind*, 297–288.

960. Throburn, W.M. (1918) The Myth of Occam's razor, *Mind*, 345–353.

961. Tosi, P., Barba, S., De Rubeis, V. and Di Luccio, F. (1999) Seismic signal detection by fractal dimension analysis, *Bull. Seism. Soc. Am.* **89**, 970–977.

962. Tsallis, C. (1988) Possible generalization of Boltzmann-Gibbs statistics, *J. Stat. Phys.* **52**, 479–487; for updated bibliography on this subject, see http://tsallis.cat.cbpf.br/biblio.htm.

963. Tsallis, C. (1999) Nonextensive statistics: Theoretical, experimental and computational evidences and connections, *Brazilian Journal of Physics* **29**, 1–35.

964. Tsallis, C., Levy, S.V.F., Souza, A.M.C. and Maynard, R (1995) Statisticalmechanical foundation of the ubiquity of Lévy distributions in Nature, *Phys. Rev. Lett.* **75**, 3589–3593; Erratum (1996) Ibid **77**, 5422–5422.

965. Tsallis, C. and Souza, A.M.C. (2003) Constructing a statistical mechanics for Beck–Cohen superstatistics, *Phys. Rev. E* **67**, 026106.

966. Tsypin, M.M. (1997) Effective potential for a scalar field in three dimensions: Ising model in the ferromagnetic phase, *Phys. Rev. B* **55**, 8911–8917.

967. Tsypin, M.M. and Blöte, H.W.J. (2000) Probability distribution of the order parameter for the 3D Ising model universality class: a high precision Monte Carlo study, *Phys. Rev. E* **62**, 73–76.

968. Tumarkin, A.G. and Shnirman, M.G. (1992) Critical effects in hierarchical media, *Computational Seismology* **25**, 63–71.

969. Turcotte, D.L. (1986) Fractals and fragmentation, *J. Geophys. Res.* **91**, 1921–1926.

970. Turcotte, D.L. and Schubert, G. (1982) *Geodynamics – Applications of Continuum Physics to Geological Problems* (John Wiley & Sons, New York).

971. Turcotte, D.L., Smalley, R.F. and Solla, S.A. (1985) *Nature* **313**, 671.

972. United States Space Command (USSPACECOM), http://www.spacecom. af.mil/usspace/

973. Utsu, T., Y. Ogata and S. Matsu'ura (1995) The centenary of the Omori Formula for a decay law of aftershock activity, *J. Phys. Earth* **43**, 1–33,

974. Van Atta and Chen (1969) Measurements of Spectral Energy Transfer in Grid Turbulence, *J. Fluid Mech.* **38**, 743–763.

975. Van der Ziel, A. (1950) On the noise spectra of semi-conductor noise and of flicker effect, *Physica* **16**(4), 359–372.

976. Van Dongen, P.G.J. and Ernst, M. (1988) Scaling solutions of Smoluchowski's coagulation equation, *J. Stat. Phys.* **50**, 295–329.

977. Vanneste, C., Gilabert, A. and Sornette, D. (1991) Finite size effects in line percolating systems, *Phys. Lett. A* **155**, 174–180.

978. Vanneste, C. and Sornette, D. (1992) Dynamics of rupture in thermal fuse models, *J. Phys. I France* **2**, 1621–1644.

979. Varnes, D.J. and Bufe, C.G. (1996) The cyclic and fractal seismic series preceding and m_b 4.8 earthquake on 1980 February 14 near the Virgin Islands, *Geophys. J. Int.* **124**, 149–158.

980. Varotsos, P., Eftaxias, K., Vallianatos, F. and Lazaridou, M. (1996) Basic principles for evaluating an earthquake prediction method, *Geophys. Res. Lett.* **23**, 1295–1298.

981. Varotsos, P., Eftaxias, K., Lazaridou, M., Antonopoulos, G. and others (1996) Summary of the five principles suggested by Varotsos et al. [1996] and the additional questions raised in this debate, *Geophys. Res. Lett.* **23**, 1449–452.

982. Veitch, D., Abry, P., Flandrin, P. and Chainais, P. (2000) Infinitely divisible cascade analysis of network traffic data, preprint in the Proceedings of the IEEE Int. Conf. on Signal Proc. ICASSP2000, Istanbul, June 2000.

983. Vere-Jones, D. (1977) Statistical theories of crack propagation, Mathematical *Geology* **9**, 455–481.

984. Vere-Jones, D. (1988) Statistical aspects of the analysis of historical earthquake catalogues, in C. Margottini, ed., Historical Seismicity of Central-Eastern Mediterranean Region, pp. 271–295.

985. Vermilye, J.M. and Scholz, C.H. (1998) The process zone: A microstructural view of fault growth, *J. Geophys. Res.* **103**, 12223–12237.

986. Vervaat, W. (1979) On a stochastic difference equation and a representation of non-negative infinitely divisible random variables, *Advances in Applied Probability* **11**, 750–83.

987. Vespignani, A. and Zapperi, S. (1997) Order parameter and scaling fields in self-organized criticality, *Phys. Rev. Lett.* **78**, 4793–4796.

988. Vespignani, A. and Zapperi, S. (1998) How self-organized criticality works: A unified mean-field picture, *Phys. Rev. E* **57**, 6345–6362.

989. Vespignani, A., Dickman, R., Munoz, M.A. and Zapperi, S. (1998) Driving, conservation, and absorbing states in sandpiles, *Phys. Rev. Lett.* **81**, 5676–5679.

990. Vidale, J.E., Agnew, D.C., Johnston, M.J.S. and Oppenheimer, D.H. (1998) Absence of earthquake correlation with Earth tides: An indication of high preseismic fault stress rate, *J. Geophys. Res.* **103**, 24567–24572.

991. Vilfan, A. and Duke, T. (2003) Two adaptation processes in auditory hair cells together can provide an active amplifier (http://arXiv.org/abs/physics/0305059)

992. Vinnichenko, N.K. et al. (1980) *Turbulence in the Free Atmosphere.* translated from Russian by Sinclair, F.L., 2d ed. (Consultant's Bureau, New York).

993. Vitulli, W.F. (1997) Beliefs in parapsychological events or experiences among college students in a course in experimental parapsychology, *Perceptual and Motor Skills* **85**, 273–274.

994. Vives, E., Rafols, I., Manosa, L., Ortin, J. and others (1995) Statistics of avalanches in Martensitic transformations. 1. Acoustic emission experiments, *Phys. Rev.* B **52**, 12644–12650.

995. Vogel, H. (1921) *Phys. Z* **22**, 645.

996. Voight, B. (1988) A method for prediction of volcanic eruptions, *Nature* **332**, 125–130.

997. Voight, B. (1989) A relation to describe rate-dependent material failure, *Science* **243**, 200–203.

998. Voight, B. and Cornelius, R.R. (1991) Prospect for eruption prediction in near real-time, *Nature* **350**, 695–698.

999. Volfson, D. and Vinals, J. (2001) Flow induced by a randomly vibrating boundary, *Journal of Fluid Mechanics* **432**, 387–408.

1000. Voss, R.F. (1992) Evolution of long-range fractal correlations and 1/f noise in DNA base sequences, *Phys. Rev. Lett.* **68**, 3805–3808.

1001. Walker, S.J. (1992) Supernatural beliefs, natural kinds and conceptual structure, *Memory & Cognition* **20**, 655–662.

1002. Wallace, D.J. and Zia, R.K.P. (1975) Gradient properties of the renormalization group equations in multicomponent systems, *Annals of Physics* **92**, 142–163.

1003. Walsh, J.B. (1986) *An Introduction to Stochastic Partial Differential Equations*, in Ecole d'Ete de Probabilite de Saint-Flour (Springer, Berlin).

1004. Walter, C. (1990) Lévy-stable distributions and fractal structure on the Paris market: an empirical examination, *Proc. of the first AFIR colloquium, Paris* **3**, 241–259.

1005. Wang, F.K., and Keats, J.B. (1995) Improved percentile estimation for the two-parameter Weibull distribution, *Microelectronics and Reliability* **35**, 883–892.

1006. Weibull, W. (1938) Investigations into Strength Properties of Brittle Materials, *Ingeniörs Vetenskaps Akademiens Handlingar, Royal Swedish Institute for Engineering Research, Stockholm, Sweden* **149**.

1007. Weierstrass, K. (1993) On continuous functions of a real argument that do not have a well-defined differential quotient, in *Classics on Fractals*, ed. by Edgar, G.A. (Addison-Wesley, Reading, MA), 3–9.

1008. Weinberg, S. (1996) *The quantum theory of fields* (Cambridge University Press, Cambridge, New York).

1009. Weinrib, A. and Halperin, B.I. (1983) Critical phenomena in systems with long-range-correlated quenched disorder, *Phys. Rev.* B **27**, 413–427.

1010. Weiss, J.B., Provenzale, A., and McWilliams, J.C. (1998) Lagrangian dynamics in high-dimensional point-vortex systems *Physics of Fluids* **10**, 1929–1941.

1011. Weissman, I. (1977) On location and scale functions of a class of limiting processes with applications to extreme value theory, *Ann. Probab.* **3**, 178–181.

1012. Weissman, M.B. (1988) 1/f noise and other slow, nonexponential kinetics in condensed matter, *Rev. Mod. Phys.* **60**, 537–571.

1013. West, B.J. and Deering, W. (1994) Fractal physiology for physicists – Lévy statistics, *Phys. Reports* **246**, 2–100.

1014. Westwood, A.R.C., Ahearn, J.S. and Mills, J.J. (1981) Developments in the theory and application of chemomechanical effects, *Colloids and Surfaces* **2**, 1–10.

1015. Wigner, E. (1960) The unreasonable effectiveness of mathematics, *Communications in Pure and Applied Mathematics*, **13** (1).

1016. Wilke, C., Altmeyer, S. and Martinetz, T. (1998) Aftershocks in coherent-noise models, *Physica D* **120**, 401–417.

1017. Wilke, C. and Martinetz, T. (1998) Simple model of evolution with variable system size, *Phys. Rev. E* **56**, 7128–7131.

1018. Wilke, C. and Martinetz, T. (1999) Lifetimes of agents under external stress, *Phys. Rev. E* **59**, R2512–R2515.

1019. Willinger, W., Taqqu, M.S., Sherman, R. and Wilson, D.V. (1997) Self-similarity through high-variability: Statistical analysis of ethernet LAN traffic at the source level, *IEEE-ACM Transactions on Networking* **5**, 71–86.

1020. Willekens, E. (1988) The structure of the class of subexponential distributions, *Probab. Th. Rel. Fields* **77**, 567–581.

1021. Wilson, K.G. (1979) Problems in physics with many scales of length, *Scientific American* **241**, August, 158–179.

1022. Wilson K.G., and Kogut, J., (1974) The Renormalization Group and the ϵ expansion, *Phys. Rep.* **12**, 75–200.

1023. Winkler, R.L. (1972) *An introduction to Bayesian Inference and Decision* (Holt, Rinehart and Winston, Inc).

1024. Witten, T.A. and Sander, L.M. (1981) Diffusion-limited aggregation, a kinetic critical phenomenon, *Phys. Rev. Lett.* **47**, 1400–1403.

1025. Wong, J.Y. (1993) Readily obtaining the maximum likelihood estimates of the three parameters of the generalized Gamma distribution of Stacy and Mihram, *Microelectron. Reliab.* **33**, 2243–2251.

1026. Wormald, N.C. (1995) Differential equations for random processes and random graphs, *Annals of Appl. Prob.* **5**, 1217–1235.

1027. Wright, S. (1982) *Evolution* **36**, 427.

1028. Wu, F.Y. (1982) The Potts model, *Rev. Mod. Phys.* **54**, 235–268.

1029. Wu, K. and Bradley, R.M. (1994) Theory of electromigration failure in polycrystalline metal films, *Phys. Rev. B* **50**, 12468–12488.

1030. Wyss, W. (1986) The fractional diffusion equation, *Journal of Mathematical Physics* **27**, 2782–2785.

1031. Xu, H.J. and Sornette, D. (1997) Non-Boltzmann fluctuations in numerical simulations of nonequilibrium lattice threshold systems, *Phys. Rev. Lett.* **78**, 3797–3797.

1032. Yale, M.M., Sandwell, D.T. and Herring, A.T. (1997) What are the limitations of satellite altimetry? *EdgeNet, http://www.edge-online.org*.

1033. Yang, C.N. and Lee, T.D. (1952) Statistical theory of equations of state and phase transitions. I. Theory of condensation, *Phys. Rev.* **87**, 404–409.

1034. Yeh, J.-C. (1973) *Stochastic Processes and the Wiener Integral* (M. Dekker, New York).
1035. Yeh and Van Atta (1970) *J. Fluid Mechanics* **41**, 169–178.
1036. Young, A.P. (1998) *Spin Glasses and Random Fields* (Singapore: River Edge, N.J.: World Scientific).
1037. Yukalov, V.I. and Gluzman, S. (1997) Self-similar bootstrap of divergent series, *Phys. Rev. E* **55**, 6552–6570.
1038. Yukalov, V.I. and Gluzman, S. (1997) Critical indices as limits of control functions, *Phys. Rev. Lett.* **79**, 333–336.
1039. Yukalov, V.I. and Gluzman, S. (1998) Self-similar exponential approximants, *Phys. Rev. E* **58**, 1359–1382.
1040. Yukalov, V.I., Gluzman, S., and Sornette, D. (2003) Summation of power series by self-similar factor approximants, submitted to *Physica A* (http://arXiv.org/abs/cond-mat/0302613)
1041. Yukalov, V.I., Moura, A. and Nechad, H. (2003) Self-similar law of energy release before materials fracture, *Journal of the Mechanics and Physics of Solids*, submitted (2003)
1042. Yule, G. (1925) A mathematical theory of evolution based on the conclusions of Dr. J.C. Willis, F.R.S. *Philosophical Transactions of the Royal Society of London (Series B)* **213**, 21–87.
1043. Zajdenweber, D. (1976) *Hasard et Prévision* (Economica, Paris).
1044. Zajdenweber, D. (1995) Business interruption insurance, a risky business – A study on some Paretian risk phenomena, *Fractals* **3**, 601–608.
1045. Zajdenweber, D. (1996) Extreme values in business interruption insurance, *J. Risk and Insurance* **63**, 95–110.
1046. Zajdenweber, D. (1997) Scale invariance in Economics and Finance, in [251], pp.185–194.
1047. Zapperi, S., Cizeau, P., Durin, G. and Stanley, H.E. (1998) Dynamics of a ferromagnetic domain wall – avalanches, depinning transition and the Barkhausen effect, *Phys. Rev. B* **58**, 6353–6366.
1048. Zapperi, S., Lauritsen, K.B. and Stanley, H.E. (1995) Self-organized branching process: mean-field theory for avalanches, *Phys. Rev. Lett.* **75**, 4071–4074.
1049. Zapperi, S., Ray, P., Stanley, H.E. and Vespignani, A. (1997) First-order transition in the breakdown of disordered media, *Phys. Rev. Lett.* **78**, 1408–1411.
1050. Zapperi, S., Ray, P., Stanley, H.E. and Vespignani, A. (1999) Avalanches in breakdown and fracture processes, *Phys. Rev. E* **59**, 5049–5057.
1051. Zasimchuk, E.E. and Selitser, S.I. (1981) Stress fields of some nonequilibrium dislocation ensembles. I. Disordered dislocation layers, *Strength of Materials* **13**, 511–517.
1052. Zasimchuk, E.E. and Selitser, S.I. (1984) Random fields of internal stresses in various dislocation ensembles, *Sov. Phys. Solid State* **26**, 695–696.
1053. Zasimchuk, E.E. and Selitser, S.I. (1985) The effect of random internal stress fields on mechanical instability and dynamics of dislocation ensembles, *Physics of Metals* **6**, 253–263.
1054. Zaslavsky, G.M. (1994) Fractional kinetic equation for Hamiltonian chaos, chaotic advection, tracer dynamics and turbulent dispersion, *Physica D* **76**, 110–122.
1055. Zhang, S.-D. (1999) Scaling in the time-dependent failure of a fiber bundle with local load sharing, *Phys. Rev. E* **59**, 1589–1592.
1056. Zhang, Y.-Z. (1987) Ground-state instability of a random system, *Phys. Rev. Lett.* **59**, 2125–2128.

1057. Zhang, Y.-C. (1988) Scaling theory of self-organized criticality, *Phys. Rev. Lett.* **63**, 470–473.
1058. Zhou, W.-X. and Sornette, D. (2002) Evidence of Intermittent Cascades from Discrete Hierarchical Dissipation in Turbulence, *Physica D* **165**, 94–125.
1059. Zhou, W.-X., Sornette, D. and Pisarenko, V. (2002) New evidence of discrete scale invariance in the energy dissipation of three-dimensional turbulence: correlation approach and direct spectral detection, *Int. J. Mod. Phys. C* **14** (4).
1060. Zhuo, J., Redner, S. and Park, H. (1993) Critical behavior of an interaction surface reaction model, *J. Phys. A* **26**, 4197–4213.
1061. Ziff, R.M. (1992) Spanning probability in 2D percolation, *Phys. Rev. Lett.* **69**, 2670–2673.
1062. Ziff, R.M., Gulari, E. and Barshad, Y. (1986) Kinetic phase transitions in an irreversible surface-reaction model, *Phys. Rev. Lett.* **56**, 2553–2556.
1063. Zipf, G.K. (1932) *Selective Studies and the Principle of Relative Frequency in Language* (Harvard University Press, Cambridge, MA).
1064. Zipf, G.K. (1935) *The Psycho-Biology of Language: an Introduction to Dynamic Philology* (Houghton Mifflin, Boston, MA).
1065. Zipf, G.K. (1949) *Human Behavior and the Principle of Least-Effort* (Addison-Wesley, Cambridge).
1066. Zolotarev, V.M. (1983) One-dimensional Stable Distributions, *Translations of mathematical monographs, American Mathematical Society* **65**, Translation from the original 1983 Russian edition, Chapter 2, Section 2.6 (Asymptotic expansions of stable laws).
1067. Zolotarev, V. M. (1986) One-dimensional Stable Distributions, *Amer. Math. Soc. Providence R.I.*, 284.

Index

$1/f$ noise, 345, 406
Π theorem, 151

Abelian sandpile model, 398
advection of passive scalars, 90
aftershocks, 278
aging, 356
anomalous diffusion, 112, 239
anti-ferromagnetic, 442
approximants, 288
ARCH: auto-regressive conditional
  heteroskedasticity, 381
Arrhenius activation law, 355, 443
asthenosphere, 411
autoregressive process, 232
avalanches, 386, 387, 401, 429
average, 13

Bak–Sneppen model, 307, 417, 427
Barkhausen noise, 386
Bayesian, 7, 70, 163
Bethe lattice, 314
bifurcation, 255
Binomial law, 63
Boltzmann formalism, 66, 358
Boltzmann function, 200, 211, 245, 378
branching, 314, 363, 407
breaking of ergodicity, 441, 450
Breiman's theorem, 375
Brownian motion, 36, 378, 411
Burgers/adhesion model, 351
Burning method, 401
Burridge–Knopoff model, 370, 405

canonical ensemble, 205
Cantor set, 124
cascade, 235
catastrophe theory, 257, 353
Cauchy distribution, 97
cavity approach, 444
central charge, 146

central limit theorem, 48, 302, 320, 330,
  356, 452, 470
chaos, 47
Chaoticity, 444
characteristic function, 16, 98
characteristic scale, 161
Charge-Density-Wave, 415
cloud, 139
clusters, 298, 368
Coast of Britain, 125
collective phenomena, 242
complex dimensions, 159
complex exponents, 157, 276
complex fractal dimension, 156
conditional probability, 7
conformal field theory, 145
conformal transformation, 145
contact processes, 305
continuous-time random walk, 121
control function, 290
control parameter, 407
convolution, 49, 57
correlation, 93, 223, 276
correlation function, 213, 223, 246
correlation length, 247
Coulomb solid friction law, 344, 413
cracks, 293, 313
Cramér, 78
Cramér function, 61
craters, 134
crisis, 255
critical exponent, 245, 275
critical phenomena, 245, 259, 294, 341
critical point, 268, 273, 330, 368, 448
cumulants, 16, 49, 54, 59

damage, 313, 335
decimation, 53, 300
decision theory, 10
density of states, 201, 445
dependence, 229
depinning, 417

detailed balance, 214, 412
deviation theorem, 60
dice game, 70
Dieterich friction law, 343
diffusion, 41
Diffusion Limited Aggregation, 139,
    148, 278
diffusion-reactions, 199
dimensional analysis, 150
directed percolation, 294, 304, 419
discrete scale invariance, 156, 276
dissipation function, 218
DNA, 225
droplets, 250
Dutch-book argument, 10

earthquake, 1, 12, 18, 19, 215, 335, 339,
    343, 344, 405, 411
effective medium theory, 296, 313, 441
Ehrenfest classification, 236, 248
entropy, 68, 72, 199, 207, 448
epicenters, 133
epidemics, 306, 381
error function, 59
exceedance, 21
exponential distribution, 167
extended self-similarity, 146
extreme deviations, 78, 362
extreme value theory, 18, 319, 331

faults, 129
feedback, 407
Fermi's theory of cosmic rays, 355
Fermi, Pasta and Ulam, 209
fiber bundle models, 318
Fick's law, 43
financial crash, 255
first order transition, 448
first-order phase transition, 248
first-order transition, 262, 263, 321
fixed point, 93, 273, 300, 330
Flinn–Engdahl regionalization, 76
floods, 107
fluctuation-dissipation, 213, 252
Fokker–Planck equation, 41, 43, 259,
    309, 376
forest fires, 306, 391, 402
Fox function, 118
Fréchet distribution, 23
fractal, 282
fractal dimension, 37, 282
fractal growth phenomena, 331
fractals, 239, 336, 470

fractional Brownian motion, 153
fractional derivative, 236
Fractional diffusion equation, 239
fractional integral, 236
fractional noise, 153
fracture, 313
fragmentation, 83, 84, 362, 381, 453,
    454
Fredholm integral equation, 471
free energy, 70
frustration, 442

Gamma law, 76, 103, 179, 315, 361
gap equation, 422, 429
Gauss distribution, 14, 21, 43, 87, 202,
    246, 302, 356, 386
Gaussian law, 93
generic scale invariance, 409
Gibbs–Duhem relation, 208
glass transition, 356
global warming, 9
Gnedenko–Pickands–Balkema–de Haan
    theorem, 29
Goldstone modes, 406
grand canonical ensemble, 205
gravity altimetry, 473
Green function, 43, 410, 466, 468
Gumbel distribution, 23
Gutenberg–Richter, 19, 68, 76, 102,
    104, 106, 164, 179, 215, 339, 412
Gutenberg-Richter, 2, 160

Harvard catalog, 77
Hausdorff dimension, 127
Hermite polynomials, 57
hierarchical network, 269, 303, 323,
    326, 341, 429
Hill estimator, 168
Holtsmark's gravitational force
    distribution, 91, 457
homogeneisation theory, 441
Hopf bifurcation, 261
hurricane, 9
Hurst, 153
Hurst effect, 153
Hurst exponent, 434
hyperbolic dynamical system, 435
hysteresis, 386

i.i.d., 33
imitation, 243
Infinitely divisible cascades, 146
infinitely divisible distributions, 65
instanton, 266

interacting particles, 243
Internet, 381
Ising model, 243, 341, 349, 368, 386
Iterated Function Systems, 435
Ito interpretation, 44

Jaynes analysis, 74
Jeffreys theory, 70

Kesten multiplicative process, 232, 374
Koch curve, 127
Kramers' problem, 266, 355
Kramers–Moyal expansion, 42
Kullback distance, 67

Lévy law, 96, 240, 365, 458, 459, 462
Lévy walk, 110
Lagrange multipliers, 72, 205, 390
lambda point, 416
Landau–Ginzburg theory, 250, 264, 414
Langevin equation, 34, 41, 213, 252,
    258, 309, 411
Laplace transform, 116, 364
large deviations, 73, 219
Ledrappier–Young, 436
Lee–Yang phenomenon, 438
Legendre transform, 80
likelihood, 9
Linear Fractional Stable Motion, 434
lithosphere, 411
localization, 332
log-normal, 94, 166, 380
log-periodicity, 152, 157, 276, 304, 341,
    368, 455

macrostate, 202
magnetization, 246
Markovian system, 47
Master equation, 41, 376
maximum value, 19
maximum-likelihood, 164, 168, 172,
    179, 181
Maxwell construction rule, 249
mean, 13, 94
mean field theory, 258, 259, 309, 315,
    341, 387, 424, 448
median, 13
Mellin transform, 119, 455
microcanonical ensemble, 205
microstates, 209
minimal paths, 339
Mittag–Leffler function, 120
Model validation, 12
moments, 15, 94

most probable value, 94
multifractal random walk, 153
multifractals, 110, 141, 332, 334, 453
multiplication, 94, 180, 380, 406
multiplicative noise, 378

Noether theorem, 149
noise, 199, 457
normal form, 257, 263
nuclear explosion, 108
nucleation, 250, 263, 264, 343

Olami–Feder–Christensen model, 425
Omori law, 385
optimization principle, 389
optimum language, 390
order parameter, 407
Ornstein–Uhlenbeck process, 36

Padé resummation, 288
parapsychology, 74
partition function, 68, 70, 201, 205,
    270, 445
parturition, 255
peaks-over-threshold, 29
percolation, 293, 332, 338, 368, 393
perturbation analysis, 55
phase space, 204
phase transition, 238
plate tectonics, 174
platoons of cars, 348
Poisson distribution, 387
Poisson equation, 465
Poisson summation formula, 464
Potts model, 243, 273, 299, 325, 341,
    400
power law, 245
power law distribution, 170
prediction, 160, 255, 294, 323, 330, 332,
    335
preferential attachment, 370
principle of Least Action, 218
probability, 1

quantile, 173
quenched randomness, 228

random directed polymer, 137, 217,
    339, 381
random energy model, 445
random field, 386
random multiplicative models, 83
random walk, 33, 307, 363, 375
rank ordering, 164

Rayleigh number, 260
Rayleigh–Bénard, 256
real-space renormalization group, 298
renormalization group, 52, 55, 93, 227, 267, 298, 323
renormalization group equation, 272, 330
replica symmetry breaking, 443
return to the origin, 364
risk, 173, 323
river networks, 135
ruler method, 125
rupture, 293, 301, 366, 389, 426

saddle node method, 447
saddle-node method, 62, 360
sandpile models, 213, 398
scale invariance, 148, 267
scaling, 99, 253
self-affinity, 39
self-averaging, 107, 449
self-organization, 309
self-organized criticality, 315, 325, 396
self-similar renormalization, 41, 287
self-similarity, 111, 267, 317, 412
Shannon information, 202
shock, 351
singularity, 275, 279, 333, 351, 387, 406
sliding, 343
social imitation, 409
solid friction, 343
spanning trees, 401
spin models, 242, 441
spinodal, 263, 264
spinodal decomposition, 251, 321, 406
spread, 14
St. Petersburg paradox, 109
stable laws, 93
standard deviation, 94
statistical mechanics, 199
stick-slip, 404
stochastic equation, 33

stochastic partial differential equation, 308
Stratonovich interpretation, 44
stress corrosion, 328
Stretched exponentials, 362
stretched exponentials, 65, 78, 80, 180
structure factor, 254
Student's distribution, 350
subcritical, 262, 307
subduction, 175
super-diffusion, 228
supercritical, 256
surface growth processes, 199
survivor function, 30
susceptibility, 213, 244, 247, 341
synchronization, 424

temperature, 199, 201, 245, 268, 349
thermal fuse model, 335
thermodynamic limit, 294
thermodynamics, 200, 358
tides, 266
time-reversal symmetry, 231
time-to-failure, 261, 336, 339
topography, 135
trapping times, 113
tri-criticality, 323
Tsallis generalized entropy, 220, 357, 360
turbulence, 83, 148, 278, 307

universality, 306, 332, 397

variance, 14
vortices, 90

wavelets, 131, 146
Weibull distribution, 23
Weierstrass function, 239, 279
Wiener process, 33, 36, 153
Wiener–Hopf integral equation, 379
Wolf's die, 206

Printing and Binding: Strauss GmbH, Mörlenbach